Springer Series in Optical Sciences   Volume 55

Edited by Arthur L. Schawlow

Springer Series in Optical Sciences

Volumes 1–41 are listed on the back inside cover

Laser
Spectroscopy VIII

Proceedings of the Eighth International Conference,
Åre, Sweden, June 22–26, 1987

Editors:
W. Persson and S. Svanberg

With 315 Figures

Springer-Verlag Berlin Heidelberg New York
London Paris Tokyo

Dr. Willy Persson
Professor Sune Svanberg

Department of Physics, Lund Institute of Technology,
P.O. Box 118, S-221 00 Lund, Sweden

ISBN 3-540-18437-6 Springer-Verlag Berlin Heidelberg New York
ISBN 0-387-18437-6 Springer-Verlag New York Berlin Heidelberg

Library of Congress Cataloging-in-Publication Data. Laser spectroscopy VIII. (Springer series in optical sciences ; v. 55) Papers from the Eighth International Conference on Laser Spectroscopy, held at the Sunwing Hotel in Åre, Sweden, June 22–26, 1987. Includes index. 1. Laser spectroscopy–Congresses. I. Persson, Willy, 1940-. II. Svanberg, S. (Sune), 1943-. III. International Conference on Laser Spectroscopy (8th : 1987 : Sunwing Hotel, Åre, Sweden) IV. Series. V. Title: Laser spectroscopy eighth. VI. Title: Laser spectroscopy 8. QC454.L3L345 1987 535.5'8 87-28687

Printing: Druckhaus Beltz, 6944 Hemsbach/Bergstr.
Binding: J. Schäffer GmbH & Co. KG, 6718 Grünstadt.
2153/3150-543210

Preface

The Eighth International Conference on Laser Spectroscopy (EICOLS '87) was held at the Sunwing Hotel in Åre, Sweden, June 22–26, 1987. Following the traditions of its predecessors at Vail, Megève, Jackson Lake, Rottach-Egern, Jasper Park, Interlaken and Maui the intent of EICOLS '87 was to provide a forum for active scientists to meet in an informal atmosphere to discuss recent developments in laser spectroscopy. The scenic and remote location of the conference venue greatly stimulated a lively and relaxed exchange of information and ideas.

The conference was attended by 227 scientists from 20 countries including Australia, Austria, Canada, the People's Republic of China, Denmark, Finland, France, the Federal Repulic of Germany, Israel, Italy, Japan, The Netherlands, New Zealand, Norway, Poland, the Soviet Union, Sweden, Switzerland, the United Kingdom and the United States.

The scientific program included 14 topical sessions with 50 invited talks, ranging in length from 20 to 40 minutes. About 70 additional invited contributions were presented in two evening poster sessions. A third evening session included 4 oral and 18 poster post-deadline presentations.

These proceedings contain oral as well as poster and post-deadline contributions. We wish to thank all participants for making EICOLS '87 a rewarding conference and, in particular, the contributors for excellent presentations.and the punctual preparation of their manuscripts. We would also like to thank the members of the International Steering Committee for their suggestions and advice. Our particular thanks go to the Program Committee members for their efforts in composing an interesting and well-balanced program.

EICOLS '87 was held under the auspices of the International Union of Pure and Applied Physics (IUPAP). We gratefully acknowledge the financial support of IUPAP, the Swedish Natural Science Research Council, The Swedish Board for Technical Development and the Swedish National Energy Administration. We are also indebted to our corporate sponsors from Sweden, the Federal Republic of Germany, France, Japan, the United Kingdom and the United States. Their support enabled us to undertake stimulating excursions to scenic spots and helped us to create a relaxed atmosphere during the post sessions.

EICOLS '87 could not have been organized without the support of a number of people. We thank the members of the Division of Atomic Physics at Lund Institute of Technology for helping in the organization of the conference. In particular we are grateful to Helen Sheppard who helped enormously in preparing for and during the conference and in the editing of the proceedings. Her devotion and enthusiasm (and genuine sense of humour) contributed a lot to the success of the conference. Finally we would like to thank the Sunwing Åre Hotel and their charming staff for providing us with ideal conference facilities and efficient and generous assistance.

Lund, July 1987

W. Persson
S. Svanberg

International Steering Committee

Program Committee

List of Sponsors

EICOLS '87 was arranged under the auspices of the International Union of Pure and Applied Physics, IUPAP.

We are also grateful to the following agencies for their support of EICOLS '87:

NFR   The Swedish Natural Science Research Council
STU   The Swedish Board for Technical Development
STEV   The Swedish National Energy Administration

The following companies kindly provided support for EICOLS '87:

ASEA	Burleigh Instruments, Inc.
HighTech Network	Coherent Ltd.
SAAB-SCANIA	Lambda Physik
Sydkraft	Lumonics Ltd.
Volvo	Marubun Corporation
	Quantel
	Sopra
	Spectra-Physics GmbH

Contents

Part II   Laser Cooling, Trapping, and Manipulation of Atoms and Ions

Part III   Quantum Jumps

Part IV   Quantum Optics, Squeezed States, and Chaos

Part V   Atomic Spectroscopy

Part VI   Molecular Spectroscopy

Part VII   Clusters, Surfaces and Solids

Part VIII   **Miscellaneous Laser Spectroscopy Experiments**

Part IX   Laser Spectroscopic Diagnostics

Part X	Spectroscopic Techniques

Part XI　　Spectroscopic Sources

Part XII   **VUV Spectroscopy**

Part I

Applications of Laser Spectroscopy
to Basic Physics

High Resolution Laser Spectroscopy of Atomic Hydrogen

T.W. Hänsch[a], *R.G. Beausoleil*[b], *B. Couillaud*[c], *C. Foot*[d], *E.A. Hildum*[e], *and D.H. McIntyre*[a]

Department of Physics, Stanford University,
Stanford, CA 94305, USA

As the simplest of the stable atoms, hydrogen permits unique confrontations between theory and experiment. Precision spectroscopy of hydrogen can be a powerful tool to determine better values of fundamental constants and to probe the limits of basic physics laws. Interest in high resolution laser spectroscopy of atomic hydrogen has been growing during the past two years, as documented by this and several following papers. At least four accurate new measurements of the Rydberg constant have been completed in 1986 [1-4]. Dramatic further improvements in spectral resolution and measurement accuracy should be achievable in the future.

Historically, the simple and regular Balmer spectrum has inspired the pathbreaking discoveries of N. Bohr, A. Sommerfeld, L. De Broglie, E. Schrödinger, P. Dirac and W.E. Lamb. However, no classical spectroscopic experiment has ever succeeded in fully resolving the fine structure of the Balmer lines. The spectra always remained blurred by Doppler broadening. Major advances in resolution became possible only in the early seventies, thanks to the advent of highly monochromatic tunable dye lasers together with techniques of Doppler-free laser spectroscopy. With the help of saturated absorption spectroscopy, our group at Stanford was able to resolve single fine structure components of the red H-α line for the first time, and we could observe the 2S Lamb shift directly in the optical spectrum [5]. In 1974, M. Nayfeh completed an absolute wavelength measurement [6] which yielded an eightfold improved value of the Rydberg constant, accurate to one part in 10^8. Refined experiments have since led to another 30-fold improvement in the accuracy of this important constant. But even the early Doppler-free spectra of H-α approached the resolution limit imposed by the natural linewidth.

At Stanford, we have long concentrated our efforts on a different transition with potentially much narrower width: the transition from the 1S ground state to the metastable 2S-state [7]. The 1/7 sec lifetime of the upper state implies a natural linewidth of only 1 Hz, or an ultimate resolution better than one part in 10^{15}. This transition is not dipole-allowed, but can be observed by two-photon spectroscopy. Doppler-broadening can be elegantly eliminated by excitation with two counterpropagating laser beams whose first-order Doppler shifts cancel.

The most serious technical obstacle to such an experiment has long been the lack of an intense and extremely monochromatic laser source at the required ultraviolet wavelength of 243 nm. Early experiments had to rely on frequency-doubled pulsed dye lasers, and the achieved resolution was entirely instrument limited. Still, C. WIEMAN [8] was able, from such relatively crude spectra, to observe a relativistic correction due to nuclear recoil in the hydrogen-deuterium isotope shift, and he could measure the Lamb-shift of the 1S ground state to a fraction of one percent by comparing the 1S-2S transition with the Balmer-β line at 486 nm.

In 1985, C. FOOT et al. [9] achieved an important breakthrough when they succeeded in observing the 1S-2S two-photon transition by continuous wave excitation. Several mW of ultraviolet light near 243 nm were generated by summing the frequencies of an argon-ion laser at 351 nm and a cw-dye laser in a 90⁰ phase-matched crystal of KDP. The intensity of the dye-laser light at the crystal was enhanced with the help of a resonant passive ring cavity. (In more recent experiments at the Max-Planck-Institute for Quantum Optics in Garching, C. Zimmermann and J. Sandberg have obtained very promising results with the much simpler scheme of angle-tuned second harmonic generation in β-Barium-Borate.)

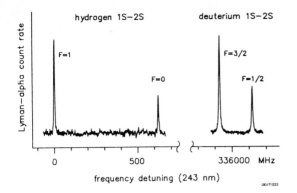

Fig. 1. CW two-photon spectra of hydrogen and deuterium 1S-2S

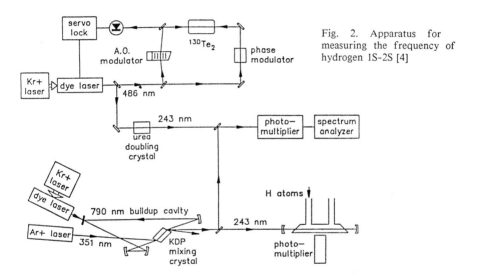

Fig. 2. Apparatus for measuring the frequency of hydrogen 1S-2S [4]

Figure 1 shows cw two-photon spectra of hydrogen and deuterium 1S-2S. The atoms are produced in a microwave discharge and simply observed in a gas cell at room temperature. This cell is placed inside a linear build-up cavity for the 243 nm radiation, and the signal is monitored by counting collision induced Lyman-α photons.

During the past year, R.G. BEAUSOLEIL et al. [4] have completed an absolute frequency measurement of the hydrogen 1S-2S transition, using the set-up illustrated in Figure 2. A cw-dye laser is stabilized to an interferometrically calibrated $^{130}Te_2$ reference line which has a frequency 65 MHz greater than one fourth of the hydrogen 1S-2S F = 1 transition frequency. An angle-tuned urea crystal produces approximately 1 nW of the second harmonic of this reference laser. The frequency of this light is compared to that of the 243 nm sum-frequency generator by overlapping the two wavefronts on a photomultiplier and observing a radiofrequency beat signal.

Collision effects in the gas cell have been carefully studied. Figure 3 shows the frequency offset (at 243 nm) of the F = 1 component of the hydrogen 1S-2S transition relative to the second harmonic of the reference laser frequency as a function of the cell pressure both for pure hydrogen and for a mixture of 0.7 % hydrogen in helium. Much smaller shifts are observed in

3

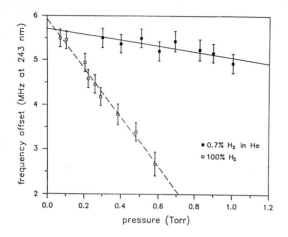

Fig. 3. Pressure shifts of the hydrogen 1S-2S F = 1 resonance relative to a reference frequency [4]

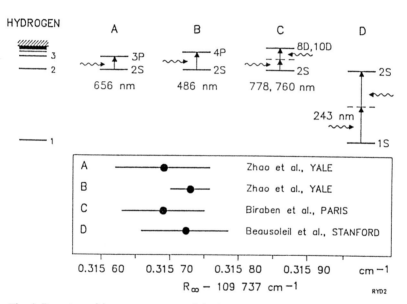

Fig. 4. Recent precision measurements of the Rydberg constant [1-4]

the hydrogen-helium mixture, and this set of data is used to arrive at the frequency value extrapolated to zero pressure. When taking the hyperfine interaction into account, we obtain the 1S-2S centroid frequency of 2 466 061 413.77(1.51) MHz.

This measured frequency can be used to determine a new value of the Rydberg constant. If we trust the theoretical determination of the 1S Lamb-shift [10], we obtain $R_\infty = 109\ 737.315\ 71(7)\ cm^{-1}$, in good agreement with the results of other recent Rydberg measurements which are compared in Figure 4 and discussed in subsequent reports. There is a disagreement with an earlier pulsed measurement of the 1S-2S frequency [11]. We believe that residual frequency chirping in that pulsed experiment was not sufficiently characterized to properly predict the resultant shift of the two-photon resonance.

4

If the Rydberg constant is regarded as known, e. g. from the Balmer-β measurement of ZHAO et al. [3], the measured 1S-2S frequency can be used to obtain an experimental value for the Lamb-shift of the 1S ground state. The measured value, 8 173.3(1.7) MHz, is in good agreement with the theoretical prediction of 8 172.94(9) MHz. This accuracy is still more than an order of magnitude behind that of the 2S Lamb-shift [12]. However, it appears very difficult to achieve further improvements of the 2S Lamb-shift since the natural linewidth of the $2S_{1/2} - 2P_{1/2}$ transition amounts to 100 MHz. The difference of 1S and 2S Lamb-shifts, by contrast, can be determined from optical transitions with much narrower widths. With future improvements, such measurements promise a most stringent test of quantum electrodynamics for a bound system.

In our new laboratory in Garching, R. Kallenbach and C. Zimmermann have begun experiments which should initially yield an at least thousandfold improvement in the resolution of the 1S-2S transition. The laser bandwidth must be reduced to only a few kHz by internal or external frequency stabilizers with fast servo response. A hydrogen atomic beam will be employed to avoid collision effects and to reach longer interaction times. The atoms will be cooled near the temperature of liquid helium by mounting the escape nozzle on a cryostat.

We are also exploring means to slow the hydrogen atoms far below liquid helium temperatures. One promising approach may be laser radiation pressure cooling [13]. An atom at 4 K can be brought to rest by resonant scattering of only 100 Lyman-α photons at 121.5 nm. A train of Lyman-α pulses could be produced by third harmonic generation, employing a recirculating ring cavity for the primary light pulse, as suggested in Figure 5. A magnetic trap could finally suspend the cold atoms in space, preventing them from falling down due to gravity.

Very recently H.F. HESS et al. [14] have successfully trapped more than 5×10^{12} atoms of hydrogen for several minutes in a static radial quadrupole trap. The atoms were cooled to a temperature of about 40 mK with the help of a dilution refrigerator. Inside the trap, the atoms are perturbed by the trapping field, but high resolution two-photon spectroscopy should still be possible since all S-states have very nearly the same magnetic moment, so that any Zeeman-shifts of $\Delta m = 0$ transitions remain small. Such studies could provide valuable information about the spatial distribution and velocity distribution of the trapped atoms, and would be a most valuable diagnostic tool once the elusive goal of Bose-condensation has been achieved.

For spectroscopy of the highest precision, the atoms could be released from the trap and observed in free fall. Two-photon optical Ramsey-spectroscopy of an atomic fountain should be able to reach the ultimate resolution limit, as given by the natural linewidth [15].

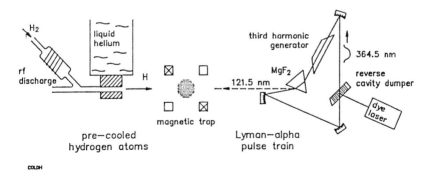

Fig. 5. Proposed experiment for laser cooling and magnetic trapping of hydrogen atoms

An elaborate frequency chain could be designed to compare the 1S-2S frequency directly with the cesium frequency standard. Such a measurement would yield a much improved Rydberg constant, improving the link between the time standard and the system of fundamental constant. However, it cannot by itself serve as a test of basic physics laws, since the cesium hyperfine splitting is not well understood theoretically.

It appears more interesting to compare the 1S-2S transition with other optical frequencies that can also be calculated from first principles, such as other transitions in the hydrogen atom itself. Very sharp resonances can be expected for two-photon transitions from the metastable 2S-state to highly excited nS or nD Rydberg states, if perturbing electric fields are avoided.

Unfortunately, a comparison with theory is still difficult. Even though quantum electrodynamics is believed to provide a very good theory for the hydrogen atom, the accuracy of computed energy levels is limited by the uncertainties of fundamental constants, unknown nuclear size and structure effects, and computational approximations. Some of the uncertainties estimated here are smaller than those given in Ref. [9], since they are based on more recent precision measurements and data analyses. The dominant contribution to the uncertainty of the 1S-2S frequency (0.7 MHz) is still the uncertainty of the Rydberg constant [3]. Precision measurements of this frequency can improve the Rydberg constant until we approach the uncertainties due to the proton/electron mass ratio [16] (2×10^{-8}: 30 kHz) and the rms charge radius of the proton [17] ($\simeq 2.5 \times 10^{-2}$: 50 kHz; estimates of the accuracy of the measured proton charge radius still differ widely). Additional uncertainties of comparable magnitude arise from approximations in the computation of the one-photon self-energy for an electron bound in a Coulomb field (35 kHz) and uncalculated higher order QED effects (50 kHz).

Fortunately, important cancellations of some of these uncertainties can occur if different transitions are compared. For instance, it is possible to construct linear combinations of two transition frequencies, such as $7 \ f(2S\text{-}nS)\text{-}(1\text{-}8/n^3) \ f(1S\text{-}2S)$, which no longer contain any terms which scale as the inverse cube of the principal quantum number. Such a frequency is independent of nuclear size, and it no longer contains many uncalculated higher order QED-corrections. It can be predicted much more easily than its individual constituents. By exploiting such cancellations of uncertainties, and by improving the accuracy of quantum electrodynamic computations, we can hope to arrive at much improved measurements of the Rydberg constant, the electron/proton mass-ratio and the charge radii of the proton and deuteron.

Interesting tests of basic physics laws will become possible, if accurate values for these fundamental constants become available from independent experiments. One promising approach towards measuring the Rydberg constant is radiofrequency spectroscopy of transitions between circular Rydberg states. Encouraging first results have very recently been achieved in experiments with lithium atoms by J. HARE et al. [18].

A direct comparison of the hydrogen 1S-2S transition with the corresponding transitions in positronium [19] or muonium [20] is also interesting. High resolution laser spectroscopy of these purely leptonic atoms avoids the complications of nuclear structure effects, but the resolution will ultimately be limited by annihilation or muon decay. An experiment aimed at two-photon spectroscopy of muonium 1S-2S is currently in progress, modelled after the earlier successful experiment in positronium [21]. Accurate calibrations of $^{130}Te_2$ reference lines for both positronium [22] and muonium [23] have recently been completed by D.H. McIntyre. The result for the positronium reference line lowers the value of the positronium 1S-2S interval by about 40 MHz and lessens the agreement between theory and experiment. Uncalculated higher order corrections could provide contributions of this order of magnitude.

Laser spectroscopy of anti-hydrogen would also be most intriguing. The successful trapping of anti-protons by G. GABRIELSE et al. [24] is a very important step towards such an experiment. A comparison with hydrogen could test with extreme sensitivity wether there are spetroscopic differences between matter and anti-matter. Even if we believe in the symmetry and simplicity of nature, it would certainly be interesting to verify our expectations experimentally. If past experience is any guide, the biggest surprise would perhaps be if we found no surprise.

This work was supported in part by the National Science Foundation under Grant No. NSF PHY86-04441 and by the U. S. Office of Naval Research under Contract No. ONR N00014-C-78-0403.

a. Max-Planck-Institute for Quantum Optics, 8046 Garching, FRG
b. Boeing Electronics Company, Bellevue, WA 98008, USA
c. Coherent, Inc., Palo Alto, CA 94303, USA
d. Clarendon Laboratory, Oxford University, Oxford, UK
e. Lawrence Livermore National Laboratory, Livermore, CA 94550, USA

1. P. Zhao, W. Lichten, H.P. Layer, and J.C. Bergquist, Phys. Rev. A34, Rapid Communications, 5138 (1986)
2. F. Biraben, J.C. Garreau, and L. Julien, Europhysics Lett. 2, 925 (1986)
3. P. Zhao, W. Lichten, H.P. Layer, and J.C. Bergquist, Phys. Rev. Lett. 58, 1293 (1987)
4. R.G. Beausoleil, D.H. McIntyre, C.J. Foot, E.A. Hildum, B. Couillaud, and T.W. Hänsch, Phys. Rev. A35, Rapid Communications (June 1987)
5. T.W. Hänsch, I.S. Shahin, and A.L. Schawlow, Nature 235, 63 (1972)
6. T.W. Hänsch, M.H. Nayfeh, S.A. Lee, S.M. Curry, and I.S. Shahin, Phys. Rev. Lett. 32, 1396 (1974)
7. T.W. Hänsch, S.A. Lee, R. Wallenstein, and C. Wieman, Phys. Rev. Lett. 34, 307 (1975)
8. C. Wieman and T.W. Hänsch, Phys. Rev. A22, 192 (1980)
9. C.J. Foot, B. Couillaud, R.G. Beausoleil, and T.W. Hänsch, Phys. Rev. Lett. 54, 1913 (1985)
10. W.R. Johnson and G. Soff, At. Data Nucl. Data Tables 33, 405 (1985)
11. E.A. Hildum, U. Boesl, D.H. McIntyre, R.G. Beausoleil, and T.W. Hänsch, Phys. Rev. Lett. 56, 576 (1986)
12. S.R. Lundeen and F.M. Pipkin, Phys. Rev. Lett. 46, 232 (1981)
13. T.W. Hänsch and A.L. Schawlow, Opt. Comm. 13, 68 (1975)
14. H.F. Hess, G.P. Kochanski, J.M. Doyle, N. Masuhara, D. Kleppner, and T.J. Greytak, to be published
15. R.G. Beausoleil and T.W. Hänsch, Phys. Rev. A33, 1661 (1986)
16. R.S. Van Dyck, Jr., F.L. Moore, D.L. Farnham, and B.P. Schwinberg, Bull. Am. Phys. Soc. 31, 224 (1986)
17. L.N. Hand, D.J. Miller, and R. Wilson, Rev. Mod. Phys. 35, 335 (1963)
18. J. Hare, M. Gross, P. Goy, and S. Haroche, to be published
19. S. Chu, A.P. Mills, Jr., and J.L. Hall, Phys. Rev. Lett. 52, 1689 (1984)
20. A.P. Mills, Jr., J. Imazato, S. Saitoh, A. Uedono, Y. Kawashima, and K. Nagamine, Phys. Rev. Lett. 56, 1463 (1986)
21. S. Chu, private communication
22. D.H. McIntyre and T.W. Hänsch, Phys. Rev. A34, Rapid Communications, 4504 (1986)
23. D.H. McIntyre and T.W. Hänsch, to be published
24. G. Gabrielse, X. Fei, K. Helmerson, S.L. Rolston, R. Tjoelker, T.A. Trainor, H. Kalinowsky, J. Haas, and W. Kells, Phys. Rev. Lett. 57, 2504 (1986)

Determination of the Rydberg Constant by Doppler-Free Two-Photon Spectroscopy of Hydrogen Rydberg States

L. Julien, J.C. Garreau, and F. Biraben

Laboratoire de Spectroscopie Hertzienne de l'Ecole Normale Supérieure, Tour 12, EO1-4, place Jussieu, F-75252 Paris Cedex 05, France

1. INTRODUCTION

The development of Doppler-free laser techniques has allowed substantial improvement of the resolution observed in optical spectroscopy. In recent years these techniques have been applied to atomic hydrogen in order to increase the precision on the Rydberg constant R_∞. Up to 1986, the most precise determination of R_∞ has been obtained in studying the Balmer α line. The measurement was limited to a precision of 10^{-9} because of the natural width of the 3P level (30 MHz) [1].

Our own way to measure the Rydberg constant is to study the Doppler-free two-photon 2S-nD transition (n=8-10). In this case, the natural width of the Rydberg level is very small. The 8D linewidth is, for example, 550 kHz and can then lead to a relative linewidth of 7.10^{-10}. In our experiment, a preliminary result gave a relative linewidth of $1.8.10^{-9}$ for the 2S-8D two-photon line [2]. We present here a first Rydberg measurement on the 2S-8D and 2S-10D two-photon lines [3].

2. OBSERVATION OF THE 2S - nD TRANSITIONS

The principle of the method is to induce the 2S-nD transitions using a metastable atomic beam collinear with two counterpropagating laser beams. Thus the line-broadening due to finite transit time of atoms in the laser beams is very small. The metastable $2S_{1/2}$ atomic beam is obtained by electronic excitation of a ground state atomic beam. The inelastic collisions with the electrons deviate the atoms and the metastable atomic beam makes an angle of 20° with the incident atomic beam. The metastable atomic beam is then collinear with the two laser beams. The optical excitation takes place in the second vacuum chamber where electric and magnetic fields are reduced at best. The metastable atoms are detected in the third vacuum chamber. An applied electric field quenches the 2S state and a photomultiplier detects the Lyman α fluorescence. Measuring the photomultiplier current, we estimate the metastable beam intensity to be about 10^7 atoms s^{-1}. To efficiently induce the two-photon transitions, the metastable atomic beam is placed inside a Fabry-Perot cavity. The cavity length (50 cm) is locked on the laser wavelength. Inside the cavity the beam waist w_0 is 570 μm and the light power is about 40 W in each propagation direction.

After a two-photon excitation from the 2S metastable state, the nD states undergo radiative cascade to the 1S ground state in a proportion of about 90 %. The two-photon transition can then be detected by observing the corresponding decrease of the 2S beam intensity. To eliminate the metastable beam intensity fluctuations, the laser frequency is modulated and the linewidth is a derivative trace (Figure 2).

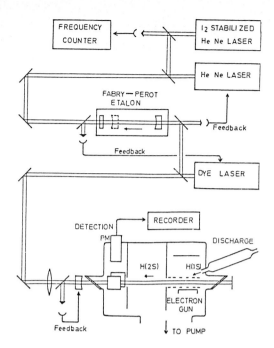

Figure 1 - Experimental set-up.

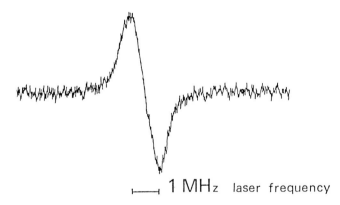

$$\longmapsto \; 1\,\text{MHz} \quad \text{laser frequency}$$

Figure 2 - Recording of the $2S_{1/2}$ - $8D_{5/2}$ two-photon transition in hydrogen.

3. MEASUREMENT OF THE TRANSITION WAVELENGTHS

Our measurement is based on the comparison between the two-photon line wavelength and the one of an iodine-stabilized helium-neon laser at 633 nm.

The key of the wavelength comparison is a nonconfocal Fabry-Perot etalon. This etalon is built with two silver-coated mirrors, one flat and the other one spherical (radius of curvature R = 60 cm). The finesse of the cavity is about 60 at 633 nm and 100 at 778 nm. The etalon is placed inside a box evacuated to less than 10^{-6} mbar. In order to eliminate the effects due to reflective phase shifts in the mirror coatings, the method of virtual mirror is used. Two etalon spacings are alternately employed (10 cm and 50 cm).

The measurement scheme is represented in Figure 1. An auxiliary He-Ne laser is mode-matched into the etalon cavity. The difference between the incident intensity and the Fabry-Perot transmission is considered to prevent a possible shift due to the variation of the laser gain with frequency. The beat frequency between this He-Ne laser and the reference one is measured by a frequency counter. The dye laser is also mode-matched into the etalon cavity and locked on it. It is brought into atomic resonance through thermal sweeping of the etalon.

For each transition, the beat frequency between the two He-Ne lasers has to be extrapolated at null light power to eliminate systematic effects due to two-photon light shift. Such an extrapolation is shown in Figure 3, where the transition involved is the $2S_{1/2}(F=1)-8D_{5/2}$ transition in hydrogen. Each dot is obtained as the average of ten measurements of the beat frequency at the centre of the atomic resonance. With a 40 W light power, the light shift is about 330 kHz. Taking into account the imprecision of the light power scale, this experimental value is in good agreement with the theoretical one (300 kHz). Figure 3 clearly shows that this extrapolation quite eliminates the two-photon light shift.

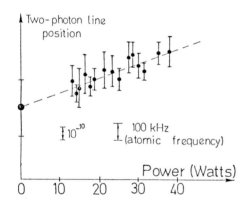

Figure 3 - Extrapolation of the two-photon line position vs. the light power.

4. RESULTS

The previous experimental method has been applied to three different transitions : $2S_{1/2} \rightarrow 8D_{5/2}$ in H and D and $2S_{1/2} \rightarrow 10D_{5/2}$ in H. The respective wavelengths in air are 777.8 nm, 777.6 nm and 759.6 nm. Table I gives the experimental frequency measurements after extrapolation to null light power and the corrections due to the second-order Doppler effect and to the hyperfine structure.

The three values obtained for R_∞ are in good agreement. Our final result is $R_\infty = 109737.315692(60)$ cm^{-1}. Figure 4 compares our result with the most recent measurements of R_∞ [4][5]. There is an excellent agreement.

Table I - Experimental results.

	Hydrogen $8D_{5/2}$	Hydrogen $10D_{5/2}$	Deuterium $8D_{5/2}$
Experimental result (MHz)	385 324 758.54(20)	394 572 421.06(19)	385 429 619.56(23)
× 2	770 649 517.08(40)	789 144 842.12(38)	770 859 239.12(46)
Second-order Doppler effect (MHz)	+ 0.044	+ 0.045	+ 0.022
$2S_{1/2}$ hyperfine splitting (MHz)	+ 44.389	+ 44.389	+ 13.641
$nD_{5/2}$ hyperfine splitting (MHz)	− 0.028	− 0.014	− 0.008
$2S_{1/2}$-$nD_{5/2}$ energy splitting (MHz)	770 649 561.49(40)	789 144 886.54(38)	770 859 252.78(46)
R_∞ − 109 737 (cm^{-1})	0.315 682(57)	0.315 711(53)	0.315 682(65)
Final result	$R_\infty = 109\,737.315\,692(60)$ cm^{-1}		

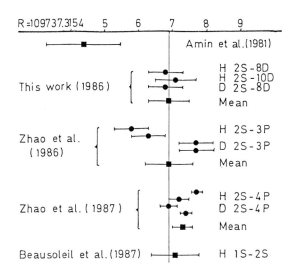

Figure 4 - Comparison of recent measurements of the Rydberg constant.

1 S.R. Amin, C.D. Caldwell and W. lichten : Phys. Rev. Lett. 47, 1234 (1981)
2 F. Biraben and L. Julien : Opt. Comm. 53, 319 (1985)
3 F. Biraben, J.C. Garreau and L. Julien : Europhys. Lett. 2, 925 (1986)
4 P. Zhao, W. Lichten, H.P. Layer and J.C. Bergquist : Phys. Rev. A34, 5138 (1986) ; Phys. Rev. Lett. 58, 1293 (1987)
5 R.G. Beausoleil, D.H. Mc Intyre, C.J. Foot, E.A. Hildum, B. Couillaud and R.W. Hänsch : Phys. Rev. A35, 4878 (1987).

Absolute Wavelength Measurements and Fundamental Atomic Physics

P. Zhao[*,1], *W. Lichten*[1], *H.P. Layer*[2], *and J.C. Bergquist*[3]

[1]Physics Department, Yale University,
 P.O. Box 6666, New Haven, CT 06511, USA
[2]The National Bureau of Standards, Gaithersburg, MD 20899, USA
[3]The National Bureau of Standards, Boulder, CO 80303, USA

1. INTRODUCTION

The Rydberg constant ties together several areas of fundamental physics: fundamental constants; atomic and molecular theory; and spectroscopy of basic systems. As determined at Yale, (Zhao et al.[1]) R is the most precisely measured, fundamental constant[2]: R = 109 737.315 73(3) [cm^{-1}].

  R enters into the least squares adjustment of other fundamental constants in which it is taken as exact.[2] In turn, other fundamental constants, such as the proton-electron mass ratio and fine structure constant are needed to interpret experimental results used to find R.

  All calculations of atomic and molecular energy levels are expressed in the atomic unit of 27.2 eV, which is twice the Rydberg constant. In turn, determination of R depends on simple systems where the theory is trustworthy. Theoretical calculations usually involve three levels of analysis: non-relativistic wave functions and energies, relativistic and quantum electrodynamical (QED) corrections.

  R is measured by means of spectra of elementary atomic systems, such as hydrogen. In turn, the interpretation of these spectra depends on a precise value of R.

  Table I shows selected measurements of R.[1,4-6]

Table I. The Rydberg constant R

Conventional Optical Spectroscopy			Dawn of the Laser Age		
Rydberg	(1890)	109 675 [cm^{-1}]	Hänsch et al.	(1978)	.3143(10)
Rydberg	later	109 734.7	Goldsmith et al.	(1978)	.31506(32)
Curtis	(1914)	109 737.7	Petley et al.	(1980)	.31529(85)
Birge	(1921)	736.9	Amin et al.	(1981)	.31544(10)
Birge	(1929)	737.42	Zhao et al.	(1986)	.31569(7)
Cohen	(1952)	737.311(7)	Hildum et al.	(1986)	.31492(22)
Martin	(1959)	737.312(8)	Barr et al.	(1986)	.3150(11)
Csillag	(1968)	.3060(60)	Biraben et al.	(1986)	.31569(6)
Masui	(1971)	.3188(45)	Zhao et al.	(1987)	.31573(3)
Kessler	(1973)	.3208(85)	Beausoleil et al.	(1987)	.31571(7)
Kibble et al.	(1973)	.3253(77)	Boshier et al.	(1987)	.31573(5)
Cohen, Taylor	(1973)	.3177(83)			

2. MEASUREMENT OF THE RYDBERG CONSTANT

"Unfortunately, preoccupation with significant figures frequently blinds participants to other aspects of the field, in particular to the distinction between significant figures, interesting figures and useful figures."

Figure 1 shows the course of the precision of the values from Table I.

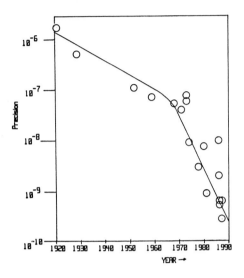

Fig. 1. Errors quoted for the Rydberg constant since 1920. The measurements around 1970 show two features characteristic of the end of an era: a frenzy of measurements, which agree with each other (see Table I) and which fail to achieve significant improvement in precision. A similar effect appears in the late 1980's. Does it also mean the end of an era?

Figure 2 shows the Yale experimental arrangement to measure R.[1]

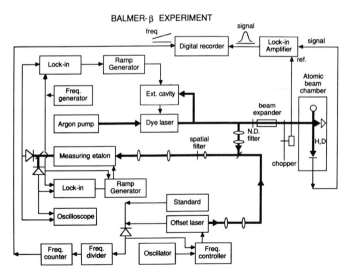

Fig. 2. The Yale measurement of the Rydberg constant

13

A dye laser, retroflected beam at 486[nm] is at right angles to the atomic beam. The laser quenches the metastable beam produced by electron bombardment of atomic hydrogen effusing from a tungsten oven at ~ 2850 K. A series of locks and offsets insures that the laser is simply related to the HeNe, I_2 stabilized laser, identical to that used to establish the meter at 633 nm.[7] Comparison with other groups is given in Table I. The Yale (Zhao et al.) measurements of the Rydberg, via Balmer-α and Balmer-β are in excellent agreement with each other and with the University of Paris (Biraben et al.).[6] Also, now that problems with "chirping" have been removed, Stanford (Hänsch and coworkers)[4] and Oxford (Boshier et al.)[5] agree. Thus, there are no longer discrepancies in this constant. Note that the quoted error in [1] is 3 parts in 10^{10}, less than twice the error of 1.6 parts in 10^{10} for the HeNe (I_2) standard (see Table II). We are up against fundamental limits of accuracy which will not change by improvements of the experiment.

Interest. One major interest in the Rydberg constant lies in measuring wavelength or frequency in a fundamental system as a test of theory. The Lamb shift of the ground state of hydrogen, was measured by subtracting the relativistically corrected, theoretical value (based on the Rydberg constant) from the experimentally measured 1S-2S wavelength. Within the experimental error of 1 parts in 10^4, there is agreement with theory.[4,5] Incidentally, this just equals the precision of Lamb's original rf measurement of his 2s shift, in over thirty years ago. The theoretical uncertainty is a factor of 10 smaller than the experimental error.[4,5] A critical test of Q.E.D. requires at least an order of magnitude improvement of the precision to which R is given.

Usefulness. **A Proposal.** Table II shows a few important points at which the meter has been realized; i.e., points on the electromagnetic spectrum where the absolute frequency/wavelength of an atomic or molecular line has been measured.

Table II. Realization of the Meter

Standard Transition	Frequency [MHz] Wavelength [µm]	Reliability (1 st. dev.)
Cs hfs	f = 9 192.631 770	few x 10^{-14}
He-Ne: CH_4	f = 88 376 181.608 λ = 3.392 231 397 0	0.44 x 10^{-10}
He-Ne: I_2 "i"	f = 473 612 214.789 λ = 0.632 991 398 1	1.6 x 10^{-10}

The last link in this long chain reaches from the methane stabilized HeNe laser in the infra-red at 3.39 µm to the iodine stabilized HeNe laser at 633 nm in the red. Two results can be compared:

wavelength ratio = 5.359 049 260 6(11) (H.P. Layer et al., 1976)[7]
frequency ratio = 5.359 049 258 9(9) (D.A. Jennings et al., 1983)[7]

The difference of 0.000 000 001 7(14) is not significant, nor is the improvement in precision. This suggests using the hydrogen spectrum to realize the meter in the optical and ultra-violet domains. Within a few years, R should be precise enough to introduce this secondary definition of the meter:

"The wavelengths of atomic hydrogen lines, as calculated by quantum electrodynamics and the Rydberg constant, realize the definition of the meter."

The Rydberg constant would act as a unique wavelength/frequency standard, which would serve throughout the entire spectrum.

3. WHERE DO WE GO FROM HERE?

Until the realization of the meter is improved in the optical domain, further measurements of R must be in the microwave or infra-red region of highly excited ("Rydberg") states of H. Ultimately, R should be measured to the same precision as time itself, currently at the 10^{-13} level (see Table II).

4. OTHER SYSTEMS. POSITRONIUM. HELIUM. MOLECULAR HYDROGEN

A test of QED is the 1S-2S transition in positronium, as fundamental an atom as any. Theory and experiment have a puzzling disagreement:

Theory: 1 233 607 202.5 MHz[8]
Experiment: 1 233 607 142.9 (10.7) MHz (8 parts in 10^9)[8]

There is mixed agreement and disagreement between theory and other experiments in positronium, presumably because of the difficulty of a complete calculation of recoil and annihilation terms.[9]

Table III. Ionization energies of helium (cm-1)

THEORY[a]	$1s^2$, $^1S_0$	1s, 2s $^1S_0$
Non-rel. energy	198 317.3865[b]	32 033.208
Rel. corr. energy	198 312.037 1	32 033.321
Lamb shift	-1.335(8) -1.377	-0.106(16)
EXPERIMENT[c]	198 310.77(15)	32 033.2325(50)
Exp. Lamb shift	1.26(15)	0.0885(50)
Possible precision	----	0.000 03

[a] Kono and Hattori, Phys. Rev. A34, 1727 (1986)
[b] Accuracy limited by Rydberg constant
[c] W.C. Martin, Phys. Rev. A29, 1883 (1984)

Table III compares theory and experiment for the helium atom.[10] Laser spectroscopy should make a two order of magnitude improvement in the precision of the Lamb shift. We can expect a similar improvement of theory in the near future.[11]

Table IV. The hydrogen molecule: theory and experiment

H_2 Dissociation energy, D_0 (cm^{-1})		Ionization Potential
Experiment	36 118.6 ± 0.5	124 417.2 ± 0.4 cm^{-1}
Theory	36 118.088	124 417.503 cm^{-1}
Discrepancy	0.5 ± 0.5	
Laser exp.	36 118.1 ± 0.2	124 417.51 ± 0.22 cm^{-1}

Table IV compares theoretical and experimental[20-22] determinations of the energy of the hydrogen molecule. The total relativistic and Q.E.D. corrections to D_0 amount to -0.76 cm^{-1}. Laser experiments by McCormack and Eyler[12] have improved the experimental precision and have removed a weak discrepancy between theory and experiment. Laser experiments by Eyler and coworkers[12] and Glab and Hessler[13] have improved the precision in the ionization potential substantially. We can expect the laser measurements to achieve a precision of ~ 0.01[cm^{-1}] and thus to provide a serious challenge to the theorists.

5. CONCLUSIONS

At least two more digits in R would be interesting. Further measurements must be of highly excited ("Rydberg") states of H, in the microwave and/or infra-red regions.

In positronium, both rf (ground state) and optical measurements disagree with theory. Higher order QED calculations are called for. In helium, non-relativistic theory and experiment are both orders of magnitude more precise than Lamb shift calculations. Better measurements and QED calculations are needed. In molecular hydrogen, laser spectroscopy is expected to improve precision by 1-2 orders of magnitude and should challenge theorists.

REFERENCES

* Current address: Harvard University, Cambridge, MA 02178. This research was aided by grant NSF-PHY-8419105.

1. P. Zhao, W. Lichten, H.P. Layer, and J.C. Bergquist, Phys. Rev. Lett. **58**, 1293 (1987); Errata, Phys. Rev. **A58**, 2506 (1987).

2. E.R. Cohen and B.N. Taylor, "The 1986 Adjustment of the Fundamental Constants," CODATA Bulletin, **63**, Nov. 1986, Table 6.

3. D. Kleppner, in Precision Measurements and Fundamental Constants, D.N. Langenberg and B.N. Taylor, Eds., Nat. Bur. Stand. (U.S.) Spec. Publ. 343 (U.S.G.P.O., Washington, DC, 1971), p. 411.

4. Talk by T.W. Hänsch at this conference.

5. Results presented at this conference by M. Boshier.

6. Talk by F. Biraben, this conference.

7. H.P. Layer, R.D. Deslattes, and W.G. Schweitzer, Appl. Opt. **15**, 734 (1976); D.A. Jennings, et al., Opt. Lett. **8**, 136 (1983).

8. S. Chu, A.P. Mills, and J.L. Hall, Phys. Rev. Lett. **52**, 1689 (1984); D.H. McIntyre and T.W. Hänsch, Phys. Rev. **A34**, 4504 (1986) (Experiment); T.J. Fulton, Phys. Rev. **A26**, 1294 (1982) (Theory).

9. S. Hatamian, R.S. Conti, and A. Rich, Phys. Rev. Lett. **58**, 1822 (1987) and C.I. Westbrook, D.W. Gidley, R.S. Conti, and A. Rich, Phys. Rev. Lett. **55**, 1328 (1987) (Experiment); W.G. Caswell and G.P. Lepage, Phys. Rev. **A20**, 36 (1979); G.S. Adkins, Ann. Phys. (N.Y.) **146**, 78 (1983) and G.P. Lepage and D.R. Yennie, in Precision Measurement and Fundamental Constants, Proceedings of the Second International Conference, Gaithersburg, MD, 1981, Nat. Bur. Stand. Publ. 617, p. 185 (Theory and summary).

10. For a review, see E.S. Chang, Phys. Rev. **A35**, 2777 (1987) and references in Table III.

11. J.D. Morgan, III and J. Baker, Bull. Am. Phys. Soc. **32**, 1245 (1987); G.W.F. Drake, ibid. Further references are given there.

12. E. McCormack and E. Eyler, Bull. Am. Phys. Soc. **32**, 1279 (1987) and personal communication.

13. W.L. Glab and J.P. Hessler, Phys. Rev. **A35**, 2102 (1987).

Precision cw Laser Spectroscopy of Hydrogen and Deuterium

M.G. Boshier, P.E.G. Baird, C.J. Foot, E.A. Hinds, M.D. Plimmer,
D.N. Stacey, J.B. Swan, D.A. Tate, D.M. Warrington, and G.K. Woodgate

The Clarendon Laboratory, University of Oxford,
Parks Road, Oxford, OX1 3PU, United Kingdom

1. Introduction

The 1S-2S transition of atomic hydrogen has attracted considerable attention because of its extremely narrow natural width (1 Hz). High resolution Doppler-free spectra of this transition, obtainable by two-photon absorption, offer the prospect of precise tests of bound-state quantum electrodynamics (QED) and more accurate values for fundamental constants /1/. The two-photon absorption was first observed by HANSCH et al. /2/ at Stanford, in the first of a series of progressively more accurate measurements /3/. Their latest result is a measurement of the 1S Lamb shift in hydrogen using continuous-wave radiation /4/.

We report here recent results from our programme of cw spectroscopy of hydrogen. We have measured the frequency of the 1S-2S transition in both hydrogen and deuterium, providing new and more precise values for the 1S Lamb shifts and the 1S-2S isotope shift. Our results can also be used to derive a new value for the Rydberg constant.

2. Description of the Experiment

The major difficulty in exciting the 1S-2S transition by two-photon absorption is the generation of sufficient cw 243 nm radiation ($\sim$1mW) in a nonlinear crystal. Our experiment uses frequency-doubling to generate the 243 nm light, in contrast to the sum-frequency mixing method employed in the recent work at Stanford. Frequency-doubling has particular advantages for this experiment; it provides a direct link to a visible wavelength (486 nm), permitting the use of accurate heterodyne calibration techniques and also facilitating a future direct comparison with the hydrogen Balmer-β transition. However, until very recently, frequency-doubling at 486 nm has been difficult because of the lack of a suitable nonlinear crystal. In our early work, an intra-cavity frequency-doubling system was developed using lithium formate monohydrate (LFM) /5/, but this was not adequate for the hydrogen experiment because of poor efficiency in LFM and damage to the crystal rapidly induced by the UV. Better results were obtained with urea, especially after it became possible to polish the crystal surfaces well enough to dispense with an index-matching fluid. Urea also suffers from UV-induced damage, but we were nevertheless able to make the initial observation and preliminary measurements of the 1S-2S transition using it. We have now obtained a crystal of β-barium borate (BBO) which is superior to urea in almost all respects: BBO is hard, has good optical quality and can be polished well. Further, it provides good conversion efficiency and, most important, it does not suffer from any UV-induced damage.

The primary quantities measured in our experiment are the frequencies of the 1S-2S transitions in hydrogen and deuterium relative to nearby accurately-

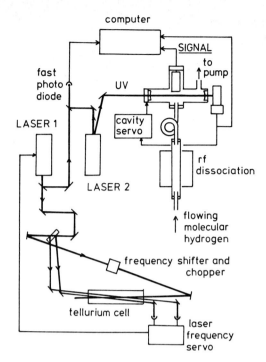

computer

fast photo diode

LASER 1

UV

SIGNAL

to pump

cavity servo

LASER 2

rf dissociation

flowing molecular hydrogen

frequency shifter and chopper

tellurium cell

laser frequency servo

Fig. 1: Schematic diagram of the apparatus. Both lasers are krypton-ion pumped Coumarin 102 dye lasers.

calibrated lines in the $^{130}Te_2$ spectrum /6/. A simplified schematic diagram of the apparatus used for these measurements is shown in Fig. 1.

The 243 nm light from the frequency-doubled ring dye laser (typically 2 mW) is enhanced by about a factor of 12 in a standing-wave enhancement cavity, which is windowless to avoid loss and scattering. The "walk-off" distortion of the uv beam profile can be fairly well approximated by an elliptical gaussian beam, and hence compensated for with an appropriate cylindrical lens. In this way more than 65% of the total uv power can be coupled into the fundamental TEM_{00} mode of the cavity. The two-photon excitation is detected by monitoring collisionally-induced Lyman-α fluorescence. A second dye laser is locked to the appropriate tellurium transition, detected by saturated absorption. Radiation from each dye laser at 486 nm is mixed on a fast photodiode and the resulting beat (∿1400 MHz for hydrogen, ∿4240 MHz for deuterium) is measured to provide frequency calibration. The measurements reported here were made with respect to the original tellurium cell calibrated by the NPL /6/. We used this cell after discovering a ∿1 MHz discrepancy between it and our own (nominally identical) cell. The reason for the difference has not yet been found.

A typical two-photon signal is shown in Fig. 2. It was recorded with an intra-cavity 243 nm power of ∿20 mW, focussed to a waist size of 100 µm in the interaction region. The cell contained 230 mTorr of a 5%H / 5%D / 90%He mixture. The linewidth at this pressure was 2.5 MHz (FWHM at 486 nm), and at lower pressures this decreased to 1.6 MHz.

The major source of systematic uncertainty in the measurement is the pressure shift in the hydrogen cell. We have made several measurements, in order to obtain an extrapolation to zero pressure. Measurements have been made both

19

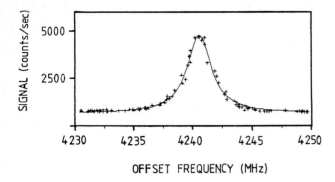

Fig. 2: F=3/2 to F=3/2 component of the 1S-2S transition in deuterium. The offset frequency is measured relative to the b_1 line of tellurium. The solid line is a least squares fit of a Lorentzian profile to the data (crosses).

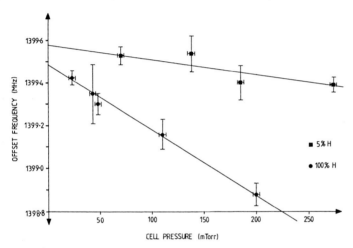

Fig. 3: Offset frequency for the F=1 to F=1 component of the hydrogen 1S-2S transition as a function of cell pressure for both 100%H and a 5% H / 5% D / 90% He mixture.

with pure hydrogen and deuterium, and also using a mixture containing a large fraction of helium; this has been found to give a smaller pressure shift /4/. The offset frequency for hydrogen as a function of cell pressure is shown in Fig. 3.

After correcting for the acousto-optic modulator shift and the hyperfine structure, we obtain 1457.21(11) MHz and 4244.64(8) MHz for the quantity $\frac{1}{4}f(1S-2S)-f(Te)$ for H and D respectively (i.e. the interval between one quarter of the 1S-2S centroid and the appropriate tellurium line). We then use the Rydberg constant /7/ and the known 2S-2P Lamb shift to extract a value for the 1S Lamb shift. We find the values 8172.93(84) MHz for H and 8183.96(80) MHz for D, in excellent agreement with theory (8173.15(10) MHz and 8184.04(9) MHz respectively /8/). Our definition of the Lamb shift is the same as that of JOHNSON and SOFF /8/, i.e. the sum of all QED terms plus the

finite nuclear size correction. Our results can also be used to provide a value
for the Rydberg constant by using the theoretical value of the 1S Lamb shift.
In this way we obtain R_∞ = 109737.31573(5) cm^{-1} from H and 109737.31573(5) cm^{-1}
from D. These compare well with other recent values /7,9/. Finally, we
obtain an H-D 1S-2S isotope shift of 670994.4(9) MHz, also agreeing well with
the theoretical value of 670994.53(12) MHz.

The precision of our determination of the 1S Lamb shift is limited by the
accuracies of the tellurium standard and the Rydberg constant. These sources
of uncertainty will be removed in the next stage of this work, which will
involve the direct comparison of the 1S-2S and 2S-4S transitions, excited in
atomic beams. The natural width of the 2S-4S transition is less than 1 MHz
and we expect to make the comparison to about 25 kHz (at 486 nm). This will
provide a value of the 1S Lamb shift with a precision that approaches that of
rf 2S Lamb shift measurements.

Acknowledgements

It is a pleasure to acknowledge the contributions to this work of Mr G. Read, Mr
C.W. Goodwin and Dr S.H. Smith. We are grateful to Dr A.I. Ferguson for lending
us the calibrated tellurium cell. One of us (MGB) acknowledges the support of
the Royal Commission for the Exhibition of 1851 and Christ Church, Oxford.
This work is supported by the Science and Engineering Research Council.

References

1. E.V. Baklanov and V.P. Chebotaev: Opt. Spectrosc. 38, 215 (1975)
2. T.W. Hänsch, S.A. Lee, R. Wallenstein and C. Wieman: Phys. Rev. Lett. 34,
   307 (1975)
3. C. Wieman and T.W. Hänsch: Phys. Rev. A22, 192 (1980)
4. C.J. Foot, B. Couillaud, R.G. Beausoleil and T.W. Hänsch: Phys. Rev. Lett.
   54, 1913 (1985)
   R.G. Beausoleil, D.H. McIntyre, C.J. Foot, B. Couillaud, E.A. Hildum and
   T.W. Hänsch: to be published in Phys. Rev. A
5. C.J. Foot, P.E.G. Baird, M.G. Boshier, D.N. Stacey and G.K. Woodgate: Opt.
   Comm. 50, 199 (1984)
6. J.R.M. Barr, J.M. Girkin, A.I. Ferguson, G.P. Barwood, P. Gill, W.R.C. Rowley
   and R.C. Thompson: Opt. Comm. 54, 217 (1985)
7. P. Zhao, W. Lichten, H. Layer and J. Bergquist: Phys. Rev. Lett. 58, 1293
   (1987)
8. W.R. Johnson and G. Soff: At. Data Nucl. Data Tables 33, 405 (1985)
9. P. Zhao, W. Lichten, J.C. Bergquist and H.P. Layer: Phys. Rev. A34, 5318
   (1986)
   F. Biraben, J.C. Garreau and L. Julien: Europhys. Lett. 2, 925 (1986)

First Antiprotons in an Ion Trap

G. Gabrielse[1], X. Fei[1], K. Helmerson[1], S.L. Rolston[1], R. Tjoelker[1],
T.A. Trainor[1], H. Kalinowsky[2], J. Haas[2], and W. Kells[3]

[1]Department of Physics, Harvard University, Cambridge, MA 02138, USA
[2]Institut für Physik, University of Mainz,
 D-6500 Mainz, Fed. Rep. of Germany
[3]Fermi National Accelerator Laboratory, Batavia, IL 60510, USA

A. The Antiproton Mass[1]

Measurements of the antiproton mass[2,3,4,5] are represented in Fig. 1. All of these
are deduced from measurements of the energy of x-rays radiated from highly excited
exotic atoms. For example, if an antiproton is captured in a Pb atom, it can make radiative transitions from its $n = 20$ to $n = 19$ state. The antiproton is still well outside
the nucleus in this case, so that nuclear effects can be neglected. The measured transition energy is essentially proportional to the reduced mass of the nucleus and hence the
antiproton mass can be deduced by comparing the measured values with theoretical
values, corrected for QED effects. The most accurate quoted uncertainty is 5×10^{-5}
and is consistent with the much more accurately know proton mass, indicated by the
dashed line. It looks like it would be difficult to extend the accuracy realized with the
exotic atom method. It might be possible, however, that proton and antiproton masses
could be compared directly in a storage ring, from the spatial separation of counter
propagating beams of protons and antiprotons at comparable or somewhat improved
accuracies.[6] Based upon precisions obtained with trapped electrons, positrons and
protons, it seems very likely that the measurement uncertainty in the ratio of antiproton to proton masses could be reduced by more than 4 orders of magnitude, to order
10^{-9} or better.

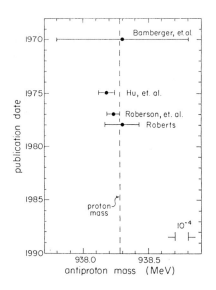

Fig. 1 Antiproton mass measurements

Comparison of the antiproton and proton masses is a test of CPT invariance since CPT invariance implies that the masses be the same. The current status of experimental tests of CPT invariance is summarized in the Particle Data Group compilation[7]. Since CPT invariance implies that a particle and antiparticle have the same magnetic moment (with opposite sign), the same inertial mass and the same mean life, the tests are so grouped for various baryons, mesons and leptons. The neutral kaon system provides a test of CPT invariance of striking precision. Equally striking, however, is that only 3 other tests exceed 1 part per million in accuracy, and these involve leptons only. In fact, there is not even a single precision test of CPT invariance with baryons. The widespread faith in CPT invariance is clearly based upon the success of field theories in general and not upon a dearth of precision measurements.

It is even conceivable that proton and antiproton masses could be different without a violation of CPT invariance. Precisely stated, CPT invariance relates the mass of a proton in a matter universe to an antiproton in an antimatter universe. A long range coupling to baryon number would not affect the kaon system but could shift differently the proton and antiproton masses, given the preponderance of baryons in our apparatus.

The scarcity of precise tests of CPT invariance makes the case for a precise comparison of proton and antiproton masses seem to be very strong to me, especially since no precise test at all involves baryons. Such a measurement also satisfies several additional criteria.

1. A big improvement in accuracy is involved, somewhere between four and five orders of magnitude.
2. A simple, basic system is involved.
3. The technique used will be convincing if the masses are found to differ.
4. The measurement will involve a reasonable effort.
5. It will be fun.

The last two criteria are more subjective than the others, but important nonetheless.

B. First Slowing and Capture of Antiprotons in an Ion Trap

Antiprotons are created at energies of several GeV. Precision experiments in Penning traps take place at millielectron volts (meV). An experimental difficulty, then, is to reduce the antiprotons kinetic energy by approximately 12 orders of magnitude. The first slowing, from GeV energies down to MeV energies takes place within LEAR. The unique capabilities of this machine are well known, so I will not discuss them further.

I am delighted to report that our TRAP Collaboration (PS196) has taken 21.3 MeV antiprotons from LEAR (200 MeV/c) and slowed them down to below 3 keV. At this energy they were caught in the small volume of an ion trap and held up to ten minutes. I should point out that this effort succeeded despite incredible time pressure. The capture of antiprotons, for example, occured during a single 24 hour period. A published account is available[8] so I will only briefly summarize.

The experiments went in two stages. In May 1986, we used a simple time-of-flight apparatus to measure the energy distribution of antiprotons emerging from a thick degrader. As degrader thickness is increased, the number of transmitted antiprotons drops as more of them are stopped in the degrader. The degrader thickness at the half intensity point is very close to the proton range which is compiled in standard tables. Most of these transmitted antiprotons have energies above 3 keV which is the highest energy we could trap. However, we were able to show that the number of antiprotons which emerge from the degrader with low kinetic energies (along the beam axis), between 2 and 8 keV, is clearly peaked at the half intensity point for antiprotons of all energies. Approximately 1 in 10^4 of the incident antiprotons emerges from the degrader with below 3 keV. These are the particles available for trapping.

In July 1986 we returned to LEAR for a 24 hour attempt to actually catch antiprotons in the small volume of an ion trap. The slowest antiprotons leaving the thick degrader are confined in 2 dimensions to field lines of the 6T superconducting magnet and are so guided through the series of 3 trap electrodes. As the antiprotons enter the trap, a first ring-shaped trap electrode (the entrance endcap) and a main ring electrode are both grounded. A third cylindrical electrode (exit endcap) is at -3 kV so that negative particles with energy less than 3 keV turn around on their magnetic field lines and head back towards the entrance of the trap. Approximately 300 ns later, before the antiprotons can escape through the entrance, the potential of the entrance endcap is suddenly lowered to -3kV, catching them within the trap. The potential is switched in 15 ns with a kryton circuit developed for this purpose and is applied to the trap electrodes via an unterminated coaxial transmission line.[9] The 3 keV potentials and 15 ns rise times contrast sharply with the several volt potentials and the 100 ns switching times used recently to capture Kr^+ in a few eV well.

After antiprotons are held in the trap between 1 ms and 10 minutes, the potential of the exit endcap is switched from -3 kv to 0 volts in 15 ns, releasing the antiprotons from the trap. The antiprotons leave the trap along respective magnetic field lines and annihilate at a beam stop well beyond the trap. The high energy charged pions which are released are detected in a 1 cm thick cylindrical scintillator outside the vacuum system. A multiscaler started when the potential is switched records the number of detected annihilations over the next $6\mu s$ in time bins of $0.4\mu s$. A second multiscaler records the pion counts over a wider time range with less resolution to monitor backgrounds.

Fig. 2 shows a time-of-flight spectrum for antiprotons kept in the trap for 100s. The spectrum includes 31 distinctly counted annihilations which corresponds to 41 trapped particles when the detector efficiency is included. We carefully checked that these counts are not electronic artifacts. When the high voltage on the exit endcap

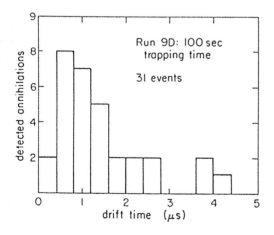

Fig. 2 Time-of-flight spectra

is switched to release antiprotons from the trap, a single count (occasionally two) is observed in the multichannel scalers. We take this to be time $t = 0$ and always remove a single count from the measured spectra. Otherwise, the background is completely negligible. When the potential of the entrance endcap is switched on just 50 ns before 3 keV antiprotons arrive in the trap, when the magnetic field is off, or when the -3 keV on one of the electrodes is adiabatically turned off and then back on during a 100s trapping time to release trapped antiprotons, no counts are observed.

The potential on the exit endcap is lowered quickly compared to the transit time of particles in the trap in order to maximize the detection efficiency. Even a small number of trapped particles can be observed above possible background rates in the $6\mu s$ window. For trapping times shorter than 100s, however, we actually released so many trapped antiprotons that our detection channel is severely saturated. For a 1 ms trapping time, we conservatively establish that more than 300 antiprotons are trapped out of a burst of 10^8, which corresponds to trapping 3×10^{-6} of the antiprotons incident at 21.3 MeV and 3% of the antiprotons slowed below 3 keV in the degrader. We observe that 5 particles remain in the trap after 10 minutes. This is actually based upon only two trials (since we were reluctant to use up our short time at LEAR holding antiprotons for long times), but both of these trials used a burst of antiprotons from LEAR of comparable intensity to that used for the 41 trapped particles of the 100s spectra in Fig. 3. If a simple exponential decay describes the number of particles trapped between 100 a and 10 minutes, the decay time is 240 seconds. An extrapolation back to the loading time $t = 0$, however, would then indicate that only 62 particles are initially trapped. We clearly observe many more for a trapping time of 1 ms, suggesting that antiprotons are lost more rapidly at earlier times.

A key point here is that the rate of cooling and annihilation via collisions with background gas will decrease with decreasing pressure. The background pressure can be made lower by orders of magnitude compared to the present vacuum by cooling a completely sealed vacuum enclosure to 4.2K. We thus expect a very significant increase in achievable trapping times.

My research group is supported by the National Science Foundation, the National Bureau of Standards (Precision Measurements Grant) and by the Air Force Office of Scientific Research. Others in the collaboration are supported by the D.O.E.

1. A more detailed review entitled "Penning Traps, Masses and Antiprotons" will appear in the Proceedings at the International School of Physics with Low Energy Antiprotons: Fundamental Symmetries, 1986, Erice.

2. A. Bamberger, et. al., Phys. Lett. **33B**, 233 (1970).

3. Hu, et. al., Nucl. Phys. **A 254**, 403 (1975).

4. P.L. Roberson, et. Al., Phys. Rev. C **16**, 1945 (1977).

5. B.L. Roberts, Phys. Rev. D. **17**, 358 (1978).

6. S. van der Meer, private communication.

7. Particle Data Group, Rev. Mod. Phys. **56**, S1 (1984).

8. G. Gabrielse, X. Fei, K. Helmerson, S.L. Rolston, R. Tjoelker, T.A. Trainor, H. Kalinowsky, J. Haas, W. Kells, Phys. Rev. Lett. **57**, 2504 (1986).

9. X. Fei, R. Davisson and G. Gabrielse, Rev. of Sci. Inst. (in press).

Antihydrogen Production

G. Gabrielse[1], L. Haarsma[1], S.L. Rolston[1], and W. Kells[2]

[1]Department of Physics, Harvard University, Cambridge, MA 02138, USA
[2]Fermi National Accelerator Laboratory, Batavia, IL 60510, USA

We know of 3 proposed methods to make antihydrogen. The first is to merge beams of antiprotons and positrons; moving at the same velocity within a storage ring.[1] The mechanism is radiative recombination,

$$p^- + e^+ \rightarrow \bar{H} + h\nu, \tag{1}$$

the extra energy being carried off by a photon. This process may be stimulated

$$p^- + e^+ + h\nu \rightarrow \bar{H} + h\nu' \tag{2}$$

with a possible increase of 100 in rate for stimulation to $n = 2$ [2]. Nonetheless, the antihydrogen production rate in a subsequent proposal is still very low, less than 1 per second. Radiative recombination is slow because the time required to radiate a photon is typically much larger than the interaction time between antiprotons and positrons.

The second method is to collide positronium e^+e^- with antiprotons stored in an ion trap:[3]

$$p^- + e^+e^- \rightarrow \bar{H} + e^- . \tag{3}$$

The cross section is much higher than for radiative recombination since the excess energy is carried off by the electron. However, the positronium is relatively hot compared to the 4.2K charged plasmas which can be realized in an ion trap. Also, positronium is short lived and neutral so it cannot be trapped in an ion trap. If we assume 10^4 antiprotons in a trap source, the anticipated rate[3] scales to 10^{-3}/sec.

We have recently suggested that another 3 body process might produce antihydrogen at a much higher rate:

$$p^- + e^+ + e^+ \rightarrow \bar{H} + e^+ . \tag{4}$$

The equivalent process with protons, electrons and hydrogen has been studied in some detail.[4] For a plasma of charged antiprotons and positrons at temperature T, this 3 body rate is proportional to the density of antiprotons, the square of the density of the positrons, the interaction volume and to $T^{-9/2}$. At 4.2K, for 10^4 antiprotons/cm^3, 10^7 positrons/cm^3 and an interaction volume of 1/cm^3, the instantaneous rate for antihydrogen production is 6×10^6/sec. The potentially high rate is very encouraging and is being investigated in more theoretical detail. including the consequences of electric and magnetic fields.

We are also beginning experimental studies with protons and electrons, to explore the possibility of producing a 4.2K plasma of antiprotons and positrons in an RF trap[5] or in a nested pair of Penning traps.[6] Recombination processes (1), (2) and (4) will be considered. The first experiment with antimatter is to measure the depletion of trapped antiprotons and positrons as they interact, along with the annihilation pions from antihydrogen hitting the walls of the trap. For the future it would be nice and perhaps necessary to capture antihydrogen as it is formed in a surrounding neutral particle trap.[7] We find it very encouraging that since we have been contemplating this difficult scenario, that copious amounts of hydrogen have been confined in a neutral particle trap[8] and that there are now intentions[9,10] to slow hydrogen and trap it as a means for doing more precise spectroscopy of hydrogen. With trapped and cooled antihydrogen, it would be possible to measure the gravitational acceleration of a neutral antihydrogen atom by observing how the gravitational force shifts the location of an antihydrogen atom in a magnetic trap. Again, we find it encouraging that such effects were recently and unexpectedly observed with trapped Na atoms.[11]

1. H. Herr, D. Mohl, A. Winnacker, *Physics at LEAR with Low-Energy Cooled Antiprotons*, Erice (1982) 659.

2. R. Neumann, H. Poth, A. Winnacker, A. Wolf, Z. Phys. A. **313**, 253 (1983).

3. B.I. Deutch, A. S. Jensen, A. Miranda, G. C. Oades, in *Proceedings of the First Workshop on Antimatter Physics at Low Energy*, edited by B. E. Bonner and L. S. Pinsky, Fermilab (1986).

4. Most recently by J. Stevefelt, J. Boulmer and J-F. Delpech, Phys. Rev. A. **12**, 1246 (1975).

5. H. Dehmelt, R. S. Van Dyck,Jr., P. B. Schwinberg, G. Gabrielse, Bull. Am. Phys. Soc. **24** 757 (1979).

6. G. Gabrielse, K. Helmerson, R. Tjoekler, X. Fei, T. Trainor, W. Kells, H. Kalinowsky, in *Proceedings of the First Workshop on Antimatter Physics at Low Energy*, edited by B. E. Bonner and L. S. Pinsky, Fermilab (1986).

7. G. Gabrielse, *Penning Traps, Masses and Antiprotons*, Invited Lecture at the International School of Physics with Low Energy Antiprotons: Fundamental Symmetries, Erice, Italy (1986).

8. H. Hess, G.P. Kochanski, J.M. Doyle, N. Masuhara, D. Kleppner and T.J. Greytak, to be published. Rev. Lett.

9. W. Phillips, et. al., (private communication).

10. T. Hänsch, et. al., (these proceedings).

11. V.S. Bagnato, G.P. Lafyatis, A.G. Martin, E.L. Raab, and D.E. Pritchard, Phys. Rev. Lett. **58**, 2194 (1987).

Excitation of the $1S$–$2S$ Transition in Muonium

S. Chu[1], A.P. Mills Jr.[1], A.G. Yodh[1], K. Nagamine[2], H. Miyake[2], and T. Kuga[2]

[1]AT&T Bell Laboratories, Holmdel and Murray Hill, NJ 07733, USA
[2]Meson Science Laboratory, University of Tokyo,
  Bunkyo-ku, Tokyo, Japan

Muonium, the bound state of a positive muon and an electron, is one of the simplest systems for testing QED. Spectroscopic measurements in muonium are particularly attractive since the atom does not possess the proton structure effects present in hydrogen. Positronium is free of nuclear structure offsets, but the relativistic two-body problem presents formidable calculational difficulties. The ultimate precision of a measurement of the muonium energy levels will be limited by the 2.2 μsec lifetime (72kHz line width) of the muon.

Our experiment was performed at the booster meson facility of the National Laboratory for High Energy Physics (KEK), in Tsukuba, Japan. 500 MeV protons incident on a Be target create a pion shower that was used to produce a low energy μ^+ beam (~100 μ^+/pulse at 27 MeV/c). The 20 Hz pulsed beam was directed onto a SiO_2 powder target where approximately one muonium per pulse was generated in vacuum at a density of ~4 X 10^{-2} atoms/cm^3.

The excitation of the 1S-2S transition in muonium follows from previous pulsed spectroscopy of positronium [2] and hydrogen [3]. The atoms were excited and ionized by counterpropagating light beams at 244nm placed ~4mm in front of the SiO_2 target. The light was produced by amplifying 50 mW of cw dye laser light at 488nm in an XeCl excimer pumped dye laser amplifier chain, and then doubling the frequency in a β-Ba_2BO_4 crystal. ~80mj/pulse at 488nm in a bandwidth of 35 MHz were converted into ~15mj/pulse at 244nm. A laser fluence of 0.25 joules/cm^2 in a fourier transform limited pulse is necessary to obtain an excitation probability of 0.5. Muons produced by the ionization of muonium were collected by an imersion lens and guided electrostatically to a microchannel plate detector. The frequency of the muonium resonance was measured relative to Te_2 line recently calibrated [4] to be at 613 881 150.89 (45) MHz. A 74.5 MHz frequency marker was used to measure the frequency interval between the reference line and the 1S-2S, F=1→1 muonium resonance.

Figure 1 is a plot of our best data as a function of laser frequency and time after the laser pulse. A time histogram of the counts for a laser frequency between 10 and 11.5 fringes to the blue of the Te_2 line is shown on the right. The peak at 1.4 μsec can be identified as muon counts caused by the laser pulse. This timing is consistent with the arrival of hydrogen ions produced in the target region by laser light. Muons created by the same light pulse should arrive $\sqrt{m_\mu/m_p}$ sooner than the protons. In figure 2 we present a plot of counts from all the data detected within 50 nsec of the expected arrival time as a function of laser frequency. Also shown in figure 2 is a fit to the data, and the expected position of the resonance after accounting for the QED prediction [5], the frequency shift of the cw dye laser with respect to the amplified pulse, the acousto-optic modulator shift used in the Te_2 spectroscopy and our best estimate of the ac Stark shift. The ac Stark shift was estimated by using the data of figure 2 to give an estimate of the

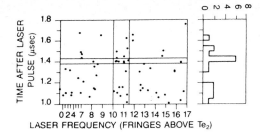

Fig. 1 A time-frequency plot of the CEMA during our best run. The scan between -2 to +7 fringes relative to the Te$_2$ line was a factor of five faster than the scan between +7 and 17 fringes.

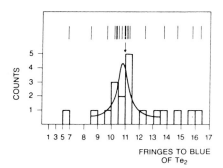

Fig. 2 Sum of all data taken in the time - laser frequency counting mode. The fit to the actual data (before binning into 0.5 fringe bins) gives a line center at 10.8 ±0.4 fringes with a fwhm of 0.9 ±0.3 fringes. Peak amplitude = 0.77 ±0.55 counts per 0.1 fringe and a background of 0.087 ±.055 counts. 1 fringe = 74.5 MHz.

width of the resonance line. By scaling the results from the positronium 1S-2S measurement where the Stark broadening to shift ratio is ~2.5[2], we estimate that the ac Stark shift is 30 ($^{+35}_{-15}$) MHz. All other uncertainties are less than 10 MHz.

We have shown the feasibility of laser spectroscopy experiments on muonium. Clearly a more intense source of thermal muonium is desirable. Active work on muon moderators, improved muon-to-muonium conversion targets, and brighter pulsed meson sources are underway in several laboratories. We anticipate a 3 order of magnitude improvement in thermal muon sources in the near future. Coupled with improvements in uv sources and detection efficiency, a high resolution measurement of the 1S-2S transition should be possible.

REFERENCES:

1. G. A. Beer, et. al., Phys Rev. Lett. 57, 671 (1986).
2. S. Chu, A. P. Mills, Jr., and J. L. Hall, Phys. Rev. Lett. 52, 1689 (1984).
3. E. A. Hildum, et. al., Phys. Rev. Lett. 56, 576 (1986); J. R. M. Barr, et. al., Phys. Rev. Lett. 56, 580 (1980) and references contained within.
4. D. H. McIntyre and T. W. Hansch, submitted to Phys. Rev. A, Rapid Communications (1987).

Atomic Physics in Confined Space:
Suppressing Spontaneous Emission at
Optical Frequencies and Measuring
the Van der Waals Atom-Surface Interaction

S. Haroche[(*)], *A. Anderson, E. Hinds, W. Jhe, and D. Meschede*
Yale University, New Haven, CT, USA

The radiative properties of an atomic system depend upon the structure of the vacuum field in the surrounding space. By confining the atoms in cavities in which the mode distribution of the field is very different from its free space configuration, it is possible to strongly alter the atomic spontaneous radiation rates and transition frequencies. These Cavity Quantum Electrodynamics effects, as they are called, have been first demonstrated in the microwave domain [1] and begin now to be observed in the optical domain [2-3]. Interest in these studies is manifold. First, the quantum noise of the electromagnetic field is a basic limitation to the precision of any optical experiment. Understanding how this noise is altered in a cavity is very important in order to analyze the ultimate precision of experiments on quantum radiators placed in the vicinity of metallic boundaries [4]. Being able to suppress spontaneous emission and to prepare quasi infinitely long-lived atomic excited states is a possibility which is certainly worth considering for ultra high resolution spectroscopy experiments. At a more applied level, measuring atomic radiative rates and transition frequencies in a small metallic structure is a new way of performing atom-surface physics. By analyzing the perturbations of the atomic radiative properties, one might get useful information on the electronic properties of the metal surfaces...

1. Experimental Arrangement

A conceptually very simple and flexible experimental arrangement for the study of these effects is sketched on Figure 1. A beam of alkali atoms is sent through a narrow tunnel of length ℓ_0, made of two gold-coated plane parallel mirrors stacked together with thin metal spacers of width w between them. Light excitation can be performed with laser beam L_1 in position A in front of the tunnel in order to prepare the entering atoms in selected excited states. State analysis is performed at the tunnel exit with the help of

---

(*) also at Ecole Normale Supérieure, Paris.

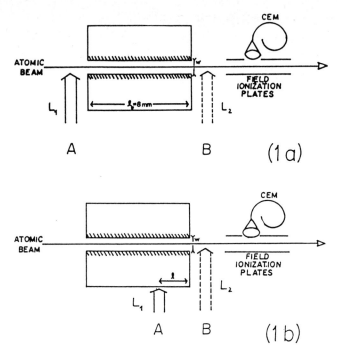

Fig. 1 : Experimental Arrangement : a beam of alkali atoms is passed through a narrow tunnel made of two gold coated mirrors stacked together with thin spacers of width w between them. Lasers are used to excite the atoms upstream and to prepare them for detection downstream (fig. 1a). The length ℓ travelled by the excited atoms inside the tunnel can be varied by focusing laser L_1 inside the tunnel at a distance ℓ from gap exit (Fig. 1b). Detection is performed by field ionization, the resulting electrons being counted by a channeltron electron multiplier (CEM)

a field ionization atomic detector : if the emerging atoms are in their ground state or in a low-lying excited state, a laser beam L_2 in position B can provide excitation from these states into a selected Rydberg level which will be subsequently field ionized and detected by the channeltron electron multiplier CEM (Fig. 1a). Alternatively the atoms can also be excited inside the tunnel structure by focusing the laser beam L_1 inside the gap through the semi transparent gold coating (Fig. 1b). In this way, the length ℓ travelled by the excited atom in the tunnel and the time spent in the gap can be continuously varied. Tunnels with a total length $\ell_o = 8$ mm, as narrow as 0.5 μm can be realized in this way, with detected atomic fluxes in the 10^3-10^{-4} s^{-1} range for w = 1 μm. The emerging atomic beams are extremely well collimated [5] (w/$\ell_o \sim 10^{-4}$ rad) and have exceedingly small kinetic energy in the direction normal to the mirrors (effective transverse temperature in the microkelvin range !).

31

2. Complete Suppression of Spontaneous Emission at Optical Frequency

This set-up has allowed us to demonstrate for the first time the complete suppression of spontaneous emission at optical frequencies [2]. We have prepared Cesium atoms in the $5D_{5/2}$ level and sent these excited atoms through an $\ell_0 = 8$ mm long tunnel of width w = 1.1 μm. The $5D_{5/2}$ state spontaneously emits radiation at a wavelength $\lambda = 3.49$ μm, down to the $6P_{3/2}$ level (the branching ratio of this transition is 1). The average tunnel crossing time for atoms in a thermal beam is ~ 20 μs, i.e. thirteen times the natural lifetime of the $5D_{5/2}$ level ($\tau_0 = 1.6$ μs). Only one excited atom out of 2.10^6 could survive such a long time in free space. We have observed however that nearly all the atoms were still in the $5D_{5/2}$ level at the tunnel exit, provided they were initially excited in a substate of maximum angular momentum along the direction normal to the mirrors. In this experiment, the mirror structure acts as a wave guide beyond cut-off for radiation polarized parallel to the mirrors with a wavelength λ larger than 2w = 2.2 μm. Since the substates of the $5D_{5/2}$ level with maximum angular momentum projection along the normal to the mirrors can radiate only light polarized parallel to the surfaces, their emission is fully inhibited. The other substates have a probability of emitting photons polarized normal to the mirrors, in an "electrostatic" mode which is not cut-off in the structure. Their emission is thus not suppressed (actually it may even be slightly enhanced) and these states do not survive the gap crossing in the excited state. The angular dependence of the emission rate in the gap is demonstrated by studying the effect of a magnetic field on the excited state transmission : a field component parallel to the surface mixes the maximum angular momentum substates with shorter lived ones and quenches the level metastability. Figure 2 shows the $5D_{5/2}$ state transmission through the tunnel as a function of the angle between the magnetic field and the normal to the mirrors. The transmission is important near $\theta = 0°$ and $180°$ and drops to zero in between. Such an experiment can be viewed as a kind of Hanle experiment on the atom + cavity system and demonstrates the anisotropy of the quantum field fluctuations in the tunnel structure.

3. Measuring the Van der Waals Forces between a Rydberg Atom and a Metallic Surface

The modification of the spontaneous emission rate in the mirror gap is due to the alteration of the density of the field modes resonant with the atomic transition. The cavity also changes the structure of the modes which are off-resonant and which are responsible for energy level shifts. In the

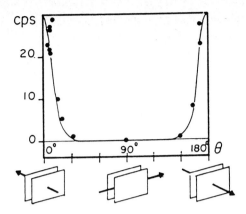

Fig. 2 : Spontaneous emission inhibition experiment at optical frequency : excited state transmission through the tunnel versus the angle θ between the magnetic field and the normal to the mirrors. Points are experimental and solid line theoretical. The angle dependence reflects the anisotropy of the vacuum field fluctuations in the confined space (from ref. 2)

vicinity of a cavity wall, this results in a position dependent shift of the atomic energy levels, proportional to z^{-3} where z is the atom-surface distance. This effect is nothing but the well known Van der Waals atom-surface perturbation, which was first analyzed in the context of QED by Casimir and Polder [6]. The derivative of this shift, scaling as z^{-4}, is the Van der Waals force pulling the atom to the nearer surface. On ground state atoms, this force is exceedingly small unless z is smaller than a few hundred Angstroem and the Van der Waals atom-surface deflection is very difficult to observe [7]. If the atom is excited into a Rydberg level of principal quantum number n, the force scaling as n^4 can however become huge and its effects are noticeable at micrometre distances from surfaces. Using the basic set-up shown on Fig. 1, we have indeed been able to observe the deflection of Rydberg atoms towards the metallic surfaces of the mirrors [8]. Cs and Na atoms prepared in states of chosen quantum number n were sent through the tunnel and the transmission ratio through the tunnel was measured for various values of n, ι and w. The atoms deflected onto the walls stick to them regardless of excitation [5] and do not emerge at the other end. The fraction of atoms escaping wall collision is thus directly related to the strength of the Van der Waals force. For given values of ℓ and w, there exists a maximum quantum number n_m above which no atom can emerge. Simple scaling arguments show that n_m, ℓ, w and the beam translation temperature T_0 are related by the relation $n_m^4 \ell^2 / w^5 T_0 =$ constant. We have plotted on Fig. 3 the experimentally measured n_m values

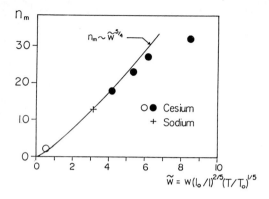

Fig. 3 : Maximum quantum number n_m of a Rydberg atom transmitted through a tunnel of "reduced" width $\tilde{w}$. The solid line is theoretical corresponding to a cut-off in transmission of 1%. Circles are experimental values for Cesium and cross for Sodium. The unshaded circle indicates the ground state Cesium cut-off, estimated from reference [7]. There is good agreement between Van der Waals theory and experiments up to $\tilde{w} = 6$ µm and $n_m \sim 25$. Deviation from theory for larger gaps is attributed to stray-field ionization in the tunnel of the very large Rydberg atom

as a function of a reduced gap width $\tilde{w} = w(\ell_0/\ell)^{2/5}(T/T_0)^{1/5}$. The solid line theoretical curve $n_m \sim (\tilde{w})^{5/4}$ is in good agreement with the experimental results, at least for gaps smaller than 6 µm and $n < 25$, above which we suspect other competing effects to take place (field ionization of very excited atoms by stray fields in the gap). This preliminary experiment demonstrates that it is possible to observe and study the Van der Waals interaction of Rydberg atoms with surfaces, at distances which are very large at the usual atomic scale.

The direct measurement of the Van der Waals energy shifts (of the order of tens of MHz for $n \sim 20$ and $z \sim 1$ µm) requires to keep the atoms at a well defined distance from the metal inside the gap. We plan to accomplish it by channelling the atoms [9] inside the tunnel using the dipole force produced by a standing light wave in the gap. It is well known that an atom interacting with a light field near resonance experiences a force which pushes it towards weaker electric fields when the light frequency is above resonance. We still pass our atomic beam through a narrow mirror tunnel as in the previous experiments, but the gap irradiated by a laser beam perpendicular to the atomic beam serves now also as Fabry-Pérot etalon with nodal plans parallel to the mirror surfaces. Atoms travelling in the structure experience a dipole force attracting them towards the nodes and competing with the Van der Waals force. This channelling process is effective without precooling stage because we are in effect selecting in our gap only those atoms which are already exceedingly cold in transverse direction. In this way, we should be able to fix the atom–surface distance to integer multiples of half the light wavelengths and hopefully we will be able to perform precise Van der Waals atom–metal spectroscopy with the help of a second laser exciting the channelled atoms into a Rydberg level. We have preliminary evidence that the channelling process is indeed working in our Fabry–Pérot tunnel, but further tests and experiments are necessary before the technique can be used for precise atom–surface spectroscopy.

[1]   P. Goy, J. M. Raimond, M. Gross and S. Haroche: Phys. Rev. Lett; 50, 1903 (1983); G. Gabrielse and H. Dehmelt: Phys. Rev. Lett. 55, 67 (1985); R. G. Hulet, E. S. Hilfer and D. Kleppner: Phys. Rev. Lett. 55, 2137 (1985)

[2]   W. Jhe, A. Anderson, E. A. Hinds, D. Meschede, L. Moi and S. Haroche: Phys. Rev. Lett. 58, 666 (1987)

[3]   D. J. Heinzen, J. J. Childs, J. E. Thomas and M. S. Feld: Phys. Rev. Lett. 58, 1320 (1987)

[4]   L. S. Brown, G. Gabrielse, K. Helmerson and J. Tan: Phys. Rev. Lett. 55, 44 (1985)

[5]   A. Anderson, S. Haroche, E. A. Hinds, W. Jhe, D. Meschede and L. Moi: Phys. Rev. A34, 3513 (1986)

[6]   H. B. Casimir and D. Polder: Phys. Rev. 73, 360 (1948)

[7]   D. Raskin and P. Kusch, Phys. Rev. 179, 712 (1969); A. Shih, D. Raskin and P. Kusch: Phys. Rev. A 9, 652 (1974)

[8]   A. Anderson, S. Haroche, E. A. Hinds, W. Jhe and D. Meschede: to be published

[9]   C. Cohen-Tannoudji, J. Dalibard, A. Heidmann, C. Salomon, A. Aspect and H. Metcalf: in these Proceedings.

Enhanced and Suppressed Visible Spontaneous Emission by Atoms in a Concentric Optical Resonator

D.J. Heinzen, J.J. Childs, C.R. Monroe, and M.S. Feld

G.R. Harrison Spectroscopy Laboratory and Department of Physics, Massachusetts Institute of Technology, Cambridge, MA 02139, USA

Recently there has been widespread interest in enhanced and suppressed spontaneous emission by atoms in cavities. Changes in spontaneous emission occur when the density of modes in the cavity differs from that of free space. One means of accomplishing this is to choose a cavity with dimensions comparable to a wavelength, so that the cavity mode spectrum is discrete, or below cutoff in the case of a waveguide. [1–3] Recently, we introduced an alternate approach and have observed enhanced and suppressed visible spontaneous emission of the resonance transition of Yb in a confocal resonator with dimensions much larger than a wavelength [4]. Our experiments make use of the fact that the transverse modes of this resonator are degenerate, so that many modes are simultaneously brought into and out of resonance as the resonator is tuned.

In the Yb experiments we observe enhancement and inhibition of the partial emission rate into the resonator modes by a factor $\sim 1/(1-R)$, where R is the mirror reflectivity. The changes in the total rate are proportional to the solid angle $\Delta\Omega$ subtended by the mirrors, and were small (+1.6%, −0.5%) because $\Delta\Omega = 0.04$ sterad was small. The maximum solid angle was limited by spherical aberration; the ideal mirror surface to maintain perfect degeneracy is parabolic, whereas the actual mirrors were spherical. However, this is not a fundamental limitation.

We have begun a series of new experiments which overcomes this limitation by use of a concentric resonator. A beam of atomic barium is collimated by a 25μm pinhole just before entering the center of the cavity and intercepted by a cw laser which is focussed to a $\sim 70\mu$m spot. The laser frequency is locked to the $^1S_0 - {}^3P_1$ transition of ^{138}Ba near 553 nm; the power is 0.1 μW. The atoms are positioned exactly in the center of the cavity. The mirrors have a radius of curvature of $a = 2.5$cm, diameter 3.5cm, and are coated with a thin bare aluminum film of reflectivity R = 0.67. This corresponds to a solid angle $\Delta\Omega = 3.6$ sterad. The spontaneous emission emerging through one of the mirrors is collected by an $f1.2$ lens system and monitored by a PMT. In addition, the light emitted out the sides of the cavity is collected by a fiber optic bundle and is monitored in a second PMT. We are thus able to monitor both the state of the photons (on–axis light) and the state of the atoms (side light) simultaneously.

Our first preliminary results are shown in Fig 1. Curve (a) shows the count rate emitted through one mirror as a function of cavity tuning. Curve (b), obtained with the back mirror blocked, provides a calibration of the free space rate into the same solid angle. As expected, the spontaneous emission into the cavity is strongly modulated by the cavity tuning. Curves (c) and (d) show the count rate recorded by the side-light detector, (c) with the cavity open and (d) with it blocked. Under condition of weak excitation, and natural linewidth limited lineshapes, it is

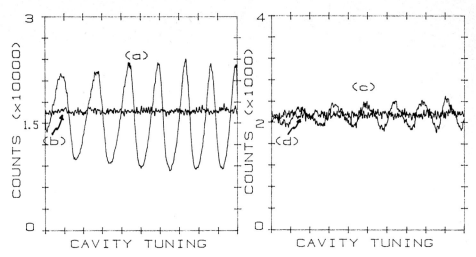

Figure 1: Concentric resonator experimental results

easy to show that $\Delta I_{side}/I_{side} = -2\Delta\Gamma/\Gamma$. We observe a 12% increase and 12% decrease in the sideways intensity, which thus corresponds to a 6% decrease and 6% increase in the total spontaneous emission rate, respectively.

Enhanced spontaneous emission as reported above is distinct from laser emission. It is a pure spontaneous emission effect, with all the characteristic properties. Thus, the emission rate is proportional only to the upper level occupation probability, N_{upper} (and not to $N_{upper} - N_{lower}$), and with increasing emission rate the linewidth of the emitted light broadens (instead of narrowing, as it does in laser amplification). However, under the appropriate conditions the number of photons in the resonator can grow large compared to unity, and stimulated emission can then occur. Thus, based on the resonator design principles discussed above, it should be possible to construct a laser with only a single atom as the gain medium. There are several interesting possibilities. In one, the atom would be optically pumped to produce a steady state inversion. The usual laser threshold condition then applies:

$$[(N_{upper} - N_{lower})/V_{eff}]\sigma L \geq 1-R \ , \tag{1}$$

with $V_{eff} = 2\lambda^2 L/\Delta\Omega$ the effective volume of the resonator, $\sigma = (3/2\pi)\lambda^2$ the stimulated emission cross section and $L=2a$ the resonator length. For complete inversion, this gives

$$3\Delta\Omega/[4\pi(1-R)] \geq 1 \ , \tag{2}$$

which is the same condition as for the enhanced emission rate into a given solid angle of the resonator to exceed the corresponding free space rate. The outcome of the current experiments will put us in a good position to determine if the necessary conditions for producing a single atom laser can be achieved.

References

1. P. Goy, S.M. Raimond, M. Gross and S. Haroche, Phys. Rev. Lett. $\underline{50}$, 1903 (1983).
2. R.G. Hulet, E.S. Hilfer and D. Kleppner, Phys. Rev. Lett. $\underline{55}$, 2137 (1985).
3. W. Jhe, A. Anderson, E.A. Hinds, D. Meschede, L. Moi and S. Haroche, Phys. Rev. Lett, $\underline{58}$, 666 (1987)
4. D.J. Heinzen, J.J. Childs, J.E. Thomas, and M.S. Feld, Phys. Rev. Lett. $\underline{58}$, 1320 (1987)

Single Atomic Particle at Rest in Free Space: Shift-Free Suppression of the Natural Line Width?

H. Dehmelt

Department of Physics, University of Washington, Seattle, WA 98195, USA

1. Introduction

An increase by a factor 20 in resolution is about as important when measuring the g-factor of the electron, as a 20-fold increase in accelerator energy is in high-energy physics. An antenna in a realistic spherical cavity resonator had been analyzed [1] by SCHELKUNOFF in '43. PURCELL showed in '46 that the life-time of atoms placed in a resonant cavity is greatly reduced. In a perfect non-resonant cavity one naively expects the opposite: no radiation can be lost, ergo no damping of an atomic oscillator. Indeed, a 10-fold increase in the lifetime was observed by GABRIELSE et al. in '84. The reduction of the natural width by tuning between cavity resonances had been proposed by DICKE in '59. After shooting an atomic beam through a parallel plate wave guide < $\lambda/2$ = 0.22 mm apart and reducing the decay rate >20-fold, KLEPPNER et al., in '85 nevertheless reached the conclusion, that cavity narrowing of spectral lines is 'fundamentally flawed by nonradiative shifts'. Such pessimism is unjustified for an individual particle closely centered in a cavity >> $\lambda/2$, VAN DYCK et al. '81. Here, the trap electrodes form a microwave cavity good enough, to greatly reduce the natural line width, and, by proper relative tuning of cavity and atomic resonances, any shift of the atomic transition may be nulled.

2. Elastically Bound Electron in Strip Line Cavity: an Exercise

Classical dispersion theory successfully describes the interaction of matter and radiation. Thus, we model the atomic transition by an elastically bound electron with adjustable charge, and represent it by an lc circuit. Free space we model as an infinite strip line of spacing a, width b, and characteristic (ohmic) impedance Z_0. When cut off and shortcircuited at both ends, waves from the electron are reflected and a symmetric 1-D 'cavity' is formed [2], see Fig. 1. Then the electron sees the impedance [1]

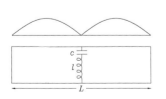

Fig. 1 Elastically bound electron in simple loss-free strip-line 'cavity' of length L. The electron is represented by the lc circuit. When excited, it produces a standing wave on the transmission line. For the critical tuning shown, an electric field node falls on the electron, the line has zero impedance and the frequency of the atomic resonance is not shifted, while its width vanishes.

$$j\omega_e L_e = \tfrac{1}{2} j Z_0 \tan Y, \quad Z_0 = \mu_0 \underline{c} a/b.$$

The eigenfrequency $\omega_e + \delta$ of the $l c L_e$ combination, the electron in cavity, is shifted vs ω_e for the $l c \tfrac{1}{2} Z_0$ combination, the free space case, due to the effective cavity inductance L_e, and given by

$$(\omega_e + \delta)^2 (1 + L_e)c = 1, \quad l = ma^2/e^2.$$

For δ, the cavity shift, then holds

$$-\delta/\omega_e = L_e/2l = (Z_0 L/8\underline{c}l)\ F(Y)$$

$$= (\pi r_e L/2ab)\ F(Y), \quad F(Y) = \tan Y\ /Y.$$

Here we have introduced $\underline{c}$, the speed of light, the shift function $F(Y)$, the normalized electron frequencies $Y = \pi X$, $X = L/\lambda = \omega_e/\Omega_D$, the "dimension" frequency $\Omega_D = 2\pi\underline{c}/L$, coinciding with the second cavity eigenfrequency, and the classical electron radius $r_e = \mu_0 e^2/4\pi m = 2.82 \cdot 10^{-13}$cm. The cavity has eigenfrequencies at $X = \tfrac{1}{2}$, 1, 1½, ..., where the electron sees alternately shunt and series resonances. We note the zero-shift of the electron frequency for $F(Y) = 0$ at the $X = 1, 2, 3, \ldots$, cavity eigenfrequencies, where the cavity looks like a short circuit. I extended this treatment to a cylindrical cavity in ´84/85, and to a spherical one in ´85. BOULWARE et al. showed in ´85, that the full QED apparatus gives the same shift as Maxwell´s equations. For the cylinder BROWN et al. in ´85 developed a more accurate image charge formalism. With this formalism, BROWN et al. [3] in ´86 confirmed my ´85 accurate spherical cavity results presented below.

3. Electron in Spherical Cavity and Zero-Shift Tuning
The task at hand is the suppression of the natural line width, and at the same time approximating the individual atom at rest in free space. Obviously, one must minimize interactions with the surroundings, and, in the face of an irreducible residue, simplify the surroundings as much as possible. To this end I have adapted the ´43 antenna-in-spherical-cavity treatment [1] to an atomic radiator. In summary: in vacuum oscillating dipole emits spherical wave. In a cavity this same wave is reflected by perfectly conducting wall, reflected wave is reflected again at center and so on. With other words, in addition to HERTZIAN vacuum wave traveling outward, a standing wave, FIG. 2, is set up, satisfying the boundary conditions at cavity wall. Only the standing wave is responsible for all cavity back-action on the physical electron of interest to us. In steady state the energy traveling outward in HERTZIAN wave is re-supplied to radiator by standing wave: atomic resonator loses no energy, radiation damping and natural line width of modified atomic transition vanish for any size cavity.

What about shifts in atomic frequency? The E-field of standing wave at site of the radiator may be decomposed into one component, of constant amplitude, $\pi/2$ out of phase with displacement z of oscillating electron. This component does net work on electron. The other, reactive component, vanishing in FIG. 2, is in phase with the displacement; it modifies the force constant (or mass, if you wish) of the electron oscillator and thereby shifts its frequency. When now radius "a" of cavity is slowly decreased,

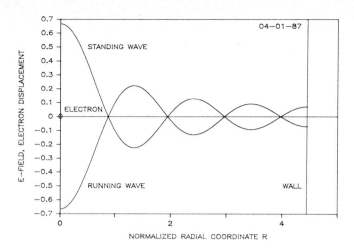

Fig. 2 E-Field distribution in spherical cavity at instant when electron displacement vanishes. Standing wave field is at maximum, π/2 out of phase with electron: no shift! Normalized radial coordinate R ≡ 2r/λ.

then, at wall, amplitude of traveling wave changes little. Likewise, at the electron the work-component of the standing wave field does not change. However, the reactive component of field there assumes huge values, when cavity wall approaches node in the standing wave. This means, of course, that an eigenfrequency of the cavity comes in resonance with the radiating oscillator. On the other hand, and most important, for tuning placing a maximum of standing wave E-field at wall, as shown in FIG. 2, the reactive component at electron and associated <u>cavity shift of its frequency vanish</u>. Then standing wave E-field at the electron and its displacement z are obviously π/2 out of phase! Figure 3 shows a shift function $F(Y) \rightarrow -Y \cot Y$, very similar to that of strip-line resonator. The function $|F| < 5$ in ±2% band around cavity eigenfrequency $\Omega(TE501)$, Fig. 4. This corresponds to a shift

$$|\delta|/\omega_e < 1.88 \cdot 10^{-13} |F(\nu)| \approx 10^{-12},$$

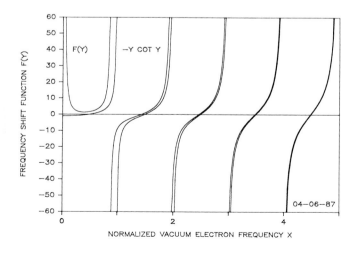

Fig. 3 Cavity shift for spherical cavity of radius a. The electron frequency is plotted as $X = 2a/\lambda$. The function $F(Y)$ quickly approaches $-Y \cot Y$, $Y \equiv \pi X$.

41

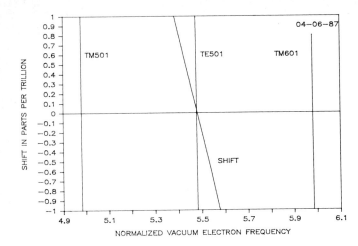

Fig. 4 Zero-shift tuning near 165 GHz in 1cm-diameter spherical cavity. As for strip line cavity, tuning the electron frequency to a non-ineracting (TEn01) eigen-oscillation produces nearly perfect zero-shift.

is ≈10X smaller than for a comparable cylindrical cavity [3] in the same band width, and follows from my general formula, valid for all frequencies, from dc to optical,

$$-\delta/\omega_e = (r_e/3a) \, F(Y)$$

$$F(Y) = Y \frac{Y \tan Y - (Y^2 - 1)}{(Y^2 - 1) \tan Y + Y} \; ,$$

with X, Y as above, L = 2a, and F(Y) = 0 for X ≈ 1½, 2½, 3½ .. For zero-shift one simply tunes ω_e to the easily measured TE501 cavity eigenfrequency. In a cloud < λ/2π of n ≈ 1000 electrons the shift is much larger, $\delta_n \approx n\delta$; at δ = 0 points the invariance of the measured $\omega_e + \delta_n$ under variation of n should be demonstrable! A more detailed paper is in preparation. The work was done under the auspices of the National Science Foundation.

1. S.A. Schelkunoff: Electromagnetic Waves (Van Nostrand, New York 1943), pp. 294 & 197
2. H.G. Dehmelt, Proc. Natl. Acad. Sci. USA 81, 8037 (1984) & 82, 6366 (1985)(erratum)
3. L.S.Brown, K.Helmerson and J.Tan, Phys.Rev. A34, 2638 (1986)

Parity Violation in Atoms

E.D. Commins

Physics Department, University of California, Berkeley, CA 94720, USA

1. Basic Features

Parity nonconservation (PNC) in an atom arises from interference between the neutral weak and electromagnetic interactions that couple a valence electron to the nucleus. The PNC neutral weak coupling is decribed by an effective zero-range pseudoscalar potential $H_p = H_1 + H_2$ that must be added to the usual atomic Hamiltonian. The dominant portion H_1 involves the axial and vector parts of the electronic and nucleonic neutral weak currents, respectively. H_1 is proportional to Fermi's constant G_F and to the "weak charge" Q_W. (In the standard model, $Q_W = Z(1-4\sin^2\theta_W)-N$ before radiative corrections, and Z=proton number, N=neutron number, while θ_W is Weinberg's angle.) H_1 connects only $s_{1/2}$ and $p_{1/2}$ orbitals, and its matrix elements vary roughly as $Z^2 Q_W$. H_2 arises from the coupling of vector electronic and axial hadronic weak currents, and its matrix elements are several orders of magnitude smaller than those of H_1. All experiments completed so far have been sensitive only to H_1; they may be regarded as measurements of Q_W, and thus of $\sin^2\theta_W$.

The first observations of PNC in atoms were reported by Barkov and Zolotorev of Novosibirsk[1]; they employed optical pumping in bismuth. Since then, optical rotation measurements have been carried out in Bi at Moscow[2], Oxford[3], and Seattle[4], and the latter group has also reported observations in Pb[5]. Stark interference experiments in ^{133}Cs have been completed at Paris[6] and Boulder[7], and somewhat similar experiments in ^{205}Tℓ have been done at Berkeley[8].

2. Stark Interference Experiments

The cases of interest are the $6^2S_{1/2}-7^2S_{1/2}$ transition (539 nm) in Cs, and the $6^2P_{1/2} - 7^2P_{1/2}$ transition (293 nm) in Tℓ, (see Fig.1). Each is a highly forbidden M1 transition, with amplitude M. However, because of PNC the 6S, 7S states of Cs are mixed with $P_{1/2}$ states; while in Tℓ the $6P_{1/2}$,$7P_{1/2}$ states are mixed with S states. Thus each transition amplitude acquires a PNC electric dipole component E_p. In an external electric field E there is also Stark mixing of 6,7S P levels in Cs, while in Tℓ, $6,7P_{1/2}$ states are mixed with $S_{1/2}$,$D_{3/2}$ states. The resulting Stark-induced transition amplitude for stimulated absorption when the incoming photon polarization $\hat{\epsilon}$ is perpendicular to E is called βE. In each Stark interference experiment one measures Im $E_p/\beta E$ and β must be determined independently for extraction of E_p. The determination of β is done in Cs by a precise semi-empirical analysis,[6,9], and in Tℓ by direct measurement[10].

Figures 2a,b,c indicate the geometrical arrangements in the Paris and Boulder Cs experiments and the Berkeley Tℓ experiment, respectively. In Fig. 2a (Paris) a circularly polarized 539 nm laser beam with wave vector k along +y passes through Cs vapor in an electric field E along x. Interference between

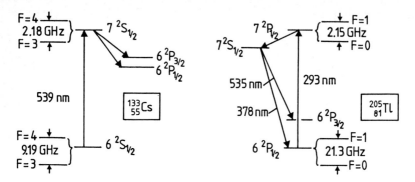

Fig.1   Energy level diagrams

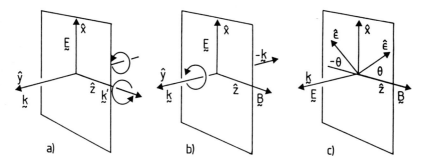

Fig. 2   Geometrical arrangement of vectors in Stark interference experiments
a) Paris[6]   b) Boulder[7]   c) Berkeley[8]

M and βE, as well as between E_p and βE results in polarization of the 7S state in the z direction. This polarization P is detected by observing the circular polarization of fluorescence accompanying the decay 7S-6P. That part of P depending on M/βE reverses sign with k, while that depending on Im $E_p/\beta E$ reverses sign with incoming helicity h. The pseudoscalar term in the transition probability arising from PNC is proportional to

$$h \, \vec{k} \times \vec{E} \cdot h'\vec{k}', \tag{1}$$

where h' is the helicity of the fluorescent decay photon and k' is its wave vector. In the Boulder experiment (Fig.2b) one has an elliptically polarized optical standing wave, indicated by vectors k,-k along y. Again E is along x. The cesium atoms are now in an atomic beam, which makes possible narrow line-widths. A magnetic field B along z splits various Zeeman sub-levels, so that the laser can be tuned to individual $F,m_F \to F',m_F'$ transitions. Thus it is not necessary to detect circular polarization in the decay fluorescence, but merely intensity. The pseudoscalar here is

$$h \, \vec{k \times E} \cdot \vec{B} . \tag{2}$$

In the Berkeley Tℓ experiment, E is along y (parallel to the 293 nm laser beam). As in the previous case a magnetic field B is imposed in the z direction to resolve Zeeman components. The incoming photons are now linearly polarized and the pseudoscalar invariant is

$$\hat{\varepsilon} \cdot \vec{B} \quad \hat{\varepsilon} \cdot \vec{E} \times \vec{B} . \tag{3}$$

One detects fluorescence at 535 nm from the decay $(7P_{1/2} \to) 7S \to 6P_{3/2}$. In the Cs experiments the contributions of M-βE interference cancel because the exciting light propagates in both directions $(y,-y)$. In the Tℓ experimental configuration of Fig.2c there is no M-βE interference.

3. Stark Interference Results

a) Cesium. The Paris group has obtained the result:

$$E_p = (-0.79 \pm 0.10) \, i|e|a_0 \cdot 10^{-11} \qquad \text{(Paris)} \tag{4}$$

while Boulder has obtained:

$$E_p = (-0.86 \pm 0.07) \, i|e|a_0 \cdot 10^{-11} . \qquad \text{(Boulder)} \tag{5}$$

These results may be compared to the many-body atomic calculations of Dzuba et al[11]:

$$\frac{E_p}{(Q_{W/-N})} = -(0.89 \pm 0.02) \, i|e|a_0 \cdot 10^{-11} \tag{6}$$

and of Martensson-Pendrill[12]:

$$\frac{E_p}{(Q_{W/-N})} = -(0.886 \pm \text{a few percent}) \, i|e|a_0 \cdot 10^{-11} . \tag{7}$$

Combination of (4) and (5) and comparison with (6) yields Q_W and thus $\sin^2\theta_W$. In Table 1. we present determinations of $\sin^2\theta_W$ from various high energy experiments, as well as the Cs result. It can be seen that the precision attained in $\sin^2\theta_W$ by the cesium results cannot rival that achieved in the deep-inelastic neutrino-nucleon scattering experiments. Nevertheless, the atomic physics results are useful in qualitative terms, because they demonstrate the validity of the standard model over an enormous range of q^2.

b) Thallium. The Berkeley group has obtained:

$$E_p = (6.9 \pm 1.3) \, i|e|a_0 \cdot 10^{-11} , \qquad \text{(Berkeley)} \tag{8}$$

which may be expressed as

Table 1. Determination of $\sin^2\theta_W$ from various reactions. Where two uncertainties are shown, the first is experimental, the second theoretical. See[13].

Reaction	$\sin^2\theta_W$	Relative Weight	q^2, GeV^2/c^2
PNC in Cs	0.209+.018+.014	.06	$5 \cdot 10^{-6}$
$\nu_\mu e \to \nu_\mu e$	0.223+.018+.002	.16	$4 \cdot 10^{-2}$
$\nu_\mu p \to \nu_\mu p$	0.210+.033	.06	
SLAC pol e/D	0.221+.015+.013	.08	7
μN	0.25 +.08	.01	90
Deep inel.νN	0.233+.003+.005	1	10^2
$\bar{p}p$ (W,Z masses)	0.228+.007+.002	.75	10^4
All Data	0.230+0.0048		

$$\frac{E_P}{(Q_{W/N})} = -0.75(1 \pm 0.19) \, i|e|a_0 \cdot 10^{-10} \qquad (9)$$

assuming $\sin^2\theta_W = 0.23$. This may be compared with a recent calculation by Dzuba et al[11]:

$$\frac{E_P}{(Q_{W/N})} = -0.81(1\pm0.06) \, i|e|a_0 \cdot 10^{-10} . \qquad (10)$$

4.Possible Future Stark Interference Experiments

In each of the three atomic PNC experiments just mentioned, considerable gains in overall precision can be realized by enlarging the data sample, since each result is statistics-limited. At Paris, a very interesting scheme for Cs is being developed[14] in which, in addition to a 539 nm laser beam with polarization $\hat{\varepsilon}$, one also applies a pulsed probe beam at 1.47μ with polarization $\hat{\varepsilon}'$. The function of the latter is to induce stimulated emission between 7S and $6P_{3/2}$ states. It can be shown that PNC gives rise to a pseudo-scalar term in the transition probability proportional to $\hat{\varepsilon}\cdot\hat{\varepsilon}' \, \hat{\varepsilon}'\cdot\hat{\varepsilon} \times E$. An interesting feature arises from the fact that polarization-dependent gain can be realized for the probe beam; hence the PNC effect can be amplified. So far, the Paris group has succeeded in observing $7S-6P_{3/2}$ stimulated emission and has detected anisotropy in the 7S state through the resulting dependence of the stimulated emission rate on probe beam circular polarization. Very large gains in detection efficiency should be realized.

At Boulder[15], improvements in statistical precision have already been obtained with the existing apparatus; in addition a new apparatus is being built in which the atomic beam will be polarized by optical pumping before it enters the interaction region. This should result in an improvement of a factor of 16 in intensity.

Finally at Berkeley we have built a new apparatus that incorporates the same geometry as in Fig.2c but replaces fluorescent detection by photo-ionization from the 7P state. The resulting gain in detection efficiency is about a factor of 100, and this should make possible a reduction of the statistical uncertainty from 15% to about 2%. With such improvements in the Cs and Tℓ experiments, it may be possible to observe a nuclear anapole moment[16] by detecting a small dependence of the PNC effect on hyperfine component.

References

1. L.M. Barkov and M.S. Zolotorev, Pis'ma Zh. Eksp. Teor. Fiz. 28, 544 (1978) (JETP Lett. 28, 503, 1978)), Phys. Lett. B85, 308 (1979)
2. G.N. Birch et al., Zh. Eksp. Teor. Fiz. 87, 776 (1984) (Sov. Phys. JETP 60(3), 442 (1984)
3. J. Taylor, PhD Thesis, University of Oxford (1984)
4. J.H. Hollister et al., Phys. Rev. Lett. 46, 643 (1981)
5. T.P. Emmons, J.M. Reeves and E.N. Fortson, Phys. Rev. Lett. 51, 2089, (1983)
6. M.A. Bouchiat, J. Guena, L. Pottier and L. Hunter, J. de Phys. 46, 1897 (1985); 47, 1175 (1986) & 47, 1709 (1986)
7. S.L. Gilbert and C.E. Wieman, Phys. Rev. A34, 792 (1986)
8. P.S. Drell and E.D. Commins, Phys. Rev. A32, 2196 (1985)
9. C. Bouchiat and C.A. Piketty, Phys. Lett. B128, 73 (1983)

10. C.E. Tanner and E.D. Commins, Phys. Rev. Lett. $\underline{56}$, 332 (1986)
11. V.A. Dzuba, V.V. Flambaum, P.G. Silvestrov and $\overline{O.P.}$ Sushkov, Inst. of Nucl. Phys. Novosibirsk Preprint 86-132
12. A.M. Martensson-Pendrill, J. de Phys. $\underline{46}$, 1949 (1985)
13. U. Amaldi et al., Univ of Pennsylvania Preprint 0331T (1987)
14. M. Lintz et al., ENS Preprint (1987), J. Guena et al., ENS Preprint (1987) & M.A. Bouchiat, Private Communication (1987)
15. C.E. Wieman, Private Communication (1987)
16. V.V. Flambaum, I.B. Khriplovich and O.P. Sushkov, Phys. Lett. $\underline{B146}$, 367 (1984)

Towards Precise Parity Violation Measurements in Cesium: Non-linear Optics Experiments in a Forbidden Three-Level System

L. Pottier, M.A. Bouchiat, J. Guéna, Ph. Jacquier, and M. Lintz

Laboratoire de Spectroscopie Hertzienne de l'Ecole Normale Supérieure, Tour 12, EO1–4, place Jussieu, F-75252 Paris Cedex 05, France

Parity violation (PV) in the Cs 6S-7S transition, measured in several experiments [1], quantitatively agrees with the prediction of the standard electroweak model within the accuracy of $\sim$ 10%. However, fundamental features remain unobserved : in particular, manifestations of electroweak radiative corrections, as well as of the vector-electronic axial-nucleonic contribution in the coupling of the Z^0 boson. Our primary goal is now to test these features through more precise measurements ($\sim$ 1%).

To surpass the statistical accuracy of our previous results [2], the novel project developed at ENS resorts to a new technology [3] to detect practically all 7S atoms, by forcing them to emit in a single transition and a single direction. This can be obtained using a probe laser beam tuned to one hfs component of the $7S-6P_{3/2}$ transition, so as to stimulate emission (or possibly trigger superradiance, depending on the experimental conditions). Sufficient 7S population is created beforehand by a laser pulse, resonant with one hfs component of the $6S \rightarrow 7S$ transition and copropagating with the probe beam. The laser polarizations $\vec{\epsilon}_1$ and $\vec{\epsilon}_2$ are chosen linear, and a dc electric field $\vec{E}$ is applied along the beams. Parity violation is expected to appear as a dependence of the stimulated emission rate on the handedness of the trihedral $\vec{E}, \vec{\epsilon}_1, \vec{\epsilon}_2$. The effect originates in atomic alignment in the 7S state due to "electroweak" interference between the weak PV amplitude and a purely electromagnetic, parity-conserving amplitude of the 6S-7S transition [3]. There is in addition an interesting possibility for the PV asymmetry to be amplified in the propagation through the vapor, optically thick at the probe wavelength.

In a preliminary step using cw excitation and detection and a transverse electric field, we have studied this unusual case of two-laser, three-level spectroscopy where one of the lasers connects two levels of same parity ("forbidden 3-level system"). As a result of this, at Cs densities of interest coherent two-photon processes are found to be negligible compared with stepwise one-photon processes. Stimulated $7S \rightarrow 6P_{3/2}$ emission has been observed, as a decrease of the fluorescence in the competing $7S \rightarrow 6P_{1/2}$ transition [4], and also as amplification of the probe intensity [5]. The Doppler-free spectra show complete resolution of the $6P_{3/2}$ hfs at the required Cs density, a prerequisite to detecting alignment. Atomic anisotropy in the 7S state (here, parity-conserving polarization), is detected through the dependence of the stimulated emission rate on the polarization of the probe beam (here, circular) [6]. Optical rotation spectra and circular dichroism spectra have been recorded with large S/N ratio and interpreted. Relaxation in the 6S-7S-6P system is understood and the parameter values are extracted from the spectra.

In a second phase the excitation was pulsed. In the intensity of the transmitted cw probe beam, immediately after the excitation pulse a gain pulse was observed. Its area indicates that practically all 7S atoms are de-

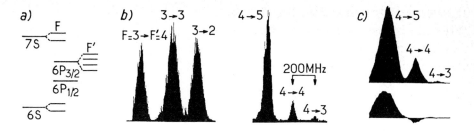

Fig.1:a) Relevant levels of Cs; b) Stimulated emission signal (photons/pulse) versus probe frequency, pump and probe polarized linearly; c) Pump polarized circularly ($\xi = \pm1$), probe detected through a fixed circular analyzer. Sum (top) and difference (bottom) of signals for $\xi = 1$ and -1

tected, as anticipated. Notwithstanding the spectral width of the pulse, the $6P_{3/2}$ hfs remains resolved (fig. 1). Atomic anisotropy in the 7S state is detected through the dependence of the transient gain on the probe polarization (fig. 1c). This dependence turns out to be amplified in the propagation of the probe beam through the optically thick vapor. At last, in sufficiently favorable conditions (regarding the electric field, the Cs density, the focussing of the excitation laser, etc...) superradiance is observed, either in absence of any probe beam, or triggered by a very weak probe intensity.

References

1. M.A. Bouchiat, L. Pottier: Science, 234, 1203 (5 Dec. 1986)
2. M.A. Bouchiat, J. Guéna, L. Pottier, L. Hunter: J. Physique, 47, 1709 (1986)
3. M.A. Bouchiat, Ph. Jacquier, M. Lintz, L. Pottier: Opt. Commun. 56, 100 (1985)
4. J. Guéna, M. Lintz, Ph. Jacquier, L. Pottier, M.A. Bouchiat: Opt. Commun. 62, 97 (1987)
5. M.A. Bouchiat, J. Guéna, Ph. Jacquier, M. Lintz, L. Pottier: to be published in the book dedicated to Pr. A. Gozzini on the occasion of his 70th birthday, Scuola Normale Superiore, Pisa, Italy
6. M. Lintz, J. Guéna, Ph. Jacquier, L. Pottier, M.A. Bouchiat: Europhys. Lett. (1987), to be published.

Intracavity Polarimetry
with a Sodium Dimer Ring Laser

A.D. May and S.C. Read

Department of Physics, University of Toronto,
Toronto, Canada, M5S 1A7, Canada

We have constructed a Na_2 ring laser which operates simultaneously on two single frequency, counterpropagating, orthogonally polarized modes. The difference in frequency between two such modes is determined by the optical anisotropy of the cavity and has been measured using a beat technique and electronic counter. Such a device forms the basis of an intracavity polarimeter. We discuss the salient features of the laser and report on preliminary results of the electric field induced birefringence (Kerr effect) in CO_2.

The system consists of a ring sodium dimer laser, pumped by a single frequency Ar^+ laser. The ring laser is bi-directionally pumped and the polarization of each pump beam is chosen to match one of the polarization modes of the ring laser. As spurious anisotropies such as strain birefringence degrade the measurement it is desirable to limit the dimer system to the minimum number of components, i.e. to a cavity, gain cell and optically anisotropic sample. Furthermore to achieve a sensitivity determined only by the Schawlow-Townes limit one wants to eliminate mode competition. Given these "boundary conditions" the following properties of the system can be appreciated: i)the narrow gain profile of Na_2 permits single longitudinal mode operation without additional intracavity elements, ii)mode matching the pump beam to the dimer cavity permits the selection of a single transverse (0,0) mode, iii)the forward/backward gain asymmetry circumvents the need for an intracavity optical diode, iv)bi-directional pumping off of line-center permits, through velocity selection, independent, countercirculating modes to operate, and v)pumping with a polarization parallel to the dimer laser selects one of the two polarization modes possible for each direction of circulation in the Na_2 ring cavity. In summary two single frequency, counterpropagating modes of selected polarization are generated in the dimer laser. (One wants selectivity to distinguish among birefringence, optical activity and Faraday rotation). For the Kerr effect measurements the modes are orthogonal and linearly polarized.

The bare laser (the Na_2 dimer laser without an anisotropic intracavity sample) runs on linearly polarized modes parallel and perpendicular to the plane of the ring, with a mode splitting of the order of 2 MHz. We have determined that this arises from the differential phase shift upon reflection from the four dielectric coated mirrors of the cavity. In our preliminary experiments we have reduced the 2 MHz beat to about 100 kHz by straining one of the windows of the sample cell. We are concerned about reducing the bare laser beat frequency too much. Frequency degenerate modes can run on any polarization and the polarization of the laser can become very sensitive to any stray polarized coupling or feedback. We want a laser with fixed polarization modes that match the anisotropy to be measured. In this way the anisotropy of the sample leads directly to a change in the beat frequency; the beat frequency would be more difficult to analyse if the polarization modes are pulled.

For a passive cavity variations in cavity length are not reflected in a variation in the difference in mode frequencies i.e. common mode rejection leads to a sharp beat even though the modes themselves may vary in frequency. However, for a laser, the dispersion of the gain medium can upset this ideal balance. The differential mode pulling depends upon the frequency of the modes relative to the gain profile and the amplitude of gain profile. Thus any fluctuations in the intensity or frequency of the Ar^+ pump laser or in the length of the ring laser or dimer density cause fluctuation in the beat frequency. This has been significantly reduced by controlling the intensity of the pump laser to about 1% and the frequency of the pump and dimer laser to about 1 MHz. Furthermore, we find empirically that single line operation is most easily obtained by using the 457.9 nm line of Ar^+ with lasing occurring at 559.3 nm.

We have measured and analysed the noise in the beat frequency of the bare Na_2 laser. The "jitter" in the beat frequency is about 2 kHz and has a Gaussian distribution with a temporal evolution indicative of thermal or acoustic noise sources. To fix an optimum modulation for the Kerr measurement we also examined the Allan variance. The variance was still decreasing with time at time intervals of 1/7 sec, a limit set by the data handling rate of our counter. We believe the use of a phase locked loop as a fast frequency to voltage converter and faster than 7 Hz modulation/demodulation is a good way around the limitation of our counter.

In a preliminary experiment to test the apparatus we have measured the d.c. Kerr effect in CO_2. Defining sensitivity as $\Delta n/n$ where Δn is the smallest detectable difference in index of refraction for two polarizations, and n as the mean index, we have achieved a sensitivity of about 10^{-15}. This corresponds to a change in the beat frequency of 1 Hz. We believe this may be improved by several orders of magnitude. Even at 10^{-15} the present polarimeter promises to be a useful tool for measuring non-linear anisotropies, anisotropies of dilute or thin samples and very weak anisotropies such as occur in parity-violation.

Fundamental Tests of Special Relativity and the Isotropy of Space

S.A. Lee[1], L.-U.A. Andersen[2], N. Bjerre[2], O. Poulsen[2], E. Riis[2], and J.L. Hall[3]

[1]Physics Department, Colorado State University,
 Fort Collins, CO 80523, USA
[2]Institute of Physics, Aarhus University, DK-8000 Aarhus C, Denmark
[3]JILA, NBS and University of Colorado, Boulder, CO 80309, USA

When one considers tests of special relativity and the isotropy of space, the experiment of MICHELSON and MORLEY [1] immediately comes to mind. These scientists might not have anticipated that one hundred years after their landmark experiment, there is an ever increasing interest in the subject. The discovery of the anisotropy in the 3K cosmic blackbody radiation and its subsequent interpretation as due to the earth's motion through the 3K radiation rest frame [2] makes it interesting to reconsider experiments that can test for the one way speed of light. Indeed, high resolution laser spectroscopy appears to have the most to offer in terms of improved techniques and increased precision for these tests.

For tests of the isotropy of space, a Michelson-Morley type experiment looks for the variation in the round-trip-averaged speed of light as the laboratory apparatus is rotated in space. The most recent and most precise experiment of this type was performed by BRILLET and HALL [3], using high resolution laser techniques. A second type of experiment uses the relativistic Doppler effect to measure how the rates of moving atomic clocks depend on their velocity through the preferred frame. If Lorentz invariance were broken, the fractional frequency difference between a moving clock and a stationary clock can be expressed as $\delta \underline{u} \cdot \underline{V}/c^2$, where $\underline{V}$ is the velocity of the moving clock in the laboratory frame, $\underline{u}$ is the velocity of the laboratory in the preferred frame, and δ is the velocity dependent anisotropy parameter. Some examples of this class of experiments are the Mössbauer rotor experiments [4], the space-borne hydrogen maser experiment of VESSOT, et al.[5], and our present experiment, to be described below. This class of experiments are sensitive to the variation in the one-way speed of light, in contrast to the "two-way" experiments which measure round trip effects.

We now describe a one-way type experiment using resonant two photon spectroscopy (TPA) in a fast beam of Ne atoms. We compare the fast beam resonant frequency to a reference transition in an I_2 cell which is at rest in the laboratory frame. For an anisotropy test, we look for a sidereal variation in the difference between the two transition frequencies when the Ne beam direction changes in space, due to the daily rotation of the earth.

The experiment is similar to the Mössbauer rotor experiment if one considers the fast beam to take the role of the moving source and the reference I_2 cell to be the fixed absorber. Although the laser beams are collinear with the Ne beam, first order Doppler shifts cancel due to the TPA process. The experiment identifies the value of the speed of light along the fast beam direction by measuring the second order (or equivalently, transverse) Doppler shift, with the velocity of the fast atoms determined by the resonant two photon process. Thus the analysis of the present experiment is similar to the Mössbauer experiments. In terms of the test theories of MANSOURI and SEXL[6], our experiments measures their time

52

dilation parameter, α. In terms of the dynamical approach of HAUGAN and WILL[7], our experiment measures a frequency difference which is dependent on u·V if Lorentz invariance is broken.

The experimental setup is shown schematically in Fig. 1. Part of the laser light was double-passed through an acousto-optic (AO) frequency shifter to be in resonance with a hyperfine component of a nearby I_2 line. A small FM was applied to the laser and the dye laser is stabilized to the saturated absorption signal of this I_2 line, shown in the upper trace of Fig.2. The lower trace of Fig.2 shows the beam TPA signal which was monitored by observing the fluorescence from the upper level. The laser was kept in resonance with Ne by using a feedback signal derived from the TPA resonance which was applied to the AO frequency shifter for the I_2 reference. A feedforward signal was applied to the laser to decouple the two servo loops. The long term velocity variations due to drifts of the acceleration and anode voltages were compensated by servoing the post acceleration voltage to maximize the TPA fluorescence.

The measured quantity in this experiment is the frequency applied to the AO shifter, which corresponds to the frequency difference between the beam TPA transition and the reference I_2 transition. Systematic effects were studied and the experimental conditions were kept fixed whenever possible. It was important to operate the source at a high Ne pressure which gave a well collimated atomic beam with a narrow and almost symmetric velocity distribution. Unfortunately this also resulted in a much reduced operating lifetime of the ion source. The change of the atomic beam divergence with aging of the ion source was a limiting factor in this experiment.

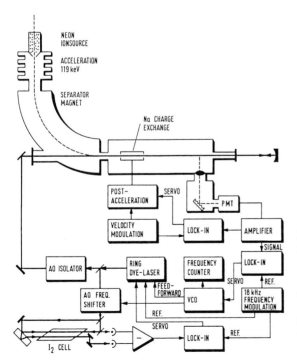

Fig.1. Schematic of the experimental setup. The cw ring dye laser was tuned to the $3s[3/2]_2-4d'[5/2]_3$ TPA transition in the fast beam. The intermediate state, $3p'[3/2]_2$, is tuned into resonance for single photon absorption by selection of the atomic beam speed.

53

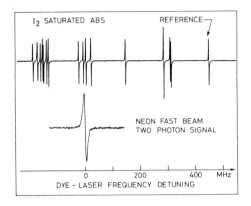

Fig. 2. Upper trace: The saturated absorption spectrum of the I$_2$ reference line.
Lower trace: The fast beam TPA signal.

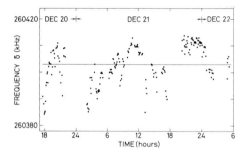

Fig. 3. Experimental data accumulated over a 36 hour period.

Fig. 3 shows a set of 198 data points accumulated over a 36 hour period. The frequency applied to the AO shifter is displayed as a function of time. Each data point consists of 25 consecutive measurements of 10 s duration each. During this run the ion source filament was replaced twice, around 20 hr on Dec.20 and 18 hr on Dec.21. The simplest way to analyze the data is to fit them with a 24 sidereal hour cosine with variable amplitude and phase, and a constant background. Using a least squares routine, with each data point weighted by the variance of its 25 individual measurements, the 24 hr component has an amplitude of 0.17+0.38 kHz, with a phase corresponding to a maximum when the fast beam direction is at a right ascension of 9.8+8.8 hr. An immediate impression from Fig. 3 is that a periodic variation of about 12 hours dominates the data. We ascribe this to be from changes in the divergence of the atomic beam resulting from aging of the ion source. To assess this systematic effect, the data were then fitted with five parameters: a constant and two cosines, each with variable amplitude and phase. The period of one cosine was fixed at 24 sidereal hours, while the other was allowed to vary, to give a first modelling of the behavior of the source. The sum-squared residuals of the fit has a minimum when the period of the second cosine is at 11.8 sidereal hours. The least squares fitting, using 24 hr and 11.8 hr periods, results in a 24 hr component of 0.81+0.26 kHz, with a phase corresponding to a maximum at a right ascension of 11.3+1.3 hr. The 11.8 hr and 24 hr results were uncorrelated. The phase of the 24 hr component is almost in the direction of the cosmic 3K radiation, which has a right ascension of 11.2+0.2 hr and a declination of -6.1^0+1.5^0 [2]. To further assess the significance of the result, we blocked out one hour's worth of data immediately before and after filament changes. If the 24 hour component is due to a real effect, its phase and amplitude should remain stable when some of the data points are removed.

54

This leaves us with 160 data points, and the results of this least squares fit has a 24 hr component with 1.54+0.25 kHz, with a phase shifted by almost 90^0 from the previous case. Therefore we must conclude that our experiment is dominated by systematic errors. This is also evident from the standard deviation obtained from the fit, 4.9 kHz, which is considerably larger than the 0.5 kHz deviation on the individual data points. We believe it is reasonable to conclude that this preliminary experiment sets a limit on the 24 hour amplitude as <2 kHz. The atomic beam axis is oriented North at a latitude of 56^0. The 24 hr component along -6^0 declination gives an amplitude of <2.5 kHz. The associated anisotropy parameter, δ, is then $\sim 10^{-6}$, using the earth's velocity of approximately 300 km/s through the cosmic microwave background and the beam velocity in the laboratory to be 1074 km/s.

Although the present experiment has uncontrolled systematic effects, it offers the potential that considerably higher precision can be obtained. An increased stability and reproducibility in the fast beam should greatly reduce the experimental uncertainties. Since the sensitity of the experiment is proportional to the beam speed, one can also use light atoms such as He to attain higher velocity in the fast beam. These steps, when implemented, should improve the present experiment by at least a factor of 10. In the future, when cold and high velocity particle beams become available in storage rings, and when combined with recent developments in ultrastable and ultra-narrowband cw dye lasers, truly exciting possibilities of anisotropy tests are foreseen.

We acknowledge the support of the Danish Natural Science Research Council, the United States National Science Foundation, the Carlsberg Foundation, and the Nordic Accelerator Committee.

References

1. A.A. Michelson and E.H. Morley, Am. J. Sci. 34, 333 (1887).
2. P. Lubin, T. Villela, G. Epstein and G. Smoot, Astrophys. J. 298, L1 (1985).
3. A. Brillet and J.L.Hall, Phys. Rev. Lett. 42,549 (1979).
4. D.C. Champeney, G.R. Isaak, A.M. Khan, Phys. Lett. 7, 241 (1963). G.R. Isaak, Phys. Bull. 21, 255 (1970). K. Turner and H.A. Hill, Phys. Rev. B134, 252 (1964).
5. R.F.C. Vessot and M.W. Levine, Gen. Rel. Grav. 10, 181 (1979).
6. R. Mansouri and R.U. Sexl, Gen. Rel. Grav. 8, 497 (1977).
7. M.P. Haugan and C.M. Will, Physics Today 40, 69 (1987).

Part II

Laser Cooling, Trapping and Manipulation of Atoms and Ions

Laser Cooling and Trapping of Atoms

S. Chu, M.G. Prentiss, A.E. Cable, and J.E. Bjorkholm

AT&T Bell Laboratories, Holmdel, NJ 07733, USA

The laser cooling, manipulation and trapping of neutral atoms has seen remarkable progress, so it would be impossible to adequately review the work done in recent years. Instead, this paper will summarize the work done on atom cooling and trapping at AT&T Bell Laboratories during the last three years, emphasizing the status of our most recent experimental results.

I. Optical Molasses and Supermolasses

We have previously reported the use of multiple laser beams to create a viscous medium of photons that both cools and confines atoms [1]. The cooling scheme [2] uses the fact that an atom moving in a light field consisting of two counter propagating laser beams tuned below resonance will Doppler shift the beam directed opposite the motion into resonance and shift the co-propagating beam out of resonance. The net result after averaging over many absorptions is that the scattered light creates a damping force directed opposite the motion of the atom.

This simple idea has remarkable consequences. For example, one can show that the damping time for slow moving atoms in this light field is on the order of $\tau \sim 4M/\hbar k^2 \approx 10$ μsec for sodium atoms using the $3S_{1/2} - 3P_{3/2}$ resonance line. The minimum temperature achievable under these conditions has been shown to be $kT \sim \dfrac{\hbar\Gamma}{2} \sim 240$ μK in the limit where the single photon recoil energy is small compared to this temperature. Also, the cooling technique provides a means of confining the atoms in space. The motion of the atoms can be modeled as a random walk in a viscous fluid of photons (hence the term optical molasses) analogous to Brownian motion. The frictional force can be calculated to be $F_{damping} = -\alpha v = 5.7\times10^{-18} gm/sec$ for $p = 1$ and $\Delta v \sim \Gamma/2$. The Einstein relation then gives a diffusion constant $D = \dfrac{<x^2>}{2t} = \dfrac{kT}{\alpha} = 5.8 \times 10^{-3}$ cm^2/sec for an infinite viscous fluid. However, the optical molasses extends as far as the overlap of the six laser beams, and an atom that randomly walks to the boundary defined by the overlapping beams will be lost. If we assume an initially uniform concentration of atoms n_0, the average concentration $\bar{n}$ has been shown to vary as $\bar{n} = n_0 \dfrac{6}{\pi^2} \sum\limits_{n=1}^{\infty} \dfrac{1}{n^2} e^{-Dt\left(\frac{\pi n}{R}\right)}$ for a spherical boundary of radius R. The presence of the spherical boundary greatly decreases the confinement time of the atoms in the molasses. For $R = 0.4$ cm, t decreases from 14 sec to 1.6 sec, and for $R = 0.1$ cm, t goes from 0.9 sec to 0.1 sec. Our

58

initial investigation of optical molasses produced results that were in reasonable agreement with calculations [1]. The measured temperature was in good agreement with the "quantum limited" temperature $\hbar\Gamma/2$ derived for a one-dimensional two-level system at low laser intensity [4]. Similarly, the confinement time achieved in our initial work was also in agreement with the random walk model with a spherical boundary. However, soon after the initial work on optical molasses, it became increasingly apparent that the behavior of the atoms was much more complicated than originally thought. Three of the more striking aspects of our more recent discoveries include the following: (i) high intensity molasses $p \geq 1$ (it should not have worked at all!) worked surprisingly well, (ii) the ideal tuning for the longest storage time was on the order of $\Delta\nu \simeq 1.5\Gamma$ rather than $\Gamma/2$, and (iii) with the proper misalignment of the nominally retroreflected laser beams. The molasses seemed to compress the atoms spatially in to a ball ~ 2 mn, in diameter, and act like a trap.

The misaligned molasses, dubbed "supermolasses" because it worked so well, was the most intriguing of these effects. The alignment of beams that produces the best result is that of a racetrack for four beams with the other two beams retroreflected. With fine tweaking of the misalignment, atoms within a 2 mm diameter region have been stored for 10 seconds, two orders of magnitude longer than expected from a simple molasses picture. We have not come up with a complete explanation for supermolasses, and the present theoretical treatments provide serious constraints for any plausible model. A careful two-level, one-dimensional treatment of the behavior of atoms in a standing wave [4] shows that the damping force exerted on the atoms for a laser tuned below resonance actually experiences a sign reversal for sufficiently large p. [5] For $\Delta\nu \sim 2\Gamma$, the upper limit on p should be $p \leq 0.3$. Furthermore, the momentum diffusion D_p grows linearly with p for $p \gg 1$. Experimentally, we find that molasses seems to work for nominally retroreflected beams up to $p \sim 1$ per beam, and for misaligned beams, intensities as high as $p = 5$ (the upper limit of our laser) could be used to produce a supermolasses trap.

There is the possibility of trapping the atoms in three dimensional standing lightwaves, as originally suggested by Letokhov, et. al.. [6] Atoms quasi-trapped in a three-dimensional checkerboard of potential wells created by the multiple standing wave interference pattern might diffuse more slowly. If the depth of the potential well U is a few times larger than kT, the diffusion of the atoms out of the molasses would be greatly reduced. Unfortunately, for pure standing waves in a one-dimensional, 2 level system, $U \leq kT$ for frequencies below resonance. The simple standing wave hypothesis also does not explain the polarization tests we have performed on supermolasses. The trap performance was insensitive to the polarization orientations (linear or circular) of the four racetrack beams. On the other hand crossed linear or opposite circular polarization on the retroreflected beams would destroy the trap.

The racetrack configuration suggests that a combination of traveling and standing waves would increase the ratio of U/kT. The damping force for a traveling wave does not undergo a sign reversal and remains fairly effective for $p>1$. Furthermore, D_p begins to saturate above $p>1$ as for the case for a traveling wave, whereas for a standing wave $D_p \propto p$ for large p. This effect was first pointed out by Cook [8]. The observation of channeling when the laser is tuned to red wavelength side of the resonance line [7] and the traveling/standing wave combination gives us hope that a sufficiently complete computer simulation can account for our discovery.

II. Optical Trapping

The optical trapping of atoms was first demonstrated with a "dipole" trap [9]. The basic physics of the trap [10] is analogous to the work done on the magnetic trapping of neutrons [11] and atoms [12]. In the magnetic trap, the magnetic moment of the neutron or atom can be in either weak-field seeking or strong field seeking states. Since it is possible to design a magnetic field with a local minimum in space, weak field seekers can be trapped. In the case of an electric dipole trap, an electric dipole moment can be induced on the atom by an external electric field. If the time varying electric field is below the natural resonant frequency of the atom, the induced dipole moment is in phase with the driving electric field and the atom will be a strong field seeker. Unlike the case of a static electric or magnetic field [13], it is possible create a local maxima with oscillating electromagnetic fields. Indeed, a single laser beam, sharply focused and tuned below resonance forms the trap successfully demonstrated by our group [9]. Although such a trap is conceptually elegant, the potential well created by the laser is shallow and limited to a small volume. In our first demonstration, 200 mW was focussed to a 20 μm diameter spot (peak power 600 kw/cm^2) to produce a trap $5 \times 10^{-3}K$ deep and 10^{-7} cm^3 in volume.

The scattering force can also be used in laser trapping. Laser intensities on the order of the saturation intensity are all that are required, so large volume traps can be constructed. We report here the first demonstration of a scattering force trap. The work was done in collaboration with E. Raab and D. Pritchard [14], was due to an unpublished suggestion by J. Dalibard. The idea can be introduced as a modification of ordinary molasses. In molasses, a red detuning of the laser creates a damping force that resists any motion of the atoms. A net restoring force can be imposed on the atoms by introducing an external magnetic field and using circularly polarized light as shown in Fig. 1. For atoms at positive z displacement, the σ^- light is more in resonance with the $S=0 \rightarrow 1$ transition than the σ^+ light, so there is a net push in the direction $-\hat{z}$. Similarly for atoms at $-z$, the σ^+ light has a higher absorption probability so there is a net force along $+\hat{z}$. Since the laser is tuned to a frequency below the resonant frequency, molasses-like damping is also present. Simulations based on the actual hyperfine structure of sodium for the $F=2 \rightarrow 3$ and $1 \rightarrow 2$ transitions in the D_2 line have been made and are reported elsewhere [14].

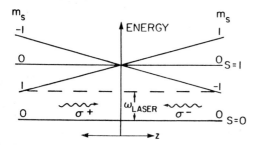

Fig. 1. Magnetic molasses for a $S=0$ to $S=1$ transition. σ^- light predominates for atoms with $z>0$ and σ^+ light predominates for atoms with $z<0$.

The three dimensional generalization to the magnetic field can be obtained by using the same quadrupole configuration as the one used by Migdall, *et al.* [12]. Less than 40 amp-turns in 5cm diameter coils were needed to produce trapping, corresponding to a magnetic field ≤ 0.2 Gauss in the central 0.8 mm region of the trap. Note that the strength of the magnetic field in this "magnetic molasses" trap is roughly 10^3 times weaker than a pure magnetic field trap of comparable depth.[12], [17]   We have trapped as many as 10^7 atoms for $1/e$ storage times of 2 minutes at initial densities of 2×10^{11} atoms/cm^3 when the laser is tuned 10-15 MHz below the $F=2-3$ and $1-2$ transitions. The temperature of the atoms in the trap, as measured by the time-of-flight method, is between 400-900 μK. Tuning the laser to the red of the $F=2\rightarrow2$ transition, increases the number of atoms by an order of magnitude, but decreases the density to $\sim 2 \times 10^{10}$ atoms/cm^3. The temperature of the atoms for the $F=2\rightarrow2$ transition is two orders of magnitude higher than for the $F=2\rightarrow3$ transition.

Fig. 2 shows an example of the decay of the atoms in the trap. It is immediately apparent that the decay of the atoms in the trap cannot be described as a single exponential decay. A good fit can be obtained if the loss of atoms is fit to the solution to the differential equation $dn/dt = -\beta n^2 - \frac{1}{\tau}n$. At high densities and early times, the loss of atoms in the trap is dominated by the βn^2 term. This fact explains why it is possible to obtain a larger number of stored atoms in the trap if it is operated at lower density. Work is now underway to determine the βn^2 loss mechanism.

Although some aspects of the trap can be understood in a quantitative way [14], there are a considerable number of details that are not understood at this time. For example, the size of the trapping region should be ≥ 1 cm in diameter, but displacement tests suggest that the size is a factor of 5 smaller for the $F=2\rightarrow3$ transition, but consistent with expectations for $F=2\rightarrow2$ transition. Also, like supermolasses, the behavior of the trap is very sensitive to small changes in the alignment of the nominally retroreflected beams. Furthermore, the trap appears to be far less sensitive to polarization than expected. Some of the behavior is probably entangled in the mysterious aspects of supermolasses, and further work is needed. It should also be noted that damping occured for $p>2$ even when the beams were retroreflected.

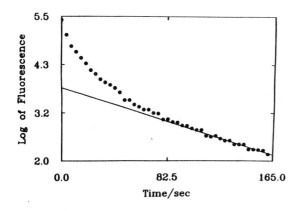

Fig. 2. Decay of atoms in the magnetic molasses trap.

III. Conclusion

Table I summarizes the atom traps demonstrated to date. As one can see, there has been remarkable progress in the two years since 1984.

Table I. Summary of Atom Traps

		$\tau_{1/e}$ (seconds)	ρ (atoms/cm^3)	N (atoms)
1)	Magnetic Trap NBS, 1985 [12]	0.8 sec	10^3	10^4
2)	Optical Molasses Bell Labs, 1985, [1]	0.2 sec	10^6	10^5
3)	Optical Dipole Trap Bell Labs, 1986 [9]	2 sec	10^{12}	10^3
4)	Supermolasses Bell Labs, 1986	10 sec	10^7	10^6
5)	Magnetic Trap II MIT, 1987 [15]	2.5 min	10^6	10^9
6)	Magnetic Molasses MIT/Bell Labs, 1987 [14]	2 min	10^{11}	10^7
7)	Magnetic Trap, III (hydrogen) MIT, 1987 [17]	20 min	10^{13}	10^{12}

The lowest temperature obtained is still $kT \sim \hbar\Gamma/2 \sim 240 \ \mu K$, [1], but many cooling schemes have been proposed to reduce the temperature by at least two orders of magnitude [16]. In addition to better traps and colder temperatures, the next few years should see an improvements in neutral atomic beam optics and the development of traps that minimize the perturbation of the internal degrees of freedom of the atom.

Clearly, the continued development of laser cooling and trapping techniques and the application of these techniques to a wide variety of physics and chemistry problems has an exciting future. Another exciting aspect of this field is the unexpected developments. Although the basic forces that radiation can exert on atoms are well understood, even seemingly simple experimental realizations of these effects have already produced surprises such as supermolasses. Perhaps a sufficiently clever quantum engineer or mechanic could have anticipated the result, but we have found it easier and more fun to discover the effects in the laboratory.

1. S. Chu, L. Hollberg, J. E. Bjorkholm, A. Cable and A. Ashkin, Phys. Rev. Lett., *55*, 48 (1985); also in *Laser Spectroscopy VII*, ed. T. Hansch, Y. R. Shen, (Springer Verlag, Berlin, 1985.) p. 14, *Methods of Laser Spectroscopy* ed. Y. Prior, A. Ben-Reuven, M. Rosenbluh (Plenum, 1986), p. 4).
2. T. W. Hansch and A. L. Schawlow, Optics Comm., *13*, 68 (1975).
3. N. A. Fuchs, *Mechanics of Aerosols*, (Pergamon, Oxford, 1964), pp. 193-200.
4. J. P. Gordon and A. Ashkin, Phys. Rev. A *21*, 1606 (1980).
5. A. P. Kazantsev, Zh. Eksp. Teor. Fiz *66*, 1599 (1974).
6. V. S. Letokhov, V. G. Minogin and B. D. Pavlik, Optics Comm. *19*, 72 (1976).
7. A. Aspect, J. Dalibard, A. Heidmann, C. Salomon and C. Cohen-Tannoudji, Phys. Rev. Lett. *57*, 1688 (1986); See also the work on channeling reported at this conference.
8. R. J. Cook, Phys. Rev. A *20*, 224, (1979).
9. S. Chu, J. E. Bjorkholm, A. Ashkin and A. Cable, Phys. Rev. Lett. *57*, 314 (1986).
10. A. Ashkin, Phys. Rev. Lett. *40*, 729 (1978).
11. K. J. Kugler, W. Paul and U. Trimks, Phys. Lett. *72B*, 422 (1978).
12. A. L. Migdall, J. V. Prodan, W. D. Phillips, J. H. Bergeman and H. J. Metcalf, Phys. Rev. Lett. *54*, 2596 (1985).
13. W. Wing, Prog. Quant. Elec. *8*, 181 (1984).
14. Details are in E. L. Raab, M. Prentiss, A. Cable, S. Chu and D. E. Pritchard, submitted to Phys. Rev. Lett.
15. V. S. Bagnato, G. P. Lafyatis, A. G. Martin, E. L. Raab, R. N. Ahmad-Bitar and D. E. Pritchard, Phys. Rev. Lett. *58*, 2194 (1987).
16. S. Stenholm, Rev. of Mod. Phys., *58*, 699 (1986).
17. H. F. Hess, G. P. Kochanski, J. M. Doyle, N. Masuhara, D. Kleppner and T. J. Graytak, submitted to Phys. Rev. Lett., (1987).

New Measurements with Optical Molasses

P.L. Gould, P.D. Lett, and W.D. Phillips

National Bureau of Standards, MET B258, Gaithersburg, MD 20899, USA

The principle of laser cooling was first set forth in 1975 [1] and was demonstrated, for trapped ions, in 1978 [2]. Nevertheless, only recently [3,4] was it appreciated that the very strong damping of atomic velocity provided by three dimensional laser cooling would lead to diffusive atomic motion and relatively long confinement times, even in the absence of restoring forces. Because of the similarity of the diffusive atomic motion to the Brownian motion of particles in a viscous fluid, the phenomenon has been called "optical molasses."

In the first demonstration of the optical molasses effect, CHU et al. [3] achieved confinement times for sodium of about 0.1 s in a molasses approximately 0.8 cm in diameter, formed at the intersection of three mutually orthogonal pairs of counterpropagating laser beams. They achieved a peak Na density of $10^6/cm^3$ with pulsed loading of the molasses from a chirp-cooled atomic beam. In a later experiment CHU et al. [5] used the optical molasses to efficiently load the first true optical trap (with a restoring force). That experiment provided one example of the great potential utility of optical molasses as a dense source of ultra cold atoms for such applications as trap loading, high resolution spectroscopy, and studies of collisions and collective phenomena.

Here we present the results of more detailed studies of optical molasses. In particular, we have accomplished the first continuous loading of optical molasses, achieving steady-state densities of $10^8/cm^3$. We have also studied the dependence of the density and the confinement time of the molasses as a function of laser power, detuning, and degree of intensity imbalance between the counterpropagating laser beams. The results of these studies have been rather surprising and are in sharp disagreement with the predictions of a simple theory.

Let us examine what we expect of optical molasses: Consider a two level atom in a one-dimensional (1-D) molasses formed by a pair of counterpropagating beams in the positive and negative x directions, having intensities I^+ and I^-. The x force due to either beam is given by

$$F^{\pm} = \pm(\hbar k\gamma/2)\ \frac{I^{\pm}/I_0}{(1 + I'/I_0) + [(\Delta \mp kv)/(\gamma/2)]^2}\ , \qquad (1)$$

where γ is the radiative decay rate, I_0 is the on-resonant saturation intensity, Δ is the detuning from resonance, $k = 2\pi/\lambda$ is the wavevector and v is the x velocity. (For Na $1/\gamma = 16$ ns and $\lambda = 589$ nm.) Saturation and extension to additional laser beams in orthogonal directions is handled naively by taking I' to be the total intensity of all laser beams (I' = 2NI for N dimensions or pairs of laser beams). Equation (1) ignores the effect of forces arising from stimulated emission (dipole forces) and is valid only for $I/I_0 \ll 1$; however, we expect it to be approximately correct up to $I' \cong I_0$.

The total force $F^+ + F^-$ for $I^+ = I^- = I$ is

$$F = 4\hbar k^2 (I/I_0) \frac{(2\Delta/\gamma)v}{[1 + I'/I_0 + (2\Delta/\gamma)^2]^2} = -\alpha v \ . \tag{2}$$

For 1-D molasses this maximizes at $\Delta = -\gamma/2$ and $I/I_0 = 1$, giving a velocity damping time $\cong 13$ µs for Na. In 3-D optimum damping is achieved at $I/I_0 = 1/3$ and the damping time is increased by a factor of 3. By equating the energy loss due to the damping in (2) with the energy gain from momemtum diffusion arising from random emission and absorption of photons, we arrive at the equilibrium temperature defined by $Nk_BT = m\langle v^2 \rangle$, where N is the number of dimensions (pairs of laser beams). Thus

$$k_BT = (\hbar\gamma/4) \frac{1 + I'/I_0 + (2\Delta/\gamma)^2}{(2\Delta/\gamma)^2} \ . \tag{3}$$

This gives the usual Doppler cooling limit of $k_BT = \hbar\gamma/2$, independent of N, for $I \ll I_0$ and $\Delta = -\gamma/2$ [6]. According the the theory of Brownian motion the positional diffusion constant D_x is k_BT/α and the mean square distance an atom diffuses in time t in 1-D is $\langle x^2 \rangle = 2D_x t/\alpha$. Thus the diffusion time is

$$t = \frac{\langle x^2 \rangle 8k^2 (2\Delta/\gamma)^2 I/I_0}{\gamma[1 + I'/I_0 + (2\Delta/\gamma)^2]^3} \ . \tag{4}$$

This maximizes for $\Delta = \gamma/2$ and, $I'/I_0 = 1$, giving $t = 7$ s for a Na atom to diffuse 0.5 cm in 1-D molasses. This is to be compared to about 15 ms if the atom moved ballistically through the same distance at the thermal velocity. In 3-D the diffusion time is reduced by 3 for the same mean square distance, and is still smaller because the optimum α is reduced. In addition, there is a further reduction from the fact that the molasses does not extend through all space in a real situation [3]. A realistic confinement time in 3-D, 0.5 cm radius molasses would be less than 1 s.

If the intensity of the laser beams is unbalanced, i.e. $I^+ = (1 + \delta)I^-$, there is a net force for $v = 0$, and the average force is zero only for a non-zero drift velocity given, for $\delta \ll 1$, by

$$v_{dr} = \frac{\delta\gamma[1 + I'/I_0 + (2\Delta/\gamma)^2]}{8k(2\Delta/\gamma)} \ . \tag{5}$$

For the parameters optimizing the diffusion time, $v_{dr} = 3\delta\gamma/8k$, with $\gamma/k = 6$ m/s for Na. This leads to a 22 cm/s drift if δ is 10%, or an effective lifetime of only 23 ms for 0.5 cm radius molasses. This high sensitivity of drift velocity to imbalance was first pointed out in [4] and led to the supposition that one needed very good intensity balance to have long molasses confinement times.

In order to test the results of the simple theory presented here, we have made detailed measurements on a 3-D optical molasses. This was formed at the intersection of three pairs of mutually orthogonal, retroreflected, counterpropagating laser beams, with $1/e^2$ diameters of about 1 cm. The laser is tuned near the F=2 → F'=3 transition of Na, with a sideband near the F=1 →

F'=2 transition to prevent optical pumping. We continuously loaded slow atoms into our molasses, using those nearly stopped atoms which we found got past the decelerating laser in our "Zeeman-tuned" atomic beam laser-cooling method [4]. The $10^8/cm^3$ densities achieved by this continuous loading are two orders of magnitude higher than those reported earlier for pulsed loading [3,5].

We determined molasses confinement times by simultaneously blocking both the atomic beam and the decelerating laser (removing the continuous source of atoms loading the molasses) and then measuring the fluorescence from the molasses as a function of time. The decay of the fluorescence is generally well described by a single exponential. We record sucessive loadings and decays of the molasses as the frequency of the molasses laser is scanned slowly. To determine this frequency we simultaneously record a saturated absorption signal in a Na vapor cell.

We obtain optimum lifetimes of about 0.5 s when the misalignment of the retroreflected beams is less than a milliradian and the polarizations of the three pairs of beams are mutually orthogonal. For deliberately large misalignments and more complicated polarizations, we sometimes see longer decays of localized regions within the molasses. These may be related to the unexplained "super molasses" effects seen by CHU et al. [7], but we have not studied these effects in detail.

Figure 1 shows the molasses lifetime vs. detuning for the experimentally determined optimum laser power at optimum detuning. Also shown is the 1-D diffusion time to 0.5 cm from (4), scaled to approximately account for the geometric factors. We assume Na to be a two-level atom with a transistion strength given by the average of all F=2 → F'=3 transitions. The average power over the $1/e^2$ diameter is used in the calculation ($I/I_0 = 0.78$ for one beam.) The contrast between theory and experiment is striking.

Figure 2 shows the increase in decay rate of the molasses fluorescence as a function of percent of intensity imbalance. For these measurements we inserted an attenuator between the molasses and the retroreflecting mirror of

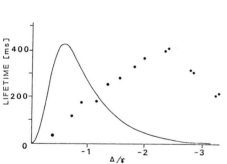 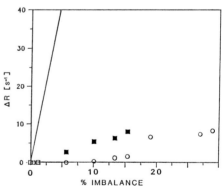

Fig. 1 Experimental molasses lifetime (points) and 1-D theoretical diffusion time (curve) vs. detuning of molasses laser. The theoretical curve has been scaled down by a factor of 15, which approximately accounts for 3-D and geometric effects.

Fig. 2 Increase in molasses decay rate as a function of percent imbalance. Squares and circles are experimental points taken with slightly different alignment. Solid line is 1-D theory.

one pair of beams. The predicted increase in decay rate was taken to be the inverse of the time required to move a distance equal to the molasses radius at the drift velocity given by (5). The predicted rate increase is at least ten times greater than the observed increase. Data presented are for the detuning and power which optimize the molasses lifetime at small imbalance, but nearby detunings and powers show similar results. For small powers, more than 100 times less than optimum, the molasses behaves approximately as expected from the simple theory.

We have obtained results from 1-D calculations which take proper account of the saturation effects and of the dipole force averaged over a wavelength. The calculations have been done for small velocity and imbalance, generalizing the approach of GORDON and ASHKIN [8], and for arbitrary velocity and imbalance, using the continued fraction method of MINOGIN and SERRIMA [9]. For the parameters used in the experiments, neither of these calculations, nor an extension of the simple theory to 3-D gave results much different from the simple approach given above. It remains to consider whether the multiple levels and sublevels of Na, multiple laser frequencies, or a consideration of the detailed motion of the atoms in 3-D can explain the surprising behavior of optical molasses.

This work was supported in part by the Office of Naval Research. P. L. Gould was supported by a NRC postdoctoral fellowship. We thank Jean Dalibard for providing the results of his continued fraction calculations.

References

1. T. Hansch and A. Schawlow, Opt. Commun. 13, 68 (1975); D. Wineland and H. Dehmelt, Bull. Am. Phys. Soc. 20, 637 (1975).
2. D. Wineland, R. Drullinger, and F. Walls, Phys. Rev. Lett. 40, 1639 (1978). W. Neuhauser et al. Phys. Rev. Lett. 41, 233 (1978).
3. S. Chu et al., Phys. Rev. Lett. 55, 48 (1985).
4. W. Phillips, J. Prodan, and H. Metcalf, J. Opt. Soc. Am. B, 2, 1751 (1985).
5. S. Chu et al., Phys. Rev. Lett., 57, 314 (1986).
6. D. Wineland and W. Itano, Phys. Rev. A25, 35 (1982), sec. II.B.2
7. S. Chu et al., presented at EICOLS 87 (this conference)
8. J. Gordon and A. Ashkin, Phys. Rev. A 21, 1606 (1980).
9. V. Minogin and O. Serimaa, Opt. Commun., 30, 373 (1979).

Optical Pumping in Translation Space

D.E. Pritchard, K. Helmerson, V.S. Bagnato, G.P. Lafyatis, and A.G. Martin

Department of Physics and Research Laboratory of Electronics, Massachusetts Institute of Technology, Cambridge, MA 02139, USA

Introduction

Perhaps the broadest overall trend in the art of experimental atomic spectroscopy has been the steady development of techniques to obtain more control of the atoms under study. After the discharge was replaced by the molecular beam to improve the control over the translational degrees of freedom, the Stern-Gerlach magnet was developed to select the internal degrees of freedom. This led immediately to the demonstration that the internal degrees of freedom were discretely quantized, and eventually to the discovery of resonance techniques. In the context of this discussion resonance may be viewed as a tool to change the internal quantum state. The development of the separated oscillatory field method opened (prematurely) the field of coherent spectroscopy, which allows, in principle, the preparation of any coherent superposition of the internal states. In addition, the coherences between internal states of different atoms may depend on the atom's spatial coordinate or velocity.

Although we can prepare the internal state as desired (and indeed are learning to slow the ravages of spontaneous decay which destroy our handiwork), we are just beginning to develop techniques to achieve comparable control of the translational degree of freedom of atoms. This explains the great amount of interest in slowing, cooling, and trapping atoms — all techniques to master the translation. In spite of this interest, only one good method has so far been demonstrated — Doppler cooling. This has been applied to both atoms and ions where milli-Kelvin temperatures have been achieved. This closely approaches the theoretical limit of Doppler cooling, $k_B T_l = 1/2\, \hbar\, \Gamma$, for strong transitions with $\Gamma \sim 10\mathrm{MHz}$. We can anticipate significant improvement when weaker transitions are used (this requires greater than current commercial dye laser stability). Unfortunately, T_l cannot be reduced arbitrarily by the use of increasingly narrow transitions: the finite momentum imparted in the last cycle of velocity selective absorption and subsequent spontaneous decay limits the temperature and velocity to the so-called recoil limit:

$$k_B T_r = E_r \equiv (\hbar k)^2/m \;, \tag{1}$$

$$v_r = \hbar k/m \;.$$

This paper explores the application of optical pumping techniques to cool the translational degrees of freedom below the recoil limit. The key idea is simple: since the final spontaneous decay of the cooling cycle has a random direction, there is a finite probability for the atom to decay to a translational state whose energy is less than $\varepsilon^2 E_r$ where ε is less than unity. If there exists a mechanism which can selectively recycle only those atoms with translational energy above $\varepsilon^2 E_r$, those may decay inside $\varepsilon^2 E_r$ on a subsequent recycling. Thus after many cycles a large fraction of the atoms will have been cooled below $\varepsilon^2 E_r$. In the following sections we shall discuss this idea more fully, derive the limits imposed by imperfections in the selectivity of the excitation, discuss solutions to the problems of systematic heating due to the repeated cycling, and give results from a computer simulation of one example of this general idea, velocity space optical pumping.

Translation Optical Pumping of Trapped Particles

Consider the case of an atom in a trap. We consider first the cooling mechanism called cyclic cooling [1], then show how this may be altered to cool below the recoil limit using translation state optical pumping. The translational degree of freedom is now quantized, as shown in Fig. 1; atoms in internal states U or L move in different potentials each with its own set of translational states. These states are densely packed, and we will have to introduce the density of states $\rho(E)$ later on. Imagine that the system starts in eigenstate $U,1$. An RF transition to internal state L provides an extremely energy selective excitation to eigenstate state $L,1'$. Let us assume that optical excitation followed by spontaneous decay now provides optical pumping back into U and a translational state in the vicinity of 2. The uncertainty results from the uncertainty of the momentum transfer in the optical pumping cycle, and is reflected in the spectrum of the fluorescence. Note that the spontaneous decay is *essential* in order to make the system cycle in the desired direction; more fundamentally it is the mechanism which removes entropy from the system and permits real cooling. The scheme described above obviously lowers the translational energy of atoms in internal state U. If the frequency of the RF is lowered slowly, atoms near translational state 2 will come into resonance and will be lowered further in translational energy [1].

A limit for this cooling process is reached as the RF frequency approaches the transition frequency at the bottom of the well. There the recoil energy becomes comparable to the amount of energy removed in the cycle and the atoms returning to state 2 after a cooling cycle are sometimes hotter than when they started!

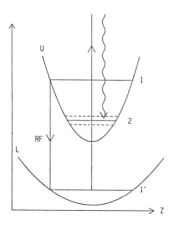

Fig. 1. Schematic diagram of translation space optical pumping mechanism. (Note: optically excited state not shown.)

69

The idea of optical pumping in translation space is now implemented by the following procedure. If the RF transition frequency is lowered only to a limit just above the transition frequency at the well bottom, atoms with energies very close to the well bottom (we call this region of phase space the *target zone*) will never come into resonance and will not be cycled (and possibly heated by the recoil). If the RF is repeatedly swept down to this limit, additional atoms will decay into the target zone so that they will not be excited again by subsequent sweeps. By repeating the sweep enough times, an arbitrarily high fraction of the atoms may be accumulated in the target zone, whose maximum energy is $\varepsilon^2 E_r$ with $\varepsilon < 1$.

Limit to pumped fraction due to finite resolution

In the preceding section we ignored the fact that the RF transition must have a finite linewidth with Lorentzian wings. Therefore atoms pumped into the target zone may possibly be excited again by the wings of this line and removed from the target zone by the recoil, even though the RF is never again in exact resonance with them. Consequently it will not be possible to pump all of the atoms into the target zone, but only that fraction for which the removal rate due to line wing excitation less than or equal to the optical pumping rate. We now calculate this fraction as a fraction of the limiting energy of the target zone, $\varepsilon^2 E_r$. This, in turn, determines the fraction which can be optically pumped into the target zone as a function of the size of this zone.

If we denote the fraction of atoms in the target zone by f_o, then

$$\frac{df_o}{dt} = + \text{optical pumping flux in} - \text{off-resonant excitation out} . \qquad (2)$$

If the excitation rate is R for resonant atoms and $R(\Gamma/\delta)^2$ for off-resonant atoms where δ is the detuning from resonance, then we approximate the rate out as the off-resonant rate for atoms with zero energy when the excitation frequency is at its lowest (where $\delta = \varepsilon k v_r$ assuming the Doppler shift dominates the shift due to the trapping potential). The rate out is then $f_o R(\Gamma/\varepsilon k v_r)^2$ ignoring the small fraction of off-resonantly excited atoms which happen to decay back inside the target zone.

In order to calculate the optical pumping flux in we need to know the integral of the density of states

$$P(E) = \int^{E} \rho(E)dE .$$

If we assume that the atoms excited resonantly have initial energy $\sim E_r$, then the excitation rate per atom is $R(\hbar\Gamma/E_r)$, because only a bandwidth Γ is excited. Assuming that the spontaneous decay distributes them evenly in phase space, a fraction $P(\varepsilon^2 E_r)/P(E_r)$ will decay into the target zone and Eq. (2) becomes

$$\frac{df_o}{dt} = (1-f_o) R\left(\frac{\hbar\Gamma}{E_r}\right) \frac{P(\varepsilon^2 E_r)}{P(E_r)} - f_o R\left(\frac{\Gamma}{\varepsilon k v_r}\right)^2 .$$

At equilibrium $df_o/dt = 0$ and this equation becomes (taking $P(E) \propto E^n$)

$$0 = R[(1-f_o)\frac{\hbar\Gamma\varepsilon^{2n}}{E_r} - f_o(\frac{\Gamma}{\varepsilon k v_r})^2].$$

Thus if it is desired to cool a fraction f_o of the atoms, they can be cooled to

$$E_{limit} = \varepsilon^2 E_r = [(\frac{f_o}{1-f_o})(\frac{\Gamma}{k v_r})]^{\frac{1}{n+1}} E_r.$$

For a three-dimensional harmonic potential n = 3, hence if the R.F. transition width is 10^5 times less than the Doppler shift due to the recoil velocity (i.e. $\Gamma/k v_r = 10^{-5}$), then 90% of the atoms can be pumped to $E_{limit} = 0.1\ E_r$. This is rather disappointing considering such good selectivity, but it nevertheless represents a substantial improvement over the recoil limit.

Velocity Space Optical Pumping

Velocity space optical pumping is another example of phase space optical pumping. It would be appropriate for particles confined to a trap with very steep walls. It is more favorable than harmonically confined atoms since the phase space exponent is only $n = 3/2$, (so that E_{limit} would be 0.02 E_r in the example above).

We have run a computer simulation of a velocity space optical pumping. Our objective was to study a simple version of the above scheme in which there were two discrete pumping frequencies rather than a continuous sweep. (It can be shown that a single frequency which excites particles inside the recoil velocity will heat the atoms; two frequencies therefore appears to be the simplest workable scheme.) Since our

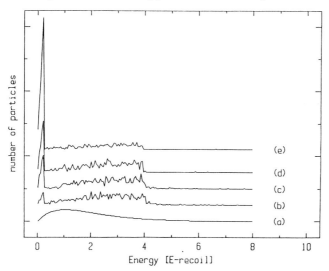

Fig. 2. Plots of energy distribution of atoms undergoing velocity space optical pumping. (Lasers tuned to 0.5 v_r and 2.0 v_r with widths 0.1 v_r and 1.0 v_r and relative intensity 1:10.) (a) Original distribution. After (b) 3,100, (c) 15,600, (d) 24,700, (e) 66,200 laser excitations.

objective was to verify that a two frequency scheme could work, as well as to study the rate of accumulation of atoms in the target zone, we did not use Lorentzian excitation functions, but rather square profile functions. This should not affect the overall feasibility or the initial accumulation rate — only the ultimate fraction (as calculated above). Figure 2 shows the results of the computer simulation where a laser tuned to 0.5 v_r is used to pump atoms into the region of energy less than 0.25 E_r, while a second laser at 2.0 v_r rescues those atoms heated by the first laser and cools them to the recoil limit.

Note. This material was not presented in the talk at EICOLS, which covered material reported in [2] and [1] on the magnetic atom trapping experiment at M.I.T.

References

[1]   D.E. Pritchard, Phys. Rev. Lett. 51, 1336 (1983).
[2]   V.S. Bagnato, G.P. Lafyatis, A.G. Martin E.R. Raab, R.N. Ahmad-Bitar and D.E. Pritchard, Phys. Rev. Lett. 58, 2194 (1987).

Ordered Structures of Ions Stored in an rf Trap

R. Casdorff, R. Blatt, and P.E. Toschek

I. Institut für Experimentalphysik, Universität Hamburg,
D-2000 Hamburg, Fed. Rep. of Germany

The demonstration of laser-cooled trapped ions[1] and atoms[2] and the recent achievement of laser confinement of neutral atoms[3] have raised the question of possible crystallization of trapped particles[4]. Numerical calculations, e.g. of a one-component plasma[5], have indicated that such a so-called "Wigner crystal" would exist[6]. An experiment[7], performed almost 30 years ago, revealed ordered structures of charged aluminum particles in an rf trap. With deep cooling of trapped ions achieved, we face the chance to observe analogous structures. Recent numerical computations of ion clouds[4,8] predict spatial order and various shapes of the cloud.

The purpose of this contribution is to communicate results of a calculation of the spatial arrangement of ions in an rf trap by means of Monte-Carlo simulation. The model allows us to include all energy contributions of the storage fields and the Coulomb interaction of the ions, as well as laser cooling.

In the simulation procedure, we compute the mechanical forces acting on the ions and (since the trap potential is explicitly time-dependent) integrate the equations of motion for time increments small enough to follow the trajectories adiabatically. Hence, the ion motion is considered free during these time intervals. This approximation enables us to apply a simulation technique which has been shown to correctly describe laser cooling of free particles[9]. Simulation of the cooling of single ions correctly shows the ion trajectories (as compared e.g. with the results in Ref. 7), and the steady-state kinetic energy of about 10 mK[1]. Without laser cooling, statistical properties of hot ion clouds, like their spatial and velocity distributions, may be derived[10]. On the other hand, laser cooling of small ion clouds indeed gives rise to ordered structures. The structural arrangement sensitively depends on the applied dc voltage in the rf trap, which determines the ratios of the radial and axial pseudo-potential well depths[11], D_r/D_z, and secular frequencies, ω_r/ω_z, where ω_r, ω_z indicate, in the approximate static model, the curvature of the pseudo-potential.

In the usual trap geometry ($r_0 = \sqrt{2}\, z_0$) $D_r = D_z$ implies $\omega_r < \omega_z$ (whereas $\omega_r = \omega_z$ corresponds to $D_r > D_z$). Fig. 1 shows an arrangement of five ions in the shape of a pentagon in the x-y plane, obtained for $\omega_r < \omega_z$ ($\vec{K}_{Laser} = (-1,-1,-1)$). Increasing the applied DC-voltage makes the structure change into a pyramidal shape, and eventually it abruptly turns into a double tetrahedron at $\omega_r = \omega_z$. This is shown in Fig. 2. Increasing the trap voltage

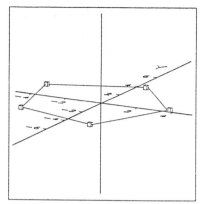
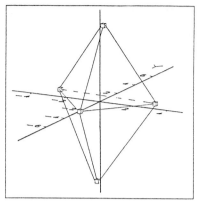

Fig. 1 Simulation of five
optically cooled ions in
an rf trap: $\omega_r < \omega_z$

Fig. 2 Same simulation
as in Fig. 1, but
$\omega_r = \omega_z$.

to $\omega_r > \omega_z$ again results in a pyramidal form, however rotated by 90° with respect to the z-axis.
Accordingly, structures of five, seven and nine ions with one centered ion are found instable and do not appear as a result of our model.

Previous computations[4,8] showed five ions at minimum kinetic energy, four of which to form a tetrahedron, with the fifth ion located at its center. Also, structures of seven and nine ions were found to include a central ion. Our result is at variance with these predictions. We think two features responsible for these findings: the micro-motion, and the presence of fluctuations, due to the diffusion of the ion momentum, from spontaneous emission during the cooling process.

References:

1. W. Neuhauser, M. Hohenstatt, P.E. Toschek, and H.G. Dehmelt
    Phys. Rev. **A 22**, 1137 (1980)
2. J. Prodan, A. Migdall, W.D. Phillips, I. So, H. Metcalf and
    J. Dalibard, Phys. Rev. Lett. **54**, 992 (1985). - W. Ertmer, R. Blatt,
    J.L. Hall, and M. Zhu, Phys. Rev. Lett. **54**, 996 (1985)
3. A. Migdall, J.V. Prodan, W.D. Phillips, T.H. Bergeman and
    H. Metcalf, Phys. Rev. Lett. **54**, 2596 (1985)
4. J. Mostowski and M. Gajda, Acta. Phys. Pol. **A 67**, 783 (1985)
5. W. L. Slattery, G. D. Doolen, and H.E. DeWitt
    Phys. Rev. **A 21**, 2087 (1980)
6. J. Javanainen, *"Fundamentals in Quantum optics II"*, F. Ehlotzky, ed.,
    Lecture Notes in Physics 282, Springer-Verlag, Berlin, 1987, p. 211
7. R.F. Wuerker, H. Shelton, and R.V. Langmuir
    J. Appl. Phys. **30**, 342 (1959)
8. E.V. Baklanov and V.P. Chebotayev, Appl. Phys. **B 39**, 179 (1986)
9. R. Blatt, W. Ertmer, P. Zoller, and J.L. Hall
    Phys. Rev. **A 34**, 3022 (1986)
10. R. Casdorff. R. Blatt, and P.E.Toschek, to be published
11. H.G. Dehmelt, Adv. At. Molec. Phys. **3**, 53 (1965)

Laser Cooling of Mg$^+$ Ions and First Experimental Observation of Resonant Particle Transport in a Penning Trap

Jin Yu, M. Desaintfuscien, and F. Plumelle

Laboratoire de l'Horloge Atomique, Equipe de Recherche du CNRS associée à l'Université Paris-Sud, Bât. 221, F-91405 Orsay, France

This experiment has been designed for the study of laser cooling of ions in either a rf or a Penning trap, and to understand the thermalization processes which the ions undergo in such a trap.

At present, our experimental study has been limited to the Penning trap. The apparatus is described in a previous paper /1/. Up to now, we have observed laser cooling from a few thousand degrees (typically 5000 K) to 1 K.

However, there are some processes which limit the cooling and the density obtained after cooling /2/.

It has been pointed out that a coupling between the axial and circular movements of an ion cloud trapped in a Penning trap could induce particle transport and limit ion density /2,3/. This coupling can be caused by a little misalignment between the magnetic field and the trap symmetry axis. Resonant coupling appears when the axial frequency ω_z and the magnetron frequency ω_m satisfy the following relation :

$$\frac{\omega_z}{\omega_m} = \ell \quad \text{where} \quad \ell = 2, 3, 4, \ldots . \tag{1}$$

In our experiment, we observe the variation of ion density n with the magnetic field B. For a simple description of the space-charge effect, we assume a uniform distribution of ions. In this case, we can write ω_z and ω_m as follows :

$$\omega_z = \sqrt{\omega_{zo}^2 - \frac{1}{3}\omega_p^2} \tag{2}$$

$$\omega_m = \frac{\Omega}{2}\left[1 - \sqrt{1 - \frac{4}{\Omega^2}\left(\frac{\omega_{zo}^2}{2} + \frac{\omega_p^2}{3}\right)}\right]. \tag{3}$$

Here $\omega_{zo} = \sqrt{eV_o/mz_o^2}$ is the axial frequency for a single trapped ion ; $\omega_p^2 = e^2 n/\varepsilon_o m$ is the plasma frequency, and $\Omega = eB/m$ is the cyclotron frequency.

Then the solutions of equation (1) give us the relation between n and B. We introduce for simplicity dimensionless variables :

$$\alpha = \frac{\omega_p^2}{\omega_z^2} \propto n , \quad (4) \qquad \gamma = \frac{\Omega}{\sqrt{2}\,\omega_z} \propto B . \tag{5}$$

The solutions of equation (1) are

$$\alpha_\ell(\gamma) = \frac{3}{(\ell^2 - 1)^2}\left(\sqrt{A_\ell^2 + C_\ell} - A_\ell\right), \tag{6}$$

where

$$A_\ell = \ell^2 \gamma^2 + (\frac{\ell^2}{2} + 1)(\ell^2 - 1),$$

$$C_\ell = 2(\ell^2 - 1)^2 \quad [\ell^2 \gamma^2 - \frac{1}{2}(\frac{\ell^2}{2} + 1)^2].$$

Figure 1 shows an adjustment between the theoretical curves according to (6) and the experimental variation of the density n of the ions versus the magnetic field. The experimental points have been obtained by measuring the fluorescence intensity when the magnetic field was swept slowly from lower to higher values. These results show that the density is limited by resonance phenomena.

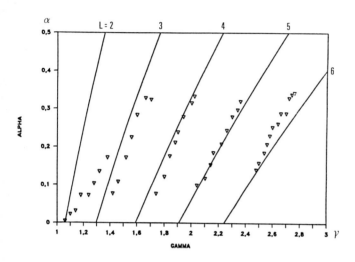

Fig. 1. Adjustment between the theoretical curves and the experimental data

References

1. F. Plumelle, M. Desaintfuscien, M. Jardino, P. Petit : Appl. Phys. B41, 183 (1986)
2. D.J. Wineland, J.J. Bollinger, W.M. Itano, J.D. Prestage : J. Opt. Soc. Am. B, Vol. 2, n° 11, 1721 (1985)
3. C.F. Driscoll : In Proc. of the Anti-Matter Workshop, Univ. of Wisconsin, Madison (1985)

Experiment to Observe a Two-Photon Transition in Stored Hg$^+$

M. Houssin, M. Jardino, and M. Desaintfuscien

Laboratoire de l'Horloge Atomique, Equipe de Recherche du CNRS associée à l'Université Paris-Sud, Bât. 221, F-91405 Orsay, France

The purpose of our experiment is to observe the two-photon transition between the fundamental state $5d^{10}6s^2S_{1/2}$ and the metastable state $5d^96s^{2\,2}D_{5/2}$ in stored Hg$^+$. This transition is attractive as an optical frequency standard for two reasons. First, the radiative lifetime of the $^2D_{5/2}$ state is about 0.1s and second, the first-order Doppler effect cancels for a two-photon absorption. Thus the transition line consists of a very narrow carrier and we can study the second-order Doppler effect (red shift and broadening, which is estimated to be a few kHz).

Hg$^+$ ions created by electron impact are confined in a cylindrical radio-frequency trap /1/. In our experiment, the radio frequency is 250 kHz, the vacuum inside the trap is $4\cdot10^{-9}$ Torr, and we are able to confine about 10^6 ions and to keep them in the trap for about 1s.

To drive the two-photon transition at 563 nm we used a dye laser which has to be very well stabilized because of the narrow linewidth of the transition. For this purpose we used a reference frequency which is a resonance frequency of a long Fabry-Perot (45 cm) interferometer. The laser was modulated at 35 MHz and we detected the beat between the central peak and the two lateral peaks reflected by the Fabry-Perot. Demodulated, this signal gives an error that permits laser frequency stabilization to ± 35 MHz /2/.

The error signal, treated by a very fast electronic system /3/, corrects the fast and medium laser frequency fluctuations using the two plates of an electro-optic modulator situated in the dye laser cavity. Using a piezocrystal on one of the laser mirrors, the same signal corrects slow fluctuations with a greater amplitude. We obtained a bandwidth of 2 MHz and a jitter lower than 140 kHz on the FP transmitted signal.

For long-term stabilization we need to control the reference Fabry-Perot length. For this purpose we used a hyperfine transition of I$_2$ near 563 nm as an error signal, resolved by a laser heterodyne saturated absorption method /4/.

A double-resonance scheme was used for the detection. Stored ions were excited, at 194 nm, from the $5d^{10}6s^2S_{1/2}$ level to the $5d^{10}6s^2P_{1/2}$ level by radiation from a mercury ion lamp, and we observed the fluorescence signal

77

at the same wavelength. If the dye laser frequency is varied a hole appears in the fluorescence intensity when the two-photon transition occurs.

The signal is expected to be weak because the relay states for the two-photon transitions are above the $^2D_{5/2}$ state. Using the results of reference /5/, we calculated $(F-F_0)/F_0$ to be $3 \cdot 10^{-13}$ P^2/sS where F_0 is the photon number off resonance, F the photon number at resonance, s and S are the sections of 194 nm and 563 nm light where the ions are located and P is the laser power. We increased the power by placing the trap in a Fabry-Perot whose finesse was 40. With our beams and a power of 0.5 W we calculated $F-F_0$ to be ~$3.6 \cdot 10^{-2}F_0$. As we had $F_0 = 10^4$ in previous experiments we expect the signal to be observable.

EXPERIMENTAL MAP

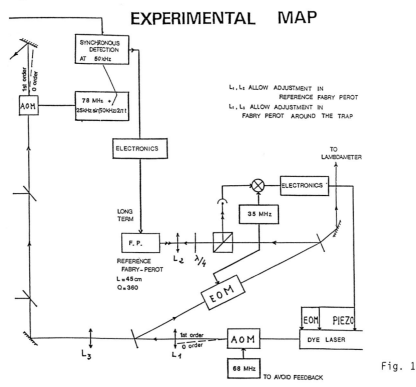

Fig. 1

References

1. W. Paul, O. Osberghaus and E. Fisher: ed. by Forschungber. Wirtsch. Verkehrsminist., 415 (Nordrhein-Westfalen, 1958).
2. J.L. Hall, L. Hollberg, Ma Long-Shing, T. Baer and H.G. Robinson: J. de Phys. C8, Suppl. No. 12, 42, 59 (1981).
3. J. Helmke, S.A. Lef and J.L. Hall: Appl. Optics 21, 1686 (1982).
4. G. Camy, D. Pinaud, N. Courtier and Hu Chi Chuan: Rev. Phys. Appl. 17, 357 (1982).
5. B. Cagnac, G. Grynberg and F. Biraben: J. de Phys., 34, 845 (1973).

High Accuracy Measurment of the g_J Factor of the ^{40}Ca Metastable Triplet Levels

N. Beverini, M. Inguscio, E. Maccioni, F. Strumia, and G. Vissani

Dipartimento di Fisica dell'Università, and INFN-Pisa,
Piazza Torricelli 2, I-56100 Pisa, Italy

The transitions between the fine structure calcium metastable levels are of great interest for metrology and astrophysics. However, the remarkable properties of these transitions can be fully utilized only if the spectroscopic constants of the levels are measured with the best available accuracy. In this paper a determination of the g_J factors of the $^3P_1$ and $^3P_2$ metastable levels is reported with an accuracy limited mainly by the actual knowledge of the proton gyromagnetic factor.

The Hamiltonian of an atom without nuclear spin, like ^{40}Ca, in an external magnetic field and in the Russel-Saunders approximation, can be written as

$$H = H_0 + A\vec{L}\cdot\vec{S} + \mu_0 g_J \vec{J}\cdot\vec{B} + \frac{e^2}{8m}\sum_a (\vec{B}x\vec{r_a})\qquad(1)$$

where the second term describes the fine structure coupling, the third the Zeeman effect and the last the diamagnetic correction, the sum being performed over all the electrons.The last term is usually smaller than the experimental sensitivity and neglected. In the case of the ^3P levels of Ca its contribution can be estimated at the level of about 10^{-6}, hence not completely negligible in our case.The level energies have been considered up to the second order as shown in Fig. 1.

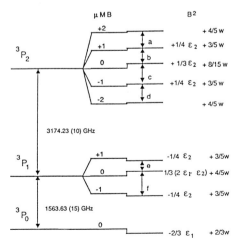

Fig. 1 - Energy level diagram of Ca ^3P levels in a magnetic field.

The third order terms are identically zero and the fourth ones are negligible in our experimental conditions.

The quadratic terms of the Zeeman transitions are symmetric with respect to the m_J quantum numbers and can be separated from the linear ones in the analysis of the experimental data. In fact we have, with reference to Fig. 1

$$\frac{f_a + f_d}{2\,\mu_0\,B} = g_2 \quad ; \quad \frac{f_b + f_c}{2\,\mu_0\,B} = g_2 \quad ; \quad \frac{f_e + f_f}{2\mu_0\,B} = g_1 \qquad (2)$$

and

$$f_f - f_e = (\frac{4}{3}\varepsilon_1 - \frac{1}{6}\varepsilon_2 + \frac{2}{5}w)\,B^2 \quad ; \quad f_b - f_a = (\frac{1}{6}\varepsilon_2 - \frac{2}{15}w)\,B^2 = f_d - f_c \ . \ (3)$$

The experimental apparatus is shown in Fig.2. The beam of metastable atoms travels in a magnetic field orthogonal to its direction and with a homogeneity of $\pm\,2$ mgauss in the interrogation region. The atoms are optically pumped before entering the magnet with a laser beam resonant with one of the $\Delta m=0$ Zeeman components of the $^3P_1 - {}^3S_1$ (or $^3P_2 - {}^3S_1$) transition. A microwave field induces Zeeman transitions at the magnet center that are detected as a fluorescence change by a second laser beam. The FWHM of the microwave transitions was measured to be about 70 kHz at a typical frequency of about 8 GHz ($Q \geq 10^5$). Four independent experimental runs have been performed giving the following result:

$$g_1 = 1.5010834 \ (21),$$
$$g_2 = 1.5011313 \ (45),$$

where the quoted errors are the statistical ones (one standard deviation). The accuracy of the measurements is also limited by the actual value of the proton gyromagnetic ratio which has an uncertainty of 3×10^{-6} .

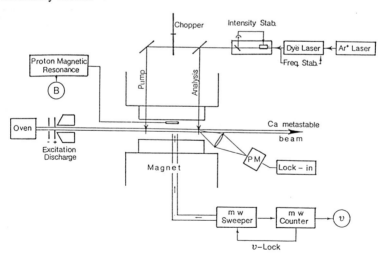

Fig.2 - Experimental apparatus.

Atomic Motion in a Laser Standing Wave

J. Dalibard, C. Salomon, A. Aspect, H. Metcalf(*), A. Heidmann, and C. Cohen-Tannoudji

Laboratoire de Spectroscopie Hertzienne de l'ENS et Collège de France, 24 rue Lhomond, F-75231 Paris Cedex 05, France

Atomic motion in a laser standing wave is a problem which has been extensively studied because of its importance for trapping and cooling. The first suggestion to trap atoms near the nodes or near the antinodes of a non-resonant laser standing wave was made about twenty years ago by LETOKHOV [1]. A few years later, KAZANTSEV predicted the existence of velocity dependent forces acting upon atoms moving in a standing wave and he proposed to use these forces to accelerate atoms [2]. At about the same time, the idea of radiative cooling was put forward by HANSCH and SCHAWLOW for neutral atoms [3] and by WINELAND and DEHMELT for trapped particles [4]. In such a scheme, one supposes that the forces due to the two counterpropagating waves can be added independently, which means that the intensity of the standing wave cannot be too high. During the last ten years, the number of theoretical and experimental papers dealing with atomic motion in a standing wave has increased considerably and it would be impossible to review here all these works.

The purpose of this paper is to present simple physical pictures based on the dressed atom approach [5] for understanding atomic motion in a standing wave. We would like also to apply these pictures to the interpretation of new experimental results obtained recently at Ecole Normale on cooling and channeling of atoms in an intense standing wave.

1. DRESSED ATOM APPROACH [5]

We consider a two-level atom with a ground state g and an excited state e, separated by $\hbar\omega_A$. We call Γ the spontaneous radiative linewidth of e. The laser field is single mode with a standing wave structure along the $0z$ direction and with a frequency ω_L. We denote $\delta = \omega_L - \omega_A$ the detuning between the laser and atomic frequencies (we suppose $|\delta| \ll \omega_A$).

The uncoupled states of the atom + laser photons system can be written |e or g,n⟩. They represent the atom in e or g in presence of n laser photons. We call $\mathcal{E}(n)$ the manifold of the two unperturbed states |g,n+1⟩ and |e,n⟩. These two states are separated by $\hbar\delta$ and they are coupled by the interaction

---

(*) Permanent address : Department of Physics, State University of New York, Stony Brook, NY 11790 USA.

hamiltonian V_{AL} describing absorption and stimulated emission of laser photons by the atom. The corresponding matrix element can be written

$$\langle e,n|V_{AL}|g,n+1\rangle = \hbar\omega_1(z)/2 \qquad (1)$$

where $\omega_1(z) = \omega_1 \sin kz$ is a z dependent Rabi frequency.

When the coupling (1) is taken into account, the two unperturbed states of $\mathcal{E}(n)$ transform into two perturbed states $|1(n)\rangle$ and $|2(n)\rangle$. These so called dressed states are linear combinations of $|e,n\rangle$ and $|g,n+1\rangle$ and their splitting $\hbar\Omega(z)$ is equal to

$$\hbar\Omega(z) = \hbar\left[\delta^2 + \omega_1^2 \sin^2 kz\right]^{\frac{1}{2}}. \qquad (2)$$

The full lines of Fig. 1 represent the variations with z of the energies of the two dressed states $|1(n)\rangle$ and $|2(n)\rangle$. We take $\delta > 0$, so that the unperturbed state $|g,n+1\rangle$ is above $|e,n\rangle$.

At a node ($\omega_1(z) = 0$), the two dressed states coincide with the unperturbed ones and the splitting $\hbar\Omega$ reduces to $\hbar\delta$. At an antinode, the splitting takes its maximum value and, if $\omega_1 \gg \delta$, $|1(n)\rangle$ and $|2(n)\rangle$ are approximately equal to the symmetric and antisymmetric linear combinations of $|g,n+1\rangle$ and $|e,n\rangle$.

We introduce now spontaneous emission. Since both states $|1(n)\rangle$ and $|2(n)\rangle$ of $\mathcal{E}(n)$ contain admixtures of $|e,n\rangle$, they can decay radiatively towards the two states $|1(n-1)\rangle$ and $|2(n-1)\rangle$ of $\mathcal{E}(n-1)$ which both contain admixtures of $|g,n\rangle$. It is important to note that the corresponding radiative linewidth of $|1(n)\rangle$ and $\cdot|2(n)\rangle$ depends on z. Consider for example the dressed state $|1(n)\rangle$ for $\delta > 0$. At a node, it reduces to $|g,n+1\rangle$ which is radiatively stable, so that the natural width of $|1(n)\rangle$ is then equal to zero. At an antinode and for $\omega_1 \gg \delta$, the weight of the unstable excited state e in $|1(n)\rangle$ is $\frac{1}{2}$ so that the natural

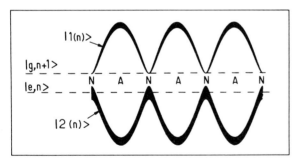

Figure 1   Spatial dependence of the dressed state energies in a standing wave (full lines). The thickness of the lines is proportional to the radiative linewidth of the levels. The dotted lines represent the unperturbed energy levels.

width of $|1(n)\rangle$ is equal to $\Gamma/2$. Similarly, the natural width of $|2(n)\rangle$ is equal to Γ at a node (since $|2(n)\rangle$ reduces then to $|e,n\rangle$) and to $\Gamma/2$ at an antinode.

2. ATOMIC ANALOGUE of the "SISYPHUS MYTH" : STIMULATED COOLING

Consider an atom moving along the direction $0z$ of the standing wave and being on one of the two dressed states $|1(n)\rangle$ or $|2(n)\rangle$. We suppose that its kinetic energy $mv_z^2/2$ is large compared to the height of the hills appearing on the energy curves of each dressed state (see Fig. 1). When the atom climbs a hill, its kinetic energy decreases : it is transformed into potential energy by stimulated emission processes which redistribute photons between the two counterpropagating waves.

Consider now the effect of spontaneous emission for $\delta > 0$ (blue detuning). The analysis of the previous section shows that, for each dressed state, the spontaneous emission rate is always maximum at the tops of the hills (antinodes for levels 1, nodes for levels 2). This means that the atom will leave preferentially a dressed state at the top of a hill. It follows that, during the time spent on a given dressed state, the atom sees on the average more uphill parts than downhill ones. Consequently, it is slowed down.

This new cooling mechanism is quite different from the usual one occuring for red detuning and at low intensity [6]. The mean energy loss per fluorescence photon is of the order of the height of the hills of Fig. 1. It scales as $\hbar\omega_1$ and does not saturate at high intensity : the velocity damping time of these "stimulated molasses" can therefore be much shorter than the one of usual molasses (by a factor Γ/ω_1). Experimental evidence for such a cooling mechanism has been obtained on Cesium atoms and is discussed in detail in reference [7].

3. CHANNELING of ATOMS

We suppose now that the kinetic energy of the atom along the standing wave, $mv_z^2/2$, is smaller than the height of the hills of Fig. 1 ($mv_z^2/2 \lesssim \hbar\omega_1$). This can be achieved for example by crossing at a right angle a one dimensional standing wave by a sufficiently well collimated atomic beam.

Figure 2 represents the energy surface of the dressed state $|1(n)\rangle$ which, for a blue detuning, connects to $|g,n+1\rangle$ out of the laser beam. Along the direction $0z$ of the standing wave, the rapid variation of $\sin^2 kz$ in (2) leads to a periodicity $\lambda/2$. Along the mean direction $0x$ of the atomic beam, the variations are much smoother since they are determined by the laser beam radius w_0. Because of the wings of the gaussian beam profile, the atoms which have a very small velocity spread along $0z$ enter the standing wave "adiabatically" and are thereby guided into

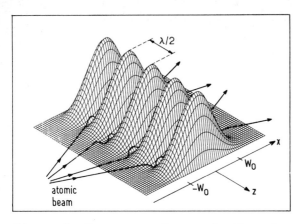

*Figure 2   Energy   surface   of   the   dressed   state $|1(n)\rangle$ in a
gaussian   standing wave   propagating along Oz. Atoms with a very
small transverse velocity are channelled near the nodes.*

the channels where they oscillate in the transverse direction. A
numerical   calculation   of   the atomic   trajectories shows that a
channeling   of the   atoms takes   place near the nodes within the
standing   wave. A similar   calculation predicts a channeling near
the   antinodes   for   a   red   detuning. We   have   checked   that
spontaneous   transitions   between   the   dressed   states   do   not
drastically   reduce   the   channeling because of the short passage
time of the atoms through the laser beam.

To   detect this   channeling, we have chosen to use the atoms
themselves as local probes of their own position. Because of the
spatially   varying light intensity, there are position dependent
light-shifts $\Delta(z)$ which are zero at the nodes (where $\omega_1(z) = 0$)
and   take their maximum value Δ_M at the antinodes. The center of
the   atomic   absorption   line   is   therefore located between two
extreme values, ω_A and $\omega_A + \Delta_M$ (for $\delta > 0$, $\Delta(z) = \delta - \Omega(z)$).

We have obtained experimental evidence for such a channeling
on   Cesium atoms [8]. Figure 3a   gives the   absorption spectrum
measured   on   a   weak   probe   laser   beam   for a uniform spatial
distribution   of atoms   (no channeling)   and for a blue detuning
(Δ_M is   then   negative). It   exhibits   a   broad   structure
corresponding   to   the   range   of   frequencies   between ω_A and
$\omega_A + \Delta_M$.   The   end   peaks   arise   because   the   line position is
stationary with respect to the position z around the nodes (peak
N)   and the   antinodes (peak   A). The heights of these two peaks
are different because the oscillator strength and the saturation
factor   of   the   atomic   line   depend   on   the   intensity.   The
modification   induced by channeling clearly appears on Fig. 3.b.
Peak N, corresponding to atoms near the nodes, is enhanced while
peak   A, corresponding   to the antinodes, is weakened. The curve
with   channeling   (Fig.   3b)   has been   obtained by   adjusting the
orthogonality   between the   atomic beam   and the   laser standing
wave   to   within 5.10^{-4} rad.   The   curve   without   channeling

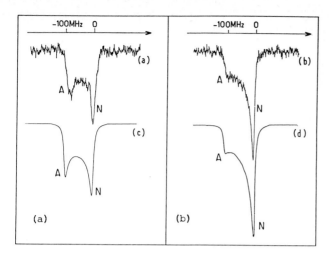

Figure 3 (a) and (b) : Experimental absorption of cesium atoms (D_2 line) at the center of a strong standing wave (Rabi frequency at the antinodes : 210 MHz ; detuning : 150 MHz ; laser beam waist : 2.3 mm), measured on a probe laser beam (power : 0,6 mW/cm² ; diameter : 1 mm). Curve (a) : no channeling. Curve (b) : channeling. (c) and (d) : calculated absorption spectra. Curve (c) : no channeling. Curve (d) : channeling.

(Fig. 3a) has been obtained by tilting the standing wave through 5.10^{-3} rad, corresponding to an average velocity along the standing wave of 1.7 m/s. With such a velocity, channeling is no longer possible.

In order to get some information on the spatial repartition of atoms $N(z)$ in the standing wave, we have calculated absorption spectra for our experimental conditions and with simple shapes for $N(z)$. Fig. 3c has been obtained with a uniform $N(z)$ and is in good agreement with Fig. 3a. Fig. 3d has been calculated with a triangular shape for $N(z)$ with a density at the nodes 5 times larger than at the antinodes. We are working on a more sophisticated deconvolution process, but the fit with the experimental curve is already of good quality. The spectra of Fig. 3 thus demonstrate the achievement of a laser confinement of neutral atoms in optical wavelength size regions.

It seems possible to improve the localization of the atoms by increasing the interaction time so that the trapped atoms could experience a further cooling for a blue detuning. Such a "dissipative channeling" has been predicted by a Monte-Carlo simulation [9] and by an analytical treatment [10]. It could be observed with laser decelerated atoms. Another attractive scheme would be to stop atoms and then to trap them at the nodes of a 3 dimensional standing wave. Finally, it would be interesting to try to observe the quantum states of vibration of the atom in the trapping light field potential. If the oscillation frequency (which in our experimental conditions is already 1 MHz) is large compared to the natural width of the dressed state, one could observe sidebands in the absorption or emission lines.

REFERENCES

1. V.S. Letokhov : Pisma Zh. Eksp. Teor. Fiz. 7, 348 (1968)
   [JETP Lett. 7, 272 (1968)]
2. A.P. Kazantsev : Zh. Eksp. Teor. Fiz. 66, 1599 (1974)
   [Sov. Phys. - JETP 39, 784 (1974)]
3. T.W. Hänsch and A.L. Schawlow : Opt. Commun. 13, 68 (1975)
4. D.J. Wineland and H.G. Dehmelt : Bull. Am. Phys. Soc. 20,
   637 (1975)
5. J. Dalibard and C. Cohen-Tannoudji : J. Opt. Soc. Am. B2,
   1707 (1985)
6. S. Chu, L. Hollberg, J.E. Bjorkholm, A. Cable and A. Ashkin :
   Phys. Rev. Lett. 55, 48 (1985)
7. A. Aspect, J. Dalibard, A. Heidmann, C. Salomon and
   C. Cohen-Tannoudji : Phys. Rev. Lett. 57, 1688 (1986)
8. A detailed description of this experiment will be published
   elsewhere
9. J. Dalibard, A. Heidmann, C. Salomon, A. Aspect, H. Metcalf
   and C. Cohen-Tannoudji : In Fundamental of quantum optics
   II, F. Ehlotzky ed., (Springer Series Lectures Notes in
   Physics), to be published
10. A.P. Kazantsev, G.A. Ryabenko, G.I. Surdutovich and
    V.P. Yakovlev : Physics Reports 129, 75 (1985)

Optical Elements for Manipulating Atoms

K. Cloppenburg, G. Hennig, A. Mihm, H. Wallis, and W. Ertmer

Institut für Angewandte Physik, Wegelerstr. 8,
D-5300 Bonn 1, Fed. Rep. of Germany

In several laboratories it has become routine by now to produce laser cooled, decelerated atomic beams of alkali elements. At present the main development and effort in this field of laser cooling points to the investigation and realisation of various trapping schemes of neutral atoms. This originates on one hand from the generic interest in understanding and optimization of these trapping schemes – e.g. to reach the sub-µK region – and on the other hand from the possibility to apply such traps to frequency standards (beyond the precision of $\Delta\nu/\nu=10^{-15}$) or other interesting future applications.

For many applications of cold atoms, however, cooled <u>beams</u> are neccessary or are in favour of trapped atoms – e.g. to avoid perturbations by the trapping fields – . Especially for frequency standard use <u>free</u> atoms (disreguarding gravity) in slow, well defined atomic beams or an atomic fountain [1] are best suited, if interaction times on the order of seconds (allowing sub-Hz precision) are sufficient. Therefore it seems to be important to study the possiblities for manipulating atoms to achieve dense, well collimated, very cold atomic beams or atomic fountains, free of hot or fast atoms. Therefore we will discuss below 1.) a new velocity filter, 2.) compression (squeezing) of atomic beams, 3.) deflection (separation) of cold atomic beams and 4.) one interesting application for the optical manipulation of atoms for the generation of VUV.

1. <u>Laser Velocity Selector</u>

Besides the deceleration of atoms, radiation pressure provides many possibilities for manipulating atoms with laser light. As an example we propose a laser velocity selector as described in this chapter.

Atoms within an atomic beam pass through a diaphragm (or slit) d_1 and cross two antiparallel laser beams with diameter s. The laser beams are tuned to zero velocity (or slightly red) and are separated by the distance x_s as shown in Fig.3. Passing the first laser beam the atoms gain transverse momentum in the z-direction by radiation pressure. Assuming high saturation the corresponding velocity change depends only on the interaction time t_1, which is a function of the atomic velocity v and is given by $t_1 \approx s/v$. Thus slow atoms gain more transverse momentum than fast ones and thus the resulting deflection angle is velocity dependent. This angular dispersion cancels in the second interaction region by the reverse momentum transfer for all atoms and velocities.

Thus atoms are sorted along the z-axis according to their velocity in beams parallel to the atomic beam axis. Therefore a second movable diaphragm (or slit) d_2 selects the desired velocity group. Equations (1) and (2) give the relations between z (position), v(z) (selected velocity group), $\Delta v(v)$ (longitudinal velocity spread), and Δv_{rms} (transverse velocity spread) (m: atomic mass) :

$$v(z) = (\frac{a \cdot s \cdot x_s}{z})^{1/2} \quad , \quad a = \frac{\hbar \cdot k}{m} \frac{1}{2 \cdot \tau} \quad , \quad (1)$$

$$\Delta v(d_2, z) = -\frac{1}{2} \frac{v^3}{a \cdot s} (\frac{d_2 + d_1}{x_s} + \frac{2 \cdot \Delta v_{rms}}{v}) . \quad (2)$$

Table 1 shows some numerical results for sodium. They demonstrate the high selectivity, especially for slow atoms. Therefore this scheme may either serve as an all-purpose velocity selector or as a separator and final selector for the decelerated atoms of a laser cooled atomic beam.

Table 1 Values of z(v), $\Delta v(v)$ as function of v and $d=d_1=d_2$ for sodium; s=2cm and x_s=0.5m; T denotes the temperature corr. to $\Delta v(d=0.1mm)$; Δv_{rms} is set about to the quantum limit

v [m/s]	z(v) [mm]	$\Delta v(d=0.5mm)$ [m/s]	$\Delta v(d=0.1mm)$ [m/s]	T [mK]
200	230	1.7	1.3	1
400	57	8	5.7	22
600	26	22	13	117
800	14	44	25	400
1000	9	76	40	1100
1500	4	204	98	6600

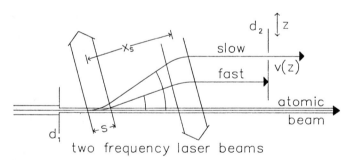

Fig.1 Experimental scheme for the laser velocity selector

2. Compression of Atomic Beams

All applications of resonance radiation pressure to free atoms lead to stochastic heating by the photon recoil. Therefore measures to compress the atomic beam in velocity as well as in position space are desirable for consecutive experiments with highly condensed, well defined atomic beams. Compression in

position space may be accomplished by the transverse dipole
force of a focused, co- or counterpropagating red shifted laser
beam [2]. As the dipole force provides a potential for the
transverse atomic motion the result is an undamped oscillation,
and the atoms leaving the focal region are radially defocussed
again. Thus an additional transverse <u>damping</u> force is needed.
It may be created by two crossed, weak standing waves orthogonal to the atomic beam, detuned one half of the natural linewidth to the red of the atomic frequency , $\delta=-\Gamma/2$. For low
saturation $s\approx1$ the characteristic transverse damping time t_d
reads $t_d\approx\hbar/E_{rec}$ ($\approx10\mu s$ for sodium). During this time a 50 m/s
(laser cooled) atom travels over a distance of about 0.5 mm.

The radial potential force near the focus O may be harmonically approximated by

$$\vec{F}_r = \hbar\delta\cdot2r/w_0{}^2\cdot\vec{e}_r = -Dr\cdot\vec{e}_r \,. \tag{3}$$

δ, w_0 denote the detuning and the waist of the copropagating
laser beam , respectively. For a waist of 500 μm, a power of
200 mW, a detuning of 130 MHz the resulting oscillation period
$t=2\pi(m/D)^{1/2}$ is on the order of 1,5 ms $>>$ t_d. Thus the oscillation is highly overdamped and atoms with the same longitudinal,
but different transverse velocities will reach the beam axis
nearly simultaneously. The resulting radial temperature can be
estimated as $k_B T\approx\hbar\Gamma$, if the induced fluctuations do not exceed
the spontaneous ones . This condition reads approximately [3]

$$D_{ind} < (\hbar\delta/w_0)^2/\Gamma < \Gamma(\hbar k)^2/4 \,. \tag{4}$$

This reduces to the experimental condition (z_R =Rayleigh-range):

$$\delta/\Gamma \le \pi w_0/\lambda =z_R/w_0 \,. \tag{5}$$

(5) will be fulfilled in situations of practical interest for
atomic beam compression. Assuming a Brownian random process the
mean square radius of the atomic beam is limited to

$$\Delta x = (2k_B T/D)^{1/2} \,. \tag{6}$$

With $D = 2\hbar\delta/w_0{}^2$ and $k_B T=\hbar\Gamma$ we obtain

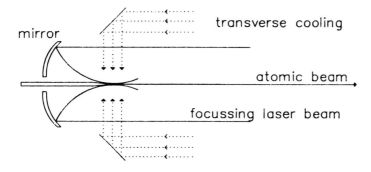

Fig.2 Experimental scheme for atomic beam compression

$$\Delta x = w_0 \; (\Gamma/\delta)^{1/2}. \qquad\qquad (7)$$

In accordance with (5) $\Delta x \approx \lambda$ could be reached for $w_0 \approx \lambda$.

3. Atomic Beam Deflection

Another method to separate laser cooled atoms from the other beam parts to form a cold beam makes use of radiation pressure from a transverse laser beam as shown in Fig.4. If all parameters are chosen correctly, the k-vector of this beam will be always orthogonal to the direction of motion of the slow atoms, thus changing only their direction. This radiation pressure has to compensate the centrifugal force, which is proportional to $1/r$ (r: bending radius); the saturation parameter $s(r)$ should read as (v_T: initial velocity, v_r: radial velocity, τ: nat. lifetime for the upper cooling level)

$$s(r) \approx \frac{v_T{}^2}{a_0} \cdot \frac{1}{r} \quad , \quad (s(r) \ll 1) , \qquad\qquad (8)$$

$$a_0 = \frac{\hbar k}{m\tau} .$$

This requires a radial intensity dependence as given by a cylindrical lens. Considering only the radial motion, the two forces form a potential $U(r)$, in which the atoms oscillate in a damped motion (damping v_r), when the laser is tuned about half a natural linewidth red (l: angular momentum):

$$U(r) = \frac{l^2}{2 \cdot m \cdot r^2} + C \ln r . \qquad\qquad (9)$$

The atoms oscillate through the potential minimum r_0 with the frequency ω_0, which is connected with the angular velocity $\dot{\varphi}$:

$$r_0 = 1/(m \cdot C^{1/2}) \quad \text{and} \quad \omega_0{}^2 = \partial^2 U(r_0)/\partial r^2 = 2 \cdot \dot{\varphi}^2 . \qquad (10)$$

So the atoms are refocussed again after $\varphi = \pi/(2^{1/2})$. The saturation necessary for a radius of 3cm and 50m/s atoms is ≈ 0.05. The deflected atoms are cooled to the quantum limit ($\hbar\Gamma$) in the plane of deflection and only slightly heated in the orthogonal direction. If the chirp method is used for deceleration, part of the cooling beam can be used for deflection [4].

Instead of the relatively weak spontaneous force at low saturation , also the dissipative dipole force, in a similarly shaped, blue shifted intense standing wave could be used

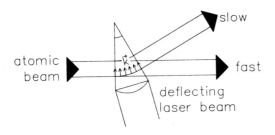

Fig.3 Optical arrangement for the deflection (separation) of laser cooled atoms. The deflecting laser beam has to be prepared to avoid possible optical pumping.

for deflection of an atomic beam on a circular orbit. The necessary field configuration is e.g. in a concentric resonator.

The action of the light force [5] is based on the kinetic
energy loss of the dressed atom in state 1(2) (notation as
given in [5]) approaching an antinode (node) of the standing
wave , combined with the enhanced relaxation rate (1⇒2) (resp.-
(2⇒1)) at the antinodes of the field. It results in a strong
damping and in a channeling of the atoms in the intensity
minima of the field. Our Monte-Carlo-simulations , performed
analogously to that mentioned in [5], show that this force
applied to Cs-atoms with λ=852nm, τ=32ns v_z= 15m/s, makes a
deflection on a circular orbit of radius R_0=5mm possible , if a
Rabi-frequency of ω=100Γ (at $r=R_0$) and a detuning of δ=35Γ is
chosen. For these conditions the angular width slightly de-
teriorates as the radial velocity width approaches the equivalence of about one half natural linewidth.

Nevertheless this proposal points into a vast field of
possibilities of manipulating atoms with the help of optical
channeling.

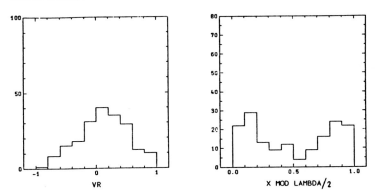

Fig.4 Monte Carlo simulation of channeling on a circle; x:
radius modulo λ/2 in units λ/2, v_r in 2k/Γ.

4. Generation of VUV-Radiation by Optical Manipulation of Atoms

Neon or the other noble gases offer a pure two-level system
between the metastable $(2p^5 3s)^o$ J=2 state (134,043.8cm^{-1}) and
the (3p) J=3 state (149,659cm^{-1}) allowing easy deceleration and
manipulation of the metastable atoms. This gives the chance for
separating the metastable atoms from a mixed atomic beam of
ground state atoms and metastable atoms by the methods described above and to increase their density.

With π-pulses from two pulsed dye-lasers the population in
the metastable state can be transferred to the $(2p^5 3s)$ J=1
states, which are population inverted in respect to the ground
state. Thus in case of sufficient density and an appropriate
geometry superradiance of $\approx$74nm can be achieved.

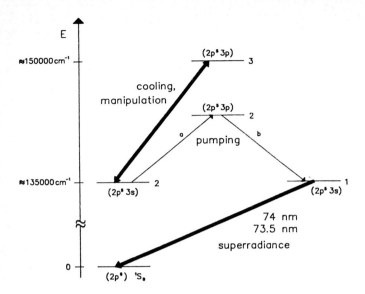

<u>Fig.5</u> Part of the neon level scheme with the transitions for optical manipulation and superradiance of 74nm

<u>References</u>

[1] J.R.Zacharias, Phys.Rev.<u>94</u>,751 (1954)
[2] J.E.Bjorkholm,R.R.Freeman,A.Ashkin,D.B.Pearson,
      Phys.Rev.Lett.<u>41</u>,1361(1978)
[3] J.Dalibard,S.Reynaud,C.Cohen-Tannoudji,J.Ph.B.<u>17</u>,4577(1984)
[4] W.Ertmer, S.Penselin, Metrologia <u>22</u>,195 (1986)
[5] A.Aspect,J.Dalibard,A.Heidmann,C.Salomon,C.Cohen-Tannoudji,
      Phys.Rev.Lett.<u>57</u>,1688(1986)

Acceleration of a Fast Atomic Beam by Laser Radiation Pressure

E. Riis, L.-U. A. Andersen, O. Poulsen, H. Simonsen, and T. Worm

Institute of Physics, University of Aarhus, DK-8000 Aarhus C, Denmark

The finite velocity distribution of atoms imposes severe restrictions on the resolution obtainable in laser spectroscopy. These problems can be eased by applying non-linear spectroscopic techniques such as Raman processes or two-photon absorption, but only to a certain extent. The second order Doppler broadening and transit time broadening plays an ever increasing role in the continuing quest for higher optical resolution.

Laser cooling has during the last few years proven to be a very efficient way to narrow an atomic velocity distribution [1]. Although the kinematic velocity compression eases the job in a fast beam experiment, it is, compared to the case of thermal atoms, a considerably more difficult task to cool a fast beam due to the short interaction time. Typically an atom scatters in the order of 100 photons, each transfering on the average a momentum of $h\nu/c$, corresponding to a change in Doppler shift of about 40 kHz.

As a demonstration of these concepts we here report having accelerated a 100 keV fast beam of metastable Ne atoms by laser radiation pressure. The experimental setup is shown schematically in Fig 1. In order to avoid loosing the atoms to the other fine-structure levels by optical pumping, the cooling laser was tuned to the closed transition from the metastable $3s[3/2]_2^O$ level to the $3p[5/2]_3$ level which has a lifetime of 19.4 nsec.

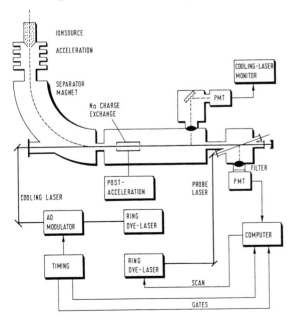

Fig. 1: The experimental setup. Ne^+ ions were accelerated to 100 keV in an electrostatic accelerator, mass separated, and charge exchanged in a Na cell. A well collimated dye-laser beam overlapped the atomic beam collinearly, providing the 4 m long interaction region for cooling. The velocity distribution at the end of the beamline was monitored with a probe laser, crossing the atomic beam at an angle of 5° and tuned to the $3s[3/2]_2^O$-$3p'[1/2]_1$ transition.

To detect the small changes in the velocity profile due to the radiation pressure and to eliminate possible non-linear laser interactions and effects of optical pumping, the cooling laser was chopped with an acousto-optic modulator. The probe detection electronics was gated accordingly to detect the cooled velocity distribution and a non-cooled reference distribution in two different spectra. Fig. 2a shows the non-cooled Doppler profile. The cooled profile looks similar, however it is shifted about 2 MHz to higher frequencies, corresponding to a velocity change of 1.2 m/s or a 0.25 eV increase in beam energy. The difference between the two distributions is shown in Fig. 2b.

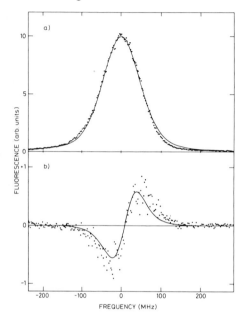

Fig. 2: The cooled velocity profile was recorded in a 2 µsec window after the cooling laser had been on for 5 µsec. When the cooled particles had passed the probe region, a 2 nsec cooling laser pulse optically pumped the new atoms into the same magnetic sublevels as previously and a non-cooled reference distribution was recorded in another 2 µsec window.
a: The experimentally detected reference spectrum (points) and the numerically evaluated distribution (solid curve).
b: The difference between the cooled and non-cooled distributions.

The modification of the velocity distribution due to the photon recoil is described by the Fokker-Planck equation [2]. In the comparison with the experimental data (Fig. 2) the velocity distribution obtained from solving this equation is convoluted with the transit time limited detection lineshape. The only adjustable parameters are the signal amplitude and the cooling laser detuning (15 MHz above resonance).

The most interesting aspect of this first demonstration of the mechanisms in laser cooling of a fast beam is the potential application of the technique in heavy ion storage rings. A dramatic increase in the influence and capability of the laser cooling will be achieved in these facilities due to the vast increase in the interaction time. The theoretical model shows that the acceleration rate is faster than any known heating process. Thus the prospects for producing a cold one-component plasma are promising and laser cooling might eventually lead to condensation of the plasma.

REFERENCES

1. W.D. Phillips, J.V. Prodan, and H.J. Metcalf, J.Opt.Soc.Am. B, 2, 1751 (1985).
2. J. Javanainen, M. Kaivola, U. Nielsen, O. Poulsen, and E. Riis, J.Opt. Soc. Am. B, 2, 1768 (1985).

Superhigh Resolution Laser Spectroscopy with Cold Particles

S.N. Bagayev, A.E. Baklanov, V.P. Chebotayev, A.S. Dychkov, and P.V. Pokasov

Institute of Thermophysics, Siberian Branch of the USSR, Academy of Sciences, SU-630090 Novosibirsk, USSR

The results are first reported of investigations of the behaviour of a nonlinear resonance in methane at 3.39 μm in the transit region depending on the saturating field intensity. It has been shown experimentally that with small fields the size of the resonance derivative $\tilde{\gamma}$ is determined only by a homogeneous halfwidth Γ ($\tilde{\gamma}$ =1.4 Γ). The main contribution to resonance formation is provided by "cold" particles whose velocities are much less than an average thermal one. This permits elimination of the influence of transit broadening and of the second-order Doppler effect.

1. Production of ultranarrow nonlinear resonances with a relative width of about 10^{-13} and the corresponding rise in resolution of optical spectroscopy require the methods that allow an increase of the time of interaction of particles with field. Particle cooling permits this without considerable increase in field dimensions. An important item in use of "cold" particles is the reduction of the second-order Doppler effect that leads to resonance broadening and shift. In this paper we report on first spectroscopic studies of saturated absorption resonances by using "cold" particles with an effective temperature of about 10^{-1} K that is lower than the working gas temperature by a factor of 10^{-3}. The width of nonlinear resonances is associated with effective selection of "cold" particles for which the interaction time is determined by a homogeneous width.

2. With $\Gamma \ll k \upsilon_0$ (Γ is the collisional line halfwidth, k is the wave number, υ_0 is an average thermal velocity of particles) the Lamb dip is due to saturation of particles whose velocity projection onto the field axis is $\upsilon_z \simeq 0$. Selection of the transverse velocity of particles υ_r is performed owing to saturation effects. In the transit region, where $\Gamma \tau_0 \ll 1$ (τ_0 is the time of flight of a particle through the field with radius a at an average thermal velocity) we meet with a new mechanism of inhomogeneous saturation. The saturation depends on the time of interaction of particles with field τ ($\tau = a/\upsilon_r$). For particles whose velocities are $\upsilon_r < a\Gamma$, the saturation parameter may be considered to be constant, independent of velocity and equal to $\varkappa = 4d^2 E^2 / \hbar^2 \Gamma^2$ (d is the matrix dipole element of

the transition, 2E is the field amplitude). With velocities $\upsilon_r > a\Gamma$ the saturation parameter depends on transverse velocity $\mathfrak{æ}(\upsilon_r)=4d^2E^2/\hbar^2(a^2/\upsilon_r^2)$ and increases with velocity decrease. A dip in the medium is the totality of dips with different intensities and widths in accordance with velocity υ_r. An important motive causing us to make detailed studies is the fact that the slope of the frequency derivative of a resonance is determined by a homogeneous width at $\mathfrak{æ} \lesssim 1$, its maximum is at a distance $\gamma =1.4\Gamma$ away from the line centre /1/. The main contribution here is provided by the particles whose velocities are $\upsilon_r \sim a\Gamma$. The corresponding temperature of particles is $T_{eff} \sim (\Gamma\tau_0)^2 T_0$, where T_0 is the gas temperature.

For the structures with halfwidth $\sim \Gamma$ at $\Gamma\tau_0 \ll 1$ to be obtained the saturation parameter should be $\mathfrak{æ} \lesssim 1$. For a Gaussian beam this corresponds to the saturation power of one travelling wave $P \lesssim (\Gamma\tau_0)^2 c(\hbar\upsilon_0/4d)^2$, where c is the light speed.

An experimentally significant feature of saturation in the transit region is associated with a great difference in saturation for the interacting particles and the medium. This is due to the fact that a contribution to the linear absorption factor is provided by the particles with all velocities, the saturation resonance is determined by their small fraction $\sim (\Gamma\tau_0)^2$.

In the microwave range a saturation line is observed on the background of a linear contour with width of about $1/\tau_0$, which makes it difficult to detect a narrow saturation line. In the optical range the saturation line is observed in a "pure" form on the background of a linear contour with a Doppler width. The saturation in the medium will be $\mathfrak{æ}_{med} \sim (\Gamma\tau_0)^2 \mathfrak{æ} = (4d^2E^2/\hbar^2)\tau_0^2$ (more exactly $\mathfrak{æ}_{med} = \mathfrak{æ}(\Gamma\tau_0)^2 \cdot \ln(1/\Gamma\tau_0)$ /1/). With low saturations ($\mathfrak{æ}_{med} \ll 1$) and absorptions the resonance intensity becomes very small. In early works the study of resonances was restricted by the region of relatively high saturation /2/ and by the qualitative observation of line narrowing /3, 4/.

3. To study the resonance and its derivative we developed the special spectrometers based on a He-Ne laser at 3,39 μm with an internal methane absorption cell ($F_2^{(2)}$ line of the P(7), υ_3 transition in methane). These spectrometers are described in detail in /5/. The studies were carried out in a wide range of pressures at small fields inside a cavity.

A qualitatively new item in these studies in the transit region is field broadening of the resonance and its derivative (Fig. 1). With $\Gamma\tau_0 > 1$ $\tilde{\gamma} \sim \tilde{\gamma}_0(1+\mathfrak{æ})^{1/2}$, where $\tilde{\gamma}_0$ is the halfwidth at $\mathfrak{æ} \to 0$. With $\Gamma\tau_0 \ll 1$ three peculiar parts can be distinguished in the dependence $\tilde{\gamma}(P)$. The initial linear part corresponds to the saturating power $P \lesssim 20$ μW at which $\mathfrak{æ} \lesssim 1$. With the saturating power P=15 μW and $\Gamma\tau_0=0.1$ ($a =0.08$ cm) we

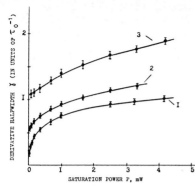

Fig. 1. Field dependence of $\tilde{\gamma}$ (in units of τ_0^{-1}) for one component of the hyperfine structure in methane, $a = 0.08$ cm, 1 – $\Gamma\tau_0 = 0.13$, 2 – $\Gamma\tau_0 = 0.54$, 3 – $\Gamma\tau_0 = 1.3$.

had in experiment $\tilde{\gamma} \approx 14$ kHz. The values of widths obtained in the linear extrapolation of $\tilde{\gamma}(P)$ and $\gamma(P)$ to zero power were used to plot the dependences of $\tilde{\gamma}_0$ and γ_0 on gas density. In the second region (P ~100 μW) $\Gamma \ll dE/\hbar \ll 1/\tau_0$ and the Rabi frequency $dE/\hbar$ serves as a homogeneous linewidth. In the region of relatively great fields (P ~1 mW) when $dE/\hbar \sim 1/\tau_0$ (the third region), the contribution of slow particles is suppressed because of strong saturation. The resonance width here is determined as $\gamma \sim \tau_0^{-1}(1+4d^2E^2\tau_0^2/\hbar^2)^{1/2} = \tau_0^{-1}(1+P/P_0)^{1/2}$, where $P_0 = c(\hbar\upsilon_0/4d)^2$ is the saturation parameter in mW. The calculated value of $P_0 \approx 1$ mW coincides with that obtained in our experiment /5/ and in ref. /2/.

The obtained dependences of γ_0 and $\tilde{\gamma}_0$ on gas pressure are given in Fig. 2. A good agreement of experimental data with theoretical dependences is observed. Some discrepancy at resonance widths of above 500 kHz ($\Gamma\tau_0 > 2$ is marked with light circles) is, as we think, due to the nonlinear dependence of Γ on methane pressure. Unlike $\tilde{\gamma}_0$ the halfwidth γ_0 at $\Gamma\tau_0 \ll 1$ depends on an average thermal velocity. In accordance with the theory /6-7/ $\gamma_0 \sim (1/\tau_0)(\Gamma\tau_0)^{1/2}$, which is confirmed by experiment. The obtained experimental results do not confirm the theoretical conclusions of /8/ and disagree with the experimental data of /9/. In /8, 9/ a rough physical error was made, which influenced all the results and conclusions (see /10/).

4. The minimum value of $\Gamma\tau_0$ at which the resonance shape was studied was $\sim 10^{-2}$, which corresponded to $T_{eff} \sim 3 \times 10^{-2}$K (Fig.3). The second-order Doppler resonance shift associated with "cold" particles is $\Delta \sim (\Gamma\tau_0)^2 \Delta_0$, where Δ_0 is the shift corresponding to the gas temperature. The preliminary gas cooling down to T=78 K at $\Gamma\tau_0 = 10^{-2}$ allows achieving effective temperatures $T_{eff} = 10^{-2} - 10^{-3}$ K. The corresponding second-order Doppler shift will be $10^{-2} - 10^{-3}$ Hz. This enables one even now

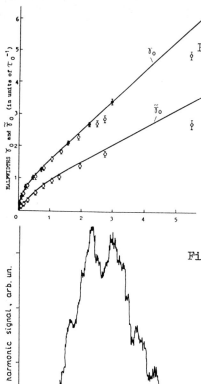

Fig. 2. Dependences of the resonance halfwidth γ_0 in methane and its derivative $\tilde{\gamma}_0$ in units of τ_0^{-1} on gas (CH_4, He) pressure in units of $\Gamma\tau_0$. For CH_4 $\Gamma = 15[\text{kHz/mTorr}] \times P_{CH_4}[\text{mTorr}]$, for He $\Gamma = 8$ $[\text{kHz/mTorr}] \times P_{He}[\text{mTorr}]$. Points $a = 0.25$ cm, circles $a = 0.08$ cm. Continuous curves are theoretical dependences /5/.

Fig. 3. Record of the recoil doublet of the hyperfine component $8 \rightarrow 7$ of the $F_2^{(2)}$ line in methane $[P(7), \nu_3]$ obtained with cold particles. $a = 0.25$ cm, $(2\pi\tau_0)^{-1} = 35$ kHz, $\Gamma\tau_0 \approx 10^{-2}$, $T_0 = 300$ K, $T_{eff} \approx 3 \times 10^{-2}$ K, saturating power ≈ 1 µW, frequency modulation $f = 600$ Hz, integration constant 8 s, recording time 30 min.

to start with production of a new generation of relatively simple lasers with frequency reproducibility better than 10^{-16}.

References

1. E.V.Baklanov et al.: Sov.J.Quant.Electr. 5, 1374 (1975)
2. J.L.Hall: In Fundamental and Applied Laser Physics, ed. by M.S.Feld, A.Javan, N.Kurnit (Wiley, New York 1973) p. 463; Colloques Internationaux du C.N.R.S. n°217, p. 105 (1973)
3. S.N.Bagayev et al.: Sov.Phys.JETP Lett. 23, 360 (1976)
4. V.A.Alekseyev et al.: Zh.Eksp.Teor.Fiz. 84, 1980 (1983)
5. S.N.Bagayev et al.: Preprint No. 125, Institute of Thermophysics (Novosibirsk 1985)
6. C.J.Borde et al.: Phys.Rev. A14, 236 (1976)
7. J.E.Thomas et al.: Phys.Rev. A15, 2356 (1977)
8. A.N.Titov: Opt.Comm. 51, 15 (1984); Sov.J.Quant.Electr. 15, 698 (1985)
9. A.Titov et al.: J.Physique 47, 2025 (1986)
10. B.Ya.Dubetsky: Kvant.Elektron. 5, 1088 (1987)

Light-Induced Drift and Isotope Separation in Alkali-Noble Gas Systems

E.R. Eliel, H.G.C. Werij, A.D. Streater, and J.P. Woerdman

Huygens Laboratory, University of Leiden, P.O. Box 9504,
NL-2300 RA Leiden, The Netherlands

1 Introduction

In a dilute mixture of optically absorbing atoms or molecules and a buffer gas the phenomenon of Light-Induced Drift (LID) can occur when: (i) the optically absorbing particles are excited in a velocity selective way (Doppler effect) and (ii) the collision rates with the buffer gas are different for particles in the ground- and excited-states, respectively [1]. The velocity selective excitation sets up antiparallel fluxes of ground- and excited-state particles, which cancel in the absence of collisions. When condition (ii) is met the absorbing particles acquire a net drift velocity.

Dramatic manifestations of LID occur in alkali-noble gas systems where the gas mixture is contained in a capillary cell. In an optically thick vapor an Optical Piston is created, as has been demonstrated for both the Na-Ar and Rb-Ar systems [2-5]. When the system is optically thin LID has, under the proper experimental conditions been applied to efficiently separate the isotopes of Rubidium [6]. Also, in the optically thin regime, quantitative measurements of the drift velocity under well-defined experimental conditions have been made and the obtained results have been compared with a recently developed numerical model for LID, in which the multi-level character of the alkali atom under study has been accounted for properly for the first time [7-9].

2 Drift Velocities in Na-Noble Gas Systems

The Optical Piston, though a remarkable demonstration of the phenomenon of LID, allowed for only qualitative determination of drift velocities, essentially due to adsorption of the alkali atoms on the capillary walls [3,4,5,10]. An experimental breakthrough was realized by coating the walls of the capillary with paraffin [11], which is known to provide an almost ideally nonsticking surface to Rb-atoms [12]. The accompanying requirement to work at much lower temperatures forces us to investigate LID effects in the optically thin regime.

An elegant tool for the study of LID under these conditions is the "Optical Machine Gun" [7,11] shown in fig.1. This device derives its name from the fact that small amounts (bullets) of alkali vapor are pushed or pulled by LID through the main capillary. They are repetitively introduced there due to the fact that a second laser beam that LID-pulls on the absorbing atoms is periodically ($8s^{-1}$) intercepted. During such a 5 ms interception period optically thin amounts of alkali vapor diffuse into the main tube. As the bullets are optically thin the ensemble has uniform drift velocity and, apart from diffusive spreading, the cloud remains together. A simple time-of-flight technique using photodetectors along the capillary allows for a direct determination of the drift velocity.

A typical example of the evolution of a cloud being driven through the capillary with a velocity of 3.9 m/s is shown in fig.2. The broadening of the cloud is due to diffusion; the extracted diffusion coefficient agrees well

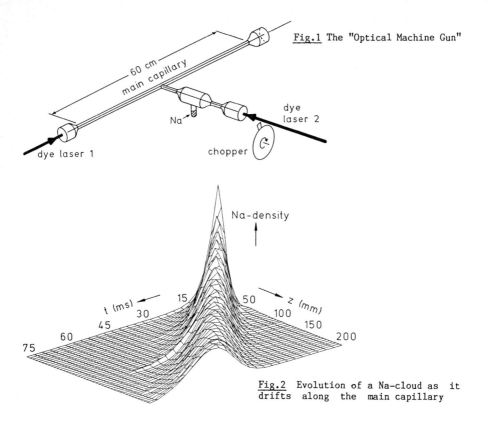

Fig.1 The "Optical Machine Gun"

60 cm
main capillary

dye laser 1

Na

chopper

dye
laser 2

Na-density

t (ms)   15   50   z (mm)
30   100
45   150
60   200
75

Fig.2 Evolution of a Na-cloud as it drifts along the main capillary

with the known literature value, confirming that wall effects are indeed absent. With this setup we can directly measure the drift velocity as a function of various parameters, such as light intensity, laser frequency, buffer gas species and buffer gas pressure. Examples thereof are shown in fig.3a for the drift velocity of Na in xenon as a function of laser frequency and in fig.3b for the drift velocity of Na in xenon as a function of laser intensity. In both graphs we also display the results of our numerical model [8,9]; the agreement is seen to be excellent.

The level of agreement is indeed impressive as the numerical model does not contain any adjustable parameters. It is based on a realistic rate-equation model, derived from the generalized Bloch equation, and includes the four relevant energy levels of the Na atom i.e. the F=1 and F=2 hyperfine levels of the 3s ground-state and the $3p\,^2P_{3/2}$ and $3p\,^2P_{1/2}$ fine-structure levels. Accounting for very considerable hyperfine pumping, fine-structure state mixing and transit relaxation due to the finite residence time of the atoms in the illuminated volume, it allows the calculation of the velocity distributions in all four atomic levels. The velocity distributions have been measured and very good agreement with the calculations exists. The large body of data, available in literature on Na-noble gas collisions provides essential input for the description of the collision processes.

The availability of the model allows us to predict the dependence of the drift velocity on experimentally accessible parameters and to investigate which physical mechanisms limit the drift velocity to be obtained. The simplest 2-level atom model of LID predicts attainable values of 50-100 m/s,

100

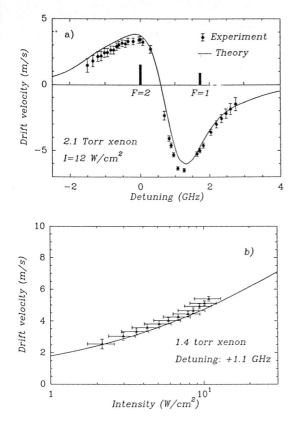

Fig.3 Experimental and calculated data for the drift velocity, (a) shows the results for Na in xenon as a function of laser frequency (D2-line). The vertical bars indicate the positions of the transitions connecting with the F=2 and F=1 hyperfine levels of the groundstate. (b) shows the drift velocity of Na in xenon as a function of laser intensity.

more than an order of magnitude larger than the value shown in fig.2. In the present model the maximum of the drift velocity lies at v_{dr}=14 m/s for Na in xenon at I=400 W/cm2 [7]. Power broadening and, more importantly, hyperfine pumping determine to a large extent this limiting value. Using two-frequency excitation with counterpropagating laser beams the effect of hyperfine pumping is largely eliminated and a maximum for v_{dr} has been predicted to occur at v_{dr} = 26 m/s for I=240 W/cm [7]. Extending this idea we believe that the ultimate limit will be v_{dr}=42 m/s at I=90 W/cm when suitably tailored broadband lasers are used for excitation [7]. A preliminary experiment to investigate the effect of a counterpropagating second laser beam through the main capillary confirmed the predictions of the model both in a qualitative and quantitative way.

3 Isotope Separation in Rb-Ar Mixtures

Naturally occurring Rb contains two isotopic species: 85Rb (72%) and 87Rb (28%),with quite different hyperfine (hf) splittings in the ground-state (3.0 and 6.8 GHz, respectively). Due to the larger ground-state hf splittings in the

101

Rb—isotopes compared to Na the drift velocities will be much smaller, as an even larger fraction of the atoms resides in the nonresonant hf level, as a result of optical pumping. Apart from a smaller numerical value of the drift velocity for the Rb—isotopes the spectral dependence of v_{dr} is predicted to be quite different, allowing separation of the isotopes [6]. Contrary to the case of Na, the spectral dependence of v_{dr} is predicted to resemble the derivative of the small—signal absorption spectrum. This was experimentally verified with the setup shown in fig.4, a setup very similar to the Optical Machine Gun. In a paraffin coated cross—shaped cell filled with a Rb—Ar mixture the LID effects are studied through the action of the optical shutter. Light from a ring dye laser operating on LD700, tunable through the Rb D2—line (780 nm), runs through the side arm of the cell and creates the optical shutter through LID. The Rb—density and isotopic composition of the vapor entering into the main capillary is determined by measuring the absorption spectrum with a frequency—swept single—mode diode laser.

When the ring dye laser is blocked a spectrum displaying the natural isotopic abundancies and the signature of the weak—field absorption spectrum is observed (fig.5a). Tuning the ring dye laser 520 MHz below the transition in ^{85}Rb starting from the F=3 hyperfine level, we push back the ^{85}Rb and we weakly pull on the ^{87}Rb. The resultant spectrum is shown in fig.5b, where the ^{87}Rb is enriched to 83%. Alternatively, tuning the ring dye laser 580 MHz above the same transition we LID—pull on both isotopes, but much more effectively on ^{85}Rb than on ^{87}Rb. Consequently the ^{87}Rb is depleted to 16% (fig.5c). For slightly different experimental conditions (e.g. buffer gas pressure and laser intensity) we have measured the frequency dependence of the enrichment. The results are shown in fig.6. The best we have been able to achieve so far is an isotopic mixture containing at least 94% ^{87}Rb.

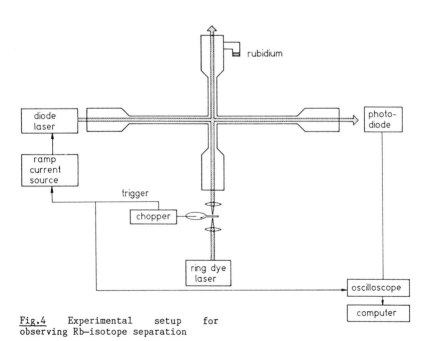

Fig.4 Experimental setup for observing Rb—isotope separation

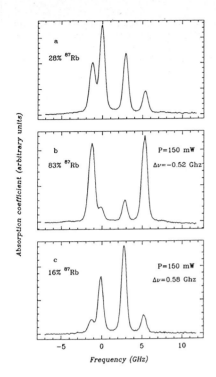

Fig.5 Rb absorption spectra. (a) shows the "normal" spectrum, (b) a spectrum enriched in ^{87}Rb and (c) a spectrum depleted of ^{87}Rb

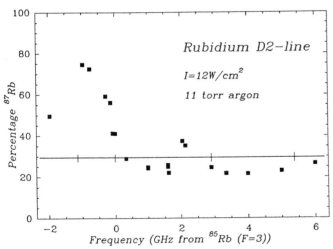

Fig.6   Frequency dependence of isotopic enrichment on the Rb D2-line

This work is part of the research program of the Foundation for Fundamental Research on Matter (FOM) and was made possible by financial support from the Netherlands Organization for the Advancement of Pure Research (ZWO).

References

1. F.Kh. Gel'mukhanov and A.M. Shalagin, Pis'ma Zh. Eksp Teor. Fiz. $\underline{29}$, 773 (1979) [JETP Lett.$\underline{29}$, 711 (1979)]
2. H.G.C. Werij, J.P. Woerdman, J.J.M. Beenakker and I. Kuščer, Phys. Rev. Lett. $\underline{52}$, 2237 (1984)
3. H.G.C. Werij, J.E.M. Haverkort and J.P. Woerdman, Phys. Rev. $\underline{A33}$, 3270 (1986)
4. G. Nienhuis, Phys. Rev. $\underline{A31}$, 1636 (1985)
5. W.A. Hamel, A.D. Streater and J.P. Woerdman, Opt. Comm. in press
6. A.D. Streater, J. Mooibroek and J.P. Woerdman, submitted to Opt. Comm.
7. H.G.C. Werij, J.E.M. Haverkort, P.C.M. Planken, E.R. Eliel, J.P. Woerdman, S.N. Atutov, P.L. Chapovskii and F.Kh. Gel'mukhanov, Phys. Rev. Lett. $\underline{58}$, 2660 (1987)
8. J.E.M. Haverkort, J.P. Woerdman and P.R. Berman, to be published
9. J.E.M. Haverkort and J.P. Woerdman, to be published
10. G. Nienhuis, Opt. Comm. $\underline{62}$, 81 (1987)
11. S.N. Atutov, St. Lesjak, S.P. Podjachev and A.M. Shalagin, Opt.Comm. $\underline{60}$, 41 (1986)
12. M.A. Bouchiat and J. Brossel, Phys. Rev. $\underline{147}$, 41 (1966)

Wall Frictionless Light-Induced Drift

J.H. Xu[*1], *S. Gozzini*[1], *M. Allegrini*[1], *G. Alzetta*[2], *E. Mariotti*[1],
and L. Moi[1]

[1]Istituto di Fisica Atomica e Molecolare del CNR,
  Via del Giardino, 7, I-56100 Pisa, Italy
[2]Dipartimento di Fisica dell'Università,
  Piazza Torricelli, 2, I-56100 Pisa, Italy
*ICTP Fellow Permanent address: Institute of Mechanics,
  Chinese Academy of Science, Beijing, China

During the last few years many experiments have dealt with the manipulation
of atomic velocities using laser sources. Mechanical action of light on the
atoms has been extensively studied using different approaches: radiation pressure has been applied in atomic beam cooling; induced dipole force has been
successfully used in collimating atomic beams and a new effect, Light-Induced
Drift (LID), has been proposed and its effectiveness shown.

LID, predicted by Gel'mukhanov and Shalagin /1/, is not based on a direct transfer of the photon momentum to the atoms but is due to the combined
effect of laser excitation and atomic collisions: the laser excited atoms are
immersed in a buffer gas in which the atoms in the ground or in the excited
state manifest a different diffusion coefficient. This difference induces
a loss of symmetry in the Maxwell velocity distribution and a drift of the
vapor inside the cell. The atoms may drift both in the direction of the laser
beam and against it, depending on the laser detuning.

The first experimental evidence for LID was obtained by Antsygin et al.
/2/, while Werji et al. /3/ reported the observation of a particular regime
of LID, i.e. the optical piston.

LID is strongly affected by the atom-wall collisions that may even totally inhibit it. In the previous experiments velocities hundred times lower
than the expected ones have been in fact measured because of this effect.
Quite recently Atutov et at. /4/ and Xu et al. /5/ have performed experiments
in which the LID evolves in a regime free of wall interactions. This is
obtained by coating the internal surfaces of the cell with paraffines or
silane compounds.

Our experiment is run in capillary cells having a diameter of the order
of 1-2mm, a length of about 15cm and an internal coating obtained after a
proper treatment with a silane-ether-tethrafluoride solution. The effect of
this coating is to reduce the adsorption time of the atoms at the cell walls
by many orders of magnitude (from 10^{-3} s for uncoated pyrex to 10^{-11} s).
Under these conditions a dramatic increase of the drift velocity has been
observed and values up to v=10m/s have been obtained for Na+Ne under single
mode laser excitation.

As LID is due to a modification of the atomic velocity distribution
through the laser excitation-collision mechanism, it depends also on the
laser bandwidth. A qualitative picture shows that, upon proper laser tuning,
the broader the range of velocities involved, the faster the drift.

Quite recently, by using a particular broadband dye laser (lamp-laser /6/) we have observed in both uncoated and coated capillary cells very fast LID. In uncoated cells we have measured velocities of 3m/s, i.e. a hundred times faster than previously observed, and velocities of the order of 100m/s in the coated ones.

In Fig. 1 the transient fluorescence signals obtained for three different tunings $\Delta\nu_L$ of the broad-band dye laser are shown. The laser is switched on at t=0 and the drift of the atoms is detected through the induced fluorescence that is collected by an optical fiber. The transients are ten times faster than those obtained by single mode laser excitation /5/.

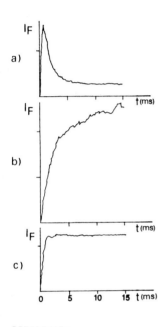

Fig. 1 Fluorescence signals obtained as a function of time for three different tunings $\Delta\nu_L$ of the broad-band dye laser

a) $\Delta\nu_L < \Delta\nu_0$

b) $\Delta\nu_L > \Delta\nu_0$

c) $\Delta\nu_L \simeq \Delta\nu_0$

REFERENCES

1. F.Kh. Gel'mukhanov and A.M. Shalagin, Sov. Phys. JEPT Lett. 29, 711 (1979)
2. V.D. Antsygin, S.N. Atutov, F.Kh. Gel'mukhanov, G.G. Telegin, A.M. Shalagin Sov. Phys. JETP Lett. 30, 243 (1979)
3. H.G.C. Werji, J.P. Woerdman, J.J.M. Beenakker, I. Kusher, Phys. Rev. Lett. 52, 2237 (1984)
4. S.N. Atutov, St. Lesjak, S.P. Podjachev, A.M. Shalagin, Opt. Commun. 60, 41 (1986)
5. J.H. Xu, M. Allegrini, S. Gozzini, E. Maariotti, L. Moi, Opt. Commun. in press
6. L. Moi, Opt. Commun. 50, 349 (1984); J. Liang, L, Moi, C. Fabre, Opt. Commun. 52, 131 (1984)

Part III

Quantum Jumps

Macroscopic Quantum Jumps in a Single Atom

R.G. Brewer[1] *and A. Schenzle*[2]

[1]IBM Research, Almaden Research Center, 650 Harry Road,
San Jose, CA 95120, USA
[2]University of Essen, D-4300 Essen, Fed. Rep. of Germany

There has been considerable interest recently in the theory [1-7] and measurement [8-10] of optical quantum jumps in a single atom. The novelty is in detecting a single atom event since in a group of atoms only some average emission is observed and the quantum jump effect is erased. The problem is a fundamental one, partly because it raises once again the measurement paradox of quantum mechanics and partly because the quantum statistics of spontaneous emission in a three-state atom had never been treated before.

The curious thing, which often is not realized, is that quantum mechanics does not predict a measurement. The wave function of an object such as an atom is described by a superposition of states and is smoothly varying. There are no discontinuities, and hence, events do not happen, whereas in the everyday world, we know they do. It is for this reason that Schrödinger [11] did not believe in quantum jumps, because it is not predicted by his wave equation.

What Schrödinger chose to ignore is the measurement process where the microworld is detected by its interaction with the macroworld or apparatus. Rather remarkably, the act of a measurement projects out of the atomic superposition state a single state, also known as the collapse of the wave function − an irreversible act. If quantum mechanics is a universal theory, it should apply to both the micro- and macroworlds. But again, rather curiously quantum mechanics predicts that macroscopic objects can also exist in superposition states. An example is Schrödinger's cat which can be partially alive and dead, an unphysical situation. This kind of predicament is the quantum measurement paradox.

The orthodox solution, which is not the only one proposed, is the Copenhagen formulation as advanced by Bohr. It states that quantum mechanics describes the microworld but that we have no knowledge of it until a measurement is performed, and, furthermore, that the measuring instruments must be described in classical terms, not quantum mechanically. Therefore, a quantum jump is the interaction of an atom with an apparatus, and it is meaningless to say that the atom executes a jump in the absence of an observation. The quantum jump reduces the atom to a single eigenstate. Bohr's view, however, evades the question of where a cutoff between the quantum description of the microworld and a classical description of the macroworld begins.

Schrödinger's 1952 paper, "Are There Quantum Jumps?" [11], was recently discussed by J. S. Bell [12], who believes that quantum mechanics itself must be modified since it fails to describe the macroscopic world.

The quantum theory I will discuss here assumes the Cophenhagen view, and it utilizes an approximate quantum mechanical treatment of the photodetection process, due to Glauber [13], which can be quite accurate and useful, even if it is not a formal quantum mechanical theory of measurement. I turn now to the main part of my talk, spontaneous emission in a three-level atom, giving only the results of the calculation. This problem originated with a proposal by Dehmelt [14] for an optical frequency standard that utilizes a single atom. Since the ultimate precision of such a standard is limited by the width of the atomic level involved, metastable states with a forbidden transition to the ground state are favorable configurations. However, atomic transitions that couple weakly to vacuum fluctuations also couple weakly to external fields which are used to monitor the resonance frequency.

A possible solution suggested by Dehmelt incorporates a three-level system, a "V" configuration, where a dipole allowed and a forbidden transition are coupled, as in Fig. 1.

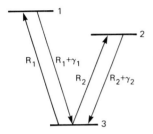

Fig. 1. The level structure considered consists of a ground state connected with two excited states via an allowed transition $|3\rangle \leftrightarrow |1\rangle$ and a forbidden transition $|3\rangle \leftrightarrow |2\rangle$. The lifetimes of spontaneous emission are γ_1^{-1} and γ_2^{-1} where $\gamma_2 << \gamma_1$. The two transitions are driven by their respective fields at rate R_1 and R_2. The observable is spontaneous emission of the allowed transition.

When each transition is driven resonantly by a laser field, one may expect on intuitive grounds that the two transitions compete in a unique way. It is evident that when the electron is driven in the allowed transition, strong spontaneous emission (frequency ω_1) will be emitted. But occasionally, the weak transition (frequency ω_2) will be excited causing the atom to be shelved in the metastable state for its radiative lifetime, and then the strong transition will be extinguished. Following the radiative decay from the metastable state to the ground state, spontaneous emission of the strong transition reappears. Thus, although the system is driven continuously, it does not respond continuously, *but shows fluorescence from the dipole allowed transition, interrupted randomly by long periods of darkness.* Consider for a moment an allowed transition with a lifetime of the order of 10^{-8} seconds and a metastable state that lives for a second. Resonance fluorescence from the forbidden state can never emit more than a single photon of frequency ω_2 per second, i.e., a single "quantum jump" creates a single photon in the interval of a second. This signal is unquestionably well below the noise limit of any photon counting experiment. When we consider, however, the proposed three-level configuration, we expect a strong fluorescence signal of frequency ω_1 with an intensity of 10^8 photons per second, which is turned on and off randomly in time. A single "quantum jump" in this picture would manifest itself in an almost macroscopic signal, which is well within the limits of experimental observation. The single event of an electron transiting from the metastable to the ground state creates a single photon of frequency ω_2 while in the three-level system, it triggers the emission of 10^8 photons per second, *a macroscopic quantum jump*, corresponding to an amplification of 10^8. It is an intriguing idea that the flickering of a quasi-macroscopic intensity scattered by a single atom could illustrate to the naked eye the nature of a quantum mechanical time evolution.

Let us begin our discussion by considering two intuitive arguments as they allow us to understand the nature of the problem without a detailed derivation. We assume that a single atom is excited by two fields, one for the allowed (1-3) and the other for the forbidden (2-3) transition. The fields are assumed to be sufficiently intense that each transition approaches saturation under steady-state conditions.

The first intuitive argument assumes that the atom is always in an eigenstate. Since the time-averaged probability of occupying any one state is 1/3, there is a 2/3 probability of occupying two states, such as $|1>$ and $|3>$. Hence, 2/3 of the time strong spontaneous emission from $|1>$ will be detected. On the other hand, there is a 1/3 probability of the atom being shelved in state $|2>$, and thus, 1/3 of the time there will be darkness. We conclude that the average ratio of the bright to the dark period is $\tau_B/\tau_D = 2$. Furthermore, during the bright period, the intensity of spontaneous scattering, $I_{sp} = \gamma_1 \rho_{11}^{ss} = 1/2\gamma_1$, will be the same as for the two-level problem where γ_1 is the spontaneous decay rate, ρ_{11}^{ss} is the steady-state diagonal density matrix or population of state $|1>$, and the atom behaves as if state $|2>$ did not exist. However, the time-averaged intensity over bright and dark periods must obey $< I_{sp}^{ss} > = (1/2)\gamma_1(\tau_B/(\tau_D + \tau_B)) = (1/3)\gamma_1$, which we see is consistent with $\tau_B/\tau_D = 2$.

The second intuitive argument assumes that all three states are in superposition, for example, when laser fields are used. One could imagine that the scattering intensity of $|1>$ is constant in time and its intensity would be only slightly reduced due to the fact that the population is shared among the three states.

In the following, we show that the first argument and its conclusions are reproduced in a quantum statistical theory. Additional details of the derivation can be found elsewhere [2]. In order to describe the photon statistics of this problem completely, we must include the entire hierarchy of multiphoton correlation functions. As in Fig. 1, we assume that both transitions are driven continuously by two resonant fields, acting independently, and the spontaneous decay rates are γ_1 and γ_2, where $\gamma_1 >> \gamma_2$. Our approach is to consider the photon counting probability $W(n,T)$, the probability of observing precisely n photon counting events in a given time T. This photon counting probability contains all orders of multiphoton correlations and allows predicting all details of a fluorescence counting experiment. According to our intuitive understanding, we expect an approximate Poissonian distribution around the average photon number, which for a collection time of the order γ_2^{-1} predicts an astronomically small probability of observing no event, i.e., a period of darkness. Alternatively, we could find that no photon in a time γ_2^{-1} is a rather frequent event and occurs on the average every third time, as suggested by our previous second-order correlation calculation [3], demonstrating that fluorescence is emitted intermittently. The zero count rate $W(0,T)$ for spontaneous emission of the 1-3 transition has been plotted in Fig. 2 as a function of the photon collection time T and for different values of the saturation parameter $S \equiv R_2/(R_2 + \gamma_2)$ of the weak transition. This figure demonstrates clearly that there are three characteristic time domains. For very short times $t << \gamma_1^{-1}$, $W(0, T)$ differs little from unity indicating that photons are emitted with an average time separation of at least $\tau = \gamma_1^{-1}$. In the intermediate time regime $\gamma_1^{-1} << T_0 << \gamma_2^{-1}$, $W(0,T)$ assumes a plateau value, independent of time, which is within the range 0.05 to 1/3 when the weak transition is driven sufficiently strongly, and for longer times $T >> \gamma_2^{-1}$, $W(0,T)$ drops sharply to a very small value.

The general photon counting probability is sketched schematically in Fig. 3 and reveals a bimodal distribution about $n = 0$ and $n = 10^8$ for $T = \gamma_2^{-1}$ where the latter

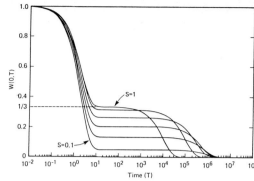

Fig. 2 (left). Probability $W(0,T)$ of observing no photon in a given interval T. The excitation rate of the weak transition has been varied from well below saturation to full saturation, $S = 0.1$ to 1, where $S = R_2(\gamma_2 + R_2)$. The spontaneous lifetimes have been assumed to be $\gamma_1^{-1} = 10^{-8}$ sec and $\gamma_2^{-1} = 1$ sec. The plateau rises up to the predicted value of 1/3 when saturation of the weak transition is approached.

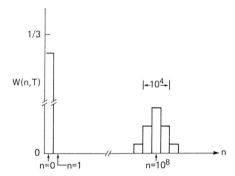

Fig. 3 (right). Photon counting distribution $W(n,T)$ for $\gamma_1^{-1} = 10^{-8}$ sec, $\gamma_2^{-1} = 1$ sec.

obeys a Poisson distribution. Since rather extreme scales are involved, the figure has not been drawn to scale. Thus, the above photon counting statistics proves in a transparent way that the first intuitive understanding of three-level fluorescence is the correct solution.

References

1. R. J. Cook and H. J. Kimble: Phys. Rev. Lett. 54, 1023 (1985)
2. A. Schenzle and R.G. Brewer: Phys. Rev. A 34, 3127 (1986)
3. A. Schenzle, R.G. DeVoe and R.G. Brewer: Phys. Rev. A 33, 2127 (1986)
4. D.T. Pegg, R. Loudon and P.L. Knight: Phys. Rev. A 33, 4085 (1986)
5. J. Javanainen: Phys. Rev. A 33, 2121 (1986)
6. C. Cohen-Tannoudji and J. Dalibard: Europhysics Lett. 1(9), 441 (1986)
7. P. Zoller, M. Marthe and D.F. Walls: Phys. Rev. A 35, 198 (1987)
8. W. Nagourney, J. Sandberg and H. Dehmelt: Phys. Rev. Lett. 56, 2727 (1986)
9. J.C. Bergquist, R.G. Hulet, W.M. Itano and D.J. Wineland: Phys. Rev. Lett. 57, 1699 (1986)
10. Th. Sauter, W. Neuhauser, R. Blatt and P.E. Toschek: Phys. Rev. Lett. 57, 1696 (1986)
11. E. Schrödinger: Brit. J. Phil. Sci. III, 109 (1952)
12. J.S. Bell, internal report CERN, June 19, 1986
13. R.J. Glauber: In Quantum Optics and Electronics, ed. by C. DeWitt, A. Blandin and C. Cohen-Tannoudji (Gordon and Breach, New York, 1965) p. 178
14. H. G. Dehmelt: Bull. Amer. Phys. Soc. 20, 60 (1975)

On the Theory of Quantum Jumps

G. Nienhuis

Fysisch Laboratorium, Postbus 80000,
NL-3508 TA Utrecht, The Netherlands and Huygens Laboratorium,
Postbus 9504, NL-2300 RA Leiden, The Netherlands

According to quantum mechanics, a measurement can only produce single eigenvalues of the measured observable, as if the system always exists in an eigenstate. For discrete states, this leads to the idea of a quantummechanical system executing jumps at random instants between eigenstates. DEHMELT [1] suggested to observe these jumps as the switching on and off of the fluorescent emission of a single ion or atom when a metastable state is weakly coupled to a strongly driven transition. This suggestion has led to several experiments [2-5] and even more theoretical treatments [6-10]. Most theoretical papers focus on the evaluation of the intensity correlation function f(t), which is the probability density for photon emission at time t, provided that another photon was emitted at time zero. Two papers [9,10] argue that the relevant quantity is rather the waiting-time distribution w(t), defined as the probability distribution for the time lapse t one has to wait for the first emitted photon after an earlier emission at time zero. Furthermore, the various treatments consider different special cases with regard to the level scheme, the degree of saturation and the degree of coherence of the driving light.

We point out that the intensity correlation function f and the waiting-time distribution w are related in a simple manner. The probability for finding at time t the n th photon emission after time 0 may be written as a repeated convolution of w, and the intensity correlation function is the summation of this expression over n. In Laplace transform this leads to a simple geometrical series with the result

$$\hat{f}(v) = \hat{w}(v)/[1 - \hat{w}(v)]. \tag{1}$$

This result allows us to calculate w once f is known. By applying standard techniques for the separation of rapid and slow time scales, we can derive an approximate expression for f, valid for arbitrary intensity and bandwidth and for any coupling scheme for three-state systems. The weakly coupled state can be coupled to the upper or the lower state of the strong transition, either by absorption or by emission. With (1), this leads to an expression for w(t) of the general form

$$w(t) = w_0(t) + R_+R_- \exp(-R_-t)/I_1 \tag{2}$$

with w_0 the waiting-time distribution and I_1 the emission rate in the absence of the weakly coupled state. The coefficient R_+ has the significance of the rate for switching off the fluorescence (corresponding to a transition to the metastable state), and R_- denotes the rate of decay of the metastable state, leading to switching on the fluorescence. Note that $w_0(t)$ decays to zero after a few radiative lifetimes of the strong transition. The long-time tail corresponding to the second term in (2) demonstrates the existence of dark periods in the fluorescence, since there is a finite probability R_+/I_1 that one has to wait a time of the order R_-^{-1} for the arrival of the next photon.

Explicit expressions for the switching rates R_+ are obtained automatically in the derivation of the explicit result for f, for any coupling scheme, and arbitrary degrees of coherence and saturation of the transitions, and general results will be published elsewhere [11]. Here we give the results in two opposite special cases. In the absence of the weakly coupled state, the populations of the upper and the lower state are given by

$$n_1 = s/(1 + 2s), \quad n_0 = (1 + s)/(1 + 2s) \tag{3}$$

in terms of the saturation parameter

$$s = \tfrac{1}{2}\pi\Omega_1^2 \, P(\omega_0)/A_1 \tag{4}$$

with Ω_1 and A_1 the Rabi frequency and the spontaneous decay rate of the strong transition, and P the convolution of the natural profile and the profile of the driving field. Likewise, Ω_2 and A_2 pertain to the weak transition. In the first case we consider, the field driving the weak transition has a large bandwidth λ (HWHM), and the other field has arbitrary properties. Then we find for the "V-configuration" (metastable state coupled by an absorptive transition to the lower state 0)

$$R_+ = Bn_0, \quad R_- = A_2 + B, \tag{5}$$

whereas we obtain in the "Λ-configuration"

$$R_+ = (A_2 + B)n_1, \quad R_- = B. \tag{6}$$

In (5) and (6), the rate $B = \Omega_2^2/2\lambda$ is the weak stimulated-transition rate.

The second special case occurs when the strong transition is driven by a monochromatic field on resonance and the field driving the weak transition is monochromatic with a frequency detuning equal to $\pm\Omega_1/2$, where the transition rate is maximal. Surprisingly, in this fully coherent case the results (5) and (6) remain valid, if we substitute the value $B = \Omega_2^2/A_1$. The saturation parameter s determining the populations (3) and (4) is equal to $s = \Omega_1^2/A_1^2$.

1. H.G. Dehmelt: Bull. Am. Phys. Soc. 20, 60 (1975)
2. W. Nagourney, J. Sandberg, H. Dehmelt: Phys. Rev. Lett. 56, 2797 (1986)
3. J.C. Bergquist, R.G. Hulet, W.M. Itano, D.J. Wineland: Phys. Rev. Lett. 57, 1699 (1986)
4. Th. Sauter, W. Neuhauser, R. Blatt, P.E. Toschek: Phys. Rev. Lett. 57, 1696 (1986)
5. Th. Sauter, R. Blatt, W. Neuhauser, P.E. Toschek: Opt. Commun. 60, 287 (1986)
6. D.T. Pegg, R. Loudon, P.L. Knight: Phys. Rev. A33, 4085 (1986)
7. H.J. Kimble, R.J. Cook, A.L. Wells: Phys. Rev. A34, 3190 (1986)
8. A. Schenzle, R.G. Brewer: Phys. Rev. A34, 3127 (1986)
9. C. Cohen-Tannoudji, J. Dalibard: Europhys. Lett. 1, 441 (1986)
10. P. Zoller, M. Marte, D.F. Walls: Phys. Rev. A35, 198 (1987)
11. G. Nienhuis: Phys. Rev. A35, 4639 (1987)

Quantum Jumps and Laser Spectroscopy of a Single Barium Ion Using "Shelving"

W. Nagourney, J. Sandberg, and H. Dehmelt

University of Washington, Department of Physics FM-15, Seattle, WA 98195, USA

We have observed quantum jumps between the $6^2S_{1/2}$ state and the $5^2D_{5/2}$ state of an individual laser cooled, radiofrequency trapped Ba^+ ion. Striking laser fluorescence "telegraph signals" appear upon incoherent excitation of the $5^2D_{5/2}$ state. Preliminary spectra of the $6^2S_{1/2} - 5^2D_{5/2}$ quadrupole transition have been obtained via direct excitation with a tunable color center laser at 1.762 μm. These spectra employed the "shelved optical electron amplification" technique, which ensures that virtually every transition will be detected.

1.0 Introduction

In 1975, one of the authors briefly described a scheme[1] in which the strong resonance fluorescence of an individual atomic ion confined in high vacuum serves as a monitor which determines whether or not the ion is in its electronic ground state. The usefulness of the scheme as a millionfold amplifier of the very weak *optical* fluorescence from one of the ion's metastable levels was pointed out. We report here the application of this scheme to the detection of forbidden *optical* transitions in a single trapped barium ion.

The technique is illustrated in Fig. 1. A red laser (650 nm) and a green laser (493 nm) cause $6^2S_{1/2} - 6^2P_{1/2} - 5^2D_{3/2}$ transitions which serve to both cool the single barium ion (confining it to <1 μm) and monitor its presence in the trap via strong fluorescence (at 493 nm) from the $6^2P_{1/2} - 6^2S_{1/2}$ transition. If the ion is excited to the $5^2D_{5/2}$ state, it will remain in this level for the $\approx$32 sec lifetime of that state. The presence of the ion in the metastable level can now be determined with unity "quantum efficiency" by the easily observed *suppression* of the 493 nm fluorescence.

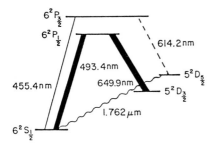

Fig. 1. Level structure of Ba^+. The red and green laser excitation is shown by the bold lines; the lamp excitation is indicated by the light solid line while the subsequent decay into the $5^2D_{5/2}$ level is indicated by the dotted line. The direct excitation of the $5^2D_{5/2}$ level by the color center laser is shown by the wavy line.

2.0 Quantum Jumps

Two methods for exciting the ion to the metastable level were used. To demonstrate "quantum jumps", a filtered barium resonance lamp was used to make relatively infrequent (roughly 1 every 10 seconds) transitions to the $6^2P_{3/2}$ state. Approximately one third of the subsequent decays will be to the metastable "shelf" ($5^2D_{5/2}$) level. A plot of the 493 nm fluorescence versus time appears in Fig. 2. The "dwell times" in the $5^2D_{5/2}$ level can be easily shown to be distributed exponentially, with an average value equal to the metastable lifetime. From a sample of 203 dwell times, we have verified the exponential distribution and measured the metastable lifetime. The result is 32 ± 5 sec, which includes a roughly 10% measured correction due to collisional quenching from the background gas. This is within one standard deviation of the previously measured[2] lifetime of 47 ± 16 sec.

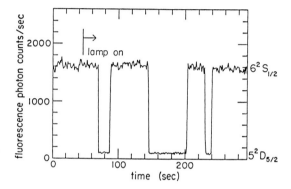

Fig. 2. A typical trace of the 493 nm fluorescence from the $6^2P_{1/2}$ level showing the quantum jumps after the barium resonance lamp is turned on. The atom is definitely known to be in the $5^2D_{5/2}$ level during the low fluorescence periods.

3.0 Laser Spectroscopy

Direct excitation of the quadrupole transition at 1.762 μm between the ground state and the metastable level was done using a tunable color center laser. The color center laser was crudely locked to a Fabry-Perot interferometer which was scanned in order to scan the laser. At each frequency in the scan, the number of abrupt changes in the strong fluorescence from the $6^2P_{1/2}$ level was counted; this number is equal to the number of transitions that the ion makes between the ground and metastable states. It was assumed that virtually all of these transitions were due to stimulated events; this is valid since the spontaneous lifetime (32 sec) is so much greater than the average time ($\approx.1$ sec) between transitions at resonance. A plot of the resulting spectrum appears in Fig. 3. The linewidth is due in part to the broadening of the ground state by the 493 nm laser and in part to the ≈10 MHz laser linewidth. When the efforts to narrow the color center laser are successful, it will then be necessary to alternate the cooling/interrogation lasers with the color center laser to eliminate the ground state broadening. The sidebands 25 MHz on either side of the main feature are due to the enhancement of the micromotion at the trapping frequency (25 MHz) by nonquadrupolar DC electric fields in the trap. These will be reduced in the future with the aid of compensation electrodes.

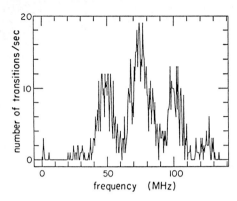

Fig. 3. Spectrum of $6^2S_{1/2}$ to $5^2D_{5/2}$ quadrupole transition at 1.762 μm made by scanning the color center laser and counting the transitions using the "shelved optical electron amplification" technique. The two sidebands are a trapping artifact and are due to the enhancement of the 25 MHz micromotion by nonquadrupoler DC fields in the trap.

4.0 Conclusion

This technique provides an essentially noiseless "front end" of a high resolution optical spectrometer. Spectroscopy of the $5^2D_{5/2}$ level of Ba$^+$ could be done with potentially 5 mHz resolution; virtually every transition to that level would be detected. Such spectroscopy would be free from Doppler shifts to all orders (due to the laser cooling) and would not be subject to collisional or transit time broadening; power broadening would be avoided by the above-mentioned sequencing of the lasers. In fact, with adequate magnetic shielding, the principal broadening mechanism for the forseeable future would be the breadth of the laser.

5.0 Acknowledgments

We would like to thank Clifford R. Pollock for kindly providing the OH-doped NaCl crystal used in the color center laser. The work was supported by the National Science Foundation under the SAPARIS PROJECT (PHY-8506028).

1. H. G. Dehmelt, Bull. Am. Phys. Soc. 20, 60 (1975).
2. F. Plumelle, M. Desaintfuscien, J. L. Duchene and C. Audoin, Opt. Commun. 34, 71 (1980).

The Observation of Quantum Jumps in Hg $^+$

W.M. Itano, J.C. Bergquist, R.G. Hulet, and D.J. Wineland

National Bureau of Standards, Boulder, CO 80303, USA

Introduction

Quantum jumps (sudden changes of quantum state) have been observed in isolated samples of one or a few atomic ions [1-4]. The method of detecting quantum jumps uses a kind of optical-optical double resonance originally proposed by Dehmelt [5]. Consider an atom, such as Hg$^+$, which has both a strongly allowed transition (at wavelength λ_1) and a weakly allowed transition (at wavelength λ_2) from the ground state (see Fig. 1). Assume that radiation is present at both λ_1 and λ_2. The atom fluoresces strongly at λ_1 until it absorbs a λ_2 photon and makes a transition to the metastable state. The λ_1 fluorescence is shut off until the atom decays to the ground state.

The first experiments were concerned mainly with verifying that quantum jumps did occur [1-3]. For a single ion, this means that the fluorescence from the strong transition switches back and forth between zero and a steady value rather than assuming some constant intermediate value. Values of metastable lifetimes were obtained from the time intervals between quantum jumps, but these values were no more accurate than those obtained previously. Here we report the use of quantum jumps in measuring radiative decay rates and branching ratios which had not been known accurately. We also report their use as a means of detection in high-resolution spectroscopy.

Hg$^+$ System

The lowest energy levels of Hg$^+$ are shown in Fig. 1. The $5d^96s^2$ $^2D_{3/2}$ and $5d^96s^2$ $^2D_{5/2}$ states are predicted to be metastable [6]. Once excited to the $5d^{10}6p$ $^2P_{1/2}$ state, the atom usually decays back to the $5d^{10}6s$ $^2S_{1/2}$

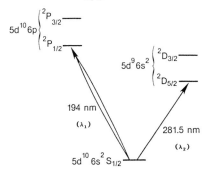

Fig. 1. Energy levels of Hg$^+$

ground state, but it has a probability, estimated to be about 10^{-7} [7], of decaying to the $^2D_{3/2}$ state. The probability is small because the transition frequency is low and because it requires configuration mixing to occur (the nominal configurations differ by two orbitals). From the $^2D_{3/2}$ state, the atom decays either directly to the ground state with probability f_1 or to the $^2D_{5/2}$ state with probability $f_2 = 1-f_1$. According to theory, $f_1 \approx f_2$ [6].

Apparatus

The apparatus has been described previously [2,8]. The Hg$^+$ ions are confined in a radiofrequency (Paul) trap under ultra-high-vacuum conditions. A single Hg$^+$ ion has been kept in the trap for over a week. The cw 194 nm radiation required to excite the $^2S_{1/2}$ to $^2P_{1/2}$ transition was obtained by sum-frequency mixing the output of a frequency doubled 514.5 nm Ar$^+$ laser with a 792 nm dye laser in a potassium pentaborate crystal [9]. The 281.5 nm radiation required to excite the $^2S_{1/2}$ to $^2D_{5/2}$ transition was obtained by frequency doubling the output of a dye laser. The 194 nm fluorescence was collected by a lens system and detected by a photomultiplier tube.

Analysis of Single-Ion Quantum Jumps - Lifetimes and Branching Ratios

Typical fluorescence data are shown in Fig. 2 for one, two, and three ions. The quantum jumps are clearly visible and occur in a time of less than 1 ms, which is the spacing between data points. For these data, only the 194 nm radiation was present. The intensity was high enough that quantum jumps due to the weak $^2P_{1/2}$ to $^2D_{3/2}$ decay occur several times per second.

Let P_1, P_2, and P_3 be the probabilities that an Hg$^+$ ion is in the $^2D_{3/2}$, the $^2D_{5/2}$, and the $^2S_{1/2}$ states, respectively. Let γ_1 and γ_2 be the total radiative decay rates of the $^2D_{3/2}$ and the $^2D_{5/2}$ states, respectively. The ensemble averaged behavior of one atom after it decays to the $^2D_{3/2}$ state can be derived from the following set of rate equations:

$$dP_1(t)/dt = -\gamma_1 P_1(t), \tag{1}$$

$$dP_2(t)/dt = f_2\gamma_1 P_1(t) - \gamma_2 P_2(t), \tag{2}$$

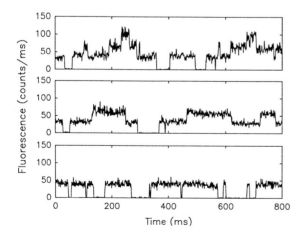

Fig. 2. Fluorescence intensity as a function of time for three ions (top), two ions (middle), and one ion (bottom), showing the quantum jumps

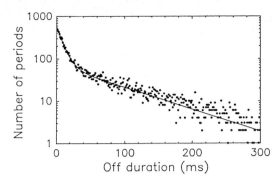

Fig. 3. Distribution of fluorescence-off periods for a single ion (dots) and least-squares fit (solid curve)

$$dP_3(t)/dt = f_1\gamma_1 P_1(t) + \gamma_2 P_2(t). \tag{3}$$

The quantity which is measured is the probability distribution $W_{off}(\tau)$ of a fluorescence-off period having a duration τ. This is proportional to $dP_3(t)/dt$ evaluated at $t=\tau$ with the initial conditions $P_1(0) = 1$ and $P_2(0) = P_3(0) = 0$. Solving Eqs. (1-3) yields

$$W_{off}(\tau) = \text{const.}[f_2\gamma_2\exp(-\gamma_2\tau) + (f_1\gamma_1-\gamma_2)\exp(-\gamma_1\tau)]. \tag{4}$$

The experimental fluorescence-off distribution was least-squares fitted to Eq. (4) to obtain values for γ_1, γ_2, and f_1. Figure 3 shows the data and the least-squares fit. The values obtained from the fit are $\gamma_1=112 \pm 7 \text{ s}^{-1}$, $\gamma_2=12.0 \pm 0.6 \text{ s}^{-1}$, and $f_1=0.498 \pm 0.031$. These values are in good agreement with calculations [6]. The value of γ_2 is in good agreement with previous measurements [2,8,10], but the value of γ_1 is about a factor of 2 higher than the only previously reported value [10]. No previous measurements of f_1 exist.

Application of Quantum Jumps to Spectroscopy - Quantized Detection

We have used the bistable nature of the single-ion fluorescence intensity (high when the ion is cycling between the $^2S_{1/2}$ and $^2P_{1/2}$ states and zero when it is in a metastable state) to detect the weak $^2S_{1/2}$ to $^2D_{5/2}$ transition [11]. The measurement cycle was as follows: If the fluorescence was high enough to indicate that the ion was cycling between the $^2S_{1/2}$ and $^2P_{1/2}$ states, the 194 nm radiation was turned off and the 281.5 nm radiation was turned on for 20 ms, which is much less than the $^2D_{5/2}$ state lifetime.

The 194 nm radiation was turned on again, and the fluorescence photons were counted for 10 ms to see whether the ion had made a transition to the $^2D_{5/2}$ state. We have used the quantum multiplication aspects of double-resonance detection previously [12]. The new feature was that the result of the measurement was a 0 or a 1, depending on whether the number of photons detected was below or above a threshold level. This eliminated certain kinds of instrumental noise, such as intensity fluctuations of the detection laser, leaving only the inherent quantum fluctuations of the atom. Resonance linewidths as small as 30 kHz were observed at 281.5 nm, and the motional sidebands were clearly resolved. Motional sidebands are caused by the frequency modulation of the laser frequency in the frame of the ion by the Doppler effect. They are offset from the central resonance by multiples of the frequencies of harmonic motion of the ion. The intensities of the motional sidebands indicated that the ion had been cooled to near the theoretical minimum of 1.7 mK.

Quantum Jumps of Two or Three Ions

We have observed quantum jumps with two and three ions (see Fig. 2). Others have reported that multiple jumps (simultaneous quantum jumps of two or more ions) occur frequently in a sample of Ba^+ ions [4]. They attribute the multiple jumps to a collective interaction of the ions with the light field. We find that the Hg^+ ions act independently. Apparent multiple jumps occur only at a rate which is consistent with the finite time resolution of the apparatus.

We gratefully acknowledge the support of the Air Force Office of Scientific Research and the Office of Naval Research.

References

1. W. Nagourney, J. Sandberg, and H. Dehmelt: Phys. Rev. Lett. **56**, 2797 (1986)
2. J.C. Bergquist, R.G. Hulet, W.M. Itano, and D.J. Wineland: Phys. Rev. Lett. **57**, 1699 (1986)
3. Th. Sauter, W. Neuhauser, R. Blatt, and P.E. Toschek: Phys. Rev. Lett. **57**, 1696 (1986)
4. Th. Sauter, R. Blatt, W. Neuhauser, and P.E. Toschek: Opt. Commun. **60**, 287 (1986)
5. H. Dehmelt: Bull. Am. Phys. Soc. **20**, 60 (1975)
6. R. H. Garstang: J. Res. Natl. Bur. Stand. Sect. A **68**, 61 (1964)
7. W.T. Silfvast: AT&T Bell Laboratories, Holmdel, NJ, U.S.A., unpublished calculation based on computer code of R.D. Cowan
8. J.C. Bergquist, D.J. Wineland, W.M. Itano, H. Hemmati, H.-U. Daniel, and G. Leuchs: Phys. Rev. Lett. **55**, 1567 (1985)
9. H. Hemmati, J.C. Bergquist, and W.M. Itano: Opt. Lett. **8**, 73 (1983)
10. C.E. Johnson: Bull. Am. Phys. Soc. **31**, 957 (1986)
11. J.C. Bergquist, W.M. Itano, and D.J. Wineland: to be published in Phys. Rev. A (July, 1987)
12. D.J. Wineland, J.C. Bergquist, W.M. Itano, and R.E. Drullinger: Opt. Lett. **5**, 245 (1980); D.J. Wineland and W.M. Itano: Phys. Lett. **82A**, 75 (1981)

Quantum Jumps and Related Phenomena of Single Ions and Small Ion Clouds

Th. Sauter, W. Neuhauser, R. Blatt, and P.E. Toschek

Universität Hamburg, I. Institut für Experimentalphysik,
Jungiusstraße 9, D-2000 Hamburg 36, Fed. Rep. of Germany

Experimentation with single "live" atomic particles - ions - has turned real a few years ago /1/. Precondition for this achievement was the development of techniques for storage and detection of small ion clouds, and in particular for damping the ion motion in the trap: optical cooling was predicted for free /2/ and bound particles /3/ in 1975 and subsequently observed /4,5/. This development has been crucial for an attempt to get evidence of and insight into a concept which is intrinsic to BOHR's successful model of the hydrogen atom /6/: *instantaneous* transitions between energy eigenstates of an atom upon interaction with light, which have been labelled "quantum jumps". The early controversy on that, as it seemed, awkward concept subsided: quantum mechanics turned out a story of overwhelming success after all, and repeatable experimenting with single atoms was considered unthinkable, anyway. As for traces in bubble chambers and on photographic plates, Erwin SCHRÖDINGER rightfully cautioned: *"In the first place it is fair to state that we are not experimenting with single particles, any more than we can raise Ichthyosauria in the zoo. We are scrutinising records of events long after they have happened..."* /7/. The issue of quantum jumps pertains to the internal dynamics of *multi-stable, quasi-stationary* systems, like an atom interacting with light, and subject to continuous observation. The question that matters is: Can a single atom undergo a continuous internal evolution, as described by an equation of motion, or do discontinuities in time occur? With spontaneous light emission, for instance, we usually acknowledge, persuaded by the click of the photo-electron counter, that the process inside the individual atom is "instantaneous" - as opposed to, say, an exponential decay - although the mere recording of the emitted photon does not permit time-resolved observation of the internal process. But what do we expect for absorption of monochromatic radiation? It seems that another channel of observation is required in order to unravel this problem.

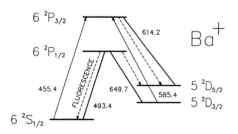

Fig. 1: Simplified energy level scheme of Ba⁺. Wavelength values in nm.

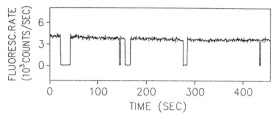

Fig. 2: Recording of the laser-excited fluorescence, at 493 nm, of a single Ba$^+$ ion

In 1975, DEHMELT suggested the detection of strongly excited resonance fluorescence of a single atom as a way of monitoring its occasional excitation on a very weak competing transition /8/. If quantum jumps existed, the observation of emissionless intervals would enable us to efficiently detect the weak excitation with intrinsic amplification up to 10^9. COOK and KIMBLE showed /9/ that the "on" and "off" time intervals would obey the statistics of a random telegraph signal, where the widths of the distributions of interval lengths are determined by the rates of the processes.

The existence of the presumptive jumps, however, is tied to whether quantum mechanics truefully describes the quasi-continuous measurement of the resonance fluorescence of an individual atom. And if so: is the result, for *short* observation times, an energy eigenstate, as opposed to the expectation value of a superposition state which is obviously observed on a long time scale? Recent theoretical work /10 - 14/ has confirmed these presumptions. With excitation on a weak line, the strong resonance fluorescence develops an extra probability for the next photon to appear after a long time interval τ. This is equivalent to the appearance of dark intervals of mean duration $<\tau>$.

We have demonstrated the existence of macroscopic interruptions in the resonance fluorescence of a single Ba$^+$ ion in a radio-frequency trap upon excitation of a metastable state /15,16/. (See also Ref. 17 for a similar observation, and Ref. 18 for a more recent experiment on Hg$^+$.) A simplified scheme of the energy levels of Ba$^+$ is shown in Fig. 1. A trapped ion's resonance fluorescence on the 493-nm line ($^2S_{1/2} - {}^2P_{1/2}$) is excited by a Coumarine-102 laser down-tuned from resonance for optical cooling /4, 19/. After some three excitation cycles, the ion drops into its $^2D_{3/2}$ level of 17 s lifetime. For its fast re-excitation, we used a cw DCM laser at 650 nm. Occasionally the laser excitation drives the atom into a dark state (which will be identified as the $^2D_{5/2}$ level) by off-resonant electronic Raman-Stokes transitions. The time of residence in that dark state is marked by the resonance fluorescence being quenched(s.Fig.2).

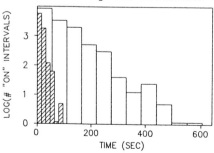

Fig. 3: Distributions of the lengths of "on"times. Shaded bars: additional incoherent excitation at 455 nm (from Ref. 16)

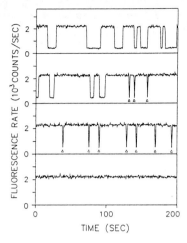

Fig. 4: Single-ion "on" and "off" intervals of fluorescence(top). Removing the ion from the "off" state ($^2D_5/_2$) by manually pulsed laser light at 614 nm (Δ, center). Coupling the $^2D_5/_2$ level to the "on" state by continuous 614 nm laser light, which results in "shelving prevention" (bottom). Full length of uninterrupted fluorescence recording: 20 min. (From Ref. 15)

The mean "on" time of the green fluorescence, 136 s $\pm$ 13 s, is characterized by the probability for Raman-Stokes transitions, the mean "off" time, 8 s, however, by Raman-anti-Stokes transitions, collisional and radiative decay of the $^2D_5/_2$ level. The faint emission of a Ba hollow cathode lamp, which contains light at 455 nm wavelength, is sometimes used to excite the ion to the $^2P_3/_2$ level and enables it to subsequent decay into the "off" state. Thus, the rate of on-off transitions is enhanced such that the mean "on" time is now reduced to 24 $\pm$ 4 s (s.Fig.3, and Fig.4, top). So far,the dark state still requires unambiguous identification. This has been done by irradiating the ion with an additional laser beam at 614 nm, which re-excites the ion to the 6 $^2P_3/_2$ level. Spontaneous decay leaves the ion in the "on" state. In Fig. 4 (center) the marks indicate, when a break in the fluorescence signal was immediately undone by a pulse of 614-nm light manually triggered by the observer, who acts as a kind of Maxwell's demon upon an internal degree of freedom of the ion. Permanent presence of this light prevents the ion from showing dark intervals (Fig. 4, bottom).

The fluorescence signal of two or three ions in the trap also evolves in a stepwise manner. An example, for three ions, is shown in Fig. 5. It is obvious even from the inspection of the traces that simultaneous jumps of two or even three ions happen much more often than can be attributed to random coincidence

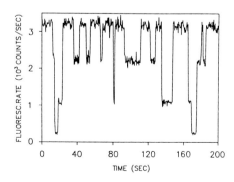

Fig. 5: Multiple jumps documented in the laser-excited 493-nm fluorescence of *three* Ba$^+$ ions. (From Ref. 15)

within the resolution time of the detection (0.4 s) /16/. This rate of random coincidence, normalized to the rate of individual jumps, is less than 1% for double events and about 10^{-4} for triple events of three ions. The recorded numbers of multiple events exceed the corresponding rate of random coincidence by more than two orders of magnitude, however. This remarkable phenomenon hints to *collective* interaction of the ions with the light field. The coupling is observable in *individual* events, unlike with macroscopic effects, where by the required ensemble averaging enhanced *probabilities* make us *infer* the collective interaction.

Our observations show that an equation of motion giving rise only to a continuous evolution of the system does not suitably describe the time evolution of an individual atomic particle, but has to be amended by taking into account the fluorescent decay on the *observed* transition. Moreover, the observations demonstrate a novel type of measurement in microphysics: the *same* system is prepared in identical conditions and detected over and over again. This is in contrast with conventional measurements, where ensemble averaging results in an expectation value, and the information on the micro-state is lost.

Finally, our observations have yielded strong evidence for the existence of cooperative interaction of atomic particles with light even in the smallest ensembles.

In 1952, SCHRÖDINGER also stated /7/:"...*we never experiment with just one electron or atom or (small) molecule. In thought-experiments we sometimes assume that we do; this invariably entails ridiculous consequences.*" - The reported single-ion observations disprove this conclusion.

1. W.Neuhauser, M.Hohenstatt, P.E.Toschek, and H.Dehmelt, Phys.Rev. A *22*, 1137 (1980)
2. T.Hänsch and A.Schawlow, Opt.Communic.*13*, 68 (1975)
3. D.Wineland and H.Dehmelt, Bull.Am.Phys.Soc.*20*, 637 (1975)
4. W.Neuhauser, M.Hohenstatt, P.E.Toschek, and H.Dehmelt, Phys.Rev.Lett. *41*, 233 (1978)
5. D.J.Wineland, R.E.Drullinger, and F.L.Walls, P.R.L.*40*,1639(1978)
6. N.Bohr, Phil. Mag.*26*, 1 and 476 (1913)
7. E.Schrödinger, Brit.J.Phil. Sci. *3*, 109 (1952)
8. H.Dehmelt, Bull.Am.Phys.Soc. *20*, 60 (1975)
9. R.J.Cook and H.J.Kimble, Phys.Rev.Lett. *54*, 1023 (1985)
10. C.Cohen-Tannoudji and J.Dalibard, Europhys.Lett.*1*,441 (1986)
11. J.Javanainen, Phys.Rev. A *33*, 2121 (1986)
12. D.T.Pegg, R.Loudon, and P.L.Knight, Phys.Rev. A *33*,4085 (1986)
13. A.Schenzle, R.G.DeVoe, and R.G.Brewer, Phys.Rev. A *33*,2127 (1986) and A *34*, 3127 (1986)
14. P.Zoller, M.Marte, and D.F.Walls,Phys.Rev. A *35*, 198 (1987)
15. Th.Sauter, R.Blatt, W.Neuhauser, and P.E.Toschek,P.R.L.*57*,1696 (1986)
16. Th.Sauter, R.Blatt, W.Neuhauser, and P.E.Toschek, |(1986) Opt.Communic. *60*,287 (1986)
17. W.Nagourney, J.Sandberg, and H.G.Dehmelt, P.R.L.*56*,2797(1986)
18. J.C.Bergquist, R.G.Hulet, W.M.Itano, and D.J.Wineland, Phys.Rev.Lett. *57*, 1699 (1986)
19. W.Neuhauser, M.Hohenstatt, P.E.Toschek, and H.G.Dehmelt, Appl.Phys. *17*, 123 (1978)

Part IV

Quantum Optics,
Squeezed States and Chaos

Quantum Devices and Measurements

R.J. Glauber

Lyman Laboratory of Physics, Harvard University,
Cambridge, MA 02138, USA

Nearly all of the measurements that detect individual events taking place at the quantum level depend in one way or another on a process of amplification. That amplification process, though essential to detection, may also impose certain limits on what can be detected. We shall discuss briefly some of the elements necessary to build a quantum amplifier and some of the features of the way in which such an amplifier works. We then discuss the application of these ideas to some experiments that show how the amplification process influences what is detected. One corollary, we shall see, is that no linear amplification process can lead to any such paradox as "Schrödinger's cat".

It may be helpful, before dealing with amplifiers, to say a few words about damping, or attenuation. There is nothing unusual about such dissipation, of course. Nearly all natural systems do it, but we may try to indicate the sort of system for which its description is simplest. Let us imagine a harmonic oscillator coupled to a large system we shall call a reservoir. The reservoir must have many degrees of freedom so that the loss of energy to it by the oscillator may be effectively irreversible. We shall assume that the coupling takes place through a complex reservoir variable we shall label $R(t)$, which has a spectrum of positive frequencies (i.e. time variations as $\exp(-i\omega_k t)$ for $\omega_k > 0$) that is essentially continuous in the neighborhood of the oscillator frequency ω. The Hamiltonian for the complete system is then taken to be

$$H = \hbar\omega a^+ a + \hbar\,(a^+ R + R^+ a) + H_R \,, \tag{1}$$

where H_R is the Hamiltonian for the reservoir system. A particularly simple example of such a damped oscillator system, which we have already discussed in detail [1], corresponds to choosing a large collection of harmonic oscillators as the reservoir,

$$H_R = \sum_k \hbar\omega_k\, b_k^+ b_k \,, \tag{2}$$

and letting the reservoir variable be the sum

$$R = \sum_k \lambda_k\, b_k \,, \tag{3}$$

where the constants λ_k are coupling constants.

One virtue of this model is its linearity. Another is that an initially coherent state for the entire system remains coherent at all later times. The linearized equations that hold for a much broader class of nonlinear systems, over a certain dynamical range, can often be cast into a form quite similar to those of the oscillator model, and yield equivalent results.

If the reservoir is at a temperature $T \neq 0$, the coupling to the variable R will feed some random excitation into the a-oscillator, and the damping process will then be accompanied by noise. That noise can be avoided altogether, in the present model, by taking the reservoir to be at temperature $T = 0$. It follows then that a coherent state for the a-oscillator, taken by itself, remains coherent at all times. If the amplitude of that state is initially α then at any later time it becomes $\alpha u(t)$ where $u(t) =$ $\exp[-(\varkappa + i\omega')t]$, $\varkappa$ is the damping constant and ω' is a slightly displaced frequency for the a-oscillator. If, more generally, the initial reduced density operator for the a-oscillator takes the form of a P-representation,

$$\rho(0) = \int P(\alpha,0) |\alpha\rangle \langle\alpha| d^2\alpha , \qquad (4)$$

then at any later time the density operator $\rho(t)$ is specified by a function P arrived at by the simple scale transformation

$$P(\alpha,t) = \frac{1}{|u(t)|^2} P \left(\frac{\alpha}{u(t)} , 0\right) . \qquad (5)$$

If the a-oscillator were to begin, for example, with an excitation so strong it could be considered classical, then $P(\alpha,0)$ would be, in effect, a classical probability density. It would be positive-valued and smoothly varying and those properties would be shared by the P functions for all the attenuated forms of the field. In fact, most of the optical fields investigated until very recently did have well-behaved and positive-valued P-functions and in that sense could be thought of as attenuated classical fields. But, as we have noted long ago [2], there exists a broad variety of other excitations for which the P-representation behaves quite non-classically, or even fails to exist altogether. Those excitations include n-quantum states and squeezed states. They cannot be reached by attenuating classical fields, but we are now learning how to reach them quite directly at the quantum level. It is especially interesting to see what forms such fields take when they are amplified.

How can we construct an amplifier? We must have a reservoir full of energy that it is ready to supply to the a-oscillator. But an ordinary reservoir that is simply hot will not do; it would just feed thermal noise into the a-oscillator. To avoid that we must require that the reservoir variable R not be highly excited, i.e. that the expectation values $\langle RR^+\rangle$ and $\langle R^+R\rangle$ remain minimal. So the requirement is the strange one that the reservoir be extremely hot, but that the reservoir variable R remain nevertheless unexcited. The practical solution to that problem used in the laser is the coupling of a field mode oscillator to a medium of atoms with an inverted population of excitations. A considerably more idealized and simple model can be obtained by making the reservoir out of a collection of inverted harmonic oscillators,

$$H_R = - \sum_k \hbar\omega_k b_k^+ b_k . \qquad (6)$$

The complete Hamiltonian can then be taken to be

$$H = \hbar\omega a^+ a + \hbar (a^+ R^+ + Ra) + H_R , \qquad (7)$$

where the reservoir variable is again given by (3). Note that the roles of the variables R and R^+ have been interchanged in the coupling term in order to permit energy conservation in the lowest order interactions, (i.e. the b_k now have the time dependence $\exp(+i\omega_k t)$).

The reservoir of inverted oscillators has no ground state, of course, but the state that interests us is that of maximum excitation, since it does indeed put all the oscillators in states for which the coordinate R has minimal excitation. The states of the inverted oscillator reservoir, having upper bounds to their energy spectra, can be spoken of as being in states of negative absolute temperature. The temperature that minimizes the noisiness of the amplifier corresponds to the state of maximum excitation, $T = -0$.

A second sort of amplifier model [3] is also worth mentioning. Though its structure is a bit different, it behaves in a way quite similar to the one we have discussed. If we take the a-oscillator to be inverted, then we can still secure amplification while taking the reservoir to be a normal, uninverted set of oscillators. The Hamiltonian is then

$$H = - \hbar \omega a^+ a + \hbar (a^+ R^+ + Ra) + H_R , \qquad (8)$$

where H_R is given by (2) and R by (3). This model is inspired by the way in which a set of N fully excited atoms $(N \gg 1)$ in a Dicke state, (corresponding to the inverted oscillator,) begins to radiate into the electromagnetic field or, in other words, by the initial stages of superfluorescent radiation [4]. We have discussed its properties at length in previous work [3].

In the first model the energy flows from the reservoir to the a-oscillator and in the second it flows the other way, but the two models are, in fact, mathematically quite similar in their behavior and differ essentially only in the signs of the frequencies ω and ω_k. In both models there is an exponential growth of whatever excitation (or de-excitation), i.e. whatever "signal" is initially present in the a-oscillator. But the zero-point displacement the oscillator has at the moment the coupling begins is also amplified, even in the absence of any signal, and that inescapable element of noise also grows exponentially along with any signal.

Let us ask what happens, for example, when the a-oscillator begins in its n-quantum state. The P-distribution in that case is initially singular, but it immediately becomes quite a well-behaved function for t > 0. With the reservoir temperature fixed at $T = \pm 0$ to minimize noise, we find the time-dependent P-distribution

$$P_n(\alpha,t) = \frac{(-1)^n}{\pi \, M^{n+1}} \, L_n [(1 + \frac{1}{M}) |\alpha|^2] \, \exp(- \frac{|\alpha|^2}{M}) , \qquad (9)$$

where L_n is the n-th Laguerre polynomial and $M(t) = e^{2 \varkappa t} - 1$. In particular, for n = 0 we see the simple Gaussian distribution that represents pure noise. It is interesting to note that the distributions P_n for $n \geq 1$ all have regions in which they take on negative values, but these remain confined to the small area $|\alpha| \sim \mathcal{O}(1)$ while the domain in which P_n is positive-valued spreads out rapidly over the entire remainder of the complex α-plane.

In the foregoing example the initial quantum state happens to have a singular P-function, but we have already noted that other initial states such as squeezed states need not have any P-representation at all. It is interesting to note therefore that the noise added to the oscillator excitation by the amplification process always has the effect of "spreading out" the excitation, so that a P-representation presently comes into existence, even if none existed initially. In fact we can state as a theorem, that for an amplifier coupled in either of the ways indicated, a quadratically integrable P-representation for the reduced density operator of the α-oscillator will always exist once the power gain $G(t) = e^{2\varkappa t}$ exceeds the value 2. A related, but less general result, that squeezing is always wiped out by more than two-fold gain, has already been noted [5].

Of course the existence of such a P-representation does not guarantee that it will be positive-valued or interpretable in classical terms. Some strong leanings in those directions can be seen, however, by making use of the function

$$Q(\alpha,t) = \frac{1}{\pi} <\alpha|\rho(t)|\alpha> , \qquad (10)$$

which was originally defined long ago [6], and furnishes an alternative specification of the density operator. The function Q is obviously positive and bounded, $0 \le Q \le \frac{1}{\pi}$, and in fact smoothly behaved for all states of the oscillator. It may be found from P, where the latter function exists, by means of the relation

$$Q(\alpha) = \frac{1}{\pi} \int e^{-|\alpha-\beta|^2} P(\beta)d^2\beta . \qquad (11)$$

Since the amplification process tends to spread the function P smoothly over the entire area of the complex plane, it is clear that the averaging implicit in (11) makes little change from the function P, except in those small areas in which P retains negative values. In this sense the functions P and Q tend, as the amplification process continues, to become asymptotically identical; both tend to become the same classical probability density.

It is somewhat simpler, in fact, to base an analysis of the amplification process on the function $Q(\alpha,t)$ than on the function $P(\alpha,t)$. The function Q not only behaves less temperamentally than P, it also obeys the simplest of scaling laws under amplification. We find then that the function $Q(\alpha,t)$ is given in terms of its initial value by the relation

$$Q(\alpha,t) = \frac{1}{|U(t)|^2} Q(\frac{\alpha}{U(t)} , 0) , \qquad (12)$$

where $U(t) = \exp(\varkappa \pm i\omega')t$, $\varkappa$ is the amplification constant and $\pm \omega'$ is a slightly displaced frequency for the oscillator. An elementary illustration of this scaling rule is given by the initial n-quantum state we have considered earlier. For this case we have

$$<\alpha|\rho(0)|\alpha> = <\alpha|n><n|\alpha> = \frac{|\alpha|^{2n}}{n!} e^{-|\alpha|^2} \qquad (13)$$

and hence

$$Q_n(\alpha,t) = \frac{1}{\pi n!|U|^2} |\frac{\alpha}{U}|^{2n} e^{-|\frac{\alpha}{U}|^2} , \qquad (14)$$

a positive-definite expression that is much simpler and more transparent in form than (9).

The principal advantage of using amplifiers in the measurement process is that they can bridge the gap between the quantum theory and classical theory; they can strengthen a quantum signal sufficiently to bring it into the classical domain and allow its measurement without any significant disturbance. They solve, in other words, one of the more awkward problems of measurement theory, the description of the means by which quantum signals are conveyed to essentially classical instruments of laboratory apparatus. They do it however at the expense of introducing noise, and that noise, which is unavoidable in quantum mechanical terms, can have a significant effect on the outcomes of many experiments.

If one imagines an optical amplifier, for example, to be a device that multiplies identical photons, then it is easy to imagine a pair of such amplifiers connected to a polarizing prism in such a way that they will amplify and "clone" photons of arbitrary polarization [7]. If someone were to provide us with only a single photon of some polarization known to him, but not to us, we would encounter a familiar sort of ambiguity in trying to establish by using the familiar types of prisms or filters, just what that polarization was. If the compound amplifier scheme could in fact provide us with arbitrarily many "clones" of the original photon we would have little difficulty in determining that polarization to arbitrary precision (and such a result could lead to miraculous things like communication faster than light via the EPR paradox [7].) So it is quite important to understand that the spontaneous noise output of our amplifiers will see to it that we do not secure any pure strain of "cloned" photons. The compound amplifier does indeed amplify photons of arbitrary input polarization, but its output includes a large proportion of unpolarized "noise photons". The output polarization will in fact be mixed in a way that imposes on our amplification scheme precisely the same limitations in determining polarization that we find by simply using prisms or filters in the most obvious way [7].

There are many other uses possible for our double or compound amplifier. We can use it for example to perform an interesting single-photon version of Young's two-pinhole interference experiment by placing one amplifier behind each of the two pinholes. These amplifiers can then be assumed to project amplified or classical beams at the distant screen and the region of overlap of the two beams does indeed show a pattern of parallel fringes. These fringes, which are due to a single incident photon (plus noise) are somewhat random in position, but average out in an ensemble of single-photon experiments to a set of Young's fringes seen against a constant background provided by the uncorrelated noise outputs of the two amplifiers.

We may alternatively mount one of our amplifiers in each arm of an interferometer and show that the system then produces amplified interference fringes that are somewhat random in phase [7]. That is true even though the system begins in a pure single-photon state and only part of the single photon amplitude passes through each amplifier.

While these experiments illustrate the preservation of coherence and interference through independent amplification processes, they cannot lead to states that are quantum mechanically "pure". In this sense both their analyses and their physical results differ strikingly from those imagined by Schrödinger in 1935 in stating his famous "cat paradox". There he assumed that a microscopic state that could be described as a coherent superposition of two alternative states (in which a nucleus either has or has not undergone α-decay) could somehow be transformed directly into a coherent superposition of two very different macroscopic states, one in which a cat remained alive and another in which it was dead. We can show that the amplification

130

process required to carry out those changes in the large-scale world can never lead to any such pure state, even of systems far less animate than a cat.

There is a simple way of seeing, in fact, that the amplification process, even when it begins with the purest of quantum mechanical states, always leads to infinitely mixed states as outputs. The trace of the square of a density operator always obeys the inequality $0 < \text{Tr}(\rho^2) \leq 1$. It can only attain the value one for pure states, while it goes to zero as a state becomes progressively more mixed. For the output field of an amplifier the scaling law noted earlier gives us the inequality

$$\text{Tr} \, \rho^2(t) \leq \frac{1}{G(t)} \, \text{Tr} \, \rho^2(0) \, , \tag{15}$$

where $G(t) = \exp(2\varkappa t)$ is the power gain. Thus an initially pure state inevitably degenerates into an infinitely mixed one. That is equally evident in the entropy, $S = - \text{Tr}(\rho \log \rho)$, of the output state. For a pure initial state the entropy is zero, but as the amplification process proceeds, the entropy begins to grow. Eventually the function $Q(\alpha, t)$ becomes, in effect, a probability density, and the scaling law then gives the asymptotic result that S increases linearly with time,

$$S(t) \sim S(0) + 2\varkappa t \, . \tag{16}$$

For the double or compound amplifier these results are all the stronger; $2\varkappa t$ must be replaced by $4\varkappa t$ in both (15) and (16).

It is worth emphasizing that nothing in the present analysis denies the existence of superpositions of macroscopically distinguishable states; it simply says that such states are not to be attained by any process of linear amplification. Not all amplifiers are linear of course, but they all do play a characteristic role in physical measurements. They quite generally exploit an instability of one sort or another to concentrate the energy supplied by the many degrees of freedom of a reservoir, or to allow a coherent emission of energy into the reservoir. In either case the very process of excitation concentrates in the amplifier a certain degree of randomness, or chaos, and causes its state to take on a highly mixed character. No quantum mechanically pure input state can survive the process of amplification as a pure state. It is largely for that reason that quantum mechanical superpositions, however common we find them to be in the micro-world, cannot be projected directly into the large-scale world -- and that the outcomes of all attempts to do that can be analyzed so universally in probabilistic terms.

References

1. R. J. Glauber: In Quantum Optics, ed. by R. J. Glauber, Proceedings of the Enrico Fermi International School of Physics, course 42, (Academic Press, New York, 1969) p. 32
2. R. J. Glauber: Phys. Rev. 131, 2766 (1963)
3. R. J. Glauber: In Group Theoretical Methods in Physics, Proceedings of the International Seminar, Zvenigorod, November 24 - 26, 1982, Vol. II, (Nauka Publishers, Moscow, 1983) p. 165. See also Group Theoretical Methods in Physics, Vol. I, (Harwood Academic Publishers, 1985) p. 137. R. J. Glauber: In Proceedings of the VI International School on Coherent Optics, Ustron, Poland, September 19 - 26, 1985, ed. by A. Kujawski and M. Lewenstein (Polish Academy of Sciences, Warsaw, 1986) p. 9, also published by D. Reidel, 1986

4. R. J. Glauber, F. Haake: Phys. Rev. A13, 357 (1976)
5. S. Friberg, L. Mandel: Optics Commun. 46, 141 (1983)
   R. Loudon, T. J. Shepherd, Optica Acta 31, 1243 (1984)
   S. Stenholm: Optics Commun. 58, 177 (1986)
6. R. J. Glauber: In Quantum Optics and Electronics, ed. by C. De Witt et al, (Gordon and Breach, New York, 1965) see p. 138
7. R. J. Glauber: In Frontiers of Quantum Optics, ed. by E. R. Pike and S. Sarkar, (Adam Hilger, Ltd., Bristol, 1986)
   R. J. Glauber: "Amplifiers, attenuators and Schrödinger's cat" in New Ideas and Techniques in the Quantum Theory of Measurement, ed. by D. Greenberger (New York Academy of Sciences, 1986)

One-Atom Oscillators for Nonclassical Radiation

F. Diedrich[1], *J. Krause*[3], *G. Rempe*[1], *M.O. Scully*[2,3], *and H. Walther*[1,2]

[1]Sektion Physik der Universität München,
 D-8046 Garching, Fed. Rep. of Germany
[2]Max-Planck-Institut für Quantenoptik,
 D-8046 Garching, Fed. Rep. of Germany
[3]Center for Advanced Studies and Department of Physics and Astronomy,
 University of New Mexico, Albuquerque, NM 87131, USA

In recent years several features of the radiation field have been found which are pure quantum effects. Among these are antibunching /1/ and squeezing properties of the radiation field /2/. In addition, in quantum electrodynamics in a cavity new phenomena in the dynamics of the atom-field interaction as e.g. the collapse and revivals in the Rabi-nutation /3,4/ have been demonstrated. In this paper recent single atom experiments will be reviewed which have been performed to demonstrate pure quantum features of the radiation field /4,6/.

1. QUANTUM COLLAPSE AND REVIVAL IN THE ONE-ATOM MASER

Experiments with the one-atom maser have demonstrated /6/ that it is possible to study the interaction of a single atom with a single mode of a resonant electromagnetic field in a cavity. The atoms used in these experiments were Rydberg atoms with a very large principal quantum number n. The probability of induced transitions between closely adjacent states becomes very large and scales as n^4. Since the lifetime for spontaneous transitions is also very large, and scales proportionally to n^3 and n^5 for low- and high-angular-momentum states, respectively, the saturation power fluxes for transitions between neighbouring states becomes extremely small so that a few photons are able to saturate a transition /7,8/.

Another important ingredient of the one-atom maser is the superconducting cavity: The quality factor is high enough for a periodic energy exchange between atom and cavity field to be observed; i.e., the relaxation time of the cavity field is larger than the characteristic time of the atom-field interaction, which is given by the reciprocal of the Rabi frequency.

The situation realized in the one-atom maser approaches the idealized case of a two-level atom interacting with a single quantized mode of a radiation field as treated by Jaynes and Cummings /9/ many years ago. It is therefore now possible to perform experiments on the dynamics of the atom-field interaction predicted in this theory. Some of the features are explicitly a consequence of the quantum nature of the electromagnetic field: The statistical and discrete nature of the photon field leads to new dynamic characteristics such as collapse and revivals in the Rabi nutation. In order to investigate the predictions of the theory, the interaction of atoms in the upper state of the maser transition with the maser field is investigated and the probability that the atoms remain in the upper state is probed behind the maser cavity. The experiments were performed with single atoms in the cavity at a time.

At a low atomic-beam flux, the cavity contains essentially thermal photons and their number is a random quantity conforming to Bose-Einstein statistics. When the velocity of the atoms is changed by means of a velocity selector, the probability that the interacting atom remains in the excited state $P_e(t)$ varies with the interaction time in an apparent random oscillation. At higher atomic-beam fluxes the number of photons stored in the cavity increases and their statistics changes. If a coherent field is prepared in the cavity the probability distribution is given by a Poissonian. As first shown by Cummings, the Poisson spread in n gives a dephasing of the Rabi oscillations, and therefore $P_e(t)$ first exhibits a collapse /10/. This is described in the resonant case by the approximate envelope $\exp(-g^2t^2/2)$ and is independent of the average photon number n (this independence does not hold for nonresonant excitation; g is the single photon Rabi-frequency). The collapse was also noted later in other work /3/. After the collapse there is a range of interaction times for which $P_e(t)$ is independent of time. Later $P_e(t)$ then exhibits recorrelations (revivals) and starts oscillating again in a very complex way. As has been shown by Eberly and co-workers the recurrences occur at times given by $t=kT_R$ (k=1,2,...), with $T_R = 2\pi(n)^{1/2}/g$ /3/. Both collapse and revivals in the coherent state are purely quantum features and have no classical counterpart.

The inversion also collapses and revives in the case of a chaotic Bose-Einstein field. Here the photon-number spread is far larger than for the coherent state and the collapse time is much shorter. In addition, the revivals completely overlap and interfere to produce a very irregular time evolution. A classical thermal field represented by an exponential distribution of the intensity also shows collapse, but no revivals. Therefore the revivals can be considered as a clear quantum feature, but the collapse is less clear-cut as a quantum effect /11/.

It is interesting to mention that in the case of two-photon processes the Rabi-frequency turns out to be $2g(n+1)$ rather than $2g(n+1)^{1/2}$, enabling the sums over the photon numbers in $P_e(t)$ to be carried out in simple closed form. In this case the inversion revives perfectly with a completely periodic sequence /12/.

For the experimental demonstration of the above effects velocity selected Rubidium Rydberg atoms were used in the one-atom maser. The evolution of the atomic inversion as atom and field exchange energy was observed. The quantum collapse and revival predicted by the Jaynes-Cummings model could be demonstrated /4/.

In the following we will focus on the discussion of the statistics of the photons in the maser cavity being strongly nonclassical. In general, two approaches to the quantum theory of the one-atom maser have been developed. In one approach Filipovicz, Javanainen and Meystre emphasize the microscopic nature of the device /13/. In the other treatment it is shown that standard macroscopic quantum laser theory leads to the same steady-state photon-number distribution /14/. The reason that the special features of the micromaser were not emphasized in the standard laser theory was due to the fact that the broadening due to spontaneous decay obscured the Rabi cycling of the atoms (see section 17.1 of ref. /15/). Similar averages in microscopic theory associated with inhomogenous (interaction time) broadening leads to an equivalent result.

The photon statistics (see Fig. 1) of the one-atom maser depends on the normalized interaction time $\tau_{int} = g\tau\sqrt{\gamma t_p}$, whereby τ is the interaction time between each atom and the mode, γ the photon damping rate and t_p the

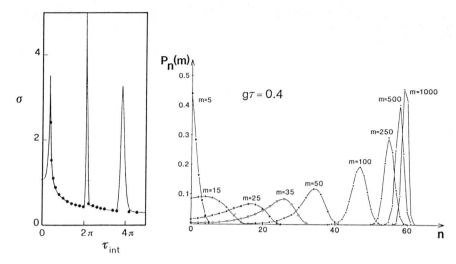

Fig. 1. Left part: Normalized standard distribution of the photon number (for details see Ref. /13/ and /14/; $\sigma = 1$ corresponds to a Poissonian distribution. Right part: Probability of obtaining a state with n photons in the cavity after m atoms have passed for $g\tau=0.4$.

average time between two atoms entering the cavity after each other. Above treshold $\tau_{int} = 1$ the photon statistics is first strongly super-Poissonian. Further super-Poissonian peaks occur at $\tau_{int} \sim 2\pi$ or at integer multiples of 2π and become less pronounced for increasing τ_{int}. In the remaining intervals at τ_{int} the field is sub-Poissonian.

The experimental results are in very good agreement with theory /4/ therefore it can be assumed that the photon statistics in the one-atom maser is sub-Poissonian for most of the values of τ_{int}.

In the near future a new cavity with a $Q\sim5\cdot10^{10}$ will be available. In this case theory predicts that a number state maser field will be generated /16/. The theoretical result for the build-up process of the maser field is shown in Fig. 1.

2. PHOTON ANTIBUNCHING OBSERVED WITH A SINGLE STORED ION

Previous experiments to investigate antibunching in resonance fluorescence have been performed by means of laser-excited collimated atomic beams. The initial results obtained by Kimble, Dagenais and Mandel /17/ showed for the second-order correlation function $g^{(2)}(t)$ a positive slope characteristic of photon antibunching, but $g^{(2)}(0)$ was larger than $g^{(2)}(t)$ for $t\rightarrow\infty$. This was due to number fluctuations in the atomic beam and to the finite interaction time of the atoms /18,19/. Later the analysis of the experiment was refined by Dagenais and Mandel /19/. Another experiment with a longer interaction time was performed by Rateike and Walther /20/. In the latter experiment the photon correlation was also measured for very low laser intensities.

The fluorescene of a single stored ion should also display the following property: The probability distribution of the photon number recorded in a finite time interval t is narrower than Poissonian, which means in other words that the variance is smaller than the mean value of the photon number. This is because the single ion can only emit a single photon. Antibunching and sub-Poissonian statistics are often associated. They are, nevertheless, distinct properties and need not necessarily be simultaneously observed /21/, as is the case in the experiment described here. Although there is evidence of antibunching in the atomic-beam experiments, the photon counts were not sub-Poissonian as a result of fluctuations in the number of atoms. In the experiments by Short and Mandel /21/ this effect was excluded by use of a special trigger scheme for the single-atom event. In the setup described here these precautions are not necessary since we have no fluctuations in the atomic number.

The centerpiece of the experiment is a Paul radio-frequency trap mounted inside a stainless-steel ultrahigh-vacuum chamber. The trap has a ring diameter of 5 mm and a pole cap separation of 3.54 mm. It is thus much larger than other single-ion radio-frequency traps /22-24/. Close confinement of a single ion at the center of the trap is achieved by photon recoil cooling. The large size of the trap allows a large solid angle for detection of the fluorescene radiation. The resonance fluorescene is transmitted through a molybdenum mesh covering a conical bore in the upper pole cap.

In order to investigate the intensity correlations in the fluorescene, an ordinary Hanbury-Brown-Twiss setup with two photomultipliers and a beam splitter was used. The photon correlation signals show the nonclassical antibunching effect connected with Rabi oscillations, which are damped out during the excited-state lifetime and a periodic feature resulting from the oscillation of the ion in the trap /5/.

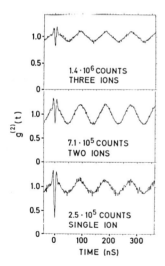 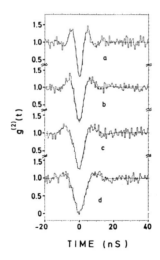

Fig. 2. Left side: Intensity correlation for one, two, and three ions. The antibunching signal occurs around t=0 and decreases with increasing ion number. The periodic signal at larger t is a result of the micromotion of the ions. Right side: Antibunching signal around t=0 for different laser intensities. The data have been corrected for the micromotion; in addition, the background due to accidental coincidences was subtracted.

136

The size of the antibunching signal around t=0 shows a strong dependence on the ion number. A single atom gives $g^{(2)}(0)=0$, whereas photons emitted by independent atoms are not correlated, thus leading to a classical correlation function as the number of photons in the field increases. If there are more than one ion stored in the trap, the Coulomb repulsion keeps them at a distance of several micrometers and the fluctuation in their position is of the same order of magnitude. Therefore coherent beating in the light emitted by different ions is negligible. The results for $g^{(2)}(t)$ measured with four different intensities are shown in Fig. 2 together with the corresponding fits. For more details see /5/. With our setup we can easily demonstrate that the fluorescene photons show sub-Poissonian statistics. The setup used for this purpose is the same as that used for the correlation measurements.

To compare this result to previous work the normally ordered variance :Q: was calculated /21/. The result obtained is :Q: = $-7 \cdot 10^{-5}$. This value is smaller than that achieved in Ref. /8/ because of the smaller overall detection efficiency in the present experiment being about $4 \cdot 10^{-4}$.

3. OBSERVATION OF A PHASE TRANSITION OF LASER-COOLED IONS

Another experiment performed recently with the above mentioned Paul trap will be described briefly in the following /26/.

Clouds of 2 to about 10 simultaneously stored laser-cooled Mg^+-ions in a Paul trap were observed in two phases, which are clearly distinguishable by their excitation spectra. Transitions between these phases can be induced by varying either the power of the laser radiation used to cool the ions or by changing the size of the radio frequency voltage applied to the trap. By the variation of the appropriate parameters transitions between a "crystalline" and a "gaseous" phase can be observed repeatedly.

The trapped ions represent a very nice model system for the study of phase transitions. It is e.g. of considerable interest to investigate in detail how the phase transitions depend on the stored ion number. It can be expected that certain ion configurations are more stable than others and therefore need a stronger heating for "evaporation". In this way the stability of the respective ion arrangement can be studied. The fact that the condensed phase exhibits linewidths in the excitation spectra not very much different from that of a single ion opens the possibility to use several ions for the frequency standard instead of a single one with all the advantages for a better signal to noise ratio.

References

1. D.F. Walls, Nature 280, 451 (1979)
2. D.F. Walls, Naure 306, 141 (1983) and the paper by H.J. Kimble, L.A. Wu, M. Xiao in this volume.
3. J.H. Eberly, N.B. Narozhny, J.J. Sanchez-Mondragon, Phys. Rev. Lett. 44, 1323 (1980) and H.I. Yoo, J.H. Eberly, Phys. Rep. 118, 239 (1985)
4. G. Rempe, H. Walther, N. Klein, Phys. Rev. Lett. 58, 353 (1987)
5. F. Diedrich, H. Walther, Phys. Rev. Lett. 58, 203 (1987)
6. D. Meschede, H. Walther, G. Müller, Phys. Rev. Lett. 54, 551 (1985)
7. S. Haroche, J.M. Raymond, in Advances in Atomic and Molecular Physics, edited by D. Bates and B. Bederson (Academic, New York, 1985), Vol. 20, p. 350
8. J.A.C. Gallas, G. Leuchs, H. Walther, H. Figger, in Advances in Atomic and Molecular Physics, edited by D. Bates and B. Bederson (Academic, New York, 1985), Vol. 20, p. 414

9. E.T. Jaynes, F.W. Cummings, Proc. IEEE 51, 89 (1963)
10. F.W. Cummings, Phys. Rev. 140, A 1051 (1965)
11. P.L. Knight, P.M. Radmore, Phys. Lett. 90A, 342 (1982)
12. P.L. Knight, Phys. Scri. T12, 51 (1986)
13. P. Filipowicz, J Javanainen, P. Meystre, Phys. Rev. A34, 3077 (1986)
14. L.A. Lugiato, M.O. Scully, H. Walther, Phys. Rev. July 1987 to be published
15. M. Sargent III, M.O. Scully, W.E. Lamb, jr. Laser Physics Addison Wesley Publishing Company, Reading, Mass (1974)
16. J. Krause, M.O. Scully, H. Walther, to be published
17. H.J. Kimble, M. Dagenais, and L. Mandel, Phys. Rev. Lett. 39, 691 (1977)
18. E. Jakeman, E.R. Pike, P.N. Pusey, and J.M. Vaugham, J. Phys. A 10, L257 (1977)
19. H.J. Kimble, M. Dagenais, and L. Mandel, Phys. Rev. A 18, 201 (1978); M. Dagenais and L. Mandel, Phys. Rev. A 18, 2217 (1978)
20. F.-M. Rateike, G. Leuchs, and H. Walther, results cited by J.D. Cresser, J. Häger, G. Leuchs, F.-M. Rateike and H. Walther, in Dissipative Systems in Quantum Optics, edited by R. Bonifacio and L. Lugiato, Topics in Current Physics Vol. 27 (Springer, Berlin, 1982)
21. R. Short and L. Mandel, Phys. Rev. Lett. 51, 384 (1983), and in Coherence and Quantum Optics V, edited by L. Mandel and E. Wolf (Plenum, New York, 1984), p. 671.
22. W. Neuhauser, M. Hohenstatt, P. Toschek, and H. Dehmelt, Phys. Rev. A 22, 1137 (1980)
23. W. Nagourny, G. Janik, and H. Dehmelt, Proc. Natl. Acad. Sci. USA 80, 643 (1980)
24. J.C. Bergquist, R.G. Hulet, W.M. Itano, and D.J. Wineland, Phys. Rev. Lett. 57, 1699 (1986)
25. D.J. Wineland and W.M. Itano, Phys. Rev. A 20, 1521 (1979)
26. F. Diedrich, E. Peik, J.M. Chen, W. Quint, and H. Walther, to be published

Tests of General Relativity and the Correlated Emission Laser

J. Gea-Banacloche, W. Schleich, and M.O. Scully

Max-Planck-Institut für Quantenoptik,
D-8046 Garching, Fed. Rep. of Germany and
Center for Advanced Studies and Department of Physics,
University of New Mexico, Albuquerque, NM 87131, USA

The arena of space-time and metric gravity is a grand playground for modern quantum optical scientists. Work in this field defines the cutting edge of technology, from precision interferometry to the quantum "limits" of measurement.

However (or perhaps "therefore"!) there have been and continue to be many spirited debates and conceptual mistakes in this field, which frequently lead to deeper understanding. In this paper we review the use of active laser interferometric devices as probes of metric gravity. In particular, we discuss such systems operating in the correlated spontaneous emission (quantum-noise-quenched) mode. The format will be that of challenge/question - reply/answer.

The theory of measurement in general relativity is a tricky and subtle business. For example, in the "old" days it was often said (incorrectly) that gravitational waves would never be seen because the meter sticks could change just as the dimensions of the apparatus so as to render the effects of gravitational radiation unobservable. Similarly, when the laser gyro was being investigated as a possible method of measuring Machian frame-dragging the results were criticised (incorrectly) by some theorists since they did not appear in the conventional (tidal force) form.

Therefore the challenge is: how can we develop a method for dealing with optical tests of general relativity so as to avoid these problems from the outset? Answer: start with the coupled Einstein-Maxwell equations, which include the effect of metric gravity from the beginning. This is developed in our Les Houches lectures [1] in some detail and is presented in a form useful to the present paper in Fig. 1.

An interesting situation is encountered when we turn to the optical side of the problem and compare two general types of detectors: "active" and "passive". As shown in Fig. 2 the active devices, in which the light is generated inside the optical cavity and which rely on the measurement of a frequency difference (i.e., a phase difference that grows linearly with time) appear to have a distinct advantage in terms of signal size. Yet the noise in active devices (which is due to spontaneous emission in the active medium) is also seen, from Fig. 2, to be larger [2], and when the signal-to-noise ratio is calculated one finds, rather surprisingly, that the two effects (larger signal and larger noise) combine to yield basically the same sensitivity as in a passive interferometer.

This remarkable coincidence was eventually explained [3], but, in the meantime, the search for a way to reduce (or even to eliminate) the spontaneous emission noise in an active device had led to developments which were interesting in their own right. The correlated (spontaneous) emission laser(CEL) arose as an answer to the question, could the phase diffusion due to spontaneous emission be eliminated from the relative phase of two laser modes? The answer turned out to be affirmative [4].

In a CEL, a correlation is imposed, by varying methods, in the atomic medium that generates the two light modes to be used in the experiment. As a result, the spontaneous

Figure 1(a)

$$\frac{\partial^2 E}{\partial y^2} - \frac{1+h(x,t)}{c^2}\frac{\partial^2 E}{\partial t^2} = 0$$

$$\frac{\partial^2 E}{\partial x^2} - \frac{1}{c^2}\frac{\partial^2 E}{\partial t^2} = 0$$

g-wave

DETECTOR

SAGNAC INTERFEROMETER

$$\Box^2 \vec{E} = -2\vec{h}\cdot\vec{\nabla}\frac{1}{c}\frac{\partial\vec{E}}{\partial t}$$

$$\vec{h} = \frac{2GS_z}{c^3 r^3}\ (-y,x,o)$$

$$\Delta\omega = \frac{4A}{\lambda P}[\Omega_\oplus + \Omega_{AETHER} + \Omega_{CURVATURE} + \Omega_{LENSE-THIRRING}]$$

Figure 1(b)

emission noise in the relative phase disappears. The theory of these devices is now well understood [5] and an experiment was performed recently [6] showing a reduction of the relative phase diffusion rate between the two modes two orders of magnitude below the Schawlow-Townes linewidth.

The question has arisen as to whether the CEL is not, after all, equivalent to an injection-locked (phase-locked) laser. The answer is "no". As discussed in a recent paper [7], there is a fundamental difference between the phase-locked laser (PLL) and the CEL. In the PLL, the diffusion coefficient for the relative phase does not vanish: the diffusion is only "quieted" by the locking. In the CEL, on the other hand, the diffusion coefficient can be made zero, i.e., the noise is "quenched" (Fig. 3). Another way to express the difference between the two systems is by the observation that the CEL has multiplicative noise, while the PLL has additive noise.

140

	SIGNAL	NOISE	SENSITIVITY
PASSIVE G-WAVE DETECTOR	$vh\,\gamma^{-1}$	$\sqrt{\dfrac{\hbar v}{P t_m}}$	$h \cong \dfrac{\gamma}{v}\sqrt{\dfrac{\hbar v}{P t_m}}$
ACTIVE G-WAVE DETECTOR	$vh\,t_m$	$\gamma t_m \sqrt{\dfrac{\hbar v}{P t_m}}$	$h \cong \dfrac{\gamma}{v}\sqrt{\dfrac{\hbar v}{P t_m}}$
PASSIVE LASER GYROSCOPE	$S\Omega\,\gamma^{-1}$	$\sqrt{\dfrac{\hbar v}{P t_m}}$	$\Omega \cong \dfrac{1}{S}\,\gamma\sqrt{\dfrac{\hbar v}{P t_m}}$
ACTIVE LASER GYROSCOPE	$S\Omega\,t_m$	$\gamma t_m \sqrt{\dfrac{\hbar v}{P t_m}}$	$\Omega \cong \dfrac{1}{S}\,\gamma\sqrt{\dfrac{\hbar v}{P t_m}}$

Figure 2

PLL:
$$\dot\psi = a - b\sin\psi + F(t)$$
$$\langle F(t)F(s)\rangle = 2D\,\delta(t-s)$$

CEL:
$$\dot\psi = a - b\sin\psi + \sin\left(\frac{\psi}{2}\right)\mathcal{F}(t)$$
$$\langle \mathcal{F}(t)\mathcal{F}(s)\rangle = 2(2D)\,\delta(t-s)$$

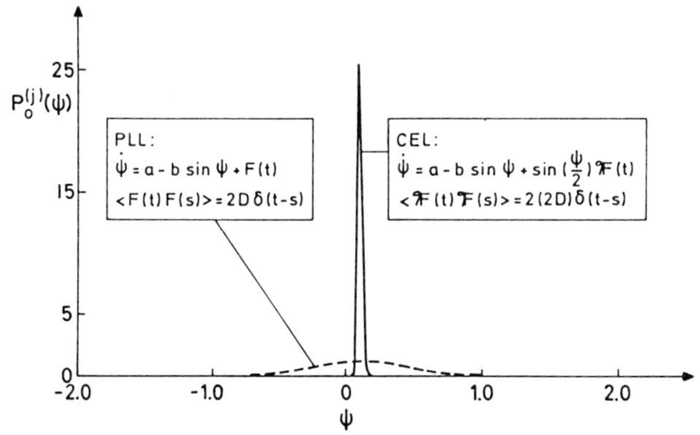

Figure 3

However, it is true that, under ordinary circumstances, the two modes in the CEL will lock. How, then, can a measurement be performed? Several alternatives are available: see, e.g., Fig. 4 for the case of a CEL gyroscope [8]. First, one can measure a signal even in the locked regime, since the perturbation to be measured (e.g., a "Machian" gravitationally-induced rotation rate Ω) will change the value to which the phase difference locks. The minimum detectable Ω in that case is indicated in Fig. 4 (under the heading "without mirrors m_1, m_2"): it is seen to be the same as for an ordinary passive device (Fig. 2), but obtained here with a device which is potentially simpler. Note that the dead-band problem is overcome in the CEL gyro.

Secondly, one can make use of a technique which extracts some light from each mode and couples it, with an appropriate phase, to the other one, to partially unlock the device. This controlled backscattering method, illustrated in Fig. 4 (mirrors m_1 and m_2 and dashed lines), allows one to, in principle, eliminate from the sensitivity formula that part of the cavity losses which is due to transmission losses, leaving only the part (γ_a in Fig. 4) which is due to absorption and other irreversible losses. This results in an increase in sensitivity. The same technique, as has been pointed out by the authors [9], could be used in a conventional passive device as well, with analogous results.

141

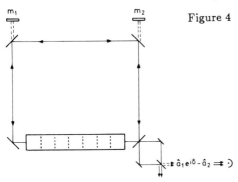

Figure 4

WITHOUT MIRRORS $m_1 + m_2$:

$$\Omega^{min} = \frac{1}{S} (\gamma_a + \gamma_t) \sqrt{\frac{\hbar\nu}{Pt_m}}$$

WITH MIRRORS $m_1 + m_2$:

$$\Omega^{min} = \frac{1}{S} \gamma_a \sqrt{\frac{\hbar\nu}{Pt_m}}$$

In conclusion, the search for optical tests of general relativity has led to a deeper understanding of fundamental physical concepts (e.g., squeezed states [10], quantum nondemolition techniques [11], the CEL, etc.) and generated a number of alternative measurement schemes whose relative merits remain, by and large, to be explored. The marriage of the macro- and microcosmos promises to yield many fascinating problems and insights.

References

1. W. Schleich and M. O. Scully, in *New Trends in Atomic Physics* (Les Houches Summer School Lecture Notes, session XXXVIII), edited by G. Grynberg and R. Stora (North-Holland, 1984), 995.
2. J. Gea-Banacloche, M. O. Scully and D. Z. Anderson, Opt. Commun. **57**, 67 (1986).
3. J. Gea-Banacloche, Phys. Rev. A **35**, 2518 (1987).
4. M. O. Scully, Phys. Rev. Lett. **55**, 2802 (1985).
5. M. O. Scully and M. S. Zubairy, Phys. Rev. A **35**, 752 (1987); J. Bergou, M. Orszag and M. O. Scully, to be published.
6. P. E. Toschek and J. L. Hall, in *XV International Conference on Quantum Electronics Technical Digest Series 1987, Vol. 21* (Optical Society of America, Washington, DC 1987), 102.
7. W. Schleich and M. O. Scully, to be published.
8. M. O. Scully, Phys. Rev. A **35**, 452 (1987).
9. M. O. Scully and J. Gea-Banacloche, to be published.
10. D. F. Walls, Nature **306**, 141 (1983).
11. C. M. Caves, K. S. Thorne, R. W. P. Drever, V. D. Sandberg and H. Zimmermann, Rev. Mod. Phys. **52**, 341 (1980).

Generation and Application
of Squeezed States of Light

H.J. Kimble, R.J. Brecha, L.A. Orozco, M.G. Raizen, Ling-An Wu, and Min Xiao

Department of Physics, University of Texas at Austin, Austin, TX 78712, USA

Squeezed states of light have recently been generated in several laboratories by a variety of techniques [1]. As the name implies, squeezed states are states for which the variance in one of two quadrature phase amplitudes is reduced (squeezed) below the level associated with the zero-point or vacuum fluctuations of the electromagnetic field [2]. Because of an uncertainty relation constraining the product of variances, this reduction is necessarily accompanied by an increase in fluctuations above the vacuum level for the orthogonal quadrature-phase amplitude. Squeezed states are of great intrinsic interest from the perspective of quantum optics because of the nonclassical nature of the radiation field in these states. There is as well a growing interest in the potential that squeezed states offer for improvements in the precision of optical measurements and for fundamental spectroscopic investigations. In this brief contribution we review the research program at the University of Texas at Austin relating to the generation and application of squeezed states.

The first experiment that we describe involves an optical parametric oscillator (OPO) pumped by a field of frequency ω_2 to generate a subharmonic field at $\omega_1 \approx \omega_2/2$. As shown by YURKE [3] and by COLLETT AND GARDINER [4], this device should in principle be capable of producing arbitrarily large degrees of squeezing over a narrow bandwidth as the pumping level is increased to the point of the threshold for oscillation of the OPO. Our research with the OPO is described in detail in Reference [5]. The basic experimental arrangement consists of a frequency-stabilized Nd:YAG laser that is intracavity frequency doubled to generate a pump beam at $\lambda_2 = 0.53$ μm. The pump is incident upon an OPO formed by mirrors of high reflectivity at both ω_1 and ω_2 and containing a crystal of MgO:LiNbO$_3$. The squeezed radiation at $\lambda_1 = 1.06$ μm emitted through the output coupler of the OPO is detected with a balanced homodyne detector, with the original laser emission at 1.06 μm serving as the local oscillator.

With this apparatus we have observed noise reductions of 63% relative to the vacuum noise level in the photocurrent from the balanced homodyne detector. (To a very good approximation, the vacuum noise level in our system is obtained simply by providing no input (a vacuum state) to the signal port of the homodyne detector). More importantly we have shown that the field emitted by the OPO would in fact be squeezed more than ten-fold in the absence of a variety of linear loss mechanisms that degrade the squeezing. These losses include the detector quantum efficiency, the homodyne efficiency, and the escape efficiency for light from the cavity of the OPO to the input of the balanced detector. In all cases the dominant losses are not associated with the nonlinear processes that generate the squeezing, and hence can in principle be eliminated with a greater investment in low loss optical elements and with improved detector quantum efficiency. Even with the current cavity losses, we infer that a field with more than five-fold squeezing is generated at the output coupler of the OPO for operation at roughly 50% of the threshold pumping power. While we cannot at present directly detect this large degree of squeezing, it is nonetheless a substantially squeezed field that could be profitably employed in some experiments.

Having made a quantitative characterization of the field produced by the OPO, we have recently used the squeezed radiation in two experiments to demonstrate improvements in sensitivity beyond the shot-noise limit. This limit is associated with the vacuum fluctuations of the electromagnetic field and has been the practical limit on the sensitivity of optical measurements. Squeezed states as we have applied them improve the sensitivity by reducing the noise level below the vacuum-state limit in the detection of a coherent optical signal. By injecting the squeezed light from the OPO into the normally open input port of a Mach-Zehnder interferometer, we have demonstrated an increase in sensitivity of 3.0 dB relative to the shot-noise limit for the detection of phase modulation [7]. By applying amplitude modulation to a squeezed beam with an acoustooptic modulator, we have achieved an improvement of 2.5 dB relative to the shot-noise limit for the detection of amplitude changes of a signal field [8].

The second experiment that has generated squeezed states in our laboratory consists of a collection of two-level atoms coupled to a mode of a high finesse optical cavity. We exploit the normal-mode structure of this coupled system to produce squeezing in the output field [9].

144

Unlike the OPO for which the spectrum of squeezing is a Lorentzian centered at the carrier frequency ω_1, the frequency of optimum squeezing for the atom-cavity experiment is offset from the carrier by approximately g√N, where g is the single-atom coupling coefficient (single-photon Rabi frequency) and N is the number of atoms. In our experiments two-level atoms are obtained by optically prepumping a well-collimated beam of atomic sodium (F=2, m_F = 2 → F = 3, m_F = 3 transition of the D_2 line). The atomic beam passes through a standing-wave cavity of length 0.83 mm which is excited by the light from a frequency-stabilized dye laser. The cavity emission is analyzed with a balanced homodyne detector in a fashion similar to that in the experiments with the OPO. With this arrangement we have observed noise reductions of 30% relative to the vacuum noise level, corresponding to 42% squeezing for the field emitted by the cavity. Although this level of squeezing is somewhat less than that predicted by our theoretical analysis, the frequency offset at which optimum squeezing occurs ($\Omega/2\pi$ = 280 MHz) agrees very well with our analysis and with the simple prediction g√N/2π = 265 MHz.

This work was supported by the Office of Naval Research and by the Venture Research Unit of British Petroleum.

References:

1. "Squeezed States of the Electromagnetic Field", Feature Issue of JOSA B, eds. H.J. Kimble and D.F. Walls, October (1987).
2. D.F. Walls, Nature (London) 306, 141 (1983).
3. B. Yurke, Phys. Rev. A 29, 408 (1984).
4. M.J. Collett and C.W. Gardiner, Phys. Rev. A 30, 1386 (1984).
5. L.A. Wu, H.J. Kimble, J.L. Hall, and Huifa Wu, Phys. Rev. Lett. 57, 2520 (1986); L.A. Wu, Min Xiao, and H.J. Kimble, Ref. 1 above.
6. H.P. Yuen and V.W.S. Chan, Opt. Lett. 8, 177 (1983); B.L. Schumaker, Opt. Lett. 9, 189 (1984).
7. Min Xiao, L.A. Wu, and H.J. Kimble, Phys Rev. Lett. (1987).
8. Min Xiao, L.A. Wu, and H.J. Kimble, in preparation (1987).
9. M.G. Raizen, L.A. Orozco, Min Xiao, T.L. Boyd, and H.J. Kimble, Phys. Rev. Lett. (1987); L.A. Orozco, M.G. Raizen, Min Xiao, R.J. Brecha, and H.J. Kimble, Ref. 1 above.

Application of Squeezed-State Light to Laser Stabilization

C.M. Caves

Theoretical Astrophysics 130–33, California Institute of Technology, Pasadena, CA 91125, USA

Squeezed-state light having now been generated in several laboratories [1], a variety of exciting applications can be pursued. Improving the phase sensitivity of an interferometer [2], recently demonstrated [3], requires only squeezed vacuum (no mean field). In contrast, exploring the full range of issues in atomic spectroscopy requires squeezed light with a coherent excitation.

Theoretical analysis of the interaction of atomic systems with squeezed light is just beginning. MILBURN [4] analyzed collapses and revivals in the Jaynes-Cummings model when the the field mode is initially in a squeezed state. GARDINER [5] considered a two-level atom immersed in a squeezed vacuum; he found suppression and enhancement of atomic phase decay and corresponding narrow and broad spectral features in the fluorescent light. CARMICHAEL, LANE, and WALLS [6] generalized GARDINER's analysis to include coherent excitation; they found spectral features that depend on the strength of the coherent excitation and on its phase relative to the phase of the squeezing.

YURKE and WHITTAKER [7] have recently noted that squeezed-state light can improve the sensitivity of frequency-modulation spectroscopy [8]. Their proposal uses squeezed-state light that has a large coherent excitation and reduced amplitude fluctuations at the modulation frequency ("two-mode" amplitude-quadrature squeezing [9]). Similar improvements might be possible in absorption spectroscopy.

The successful experiment by SHELBY *et al.* [1] generates high-power squeezed light, and one other experiment [10], so far unsuccessful, has aimed specifically at generating squeezed-state light with a large coherent excitation. The other experiments to date [1] generate squeezed vacuum. Ideally one desires techniques for generating squeezed light with a coherent excitation that has adjustable amplitude and adjustable phase relative to the phase of the squeezing. A variety of techniques could be used. One could "seed" one of the existing experiments with a coherent field, instead of vacuum; this would work better for generating reduced phase fluctuations (phase-quadrature squeezing) than for generating reduced amplitude fluctuations (amplitude-quadrature squeezing).

Related techniques include suppression of pump-amplitude fluctuations in a highly saturated laser [11] and photon twinning, with feedback from one twin, in a nondegenerate optical parametric oscillator [12]. Both techniques produce amplitude (or intensity) squeezed states. The former technique has been demonstrated in a constant-current-driven semiconductor laser [13], and the latter experiment seems on the verge of success [12].

Perhaps the most practical methods of generating high-power squeezed light separate production of power from generation of squeezing. Power is produced by a laser, a small portion of whose light pumps a nonlinear medium—a "squeezer"—that generates squeezed vacuum. The job is to put the squeezed vacuum fluctuations on top of the laser's large mean field.

The simplest method, demonstrated at the University of Texas [14], combines the coherent laser light with squeezed vacuum at a beamsplitter. Both outputs are squeezed, but both have worse signal-to-noise for the squeezed quadrature than does the laser, because both carry less power than the laser.

A more sophisticated approach employs an optical cavity in place of the beamsplitter. The laser's frequency is stabilized to a cavity resonance by an RF phase modulation technique [15], and squeezed vacuum light is directed onto the far end of the cavity. Feedback ensures that the cavity transmits essentially all the laser power. The crucial element is the optical cavity, acting as a dispersive beamsplitter. Outside the cavity resonance the fluctuations of the transmitted light are the squeezed fluctuations reflected off the far end of the cavity. Thus, for fluctuations at frequencies outside the cavity resonance, this technique achieves the ideal of a coherent excitation with adjustable amplitude and phase.

Feedback plays a more important role in other techniques. Consider the scheme sketched in Fig. 1 [16]. The laser's intensity is stabilized by negative feedback from a photodetector with quantum efficiency η; the feedback is modeled by a beamsplitter with variable transmissivity $T(t)$. Light is extracted from the feedback loop at a beamsplitter with transmissivity χ. Feedback suppresses the laser's shot noise within the feedback loop, but in doing so it imposes noise on the extracted light, which thus has super-shot-noise statistics. In a semiclassical description the imposed noise arises from photodetector shot noise; in a quantum-mechanical description, if $\eta=1$, it arises from vacuum fluctuations that enter the extraction beamsplitter's normally open port.

In Fig. 1 the normally open port is illuminated not by vacuum, but by squeezed light generated by a squeezer that is pumped by a fraction ξ of the laser light. For perfect feedback the extracted light has intensity noise-to-signal spectrum

$$\frac{2}{\mathscr{P}_2}\left[\frac{R(\varepsilon)}{\chi}+\frac{1-\chi}{\chi}\frac{1-\eta}{\eta}\right]\underset{\eta=1}{\longrightarrow}\frac{2}{\mathscr{P}_2}\frac{R(\varepsilon)}{\chi} \tag{1}$$

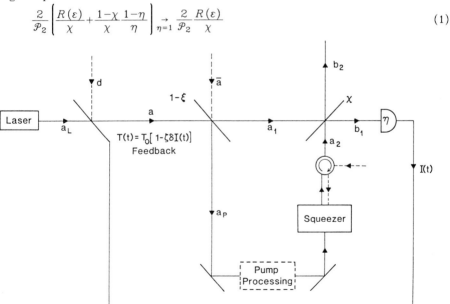

Figure 1. Laser intensity stabilization using squeezed-state light

as a function of RF frequency ε. Here $\mathscr{P}_2=(1-\chi)(1-\xi)T_0\mathscr{P}_L$ is the photon intensity of the extracted light, $\mathscr{P}_L$ being the photon intensity of the laser and T_0 being the fiducial transmissivity of the variable beamsplitter. The quantity $R(\varepsilon)$ is the spectral density of the appropriate quadrature of the squeezed light, measured in units of the vacuum level. For $\eta=1$ the extracted light displays sub-shot-noise behavior if $R(\varepsilon)<\chi$ and has better noise-to-signal spectrum than a shot-noise-limited laser with intensity $\mathscr{P}_L$ if $R(\varepsilon)<\chi(1-\chi)(1-\xi)T_0$.

In this scheme the role of squeezing is to reduce fluctuations that feedback imposes on the extracted light. What lesson does it teach? *Reduce inefficiencies and then cover open ports with squeezed light.*

How does this lesson apply to frequency stabilization? Figure 2 shows a variant of the fringe-side technique for stabilizing a laser's frequency to a resonance of an optical cavity [17]. (The difference from the usual fringe-side technique minimizes the number of open ports.) Laser light illuminates the cavity, whose two outputs (reflected and transmitted light) are directed onto photodetectors. The differenced photocurrent is integrated and fed back to an electro-optic modulator that adjusts the phase of the laser. The technique locks the laser's frequency halfway up the transmission fringe of the cavity resonance. Useful light is extracted at a beamsplitter with transmissivity χ.

The crucial open port is not the open port of the extraction beamsplitter, but rather the far end of the cavity. In Fig. 2 this cavity port is covered by squeezed light. For RF frequencies $\varepsilon\ll\beta_{cav}=$ (half-width of the cavity resonance), one finds in the case of perfect feedback that the extracted light has a spectral density of frequency noise

$$S_{\Omega}(\varepsilon)=\frac{(2\beta_{cav})^2}{2\mathscr{P}_{out}}\frac{1-\chi}{\chi}\left[R(\varepsilon)+\frac{1-\eta}{\eta}\right]\underset{\eta=1}{\rightarrow}\frac{(2\beta_{cav})^2}{2\mathscr{P}_{out}}\frac{1-\chi}{\chi}R(\varepsilon) .\tag{2}$$

Here $\mathscr{P}_{out}=(1-\chi)(1-\xi)\mathscr{P}_L$ is the photon intensity of the extracted light, and $R(\varepsilon)$ is the spectrum of the appropriate quadrature of the squeezed light. If $R(\varepsilon)$ is constant, the flat frequency spectral density corresponds to phase diffusion with diffusion constant $S_{\Omega}/2$. There is no question here of sub-shot-noise frequency fluctuations, but if the photodetectors have good quantum efficiency, squeezing allows in principle an improvement in frequency stability.

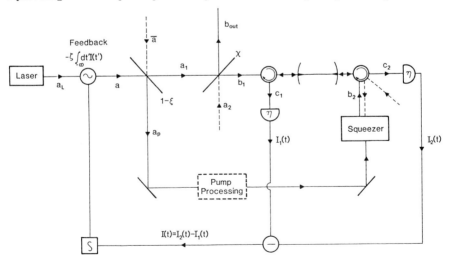

Figure 2. Fringe-side laser frequency stabilization using squeezed-state light

As mentioned above, squeezed light can also be used in the RF phase modulation technique for stabilizing a laser's frequency [15]. Within the cavity resonance, however, squeezed light incident on the far end of the cavity does little good, because there is another important, but subtle open port. The feedback loop senses vacuum phase fluctuations in the laser light at twice the modulation frequency, and it imposes these vacuum fluctuations on the transmitted light as phase fluctuations within the bandwidth of the cavity resonance. (A similar effect has been discussed for an interferometer [18].) Unless these fluctuations, too, are squeezed, the RF modulation technique does not stabilize the frequency as well within the cavity resonance as does the fringe-side technique.

This work was supported in part by the Office of Naval Research.

1. R. E. Slusher, L. W. Hollberg, B. Yurke, J. C. Mertz, and J. F. Valley: Phys. Rev. Lett. **55**, 2409 (1985); R. M. Shelby, M. D. Levenson, S. H. Perlmutter, R. G. DeVoe, and D. F. Walls: Phys. Rev. Lett. **57**, 691 (1986); L.-A. Wu, H. J. Kimble, J. L. Hall, and H. Wu: Phys. Rev. Lett. **57**, 2520 (1986); M. W. Maeda, P. Kumar, and J. H. Shapiro: Opt. Lett. **12**, 161 (1987); M. G. Raizen, L. A. Orozco, M. Xiao, T. L. Boyd, and H. J. Kimble: Phys. Rev. Lett. (to be published).

2. C. M. Caves: Phys. Rev. D **23**, 1693 (1981).

3. M. Xiao, L.-A. Wu, and H. J. Kimble: Phys. Rev. Lett. (to be published).

4. G. J. Milburn: Opt. Acta **31**, 671 (1984).

5. C. W. Gardiner: Phys. Rev. Lett. **56**, 1917 (1986).

6. H. J. Carmichael, A. S. Lane, and D. F. Walls: Phys. Rev. Lett. **58**, 2539 (1987); Opt. Acta (to be published).

7. B. Yurke and E. A. Whittaker: Opt. Lett. **12**, 236 (1987).

8. G. C. Bjorklund: Opt. Lett. **5**, 15 (1980).

9. C. M. Caves and B. L. Schumaker: Phys. Rev. A **31**, 3068 (1985).

10. H. J. Kimble and J. L. Hall: In *Quantum Optics IV*, ed. by J. D. Harvey and D. F. Walls (Springer, Berlin, Heidelberg 1986) p. 58.

11. Y. Yamamoto, S. Machida, and O. Nilsson: Phys. Rev. A **34**, 4025 (1986).

12. E. Giacobino, C. Fabre, A. Heidmann, R. Horowicz, S. Reynaud, and G. Camy: this volume.

13. S. Machida, Y. Yamamoto, and Y. Itaya: Phys. Rev. Lett. **58**, 1000 (1987).

14. H. J. Kimble: private communication.

15. R. W. P. Drever, J. L. Hall, F. V. Kowalski, J. Hough, G. M. Ford, A. J. Munley, and H. Ward: Appl. Phys. B **31**, 97 (1983).

16. C. M. Caves: Opt. Lett. (to be published), Caltech preprint GRP-133.

17. R. L. Barger, J. S. Sorem, and J. L. Hall: Appl. Phys. Lett. **22**, 573 (1973); J. Helmcke, S. A. Lee, and J. L. Hall: Appl. Opt. **21**, 1686 (1982).

18. J. Gea-Banacloche and G. Leuchs: Opt. Acta (to be published).

Quantum Nondemolition Detection and Squeezing in Optical Fibers

M.D. Levenson, R.M. Shelby, and S.H. Perlmutter

IBM Almaden Research Center,
650 Harry Road, San Jose, CA 95120, USA

The nonlinear optical interactions in an optical fiber permit the amplitude of one wave to be inferred from the phase of a coupled wave, and the partial suppression of quantum noise.

Conventional light detectors must destroy the quantum state of the electromagnetic field in order to measure its amplitude. They are Quantum Demolition detectors. The nonlinear index of refraction of an optical fiber permits one to infer the amplitude of one wave by measuring the light induced phase shift of a nonlinearly coupled wave. This process has been termed Quantum Nondemolition Detection or Back Action Evading Measurement. The amplitude of the first wave is unchanged by the interaction. That amplitude is the QND variable; the phase of the second wave is the QND readout. The uncertainty which quantum mechanics requires to be added to a system being measured appears in the phase of the first beam, not its amplitude, thus the back action is evaded.

We have demonstrated this effect by correlating the QND readout with a subsequent QDD measurement of the QND variable (1). We have demonstrated back action evasion by showing that the QND variable at the output of the detector has no greater noise than at the input - which was at the vacuum noise level.

When the correlated noise on two coupled laser beams is made to subtract coherently, the net noise level can be below the vacuum noise level. This effect has been termed "four-mode squeezing" since the fluctuations involve two sidebands of each of two strong pump waves (2). The noise on each beam can separately be above the vacuum noise level, but strong four mode correlations can lower the sum of the noise at two detectors to a value below the vacuum. Indeed, the correlations can be so strong that the noise on the two detectors can be less than that on one of them (3).

It is necessary to cool the optical fiber to 2K to suppress phase noise caused by light scattering in the optical fiber. The stimulated Brillouin effect must be suppressed by phase modulating the light to broaden the spectrum. The modulation frequency must be exactly equal to the mode spacing of the Fabry Perot interferometer needed to phase shift the carrier wave. Other experimental innovations are necessary to accurately measure the vacuum level and to maximize the effective quantum efficiency.

The same nonlinear interactions in an optical fiber give rise to conventional two-mode squeezing (4). The fluctuations in the amplitude of the beam cause correlated phase fluctuations of that beam. The nonlinear index of refraction cannot alter the amplitude of a wave. The correct superposition of amplitude and phase quadratures has been shown to have a noise level 12.5% below the vacuum level.

Squeezed light has the property that the ratio of noise to intensity rises as the light is attenuated (5). The squeezed light generated in an optical fiber is not a minimum uncertainty state as light scattering adds phase noise.

References:
1. M.D. Levenson, R.M. Shelby, M. Reid, and D.F. Walls, Quantum Nondemolition Detection of Optical Quadrature Amplitudes, Phys. Rev. Lett. 57, 2473 (1986).

2. B.L. Schumaker, S.H. Perlmutter, R.M. Shelby and M.D. Levenson, Four-Mode Squeezing, Phys. Rev. Lett 58, 357 (1987).

3. M.D. Levenson and R.M. Shelby, Four-Mode Squeezing and Applications, Optica Acta (to be published).

4. R.M. Shelby, M.D. Levenson, S.H. Perlmutter, R.G. DeVoe, and D.F. Walls, Broad-Band Parametric Deamplification of Quantum Noise in an Optical Fiber, Phys. Rev. Lett 57, 691 (1986).

5. G.J. Milburn, M.D. Levenson, R.M. Shelby, and D.F. Walls, On Optical Media for Squeezed State Generation, J. Opt. Soc. Am (to be published).

Two-Photon Chaotic States of the Radiation Field

Y.Q. Li and Y.Z. Wang

Shanghai Institute of Optics and Fine Mechanics, Academia Sinica,
P.O. Box 8211, Shanghai, People's Republic of China

Many nearly monochromatic physical radiation fields, such as chaotic light (or a thermal radiation field) and fluctuating lasers, have equal quantum fluctuations in the two quadratures but much larger products than the minimum Heisenberg uncertainty products. Could these quantized fields be used to construct new fields such as squeezed states [1,2], in which their fluctuations are reduced in one quadrature? In one experiment, such an effort has been made [3]. This paper generalizes the concept of squeezing to any quantum state, and two-photon chaotic states (TCHS), also called squeezed chaotic states, are then discussed.

1 Squeezing and Quantum Fluctuation Beats in TCHS

For a single mode $(a, a+)$ mix part of the field with its phase conjugate to produce a new mode, b, such that $b = \mu a + \nu a+$, where $|\mu|^2 - |\nu|^2 = 1$ [2]. Define mode b to be a two-photon quantum state for mode a in any quantum state. From the density operator $\rho(a, a+)$, the Wigner characteristic function $C_b^{(w)}(\xi)$ of the operator b can be derived [4], and for convenience referred to as the equivalent definition of the state. For the case when mode a is a chaotic state, the variances in the two quadratures are $V(a_1) = V(a_2) = 1/4(1 + 2\bar{n})$ where n is the average photon number in the chaotic state. Then the mode b will be in a TCHS, and

$$C_b^{(w)}(\xi) = \exp\left[-1/2\eta^2 \, |\mu\xi + \nu^*\xi^*|^2 \, (1+2\bar{n})\right] \qquad (1)$$

where η is real. Using the derivation of $C_b^{(w)}(\xi)$ [4], the unequal variances in each quadrature are caluated to be

$$V(b_1) = 1/4 \, (1+2\bar{n}) \, |\mu^*+\nu|^2 \text{ and } V(b_2) = 1/4 \, (1+2\bar{n}) \, |\mu^*-\nu|^2 \qquad (2)$$

where b_1 and b_2 satisfy $b = b_1 + ib_2$.

Figure 1 shows the fluctuating electric field of the modes and their ellipses. It is clear that in the chaotic state, the single-mode electric field fluctuates around zero randomly with a constant uncertainty. However, in TCHS it fluctuates with a time-modulated variance (its modulation frequency is 2ω). This feature may be called quantum fluctuation beat of the electric field, which shows that the field in not entirely stochastic, but there is a somewhat collective coherent characteristic among the groups of photons mixing between the chaotic field and its phase conjugate.

2 Super-Bunching Effect

We have calculated the normalized second-order correlation function, $g^2(0)$ of the single mode field in TCHS

$$g^{(2)}(0) = 2 + \left[|\mu| \cdot |\nu| (1+2\bar{n}) / \langle \hat{n} \rangle_b\right]^2 \qquad (3)$$

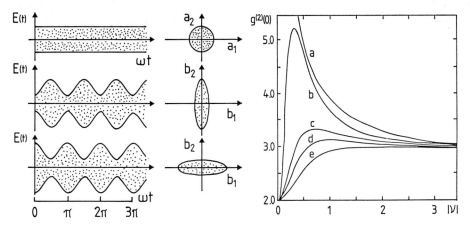

Fig. 1 Plot of single-mode electric field against time, and phase space plot showing the uncertainty in (a) a chaotic state, (b) a two-photon chaotic state with reduced fluctuation in the in-phase component, b_1 $(ang(\mu)+ang(\nu)=\pi)$, and (c) a two-photon chaotic state with reduced fluctuation on the out-phase component, b_2 $(ang(\mu)+ang(\nu)=0)$.

Fig.2 The normalized second-order correlation function $g^{(2)}(0)$ vs. $|\nu|$ the squeezing parameter of the single-mode field in two-photon chaotic states with various values of $\bar{n}$. (a) $\bar{n}=0$ (i.e. squeezed vacuum field, (b) $\bar{n}=0.1$; (c) $\bar{n}=0.5$; (d) $\bar{n}=1.0$ and (e) $\bar{n}\geqslant10$

where $\langle\hat{n}\rangle_b=|\mu|^2\,\bar{n}+|\nu|^2\,(1+2\bar{n})$. Figure 2 shows that the bunching effect of the field in TCHS is always greater than in the chaotic states in which $g^{(2)}(0)=2$, and $g^{(2)}(0)\to3$ when ν is very large. This may be called the super-bunching effect of photon statistics, caused by the partial coherence in the stochastic field.

Consequently, TCHS's of light, with many distinguishing features, are easily generated via DFWM [3,5], and will be encountered in optical fiber communication if it is necessary to reduce the enormous fluctuation of the classical noise field in one quadrature.

References

1. D.F. Walls, Nature 306, 141 (1983)
2. H.P. Yuen, Phys. Rev. A13, 2226 (1976)
3. M.D. Levenson, R.M. Shelby and S.H. Perlmatter, Opt. Lett. 10, 514 (1985)
4. W.H. Louisell, in Quantum Statistical Properties of Radiation, John Willy, (1973)
5. Y.Q. Li and Y.Z. Wang, (to be published)

153

Squeezing Intensity Noise on Laser-like Beams

E. Giacobino[1], C. Fabre[1], H. Heidmann[1], R. Horowicz[1], S. Reynaud[1], and G. Camy[2]

[1]Laboratoire de Spectroscopie Hertzienne de l'Ecole Normale Supérieure, Université P. et M. Curie, F-75252 Paris Cedex 05, France
[2]Laboratoire de Physique des Lasers, Université de Paris-Nord, F-93430 Villetaneuse, France

Parametric down-conversion generates beams having strong quantum correlations. This property can be easily understood by reasoning in terms of photons : let us consider a non-linear crystal irradiated by light at frequency ω_0. For each pump photon absorbed at frequency ω_0, it simultaneously emits two "twin" signal photons with frequencies ω_1 and ω_2 (such that $\omega_0 = \omega_1 + \omega_2$). Such non-classical correlations have been observed by photon coincidence techniques [1]. However, with the available cw lasers and non-linear crystals, the parametric process generates only very weak correlated beams (typically a few photons/second), because the pump power is spread into an infinity of twin modes. To concentrate the output energy into a few pair of modes only, the non-linear crystal can be inserted in an optical cavity having mirrors with a high reflectivity for the signal frequencies ω_1 and ω_2. Above some pump power threshold, the system can oscillate and yields a pair of intense, laser-like, "twin" beams.

Using again the picture of the twin photons, we can predict the qualitative behaviour of the correlation between the two signal beams emitted by the parametric oscillator : photons are emitted by pairs by the parametric process and then stored in the cavity during a random time of the order of the cavity storage time τ_c. Therefore counting the photons emitted in the two output signal beams during a time much longer than τ_c will give exactly equal numbers. Conversely, in the frequency domain, the noise on the intensity difference between the two beams is expected to be below shot noise for frequencies lower than τ_c^{-1} only. These arguments are fully confirmed by an exact calculation [2] : It can be shown that the noise spectrum on the intensity difference is given by

$$S_I(\omega) = S_0 \frac{\omega^2}{\omega^2 + \tau_c^{-2}}$$

where S_0 is the shot noise level.

The experimental set-up designed to demonstrate this property is shown in Fig. 1. The doubly resonant parametric oscillator is pumped by the 528 nm line of a single mode Ar^+ laser, stabilized on an external Fabry-Perot cavity. The non-linear medium is a 7 mm type II KTP crystal, inserted in a 2 cm long, highly focused cavity. Both mirrors have a high transmission for the green pump light. The front mirror has maximum reflectivity for the infrared around 1.06 μm, the rear mirror has a 0.8 % transmission for the infrared. Above a pump threshold of about 80 mW the system oscillates and two co-propagating cross-polarized beams with wavelengths 1.067 μm and 1.048 μm are emitted with intensities of a few mW. The cavity length is electronically stabilized for continuous parametric emission. The two infrared beams are separated by a polarizing beamsplitter and focused on 0.9 quantum efficiency InGaAs photodiodes. The photodiode signals are amplified, substracted and fed into a spectrum analyser.

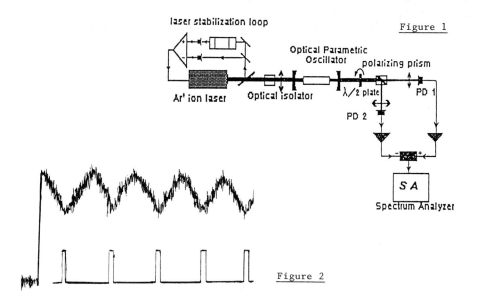

Figure 1

Figure 2

To obtain the shot noise reference level, we have used a rotating $\lambda/2$ plate before the polarizing beamsplitter in the following way : if the $\lambda/2$ plate is oriented so that the two outgoing polarizations are rotated by 45°, each beam is divided by the system $\lambda/2$ plate + polarizer in two equal parts : Since the two twin beams have very different frequencies ($\Delta\nu \sim 10^{12}$ Hz), the intensities detected by the two photodiodes are both equal to $(I_1+I_2)/2$. The system acts just like a 50 % beamsplitter ; under these conditions, it can be shown |3| that one get the shot noise level corresponding to a beam of intensity (I_1+I_2) when one monitors the noise of the difference between the intensities seen by the photodiode. This happens when the axes of the $\lambda/2$ plate are at 22.5°, 67.5°... from the axes of the polarizer. When the plate is at 0°, 45°, 90° from the axes of the polarizer, one gets the noise on I_1-I_2. Fig. 2 gives the noise amplitude obtained when spectrum analyzer is set at a fixed frequency of 10 MHz and the $\lambda/2$ plate slowly rotated. The pulses on the lower trace correspond to angles 0°, 45°, 90°... and the left part of the trace gives the electronic noise level. One observes a reduction of the noise below the shot noise level by a significant amount. We have observed this effect on a broad frequency range, from about 3 to 13 MHz. We are limited on the high frequency side by the pass-band of our detection electronics. Improvements in the amplifiers should allow us to study the noise spectrum on a wide enough frequency range to see the rise of the noise to the shot noise level which should occur at $\tau_c^{-1} \approx 20$ MHz.

In conclusion, we have observed a strong correlation between two intense, "macroscopic" beams having a pure quantum origin. Such a correlation could be used to generate a sub-shot noise intense beam ("intensity squeezed state") by feedback technique, or directly in a sub-shot noise high sensitivity absorption measurement |2| .

|1| S. Friberg, C. Hong, L. Mandel, Phys. Rev. Lett. 54, 2011 (1985)and ref in
|2| S. Reynaud, C. Fabre, E. Giacobino, JOSA B 1987 to be published
|3| H. Yuen, V. Chan, Optics Lett. 5, 177 (1983).

Probing Quantum Chaos and Localization in the Diamagnetic Kepler Problem

J.C. Gay and D. Delande

Laboratoire de Spectroscopie Hertzienne de l'Ecole Normale Supérieure, Tour 12, EO1-4, place Jussieu, F-75252 Paris Cedex 05, France

Recent studies on the diamagnetic Kepler problem have allowed greater insights into the nature of the quantum analog of classical chaos. Dynamical symmetries and the quantum/classical motions are well understood. Adding that this is experimentally realized in Rydberg atoms interacting with static magnetic fields, makes this situation one of the most promising for studying 3D-hamiltonian chaos.

For this system, the quantum statistics of energy level fluctuations was shown to evolve from Poisson to Wigner-G.O.E. when the dynamics becomes irregular |1|-|3|. This confirms the conjecture on the universality of energy level fluctuations and provides us with an unambiguous signature of quantum chaos which should be displayed by a large class of chaotic systems.

However the G.O.E. statistics is the one which the eigenvalue spectrum of a real symmetric Random matrix obeys. Certainly something beyond statistical theories should allow to discriminate in between the dynamics of a hamiltonian system and a Random matrix representation.

There is actually a key notion, which is intrinsic (not basis dependent) the one of dynamical symmetry of the hamiltonian which allows such a discrimination. By essence it completely characterizes the dynamics of the system. If the symmetry is "exact", there will be no chaos. This concept allows to go beyond the statistical Random matrix theories of quantum chaos as demonstrated below.

Extensive studies of the dynamical symmetries in the diamagnetic Kepler problem have been performed using the Coulomb dynamical group SO(2,2). In the regular region, the (approximate) dynamical symmetry is of ro-vibrational type associated with two limiting coupling schemes of SO(2,2). Classically, chaos develops near the separatrix associated with the cross-over in the symmetries. Writing $\beta = \gamma^2/(-2E)^3$ ($\gamma = B/B_c$; $B_c = 2.35.10^9$ G ; E the energy), the transition to complete chaos takes place for $\beta = 60.638$ upon which the destruction of the Coulomb dynamical symmetry is complete |2| |4|.

The study of the projections of the eigenstates onto the states with limiting rotational and vibrational symmetries allows to characterize how chaos manifests itself in the quantum system. This is shown on Fig.(1) and (2).

The destruction of the vibrational and rotational symmetries is quite clearly seen on the plots. Chaos manifests itself in the fractalization of the symmetry onto several eigenstates. From the G.O.E. statistics obeyed, the eigenstates repell each other. But they still do present a tendency to cluster. This indicates a quantum memory of symmetries and the existence

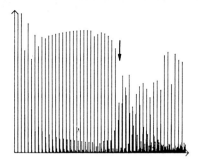 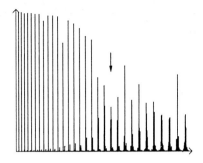

Fig. (1) - Destruction of the <u>vi-brational</u> limiting symmetry ; new spacings and <u>clustering</u> associated with islands of stability and bi-furcation appear.

Fig. (2) - Destruction of the <u>rota-tional</u> limiting symmetry. Quantum analog of a 1:1 resonance and <u>memory</u> of the symmetries beyond complete chaos.

of <u>long range order</u> in the quantum spectrum. Hence, the quantum behaviour is more regular than its classical counterpart in the chaotic region.

Such a tendency is also seen in the oscillator strengths distributions and leads to quasi-Landau resonances |4|, and to secondary series with different spacings |5|. A final confirmation can be drawn from the building of the Wigner or Q semi-classical distributions through choice of a conve-nient lattice of coherent states spanning phase space. <u>Quantal Poincaré</u> <u>sections</u> built this way, in the chaotic region, reveal the existence of a remaining localization of the quantum motion beyond complete classical chaos for β = 60.638.

It is finally worth noticing that Figure (1) illustrates the <u>quantum</u> <u>appearance</u> of a process involving <u>islands of stability and bifurcations</u>, which govern the motion along B field |6|. This is actually seen in expe-riments |5|. Figure (2) is the <u>quantum analog of a 1:1 resonance</u> alike the process in parametric oscillation and governs the usual quasi-Landau phe-nomena.

1. D. Wintgen and H. Friedrich, Phys. Rev. Lett. 57 (1986) 571
2. D. Delande and J.C. Gay, Phys. Rev. Lett. 57 (1986) 2006
3. G. Wunner et al., Phys. Rev. Lett. 57 (1986) 3261
4. D. Delande and J.C. Gay, J. Phys. B 17 (1984) L335
                    Com. Atomic Mol. Phys. 19 (1986) 35
5. K. Welge, contribution, this volume
6. M.Y. Sumetskii, JETP 56 (1982) 959.

Weak Localization of Light

M. Rosenbluh, M. Kaveh, and I. Freund

Department of Physics, Bar-Ilan University, Ramat Gan, Israel

Weak localization of light has been demonstrated in the propagation properties of electromagnetic waves in random, strongly scattering media.[1] The most striking demonstration of weak localization is shown in Fig.1, which shows the scattered intensity of a laser beam from a solid, diffuse scatterer as a function of backscattered angle. The peak represents an ensemble average of the speckle patterns of the scattered light, obtained by rotating the sample. The nearly two times stronger, narrow, backscattered peak is predicted by weak localization theories[2] and is due to the coherent superposition of time reversed loops through the random media.

When similar experiments are performed on a system of dense scatterers suspended in a liquid, the ensemble averaging is

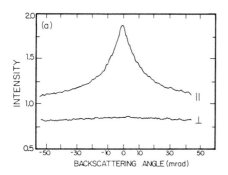

Fig. 1. Coherent backscattered peak for a rotating $BaSO_4$ scatterer at 515 nm. The backscattering is shown for polarizations both parallel and perpendicular to the incident laser polarization. The intensity scale is adjusted so that the large angle parallel polarization backscattering intensity equals one.

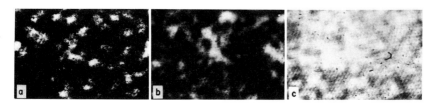

Fig. 2. Time evolution of the contrast of the backscattered speckle for a single optical pulse of width: a) 40μ sec, b) 400μ sec, and c) 4 msec from concentrated polystyrene spheres of $0.46\ \mu$m diameter, suspended in water.

automatic due to the random motion of the scatterers, but localization effects are weakened by the inelastic nature of the
individual scattering events.[3] The Doppler shift upon each scattering is random and the accumulated Doppler shift in a loop
with a constant number of scatterers is a function of time as
the scatterers move via Brownian motion. The time evolution of
the backscattered speckle patterns from such systems is strongly
affected by the evolving optical phase of the trajectories of
long loops through the sample. The time resolved intensity-
intensity correlation function of the speckle pattern drops very
rapidly for times short compared to the single particle scattering correlation time, but tails off exceedingly slowly for
long times.[4] One of the results of this behaviour is demonstrated in Fig. 2 which shows that, although the measured correlation time is about 40 μ sec, a high contrast speckle pattern is
observed for times orders of magnitude longer than this time.

The correlation of phase between long scattering loops
also has dramatic consequences for the total accumulated Doppler
shift of light propagating in the forward direction in a
strongly scattering liquid suspension.[5] In stark contrast to
the predictions of the photon diffusion approximation, which
predicts the wrong shape and scale for the frequency spectrum
and the wrong sample thickness dependence (s^2), our experiments
(Figs. 3 and 4) show that the dynamic intensity-intensity
correlation function (and thus the frequency spectrum) is
dominated by short quasi-ballistic photon trajectories which
scale linearly with the sample thickness.

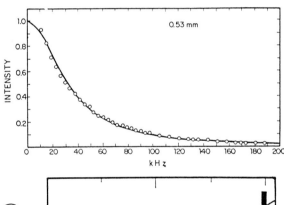

Fig. 3. Forward scattered
frequency spectrum for a
∅.53 mm thick cell con-
taining ∅.46 µm diameter
polystyrene spheres. The
solid line is a best-fit
Lorentzian.

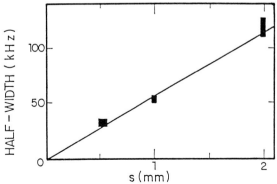

Fig. 4. Half-width of
the measured frequency
spectra as a function
of cell thickness, s,
howing linear depen-
dence.

1)M. Kaveh, M. Rosenbluh, I. Edrei, I. Freund, Phys. Rev. Lett., 57, 2049(1987), and references therein.
2)M. Rosenbluh, I. Edrei, M. Kaveh, I. Freund, Phys. Rev. A, 35, 4458(1987).
3)M. Kaveh, M. Rosenbluh, I. Freund, Nature, 326, 778(1987).
4)M. Rosenbluh, M. Hoshen, I.Freund, M. Kaveh, Phys. Rev. Lett., 58, 2754(1987).
5)I. Freund, M. Kaveh, M. Rosenbluh, to be published.

Part V

Atomic Spectroscopy

High Resolution Laser Spectroscopy of Radioactive Atoms

S. Liberman et al.

Laboratoire Aimé Cotton, CNRS II, Bât. 505,
Campus d'Orsay, F-91405 Orsay, France

1. FRANCIUM EXPERIMENTS

Contributors : J. Bauche, H.T. Duong, P. Juncar, S. Liberman, J. Pinard, from Laboratoire Aimé Cotton, Orsay (France).

C. Thibault, F. Touchard, A.Coc, from Laboratoire René Bernas, Orsay (France)

J. Lermé, J-L. Vialle, from Laboratoire de Spectrométrie Ionique et Moléculaire Lyon (France).

S. Büttgenbach, from Institut für Angewandte Physik, Universität Bonn (GFR).

A. Pesnelle, Service de Physique des Atomes et des Surfaces, Saclay (France).

A. Mueller, GANIL, Caen (France).

ISOLDE. Collaboration, CERN, Genève (Switzerland).

Until very recently no optical lines of the francium atom were known. Using basically the same techniques which had enabled us to study long series of radioactive isotopes of alkali atoms by means of the high resolution laser spectroscopy /1/, we have undertaken the systematic study of the spectral characteristics of the francium atom. Francium atoms are provided by the ISOLDE facility at CERN and formed as an atomic beam. Illuminated at a right angle by a CW single mode tunable laser, the francium atoms may undergo optical resonances which induce an optical pumping in the two hyperfine sublevels of the ground state ; this optical pumping can in turn be detected magnetically through a six-pole magnet analyser /2/. In a first series of experiments we have been able to detect the two first resonance lines connecting the 7s ground state to the 7p excited states, and to measure accurately the corresponding wavelengths which lie in the red and near infrared region /3/, and also to detect the second resonance lines connecting the ground state to the 8p excited states which lie in the blue region /4/. In a second series of experiments, we analyzed the hyperfine structure and isotope shifts of 16 isotopes of francium, and we obtained the results which are schematically reported in the figure 1. A careful study of the isotope shift results has shown us an unexpected inversion of the so-called "odd-even staggering" effect, as can be observed in the figure 2 for those isotopes having a neutron number N located between 133 and 137.

Similar behaviour has already been observed with radium isotopes by another group /5/, who suggested that such an inversion could be interpreted as an indication of the presence of a nuclear octupole electric moment.

Fr	Half-life	Spin
207	14.8s	9/2
208	58.6s	7
209	50s	9/2
210	192s	6
211	186s	9/2
212	1200s	5
213	34.6s	9/2 N=126
214	$5 \cdot 10^{-3}$ s	
215	$0.09 \cdot 10^{-6}$ s	
216	$0.7 \cdot 10^{-6}$ s	
217	$22 \cdot 10^{-6}$ s	
218	$0.7 \cdot 10^{-3}$ s	
219	$21 \cdot 10^{-3}$ s	
220	27.4s	1
221	294s	5/2
222	864s	2
223	1308s	3/2 .
224	198s	1 .
225	236s	3/2 .
226	48s	1 .
227	148.2s	1/2 .
228	39s	2 .

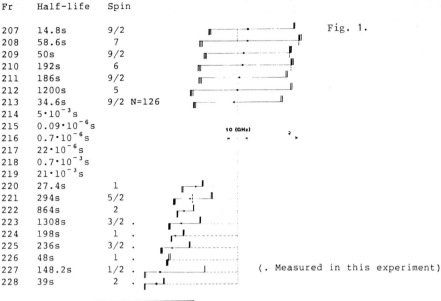

Fig. 1.

10 (GHz)

(. Measured in this experiment)

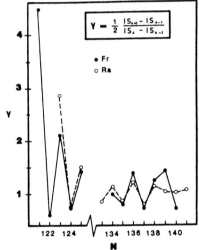

$$Y = \frac{1}{2} \frac{|IS_{A+1} - IS_{A-1}|}{|IS_A - IS_{A-1}|}$$

• Fr
o Ra

Fig. 2.

2. RADIUM EXPERIMENT

Contributors : H.T. Duong, S. Liberman, J. Pinard, from Laboratoire Aimé Cotton, Orsay (France).

E. Arnold, W. Borchers, W. Neu, R. Neugart, E.W. Otten, K. Wendt, from Institut für Physik, Universität Mainz (GFR).

M. Carré, J. Lermé, M. Pellarin, J-L. Vialle, from Laboratoire de Spectrométrie Ionique et Moléculaire, Lyon (France).

P. Juncar, from Institut National de Métrologie, CNAM Paris (France).

G. Ulm and the ISOLDE Collaboration, CERN, Genève (Switzerland).

This experiment, which aims to measure precisely the nuclear magnetic moment of Ra isotopes, is basically grounded on the collinear laser spectroscopy techniques in which a fast atomic beam interacts with the light of a collinearly propagating laser beam. The first atomic beam is produced with an accelerated ion beam which is neutralized in an efficient way through a change – exchange cell without affecting noticeably the propagating velocity of the atoms.

As shown schematically in figure 3, the atomic beam crosses 3 different zones : the zone 1 in which they interact with the laser light, the zone 2 in which they interact with a constant magnetic field B_0 oriented perpendicularly to the atomic beam, and the zone 3 in which they again interact with the laser light. As the laser light is linearly polarized, the interaction of the atoms with the resonant light in zone 1 induces an optical pumping process which gives rise to an alignment in the ground state and consequently to a weak fluorescence signal. In zone 2 the atoms experience a constant magnetic field B_0 which firstly brings them out of the optical resonance and secondly makes the aligned magnetic moments precess around the magnetic field direction. This precession leads to a periodic repopulation of the emptied magnetic sublevels according to a signal formula obtained in the simplest case of a linear polarization of the light perpendicular to $\vec{B_0}$: $S = \alpha \left\{ 2\rho + \frac{3}{4} \Delta\rho \left[1-\cos(2\omega_0\tau) \right] \right\}$ in which ρ represents populations, $\Delta\rho$ the maximum change in magnetic sublevel populations, $\omega_0 = g_I \beta_n B_0/\hbar$ and τ the time of flight duration of the atoms in zone 2. In zone 3, immediately after the magnetic zone 2, the atoms being again in resonance with the laser light, they fluoresce   afresh and the detected fluorescence si-

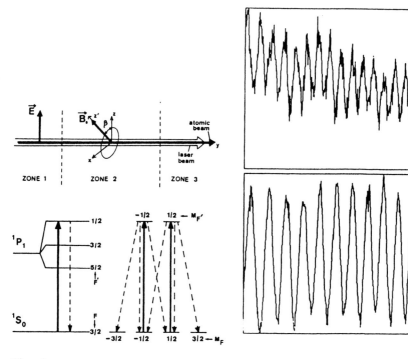

Fig. 3   .Fig. 4

gnal appears as a fringed pattern when B_0 is scanned (with τ fixed) /6/. It is then possible to measure the nuclear magnetic moments of radium isotopes by comparison of the fringes obtained in exactly the same conditions with a barium isotope whose nuclear magnetic moment is accurately known /7/. Typical recordings are shown in figure 4 , from which we have been able to deduce the values :

$$\mu(^{225}Ra) = 0.7338 \ (15) \ \mu_N ,$$
$$\mu(^{213}Ra) = 0.6133 \ (18) \ \mu_N .$$

3. GOLD EXPERIMENT

Contributors : H.T. Duong, S. Liberman, J. Pinard, from Laboratoire Aimé Cotton, Orsay (France).

F. Le Blanc, P. Kilcher, J. Obert, J. Oms, J-C. Putaux, B. Roussière, J. Sauvage, from Institut de Physique Nucléaire - ISOCELE Collaboration, Orsay (France).

J.K.P. Lee, G. Savard, J.E. Crawford, G. Thekkadath, from Foster Radiation Laboratory, McGill University, Montréal (Québec, Canada).

The radioactive isotopes provided by "isotope separator on line" facilities are generally available in the form of highly accelerated ion beams (40 to 60 keV), which makes their utilisation not so easy. In the new experiment that we have performed on gold isotopes, the radioactive ions are implanted onto a graphite foil placed on a rotating cylinder (see figure 5). The cylinder is then rotated in order to place the active region under the impact of a frequency doubled Nd-YAG pulsed laser which evaporates the deposited Au atoms ; the atoms propagate through an active region in which they are selectively excited and then photoionized according to the spectral scheme given in figure 5.

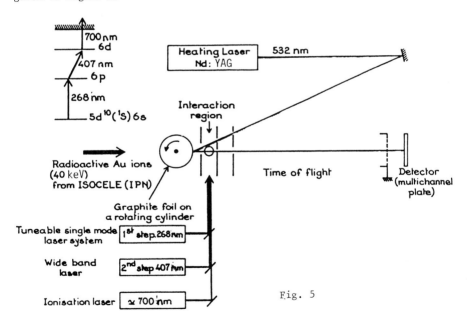

Fig. 5

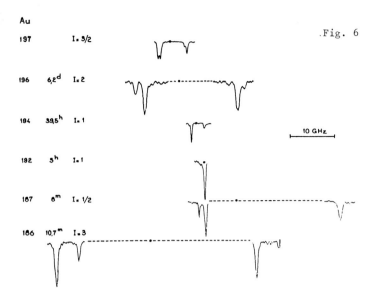

Then the produced ions are mass selected with a time-of-flight mass spectrometer and detected by a multichannel plate. It has been possible this way to study accurately both hyperfine structures and isotope shifts of the resonance line $5d^{10}6s - 5d^{10}6p$ of gold isotopes /8/ for mass numbers : 186, 187, 192, 194, 196 and 197, as is summarized in figure 6.

Original measurements are still to be obtained for a few more iso-topes, and the whole results will be exploited the same way as we did for other series of isotopes (like alkali atoms for instance).

Some of the interest of such an experimental arrangement lies in the possibility that it affords to study the descendant isotopes of primary elements such as platinum isotopes which are descendant from gold.

All these experiments are now in progress.

REFERENCES

1. S. Liberman et al. In Laser Spectroscopy IV, Proceedings of the Fourth International Conference, Rottach-Egern,June 1979. ; p. 527, ed. H. Walther and K.W. Rothe, Springer-Verlag, Berlin.

2. C. Thibault et al. ; Phys. Rev. C23, 2720 (1981).

3. N. Bendali et al. ; C.R.A.S. t. 299, série II, n° 17, 1157 (1984).

4. F. Touchard et al. In Atomic Masses and Fundamental Constants 7, ed. O. Klepper, p. 353 (1985).

5. A. Coc et al. ; Phys. Lett., vol. 163B n° 1,2,3,4,66 (1985).

6. M. Carré, J. Lermé, J-L. Vialle ; J. Phys. B. 19, 2853 (1986).

7. E. Arnold et al. ; accepted for publication in the Phys. Rev. Lett.

8. K. Wallmeroth et al. ; Phys. Rev. Lett ; 58, 1516 (1987).

Ultrasensitive Laser Photoionization Spectroscopy of Short-Lived Isotopes and Very Rare Atoms

V.S. Letokhov and V.I. Mishin

Institute of Spectroscopy, Acad. Sci. USSR, Troitzk,
SU-142092 Moscow Region, USSR

1. Introduction

In the paper "Future Applications of Selective Laser Photophysics and Photochemistry" presented at the previous conference "Tunable Lasers and Applications" held in Scandinavia (Loen, Norway, 1976), one of us discussed, among other possibilities, prospects of applying laser resonance multistep ionization to detect selectively single nuclei, atoms, molecules, and molecular bonds [1]. Now, more than ten years later, it is only natural to demonstrate how much this forecast has proved to be correct, if only for rare atoms and isotopes.

The application of various laser spectroscopy techniques in nuclear physics investigations has been described in a number of reviews [2-4] and considered in the paper [5] read at the present conference. In this paper, we present in short the results of our research on radioactive atoms by the laser resonance multistep photoionization technique described in detail in the recent monograph /6/.

Figure 1 illustrates various ways of using the photoionization technique for the spectroscopy of ultralow amounts of rare isotopic atoms. Most of our experiments with a proton accelerator [7-9] were carried out with a transverse irradiation of a beam of thermal atoms liberated from a hot sample into which fast radioactive ions were implanted after mass separation (Fig. 1a). The high sensitivity of resonance photoionization detection allowed us to measure reliably hyperfine structure (HFS) spectra and isotope shifts (IS) at radioactive ion fluxes at the exit from the mass separator as low as 10^4 - 10^5 ions/s. In experiments with still rarer isotopes, we used the method of resonance ionization of atoms moving inside a hot cavity, followed by the extraction from it and subsequent mass-selective detection of the photoions produced (Fig. 1b) [10]. In this case, a higher probability of atomic ionization is reached, the losses due to the formation of the atomic beam are absent, and, as a result, measurements can be made at a rare atom production rate as low as 10^3 s^{-1}.

Finally, for the purpose of highly selective detection of extremely rare isotopes, use can be made of resonance photoionization of accelerated atoms under a collinear irradiation (Fig. 1c) where the Doppler broadening is eliminated [11] and there occurs

167

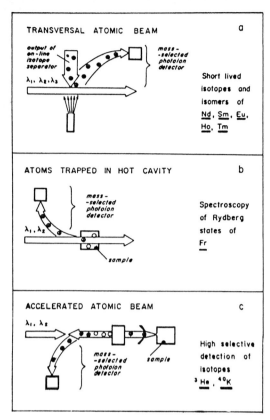

TRANSVERSAL ATOMIC BEAM   a

output of
on-line
isotope
separator

$\lambda_1, \lambda_2, \lambda_3$

mass-
-selected
photoion
detector

Short lived
isotopes and
isomers of
Nd, Sm, Eu,
Ho, Tm

ATOMS TRAPPED IN HOT CAVITY   b

mass-
-selected
photoion
detector

λ_1, λ_2

sample

Spectroscopy
of Rydberg
states of
Fr

ACCELERATED ATOMIC BEAM   c

λ_1, λ_2

mass-
-selected
photoion
detector

sample

High selective
detection of
isotopes
^{3}He, ^{40}K

Fig. 1 Various resonant photoionization techniques of ultra-sensitive spectroscopy of short-lived, rare and long-lived isotopes developed at Institute of Spectroscopy

a kinematic mass isotope shift for all spectral transitions of the atoms [12]. By employing this technique, one can easily achieve a resonance excitation selectivity of 10^5 in a single excitation stage [13] and expect a much higher spectral selectivity with multistep isotope-selective excitation and ionization [12]. This technique combines the high resolution of collinear excitation and the high sensitivity of photoionization detection. For this reason, it is especially promising for investigating radioactive isotopes far from stability [14].

In the present paper, we outline some results of investigations carried out only with thermal radioactive atoms in accordance with the schemes of Figs. 1a and b.

2. Photoionization Spectroscopy of Long Chains of Radioactive Isotopes

Information about nuclei (their charge radii, spins, and electric quadrupole and magnetic dipole moments) for long chains of isotopes is of great interest to nuclear physicists. These nuclear characteristics can be derived from the HFS and IS of optical atomic transitions [2-5]. To measure the HFS and IS of radio-

168

active atoms obtained at the exit from a mass separator operated on-line with the proton accelerator installed at the Leningrad Institute of Nuclear Physics, we used the ultrasensitive resonance atomic photoionization technique reported in [1,6]. Investigations were performed for long chains of isotopes of the rare-earth elements Eu [7,9], Tm [8,9], Sm [8,9], and Ho and Nd [9].

The ions of the isotopes under study were formed in the tantalum target of the mass separator upon its being irradiated with 1-GeV protons. These ions were directed into a hot tantalum crucible placed at the exit from the mass separator (Fig. 2), so that a collimated atomic beam of the given isotopes issued from the crucible in the opposite direction and was crossed at right angles by three matched dye-laser beams. To utilize the laser radiation more effectively, the laser beams were made to transverse the atomic beam several times. The laser frequencies were equal to the frequencies of three successive atomic transitions. As a result of this three-step excitation, the atom was raised to its autoionization state. The optical spectrum was represented in terms of the number of photoions as a function of the laser frequency at the first excitation state. The dye lasers were pumped by copper-vapor lasers. One of the pump lasers was operated as an oscillator and the other two, as amplifiers. Such an arrangement made it possible to obtain a total average power of around 1C W. The laser pulse repetition frequency was 10 kHz.

Using this setup, we managed to measure IS for isotopes of Eu (A = 138-151) and also IS and HFS for isotopes of Sm (A = 138 - 145, 147, 149, 150, 152, and 154), Nd (A = 132 and 134-142), Ho

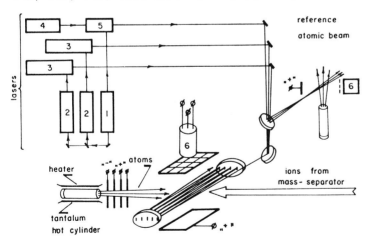

Fig. 2. Laser photoionization "on-line" spectrometer of short-lived isotopes and isomers (1,2-Cu-vapor lasers, 3-pulsed dye laser, 4-CW dye laser, 5-pulsed dye amplifier) [8]

169

(A = 152-165), and Tm (A = 156-172). Of course, optimum schemes
of photoionization through autoionization states were found for
each of these elements. We obtained for the above isotopes the
mean-square charge radii variations $\Delta \langle r^2 \rangle$ and magnetic dipole
and electric quadrupole moments, and verified the spins of some
nuclei [9]. To illustrate, Fig. 3a presents $\Delta \langle r^2 \rangle$ for the isotopes studied, with account being taken of the data reported in
[15] for stable and long-lived isotopes. In the case of Nd, Sm,
and Eu isotope chains, one can clearly observe the shell effect
— different rates of charge radii variations on both sides of
the magic number N = 82. Figure 3b shows $\Delta \langle r^2 \rangle$ for isotopes of
Eu (N > 82), Ho, and Tm. The irregular character of the isotopic
dependence of $\Delta \langle r^2 \rangle$ for Tm isotopes is clearly manifest.

Using the above technique at the LINP-ISAN laser nuclear facility, we measured during the past two years the isotopic mean-
square charge radius variations, spins, and electric quadrupole
and magnetic dipole moments of some 70 radioactive isotopes and
isomers [9].

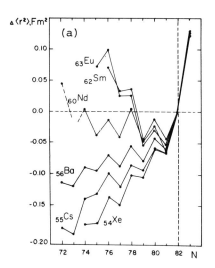 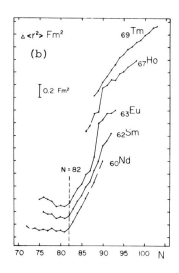

Fig. 3. Variations of mean square charge radius of long isotope chains as a function of neutron number. a) N < 82, Sm,
Eu, Nd [9], Ba, Cs, Xe [15]; b) N > 82 [9]

3. Photoionization Spectroscopy and Separation of Nuclear Isomers

Of great interest is the selective laser photoionization of atoms
with nuclei in the ground and an excited state, which makes it
possible to separate such nuclei [16]. The selective laser photoionization of nuclear Eu isomers was observed in our on-line
experiments aimed at studying IS for a chain of radioactive Eu
isotopes [7, 17]. Experiments are also known on the selective

photoionization of [137]Hg isomers [18]. Recently, in the course of experiments to measure the IS and HFS of atomic lines of Sm and Tm isotope chains, the [141]g,mSm and [164]g,mTm isomers have been separated [8]. The experiments have been conducted using the set-up illustrated in Fig. 2.

The Sm and Tm atoms were photoionized by being raised to their autoionization state in accordance with a three-step scheme. Figure 4 shows the sequence of quantum transitions for the [141]Sm

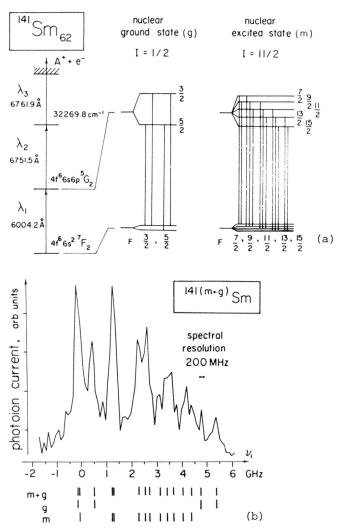

Fig. 4. Laser photoionization spectroscopy and separation of nuclear isomers. a) scheme of three-step resonant photoionization and HFS of stable and isomeric [141]Sm$_{62}$, b) photoionization spectrum as a function of ν_1.

isotope, photoionization spectrum of an isotopic mixture, and the result of its identification. The spectrum features clearly defined lines of [141m]Sm. This means that when the first-step excitation laser frequency was tuned to fall within the range 1-4Ghz, there appeared in a direction perpendicular to the atomic beam a flow of photoions with isomeric nuclei. The photoions were extracted by means of an electric field and deposited onto the cathode of a channel multiplier.

Our experimental setup operating on-line with the proton accelerator makes it possible to investigate and separate nuclear isomers having their life limited by the time it takes them to become released from the target (about a few seconds). Both the technique and setup are also applicable to other sources of isomeric nuclei.

4. Ionization Spectroscopy of the Fr Atom

The high sensitivity offered by the photoionization method makes the latter very promising for the spectroscopy of very rare elements whose spectral properties have, for all practical purposes, not yet been studied [1]. One such element is francium (Fr) the natural abundance of which is extremely low (1 Fr atom is found in 3×10^{18} atoms of natural uranium) and which is formed as a result of the decay of [235]U. The first spectral investigations of Fr were carried out at the ISOLDE facility at CERN with a flux of artificially produced Fr atoms of around 10^8 atoms /s [19]. New information on the spectrum of Fr is presented in the paper [5].

In our experiments, we managed, by using the resonance photoionization technique, to detect Fr atoms [20] and study their Rydberg states [21] at a rate of their formation of about 10^3 atoms/s in a sample containing 10^9 atoms of [225]Ra implanted into a tantalum foil. Such an ultrahigh sensitivity was achieved thanks to the two-step ionization of the Fr atoms inside the hot cavity (Fig. 1b).

The experiment is illustrated schemetically in Fig. 5. The sample in which the Fr atoms formed as a result of the radioactive decay $^{225}Ra \rightarrow \, ^{225}Ac \rightarrow \, ^{221}Fr$ (4.8 min) was placed inside the hot cavity. The quasienclosed cavity had two small holes in its wall to introduce a two-colour laser beam and extract the Fr photoions produced. Moving inside the cavity, the Fr atoms released from the sample could cross the irradiation region many times before leaving through the hole. With the laser pulse repetion frequency being sufficiently high (10^4 Hz) and the cavity geometry chosen correctly [10, 20], practically each released atom would be ionized.

The Fr atoms were ionized through two-step excitation of their Rydberg state $ns^2S_{1/2}$ and nd^2D via the known intermediate state $7p^2P3/2$ [19] which was excited by laser radiation at $\lambda_1 = 718$ nm. The dye lasers employed were pumped by the output pulses from a

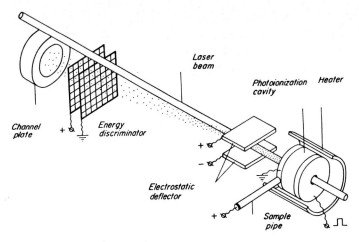

Fig. 5. Experimental scheme of laser photoionization detection and spectroscopy of very rare Fr atoms [20]

Cu-vapor laser (8.7 kHz). The photoions generated were extracted from the cavity through the hole in its wall by the electric field pulses used to ionize the Rydberg atoms and were detected by a channel multiplier. To reduce the background noise signal due to thermal ions, the ions were mass- and energy-analyzed in a time-of-flight mass-spectrometer and an electrostatic analyzer.

As an example, Fig. 6 shows an ion signal obtained when scanning the laser wavelength λ_2 at the second excitation stage. The spectral lines correspond to transitions from the state $7p^2P3/2$ into the states $nd^2D5/2$ and $nd^3D3/2$, the laser bandwidth $\Delta\nu_2 = 1 cm^{-1}$. The measured level quantum defect $\Delta = 3.42 \pm 0.1$. The data obtained yield the ionization limit $I = 32848.25 \pm 0.25$ cm^{-1} which is close to the theoretical value $I = 32841$ cm^{-1} [22].

To our view, the ultrahigh sensitivity of photoionization spectroscopy of atoms trapped in a hot cavity paves the way for systematic investigations into the optical spectra of very rare atoms with the use of low-radioactivity sources.

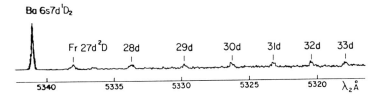

Fig. 6. Resonant ionization spectrum of Fr Rydberg states for $7p^2P_{3/2} \longrightarrow nd^2D$ transitions [21]

In conclusion, the authors express their gratitude to their colleagues from the Institute of Spectroscopy in Troitzk —
S.V.Andreev, S.K.Sekatskii, and V.N.Fedoseev — and from the Institute of Nuclear Physics in Leningrad — G.D.Alkhazov, A.E.
Barzakh, E.E.Berlovich, V.P.Denisov, A.G.Dernyatin, V.S.Ivanov, and I.Ya.Chubukov — for collaboration.

References

1. V.S.Letokhov. in Tunable Lasers and Applications, ed. by A. Mooradian, T.Jaeger and P.Stokseth, Springer Ser.Opt.Sci., vol.3 (Springer, Berlin, Heidelberg 1976) p.122
2. E.-W.Otten. Nuclear Physics, A354, 471 (1982)
3. M.S.Feld. in Lasers in Nuclear Physics, ed. by C.E.Bemis,Jr. and H.K.Carter.Nuclear Science Research Conf.,Ser.,vol.3, (Harwood Academic Publ., Chur , London, New York, 1982)
4. H.-J.Kluge. Hyperfine Interactions, 24-26, 69 (1985), C.Thibault, ibid., 95, R.Neugart, ibid., 159
5. S.Liberman. in Laser Spectroscopy VIII, ed. by S.Svanberg and W.Persson, Springer Ser.Opt.Sci., vol. (Springer, Berlin, Heidelberg, 1987), p.
6. V.S.Letokhov. Laser Photoionization Spectroscopy (Academic Press, Orlando, San Diego, New York, 1987)
7. V.N.Fedoseyev, V.S.Letokhov, V.I.Mishin, A.D.Alkhazov, A.E. Barzakh, V.P.Denisov, A.G.Dernyatin, V.S.Ivanov: Optics Comm., 52, 24 (1984)
8. V.I.Mishin, S.K.Sekatzskii, V.N.Fedoseyev, N.B.Buyanov, V.S. Letokhov, A.E.Barzakh, V.P.Denisov, A.G.Dernyatin, V.S.Ivanov, I.Ya.Chubukov, G.D.Alkhazov. Optics Comm., 61, 383 (1987)
9. G.D.Alkhazov, A.E.Barzakh, N.B.Buyanov, V.P.Denisov, V.S.Ivanov, V.S.Letokhov, V.I.Mishin, S.K.Sekatzkii, V.N.Fedoseyev, I.Ya.Chubukov. Study of charge radii and electromagnetic moments of radioactive atoms Nd, Sm, Eu, Ho, Tm by laser photoionization method. Preprint Leningrad Institute of Nuclear Physics, Gatchina, 1987.
10. S.V.Andreev, V.I.Mishin, V.S.Letokhov. Optics Comm., 57, 317 (1986)
11. K.-R.Anton, S.L.Kaufman, W.Klempt, G.Moruzzi, R.Neugart, E.-W.Otten, B.Schinzler. Phys.Rev.Lett., 40, 646 (1978)
12. Yu.A.Kudriavtzev, V.S.Letokhov. Appl.Phys., B29, 219 (1982)
13. Yu.A.Kudriavtzev, V.S.Letokhov, V.V.Petrunin. Pis'ma Zh.Eksp. Teor.Fiz.(Russ.), 42, 23 (1985)
14. CERN Project, ISOLDE-IS82, "Multiphoton Ionization Detection in Collinear Laser Spectroscopy of ISOLDE Beams", Collaboration "Mainz - CERN - Troitzk"
15. E.-W.Otten. in "International School-Seminar on Heavy-Ion Physics", Dubna, 1983, p.158
16. V.S.Letokhov. Optics Comm., 7, 59 (1973); Zh.Eksp.Teor.Fiz. (Russ.) 64, 1555 (1973)
17. G.D.Alkhazov, A.E.Barzakh, E.E.Berlovich, V.P.Denisov, A.G. Dernyatin, V.S.Ivanov, V.S.Letokhov, V.I.Mishin, V.N.Fedoseev: Pis'ma Zh.Eksp.Teor.Fiz.(Russ.), 40, 95 (1984)

18. R.Dyer. in <u>Resonance Ionization Spectroscopy 1986</u>, ed. by G.S. Hurst and C.G.Morgan, Institute of Physics Conf.Ser.,N84 (Institut. of Phys., Bristol, 1987), p.257
19. S.Liberman, J.Pinard, H.T.Duong, P.Juncar, P.Pillet, J.-L.Vialle, P.Jacquinot, F.Touchard, S.Buttgenbach, C.Thibault, M. de Saint-Simon, R.Klapisch, A.Pesnelle, G.Huber. Phys.Rev., A22, 2732 (1980)
20. S.V.Andreev, V.S.Letokhov, V.I.Mishin. Pis'ma Zh.Eksp.Teor. Fiz. (Russ.), 43, 570 (1986)
21. S.V.Andreev, V.S.Letokhov, V.I.Mishin. Phys.Rev.Lett., (1987)
22. V.A.Dzuba, V.V.Flaubaum, O.P.Sushkov. Phys.Lett., 95A, 230 (1983)

A New Sensitive Technique for Laser Spectroscopic Studies of Radioactive Rare-Gas Isotopes

W. Borchers, E. Arnold, W. Neu, R. Neugart, G. Ulm, and K. Wendt

Institut für Physik, Universität Mainz,
D-6500 Mainz, Fed. Rep. of Germany,
and the ISOLDE Collaboration, CERN, CH-1211 Geneva, Switzerland

The concept of laser ionization has been widely used in spectroscopy studies and for the detection of minute samples of atoms. Being based on ion counting, it avoids the sensitivity problems of conventional fluorescence spectroscopy, which are due to low detection efficiency and large background from scattered laser light. We report the first application of an alternative ionization scheme which we have developed for collinear laser spectroscopy on fast atomic beams /1/. Here the increase in sensitivity has considerably enlarged the range of isotopes very far from stability, for which nuclear moments and radii can be investigated in hyperfine structure and isotope shift measurements.

The scheme is based on a collisional ionization process. It utilizes the large difference in the cross-sections for electron stripping between atoms in the ground state and a high-lying metastable state. Therefore it is particularly suitable for the rare-gas elements. The relevant part of the energy level diagram for the example of radon is shown in Fig. 1.a. The metastable $6p^5 7s[3/2]_2$ state is populated in the charge-transfer neutralization of the initial Rn^+ beam with caesium vapour. Laser excitation to $6p^5 7p[3/2]_2$ at 705.5 nm depopulates the metastable level and pumps the atoms into the low-lying $6p^6$ $^1S_0$ ground state. In passing the beam through a differentially pumped gas target, one can ionize predominantly the metastable atoms and thus detect the optical pumping as a flop-out signal in the ion current.

Our experiment at ISOLDE (CERN) has complemented the earlier optical measurements on $^{202-212, 218-222}Rn$ /2/ and extended them to the extremely neutron-rich isotopes which are interesting in connection with octupole deformation effects found for the neighbouring heavier elements. The half-lives of the investigated isotopes $^{213, 223-226}Rn$ range from 1.78 h to 25 ms. Fig. 1.b shows the signal of ^{226}Rn. Recent measurements on the very neutron-rich xenon isotopes, $^{141-146}Xe$, have led to the discovery of the new isotope ^{146}Xe, and will yield the spins, moments and radii around the neutron number N=90 which in the rare-earth region is established as the borderline of strong deformation.

Here we shall discuss some technical problems inherent in the experiments on very weak beams of radioactive atoms: a) As all fast (60 keV) ions created in the gas target are deflected into the ion detector, one has to avoid isobaric contaminations of the beam from the isotope separator. We have found that beams in the investigated mass ranges are rather clean, if the ISOLDE plasma ion source is well outgassed and all less volatile reaction

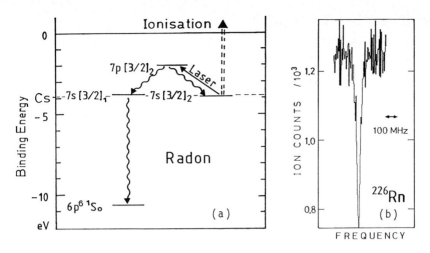

Fig. 1. a) Partial energy level diagram of radon; b) Resonance
signal of ^{226}Rn

products from the 600 MeV proton beam are trapped in a water-
cooled transfer line between the ThC$_2$ target and the ion source.
b) The radioactive species collected on the secondary-electron
ion detector give rise to a considerable background, particularly
for the α-emitters. Therefore it is essential to remove the
radioactivity by using a remote-controlled metallic tape as a
cathode of the secondary electron multiplier. Still, a constant
background of about 10 counts/s is ascribed to the general radioactivity level of the experimental area and should be removed by
more careful shielding.

Considering the sensitivity limits of our technique we can
state that 400 atoms/s from ISOLDE of ^{146}Xe were sufficient to
record a clear signal (S/N = 5) within 35 min. Involved in this
signal to noise ratio is a 50% total transmission of the beam
line and our apparatus, a 50% charge exchange and a 10% reioni-
zation efficiency with a 50% flop-out signal on the ion current.
These numbers refer to the 2x10^{-3} mbar Cl$_2$ target used in most of
our experiments. It is obvious that even without further improvements signals from less than 100 initial atoms/s can be recorded
within reasonable measuring times. This should be compared to the
fluorescence detection limit for rare gases which is about 10^6
atoms/s in the far-red transitions that are accessible from the
metastable state.

This work has been funded by the German Federal Minister of
Research and Technology (BMFT) under contract number 06 MZ 458 I.

1. R. Neugart et al., Nucl. Instr. Meth. B17, 354 (1986)
2. W. Borchers et al., Hyperfine Interactions, in press

Two-Photon Spectroscopy of Atomic Fluorine and Oxygen

W.K. Bischel, D. Bamford, M.J. Dyer, G.C. Herring, and L.E. Jusinski

Chemical Physics Laboratory, SRI International,
333 Ravenswood Ave., Menlo Park, CA 94025, USA

The technique of two-photon excited fluorescence (TPEF) was first demonstrated in 1981 for the light atomic radicals of oxygen [1], nitrogen [1] and hydrogen [2]. Of the light radical atoms, only fluorine has yet to be studied using this technique. We report here the first observation of two-photon excitation of atomic F at 170 nm followed by the detection of near ir fluorescence and ionization. In addition, we report the first Doppler-free spectra of atomic O at 226 nm using a single-frequency laser.

1. Two-photon Spectroscopy of Atomic Fluorine

The basic concept for the TPEF technique is illustrated in Fig. 1a for the fluorine atomic system. Electronic states in the region of 118,000 cm^{-1} are excited using a laser with a wavelength of 170 nm. The F atoms can then be detected by observing visible fluorescence at 776 nm, or by observing ions produced by photoionization of the excited state.

The experimental apparatus developed for these experiments is similar for the study of both atomic F and O [1,3]. For the F atom studies, the most difficult aspect of the experiment is the generation of the 170 nm radiation at sufficient intensity to produce an observable fluorescence signal. Using multiwave Raman mixing in H_2, we have generated over 10 µJ of energy at 170 nm starting with a YAG-pumped-dye laser doubled to the uv (30 mJ at 280 nm). The anti-Stokes (AS) orders are separated using a CaF_2 Pellin-Broca prism, and the 6th AS at 170 nm is focused into the experimental cell using a 5 cm focal length CaF_2 lens. Both the F and O atoms are produced using a microwave discharge. For F atoms, we have verified that over 90% of the F_2 is dissociated in the discharge by observing the decrease in the resonantly-enhanced multiphoton ionization (REMPI) signal from F_2 [4].

An example of the experimental data is given in Fig. 1b where the fluorescence intensity at 776 nm is plotted as a function of vuv laser frequency. We observe four transitions (see Fig. 1a): (1) $^2D_{3/2} \leftarrow {}^2P_{3/2}$ (58,936.5 cm^{-1}), (2) $^2D_{5/2} \leftarrow {}^2P_{3/2}$ (58,811.5 cm^{-1}), (3) $^2D_{3/2} \leftarrow {}^2P_{1/2}$ (58,734.4 cm^{-1}), and (4) $^2D_{5/2} \leftarrow {}^2P_{1/2}$ (58,609.4 cm^{-1}). Transitions 1-3 are clearly above the noise in Fig. 1b, and we have observed transition 4 with higher gain.

This data is unambigious evidence that we are indeed observing atomic fluorine, and it is currently being analyzed to determine the relative two-photon cross sections. Measurements of the radiative lifetimes and quenching rates for the ^2D_J excited states are also in progress.

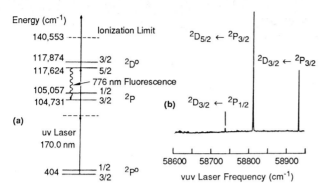

Fig. 1: (a) Fluorine excited states relevant to the two-photon excitation technique. (b) Two-photon excited fluorescence intensity at 776 nm.

2. Doppler-free Spectroscopy of Atomic Oxygen

We have obtained the first Doppler-free spectra of atomic oxygen at 226 nm. Single-frequency radiation is necessary for the Doppler-free experiment [3,5]. The 226 nm radiation is generated by first using the second harmonic of an injection-seeded, single-frequency Nd:YAG laser to pulse amplify a cw-ring dye laser tuned to 574 nm. This radiation is frequency-doubled and mixed with the residual 1064 nm YAG radiation to provide a narrow-band ($\Delta\nu < 100$ MHz) source of 226 nm radiation with pulse energies exceeding 250 μJ. Fluorescence at 845 nm is observed as the ring-dye laser is tuned by 10 GHz. The experimental details are given in Ref. [5].

Figure 2 shows typical Doppler-free data from the $^3P_2$ ground state to the $3p^3P_{2,1,0}$ excited states at approximately 88,631 cm^{-1}. These data allow a comparison with the theoretical relative fine structure two-photon cross sections and give precise relative energy measurements in the $3p^3P_J$ electronic state, while extending Doppler-free spectroscopy to shorter wavelengths than ever before.

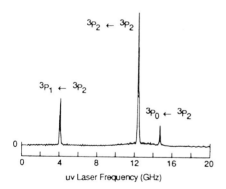

Fig. 2: Doppler-free two-photon excited fluorescence intensity at 845 nm in atomic O as a function of uv laser frequency.

This work has been supported by AFOSR and Sandia National Laboratories.

REFERENCES

1. W. K. Bischel, B. E. Perry, and D. R. Crosley: Chem. Phys. Lett. 82, 85 (1981).
2. J. Bokor, R. R. Freeman, J. C. White, and R. H. Storz: Phys. Rev. A24, 612 (1981).
3. D. J. Bamford, L. E. Jusinski, and W. K. Bischel: Phys. Rev. A34, 185 (1986).
4. W. K. Bischel and L. E. Jusinski: Chem. Phys. Lett. 120, 337 (1985).
5. D. J. Bamford, M. J. Dyer, and W. K. Bischel: "Single-frequency laser measurements of two-photon cross sections and Doppler-free spectra for atomic oxygen," submitted to Phys. Rev. Lett. (1987).

Spectroscopy on Laser-Evaporated Boron and Carbon

H. Bergström, G. Faris*, H. Hallstadius, H. Lundberg, A. Persson, and C.-G. Wahlström

Department of Physics, Lund Institute of Technology,
P.O. Box 118, S-221 00 Lund, Sweden

Laser spectroscopic studies have been performed on neutral boron and carbon. Free atoms and ions were produced by focussing a Q-switched Nd:YAG laser on a solid target of the element to be studied in a vacuum of 10^{-5} torr. The resulting pulse of atoms and ions had a duration of about 10 μs. A second delayed Q-switched Nd:YAG laser pumped one or two tunable dye lasers. These lasers were used to excite the atoms in the evaporated pulse at a distance of about 10 mm from the target. The experimental set-up is shown in Fig. 1.

Using this technique we have studied lifetimes and hyperfine structure. It was found that the rather hostile environment in the plasma did not represent a limitation for accurate determinations of atomic properties. In boron, lifetimes of $2s^2np$ ^2P states were measured. The 3p-level was populated directly by two photon excitation from the ground state. The 3s-level was reached by doubling and Raman shifting in H_2 and this level was then used as the lower level for further excitation to other higher p-levels.

For long lived states (that is $\tau > 30$ ns) the signal was captured by a transient digitizer with signal averaging. Lifetimes were determined by fitting an exponential curve plus background to the decay curve.

Because of the long duration of the exciting dye laser pulses (approx. 10 ns) shorter lifetimes were evaluated by recording the laser pulse and deconvolving the pulse profile from the fluorescence signal. This technique was used for the 3s-state, which had an experimentally measured lifetime of about 4 ns. The magnetic dipole interaction constant, $A_{3/2}$, of the 4p-level was evaluated by Fourier transforming the quantum-beat signal. Theoretical calculations of the radiative properties for this light element were performed using multi configuration Hartree-Fock wavefunctions, and the hyperfine interaction using many-body perturbation theory. Finally, we have demonstrated that this laser evaporation technique could also be used for spectroscopic studies in neutral carbon. The atomic density in the pulse was sufficiently high to allow the 2p3p ^3P-level to be excited by a two-photon transition from the ground state. The determined lifetime and hyperfine structure data is given in table 1. Properties of the plasma such as density and velocity have been investigated using tomographic techniques.

---

*Supported by NSF U.S.-Industrialized Exchange Fellowship.

Table 1: Experimental and theoretical lifetime and hyperfine structure data

			Lifetime [ns]		hfs $A_{3/2}$ [MHz]	
			exp.	theory.	exp.	theory.
Boron	$1s^2 2s^2 3s$	^2S	4(1)	4.3		
"	$1s^2 2s^2 3p$	^2P	52(5)	51		
"	$1s^2 2s^2 4p$	^2P	210(20)	220	7.4(2)	7.3
"	$1s^2 2s\,2p^2$	^2D	26(4)			
Carbon	$1s^2 2s^2 2p3p$	^3P	33(4)			

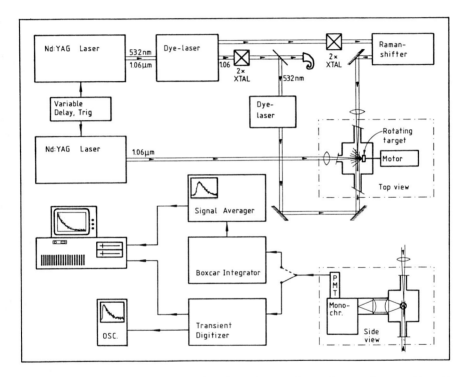

Figure 1: Experimental set-up showing the two laser beams intersecting the plasma. The fluorescence is imaged on the monochromator entrance slit oriented parallel to the expanding plasma to avoid flight out of view effects caused by the rapidly expanding plasma. The signal from the photomultiplier is captured by the transient digitizer and transferred to the computer for signal averaging and evaluation.

Highly Excited Barium Rydberg States in External Fields

H. Rinneberg, J. Neukammer*, A. König, K. Vietzke, H. Hieronymus,
M. Kohl, H.-J. Grabka, and G. Jönsson***

Freie Universität Berlin, Institut für Atom- und Festkörperphysik,
Arnimallee 14, D-1000 Berlin 33, Germany
*Present address: Physikalisch-Technische Bundesanstalt,
  Abbestr. 2–12, D-1000 Berlin 10, Germany
**Present address: Department of Physics, Lund Institute of Technology,
  P.O. Box 118, S-22100 Lund, Sweden

The distribution of oscillator strength of barium Rydberg states in external electric and magnetic fields shows conspicuous modulations. Such modulations are characteristic for non - hydrogenic Rydberg atoms.

We have populated barium Rydberg states in an atomic beam by resonant two - photon absorption ($6s^2\ ^1S_0 \rightarrow 6s\,6p\ ^1P_1 \rightarrow 6s\,ns\ ^1S_0$, $6s\,nd\ ^1D_2$), employing two cw dye lasers. The excitation was monitored by counting Ba^+ ions produced by collisional or field ionization of the Rydberg atoms. The electric and magnetic fields were applied along the atomic beam. The same experimental set - up was used previously to study Stark effects and atomic diamagnetism of barium Rydberg states in the ℓ - mixing region /1/.

In the following we discuss spectra recorded at field strengths suffi- ciently high to cause considerable mixing between Rydberg states with dif- ferent principal quantum numbers (n - mixing region). The Stark spectrum shown in Fig. 1 exhibits conspicuous modulations in the distribution of os- cillator strength. The time constant of the counting electronics was set sufficiently high to record the envelope rather than individual Stark com- ponents. The Stark spectrum (Fig. 1) corresponds to the energy range be- tween the zero field positions of the $6s\,96d\ ^1D_2$ and $6s\,210d\ ^1D_2$ Rydberg states. Initially at $n=96$ the Stark-mixed Rydberg states $6s\,(n+1)p\ ^1P_1$, $6s\,(n-3)f\ ^1F_3$, $6s\,(n+1)s\ ^1S_0$, and $6s\,nd\ ^1D_2$ - given in the order of increasing energy - can be clearly identified (ℓ - mixing region). With increasing principal quantum number the $6s\,(n+1)\,p\ ^1P_1$ and $6s\,(n-3)\,f\ ^1F_3$ states approach each other and finally coincide at $n \simeq 104$. Simultaneously the $6s\,nd\ ^1D_2$ state joins the $(n-3)$ hydrogenic Stark manifold and loses its intensity rapidly. Above $n \simeq 103$ it can no longer be identified. The Stark - mixed $6s\,(n+1)s\ ^1S_0$ and $6s\,(n+1)p\ ^1P_1$ Rydberg states exhibit a similar behaviour. In Fig. 1, the drop in intensity of the $6s\,(n+1)p\ ^1P_1$ states between $n \simeq 110$ and $n \simeq 120$ is evident. The signal corresponding to the Stark - mixed $6s\,(n+1)s\ ^1S_0$ state vanishes at $n \simeq 111$ and a new line appears at $n \simeq 113$. This line gains intensity rapidly, reaches a maximum in oscillator strength at $n \simeq 119$, and disappears at $n \simeq 129$. This pattern is repetitive and exhibits progressively shorter periods. In Fig. 1, as many as ten cycles can be identified.

We explain the observed modulations as being caused by avoided crossings between Rydberg states with non-vanishing quantum defects. Because of their proximity, $6s\,(n+1)p\ ^1P_1$ and $6s\,(n+1)s\ ^1S_0$ barium Rydberg states strongly re- pel each other in an external electric field. With increasing electric field strength - or, equivalently, increasing principal quantum number at constant field strength - these Stark - mixed states approach the $6s\,nd\ ^1D_2$ states and undergo a series of avoided crossings. The exchange of oscillator strength

183

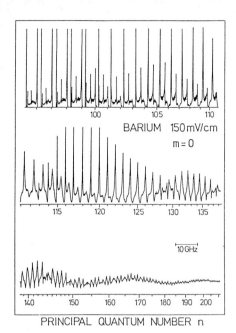

Fig.1 Low resolution Stark spectrum of barium Rydberg states. The quantum numbers refer to $6s\,nd\,^1D_2$ states in zero field

BARIUM 150 mV/cm

m = 0

10 GHz

PRINCIPAL QUANTUM NUMBER n

occurring at avoided crossings leads to the observed periodic modulation in intensity. It should be noted, that the Stark-mixed s, p, and d states are coupled to a quasi-continuum of Stark components with zero quantum defect. Hence the oscillator strength is distributed over many individual Stark components and their envelope is modulated. We have calculated the energies of Stark-mixed $6s\,n\ell\,^1L_\ell$ states ($0 \leq \ell \leq 2$, $120 \leq n \leq 130$) by diagonalization of the Stark Hamiltonian in a truncated spherical basis. Line positions observed experimentally as a function of the external electric field strength coincide with calculated ones for $F \leq 200$ mV/cm and can be followed through consecutive avoided crossings. Because of the high density of states at $n \simeq 120$, a recently published theory /2/ of the Stark effect of non-hydrogenic atoms cannot be applied easily.

Consistent with our explanation of the observed modulations, Stark manifolds of barium Rydberg states with $m = \pm 2$ do not show such intensity variations. Furthermore, even-parity $m = 0$ diamagnetic manifolds of barium exhibit similar modulations of the oscillator strength whereas those with $m = \pm 1$ do not.

References

1. H.Rinneberg, J.Neukammer, G.Jönsson, H.Hieronymus, A.König, and K.Vietzke: Phys. Rev. Lett. 55, 382 (1985)
2. D.A.Harmin: Phys. Rev. A30, 2413 (1984)

CW Laser Spectroscopy of Long-Lived 5dnl (l>2) Autoionizing States in BaI

W. Hogervorst

Subfaculteit Natuur- en Sterrenkunde, Vrije Universiteit,
De Boelelaan 1081, NL-1081 HV Amsterdam, The Netherlands

Autoionizing 5dnf J=4,5, 5dng and 5dnh J=5 series in neutral barium have been studied with CW high resolution laser spectroscopy on a collimated atomic beam. These doubly-excited states are characterized by extremely narrow autoionization linewidths down to a Doppler limited value of 8 MHz. The states are populated in one – photon transitions from the recently discovered metastable $5d^2$ $^1G_4$ state at 24.700 cm^{-1} [1] using a CW ring dye laser operating on Stilbene 3.

Many levels of the $5d_{3/2}nf$ J=4 and 5 (n=16 to 50) and $5d_{5/2}nf$ J=4 and 5 (n=11, 12, 20 to 50) have been excited. As an example in Fig.1 the excitation $5d^2$ $^1G_4 \rightarrow 5d_{3/2}41f$ is shown. A narrow J=5 (linewidth 13 MHz) and two J=4 (with the Doppler limited linewidth of 8 MHz) signals for the most abundant even isotope ^{138}Ba are identified. J=3 signals could not be observed except in the vicinity of perturbing levels. This may be ascribed to small transition probability from the $5d^2$ $^1G_4$ level as well as to an expected line broadening due to configuration interaction with $5d_{3/2}$ $np_{3/2}$ J=3 levels with large autoionization linewidths. Also some of the $5d_{5/2}np_{3/2}$ J=4, $5d_{3/2,5/2}$ nh J=5 and $5d_{3/2,5/2}ng$ states have been excited. Excitation of 5dnh states is possible because of a 5dg admixture in the $5d^2$ $^1G_4$ state. 5dng states may be observed when the atoms are excited in a weak DC electric field. The (quadratic) Stark effect is used to assign the 5dnf states from the field-induced splittings. Ab initio calculations [2] indicate jK coupling for the 5dnf J=5 states, whereas a jj assignment holds for the J=4 series.

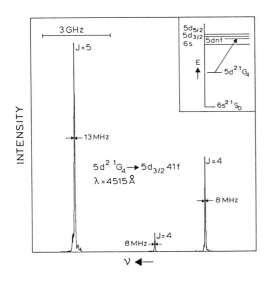

Fig.1 Excitation
$5d^2$ $^1G_4 \rightarrow 5d_{3/2}$ $41f$

Remarkable feature of the $5d_{3/2}nf$ J=4 and 5 series is the overall narrow linewidth Γ. The mean values of the product $(n*)^3$ Γ (HWHM, n* is effective principal quantum number) for these series are in the order of 500 GHz, a factor of 10 to 100 lower than the corresponding values for the $5d_{3/2}nd$ series [3] also ionizing to the 6s ion state. This large difference may be understood qualitatively by noting that in 5dnl series 5dnf is the first with a non-core-penetrating orbit. In the linewidth of the $5d_{3/2}nf$ series as a function of n pronounced stabilization effects occur, which may be ascribed to the interaction with perturbing states of the $5d_{5/2}nf$ and $5d_{5/2}np$ configurations.

Lu-Fano plots of the 5dnf J=4 and 5 series below the $5d_{3/2}$-ionization limit show the interactions between series converging to the $5d_{3/2}$- and $5d_{5/2}$-limits. A remarkable fact is that hardly any interaction between the $5d_{3/2}nf$ [9/2] J=5 and $5d_{5/2}nf$ [9/2] J=5 channels exists, whereas the $5d_{5/2}nf$ [11/2] J=5 series strongly perturbs the $5d_{3/2}nf$ J=5 series. This may be observed directly in spectra above the $5d_{3/2}$-limit. In the autoionization process of the $5d_{5/2}nf$ [9/2] J=5 states only "fast" electrons (~0.5 eV) are detected showing ionization to the 6s-continuum, whereas in the decay of the $5d_{5/2}nf$ [11/2] J=5 states only "slow" electrons (~0.1 eV) from ionization to the $5d_{3/2}$-continuum are observed [4]. Multichannel Quantum Defect analyses of these J=5 series show that a fit to the accurate level energy data (0.006 cm^{-1}) is not possible without inclusion of the 6sϵh continuum channels.

The narrow linewidths in these autoionizing series allow for the observation of isotope selective signals. Using an isotope selective detection scheme with a quadrupole mass filter signals of both odd and even Ba isotopes have been resolved and first observation of hyperfine structure in autoionization states is reported. As an example of isotope selective detection in Fig.2 the excitation $5d^2$ $^1G_4 \rightarrow 5d_{3/2}38f$ J=5 is shown. In Fig.2a the hyperfine spectrum of ^{137}Ba is displayed with the mass filter tuned precisely for the correct mass, whereas in Fig.2b a spectrum with a slightly detuned mass filter is given. The reappearance of the single ^{138}Ba-peak is

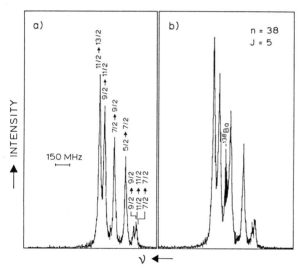

<u>Fig.2</u> Hyperfine structure in the transition $5d^2$ $^1G_4 \rightarrow 5d_{3/2}$ 38f J=5.
a. massfilter tuned for ^{137}Ba-isotope b. massfilter detuned for ^{137}Ba and ^{138}Ba

easily observed. Hyperfine structure and isotope shift data provide important information on the interactions between series converging to the same ionization limit.

References
1. W. Vassen, E.A.J.M. Bente, W. Hogervorst: J. Phys. B.: At. Mol.
   Phys. 20, 2383 (1987)
2. J.E. Hansen: private communication
3. J. Neukammer, H. Rinneberg, G. Jönsson, W.E. Cooke, H.
   Hieronymus, A. König, K. Vietzke, H. Spinger-Bolk: Phys. Rev.
   Lett. 55, 1979 (1985)
4. E.A.J.M. Bente, W. Hogervorst: accepted for publication in
   Phys. Rev. A

Laser Spectroscopy
of Double-Rydberg States of Barium

P. Camus[1], *P. Pillet*[1], *and J. Boulmer*[2]

[1]Laboratoire Aimé Cotton CNRS, Bât. 505,
  Université Paris XI, F-91405 Orsay, France
[2]Institut d'Electronique Fondamentale, Bât. 220,
  Université Paris XI, F-91405 Orsay, France

Highly excited two-electron atoms, just below the double ionization limit, are an interesting example of a three-body quantal system: the ionic core and the two Rydberg electrons. Semiclassical and quantum theory have shown that their properties can be strongly connected with the radial and angular correlations of the electron pair.

The spectroscopy of such double-Rydberg $(Nl, n'l')$ states of barium atoms has been studied experimentally using two-step laser excitation where each Rydberg electron is separately brought to Nl and $n'l'$ states. Excited barium ions, produced after autoionization of the double-Rydberg states, are detected by microwave ionization (for the highest levels) or by laser photoionization (for medium levels) and a time-of-flight mass spectrometer /1/.

Two classes of $(Nl, n'l')$ states can be considered. If $N \sim n'$, the correlation between electrons may become predominant. If $N \ll n'$, the innermost electron sees mainly the ionic core and the outermost electron as a perturbation, while the outermost electron sees the ionic core screened by the innermost electron /2/. For these states, a quantum defect theory could be adequate.

Systematic data on $9d_{5/2}$ and $9d_{3/2}$ $n'l$ series with $l=1$ and 2 and $12 \leqslant n' \leqslant 45$ have been recorded. Figure 1 shows part of the 9d n'd double-Rydberg series. The properties of most of these states have been described correctly in terms of a 2-channel quantum defect theory /2/. Nevertheless, this simple model can not explain the more intense and narrow double-Rydberg resonance shown in figure 2. This effect has been attributed to interference due to fortuitous coincidence in the energy spectra of $9d_{5/2}n'd$ and $9d_{3/2}n''d$ members. A more complete 6-channel QDT taking into account the interference between autoionization paths predicts this stabilization effect /3/ and indeed can fit our experimental data (see figure 2). A more detailed experimental analysis of this interference effect on the autoionization properties of double-Rydberg states, especially concerning the branching ratios for different autoionization paths, would be more informative.

Another interesting feature is shown in figure 1. As n' decreases and becomes closer to $9(n' \leqslant 14)$, autoionization becomes so effective that the observed spectra for $n' \leqslant 14$ are almost a continuum: interactions between electrons are probably important and a MQDT analysis is no longer appropriate. Similar behaviour has been observed for the 26d n'p and 27s n'p series with n' ranging from 39 to 47.

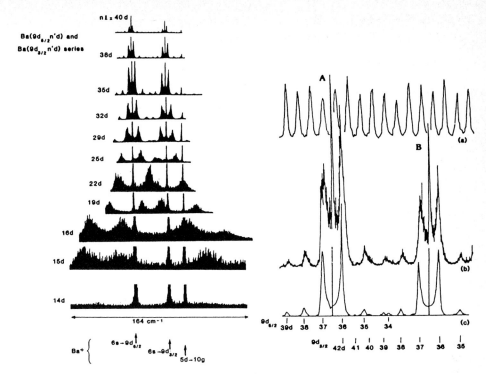

Fig.1 Measured $9d_{5/2}$ n'd and $9d_{3/2}$ n'd series of double-Rydberg states obtained by the two-step laser excitation scheme: $Ba(6s^2) \rightarrow Ba(6s\ nd) \rightarrow Ba(9d\ n'd)$ with $14 \leqslant n \leqslant 40$. The sharp lines correspond to the two-photon Ba^+ transitions indicated in the figure.

Fig.2 Double-Rydberg spectrum obtained by the two-step laser excitation $Ba(6s^2) \rightarrow Ba(6s\ 35d) \rightarrow Ba(9d\ n'd)$ (curve b) compared with the 6-channel QDT (curve c) /3/. Peaks A and B correspond to the $Ba^+(6s) \rightarrow Ba^+(9d_{5/2})$ and $Ba^+(9d_{3/2})$ 2-photon transitions. Curve a shows Fabry-Perot fringes (free space interval: $4\ cm^{-1}$).

References

1. J. Boulmer, P. Camus, J-M. Gagné and P-Pillet, J. Phys. B: At. Mol. Phys. 20, L143 (1987).
2. J. Boulmer, P. Camus and P. Pillet, J. Opt. Soc. Am. B: 4, 805 (1987).
3. J-M. Lecomte, J. Phys. B: At. Mol. Phys., to be puplished.

Some Studies on Barium Autoionization States

L. Xu, Y.-Y. Zhao, G.-Y. Wang, and Z.-Y. Wang

Department of Physics, Fudan University, Shanghai, China

In recent years, studies of the Rydberg states of barium, both bound states and autoionization states, have received great interest. In 1983, M. Aymar [1] gave a six-channel quantum defect analysis of the autoionization Rydberg series of $(5d_{3/2}nd)_2$ and $(5d_{3/2}ns)_2$ which were perturbed by $(5d_{5/2}nd)_2$ and $(5d_{5/2}ns)_2$ levels below the $5d(^2D_{3/2})$ limit. Since a complete MQDT analysis needs at least two more open channels to give the asymmetric line profile, theoretical fitting becomes quite complex due to the large number of parameters which must be fitted to the spectrum. We present a simplified analysis by adding some reasonable approximations.

The experimental spectrum of barium autoionization Rydberg series is obtained by two-photon absorption rather than by a multistep excitation approach. Spectral lines are identified in the $5d_{3/2}nd$ series ($J=0,2$, $n=16$-35) converging to the $5d(^2D_{3/2})$ limit and also some lines of the $5d_{5/2}nd$ series ($J=0$, $n=12$-14) corresponding to the $5d(^2D_{5/2})$ limit.

According to the MQDT [2], we can first neglect the interaction between two series with the same ionization limit, and then reduce the six-channel case to four channels: $5d_{3/2}nd_{5/2}$, $5d_{5/2}nd_{3/2}$, $5d_{5/2}nd_{5/2}$ and $5d_{5/2}ns_{1/2}$. Moreover, we regard $U_{i\alpha}$ as a pure LS-jj coupling transformation matrix in MQDT analysis. This approximation is acceptable here because the two 5d ionization limits lie close to each other (within about 0.1 eV). Thus the fourth channel can be neglected and the three-channel model is considered.

Figure 1 gives the calculated Lu-Fano plot of the three-channel case which shows good agreement with Aymar's six-channel case.

A fourth open channel $6s\varepsilon d_{5/2}$ is introduced as the sink for ionization. The eigen quantum defect of this open channel is fixed at the value obtained in describing the bound spectrum below the 6s ionization limit. The $U_{i\alpha}$ matrix is expressed in terms of Euler angles $V_{\bar{\alpha}\alpha} = R(\theta_{14})R(\theta_{23})R(\theta_{24})R(\theta_{34})$ [3], where θ_{14}, θ_{24} and θ_{34} are set by fitting the width of the autoionization levels and θ_{23} is fixed at the three-channel value. The oscillator strength $D_{\bar{\alpha}}$ in the autoionization region can be calculated. Finally we get the eigen phase shift of the open channel by solving the equations [4]:

$$F(-\tau,\nu_i) = \det |F_{i\alpha}| = 0$$

where

$$F_{i\alpha} = \begin{cases} U_{i\alpha}\sin\pi(\nu_i+\mu_a) & \text{for the closed channel, } i=1,2,3 \\ U_{i\alpha}\sin\pi(-\tau+\mu_a) & \text{for the open channel, } i=4. \end{cases}$$

All the results obtained are shown in the Table, and Figure 2 shows the calculated spectrum.

Table. MQDT parameters

	1	2	3	4
$\lvert i\rangle$	$5d_{3/2}nd_{5/2}$	$5d_{5/2}nd_{3/2}$	$5d_{5/2}nd_{5/2}$	$6s_{1/2}\varepsilon d_{5/2}$
$\lvert\alpha\rangle$	$^3D_2$	$^1D_2$	$^3P_2$	$^1D_2$
μ_a	0.766	0.582	0.522	0.712
	$\theta_{14}=-0.55$	$\theta_{24}=0.2$	$\theta_{34}=0.17$	$\theta_{23}=0.07$
$U_{i\alpha}$	0.756	0.421	0.454	0.217
	0.536	-0.421	-0.454	0.575
	0.094	0.562	-0.847	-0.153
	0.480	0.389	0.064	0.783
$D_{\bar{\alpha}}$	0	1	0	1

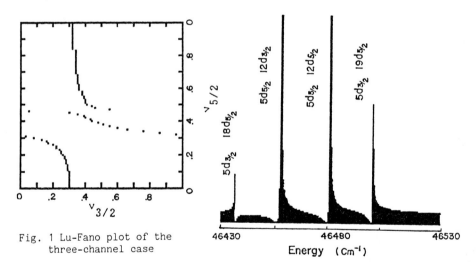

Fig. 1 Lu-Fano plot of the three-channel case

Fig. 2 Autoionization profile calculated with MQDT parameters shown in Table

References

1. M. Aymar et al., Physica Scripta, 27, 183 (1983).
2. K.T. Lu, Phys. Rev. A4, 579 (1971).
3. M. Aymar et al., J. Phys. B12, 531 (1979).
4. C.M. Lee and K.T. Lu, Phys. Rev. A8, 1241 (1973).

Part VI

Molecular Spectroscopy

Impact of Laser Spectroscopy on Chemistry

R.N. Zare

Department of Chemistry, Stanford University, Stanford, CA 94305, USA

It is not true that laser techniques affect all practitioners of chemistry, just most! To gauge some idea of how widespread laser use is in chemistry I turned to my own faculty and conducted a poll. I found that 11 of the 19 active members of the Stanford University Chemistry Department use lasers in their daily research — all physical chemists except one theorist, two of the four inorganic chemists, two of the six organic chemists, and one of the two biochemists. I believe this situation is fairly typical across the U.S. Several conclusions follow: (1) the use of lasers confers, alas, no great distinction among research chemists; and (2) lasers have become an integral part of modern chemistry finding expression in a diverse set of activities but with little application to the synthesis of chemical compounds.

Such an analysis fails to grasp how significant laser use really is in chemical research. A better measure may be obtained by examining the May 1st and the May 15th issues of the *Journal of Chemical Physics*, which is arguably the premier journal for reporting new findings in physical chemistry in the U.S., if not in the world. A superficial analysis of the contents of these articles is quite revealing. Of the 150 articles published during that period, 84 were theoretical/computational, while 66 were experimental. This already was a revelation to me. Physical chemists are fond of telling others how their discipline is an empirical one, grounded in observation, not one characterized by handwaving, speculation, and flights of fancy. However, this statistic may be interpreted in yet another way: the computer is more important than the laser to research in physical chemistry; certainly it is easier to use.

Continuing with this analysis, of the remaining 66 articles which report experimental studies, 32 of them employ lasers to carry out these investigations. This represents 48% of these articles — an amazingly high percentage — which exceeds that of a number of other techniques. Thus, it may be argued that laser techniques are more important to modern day research in physical chemistry than for example, NMR or ESR or the use of X-rays and synchrotron radiation — methods which have unquestionably changed the practice of chemistry. Moreover, these 32 articles involving laser techniques show a wide diversity in interests. For example, a sampling of the laser articles from the May 1st JCP issue includes the determination of zero kinetic energy photoelectrons accompanying the multiphoton ionization of benzene [1], the electronic spectra of weakly bound mercury atom complexes in jet-cooled beams [2], the high resolution energy level structure of the copper monoxide diatomic molecule [3], use of Raman spectroscopy for measuring small frequency differences in isotopic mixtures of liquid benzene species, pyridine species or binary mixtures of benzene and pyridine [4], the use of lasers in dynamic light scattering in order to understand relaxations in polymeric and small molecule liquids [5], the picosecond torsional dynamics of molecules in solution [6], the radiative lifetime and quenching of the 3p $^4D^o$ state of atomic nitrogen using two-photon laser induced fluorescence [7], the mechanisms of laser interaction with metal carbonyls adsorbed on Si(111) 7 x 7 surfaces [8], the rotational dynamics of electronically excited aniline in solution as directly revealed by its time-dependent fluorescence depolarization [9], the nature of the ArHCl potential surface as determined by far infrared laser spectroscopy [10], and the low-lying singlet states of a short polyacetylene oligomer [11]. I conclude that laser spectroscopy is being applied to matter in all phases and under extreme conditions in order to probe both structure and dynamics. Moreover, these applications range all over the electromagnetic spectrum covered by laser sources.

As I surveyed other fields of chemistry, often the use of lasers seems more mundane, such as to take Raman spectra, but the problems being tackled are far from the ordinary or expected. Clearly, the use of lasers pervades chemical research, and this use is primarily as a spectroscopic tool — like NMR or ESR — finding its greatest expression as a means of determining structure, following chemical change, particularly on the most rapid timescale available to chemists, and serving as a tool for chemical analysis.

Now I have a confession to make; the cosmic title for this talk is not my own choosing but instead is what the Organizing Committee suggested. As such, I feel very much like the zoology student who does not know much about elephants but a great deal about ants. When asked to speak to the class on elephants, he began "The elephant is a very large beast who walks ponderously through the jungle and sometimes steps on ants. As for ants ... ..." Therefore it is tempting to concentrate the rest of my talk on specific laser applications to chemistry, particularly those which I know best, i.e., those being carried out in my laboratory. However, let me take the more difficult path and attempt to address the question: What characteristics of laser spectroscopy compared to traditional spectroscopy make it so useful to chemists?

How do I love thee, laser spectroscopy? Let me count the ways [12]. My answer has four parts. First, *the laser source is brighter.* This not only implies the trivial but useful consequence of better signal to noise but more importantly, laser spectroscopy enables chemists to probe and record transient trace species, unstable reaction intermediates, photodissociation fragments, and reaction products under sufficiently reduced pressure conditions so that the product attributes faithfully reflect the results of single-collision processes. In addition, powerful pulsed laser sources open an entirely new dimension in molecular spectroscopy, namely, nonlinear response, such as two-photon absorption, two-photon fluorescence, multiphoton ionization, second harmonic generation, sum and difference frequency mixing, coherent anti-Stokes Raman scattering, etc. These nonlinear laser spectroscopies are vastly extending the chemist's ability to probe molecular structure and dynamics, from what happens at an interface to the hellish environment of a flame.

Second, *the laser source is tunable and highly monochromatic*. These two features work together to make laser spectroscopy the method of choice in many high resolution problems. Indeed, tunable laser sources span an incredible frequency range, which by harmonic generation and sum and difference frequency mixing cover the infrared to the extreme vacuum ultraviolet. The monochromaticity of the laser also implies that its interaction with molecules can be selective, permitting chemists to investigate whether energy in a particular motion or excitation of the molecule influences its reactivity and subsequent collisional and/or dissociative behavior. This offers the hope of carrying out photochemical processes with such control that one can separate isotopes or write fine lines on surfaces. Some of these prospects look to be of commercial interest; others are only of interest in how they elucidate the behavior of matter, such as the nature of highly vibrationally excited states of polyatomic molecules.

Third, *the laser source allows polarization control*. Because radiation- matter interactions are inherently anisotropic, involving for example the electric field of the light beam interacting with the transition electric dipole moment of the molecule, polarized beams of light are ideal for preparing and/or probing anisotropic distributions of molecules and thereby advancing our understanding of directionality, in particular, the stereochemistry of collision dynamics. Often chemists observe only scalar properties, but polarization control permits vector information as well which can afford the extra insight necessary to understand complex chemical phenomena on the microscopic level.

Fourth, *the laser source is temporally and spatially coherent*. The temporal coherence allows the control of laser pulses, permitting chemists to observe rapid change down to femtosecond timing. This offers the opportunity of watching reactions actually occur in real time. The spatial coherence permits lasers to be tightly focused, on the order of the wavelength of the light. This provides in many cases unparalleled sensitivity in laser applications, making laser methodologies one of the most powerful analytic tools chemists have of finding trace quantities of interesting molecules in real matrices. Increasingly we are finding that rational medicine is a form of applied chemistry and that progress in this subject is gated by our ability to follow and analyze the complex chemistry of life processes. Here laser techniques offer a shining hope for pursuing this most fundamental detective work.

Can there ever be too much of a good thing? Probably yes, but in closing let me remark that the trend in applying laser techniques to chemical problems is increasingly to use more than one laser. The reason for this is that laser techniques can work in concert to provide a more detailed specification of the experimental conditions. For example, one laser prepares the system for study while a second laser probes the outcome.

We can expect the impact of laser spectroscopy on chemistry to grow and deepen as new laser sources are tamed and domesticated, giving chemists a greater variety of controlled coherent radiation sources for application to current and future problems. The areas in which useful advances can be made include more brightness, extensions in frequency range and spectral purity, more control over pulse shape and pulse repetition rate, and, of course, improved cost and reliability.

Acknowledgment

Support from the U.S. Office of Naval Research under grant number N00014-87-K-0265 is gratefully acknowledged.

References

1. L. A. Chewter, M. Sander, K. Müller-Dethlefs and E. W. Schlag, J. Chem. Phys. 86, 4737 (1987)

2. K. Fuke, T. Saito, S. Nonose and K. Kaya, J. Chem. Phys. 86, 4745 (1987)

3. M. C. L. Gerry, A. J. Merer, U. Sassenberg and T. C. Steimle, J. Chem. Phys. 86, 4754 (1987)

4. N. Meinander. M. M. Strube, A. N. Johnson and J. Laane, J. Chem. Phys. 86, 4762 (1987)

5. K. L. Ngai, C. H. Wang, G. Fytas, D. L. Plazek and D. J. Plazek, J. Chem. Phys. 86, 4768 (1987)

6. D. Ben-Amotz and C. B. Harris, J. Chem. Phys. 86, 4856 (1987)

7. R. A. Copeland, J. B. Jeffries, A. P. Hickman and D. R. Crosley, J. Chem. Phys. 86, 4876 (1987)

8. N. S. Gluck, Z. Ying, C. E. Bartosch and W. Ho, J. Chem. Phys. 86, 4957 (1987)

9. A. B. Meyers, M. A. Pereira, P. L. Holt and R. M. Hochstrasser, J. Chem. Phys. 86, 5146 (1987)

10. R. L. Robinson, D.-H. Gwo, D. Ray and R. J. Saykally, J. Chem. Phys. 86, 5211 (1987)

11. B. E. Kohler and D. E. Schilke, J. Chem. Phys. 86, 5214 (1987)

12. With apologies to Elizabeth Barrett Browning, "Sonnets from the Portuguese," 1850

Rydberg States of H_2

R.D. Knight[1], *J.E. Sohl*[1], *Yang Zhu*[1], *and Liang-guo Wang*[2]

[1]Department of Physics, Ohio State University,
Columbus, OH 43210, USA
[2]Department of Physics, University of Virginia,
Charlottesville, VA 22901, USA

While Rydberg atoms have been extensively studied for well over a decade, high resolution laser studies of Rydberg molecules have only recently begun. To the extent that an electron moves far from the core, one might expect that Rydberg molecules should differ little from Rydberg atoms. However, the structure of the molecular ion core of a Rydberg molecule is not negligible. Its motion couples strongly to that of the electron via long range potentials, and the core's ro-vibrational energy levels are very closely spaced in comparison with core excitations of atoms. This provides a much higher multiplicity of channel limits than is encountered in Rydberg atoms. Finally, molecules have modes of decay, namely dissociation, not available to atoms. Thus a fundamental question to be addressed experimentally is whether or not Rydberg molecules can be understood in similar terms to their atomic counterparts.

We have begun to study this question in the simplest neutral molecule, H_2. Our choice of molecular hydrogen is based on two overriding concerns. First, it is a simple enough system that one can hope for accurate ab initio calculations. Second, the rotational spacing of H_2 is sufficiently large that only a small number of levels are populated at room temperature. Several groups have recently reported experimental studies of Rydberg states in H_2. These works have mostly reported on autoionizing Rydberg series. Some of them [1-3] have reported the observation of field ionization, but none made a separation of the field ionization spectrum from overlapping spectra due to autoionization or collisional ionization.

We here report on our recent work on the lowest Rydberg series in H_2. We have acquired simple, well-resolved spectra, to principle quantum number n>50, which can be studied in detail and which will offer opportunities for further studies of fine structure details, field effects, etc. Our procedures have been to measure the pure field ionization spectrum and also to use pure parahydrogen samples. In the case of parahydrogen, we report the observation of forced rotational autoionization in very weak electric fields.

Our experimental technique, single photon spectroscopy on a beam of metastable 2p $c^3\Pi_u$ molecules, has been previously described [4]. A frequency doubled pulsed dye laser ($\lambda \sim 340$ nm) excites triplet nd Rydberg series. Autoionization spectra are acquired by a weak DC field (≈ 10 V/cm) which extracts ions from the beam. Field ionization spectra are taken by pulsing a high voltage field 1 μsec after the laser has fired. This delay allows autoionization products, which are formed simultaneously with the laser pulse, easily to be distinguished, via time-of-flight, from field ionization products.

Although we have recorded extensive autoionization spectra over the laser frequency range 28500 - 30000 cm^{-1}, in this paper we wish to focus on the lowest Rydberg series, those converging to vibrational level v=0 of the mole-

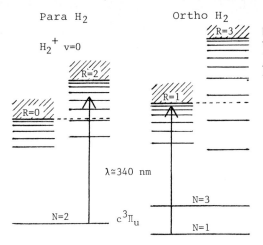

Fig. 1 Energy level diagram of the four lowest nd Rydberg series in H_2, showing observed transitions from the metastable $c^3\Pi_u$ level.

cular ion core. These series are thus energetically unable to undergo vibrational autoionization. The possible decay channels are thus radiative decay, rotational autoionization, or predissociation. Figure 1 shows the first four rotational levels of the v=0 level of H_2^+. The ortho-para interaction is extremely small in H_2, giving rise effectively to two separate species from the same gas bottle. Of the possible lower levels, rotational levels with even N (Hund's case b notation) are parahydrogen while those with odd N are orthohydrogen. As this is a Π electronic state, there is no N=0 level. In our near room temperature beam, only levels N=1-3 have significant population.

To the extent that the long-range interactions between the Rydberg electron and the core can be neglected, the allowed transitions are from lower level N to upper level with R=N. Since there is no lower N=0 level, the lowest two series which can be followed to high n are thus the nd1 series converging to R=1 of orthohydrogen and the nd2 series converging to R=2 of parahydrogen. These are Hund's case d states, and the notation employed is nlR, showing the rotational level R to which the nl series is converging. The nd2 series was studied in a pure parahydrogen sample. This eliminated overlapping series in orthohydrogen and produced a much simplified spectrum. As can be seen from Fig. 1, the upper levels of this series are above the limit of the R=0 series, which is the ionization limit of H_2, and thus can autoionize into the R=0 continuum. Figure 2 shows the autoionization spectrum of parahydrogen. It has an abrupt beginning at n=25, with lower members not seen in autoionization. The linewidths are not significantly different from that of the laser (≈ 1 cm^{-1}), indicating that autoionization rates are not large.

The transition frequencies agree to better than 1 cm^{-1} with those predicted by the long-range interaction model developed explicitly for H_2 by Eyler [5]. However, these calculations indicate that n=25 lies 1.3 cm^{-1} below the ionization limit and should not autoionize. The fact that we observe it in our spectrum is attributed to forced rotational autoionization. As noted above, a weak DC field of ≈ 10 V/cm is used to collect the photoions. While this field is not strong enough to perturb n=25, it is sufficient to lower the R=0 limit by several cm^{-1} and thus to allow n=25 to autoionize. This process, known as forced autoionization, has been extensively studied in atomic systems. We wish here to emphasize that the many closely-spaced channel limits in molecules will make forced autoionization a much more common phenomenon than in atoms and that it will often be found to occur in very weak fields.

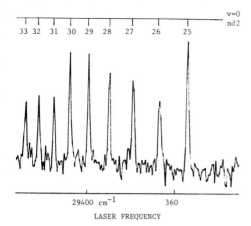

Fig. 2 Autoionization spectrum of the v=0 nd2 Rydberg levels of para-hydrogen. The n=25 level appears due to forced rotational autoionization in a weak electric field.

The upper members of the nd1 series, as seen in Fig. 1, lie above the ionization limit of H_2. However, the ortho-para coupling is so weak that autoionization into the R=0 continuum is negligible and the R=1 level of H_2^+ acts as the effective ionization limit for orthohydrogen. Thus the nd1 series cannot autoionize but can be observed by field ionization. Figure 3 shows the pure field ionization spectrum of the nd1 series for quantum numbers n=23-52. By selectively detecting just the field ionization products, we remove several overlapping autoionization series as well as a sizable photoionization continuum. The resulting spectrum is remarkably clear and resembles an atomic spectrum more than it does a molecular spectrum.

Of particular note are the strong variations in intensity throughout this spectrum, with n=23 and n=28 almost missing. Although not shown on this spectrum, the intensity does recover for n<23. The origin of these variations is still unclear. One possibility is that they arise from perturbations by lower members of the Rydberg series converging to the next higher orthohydrogen limit, R=3. This is illustrated in Fig. 3, where arrows show the positions of n=15-18 of the nd3 series lying within the energy range covered in the spectrum. In particular, the unperturbed positions of 28d1 and 16d3 are calculated to be less than 1 cm^{-1} apart. In the singlet system of H_2, the np0 series is known to exhibit intensity variations as a result of interaction with the np2 series. In that case, the level positions are shifted substantially as a result of the perturbations. Here, however, there appear to be no shifts at all, to within the present measurement accuracy of ≈ 1 cm^{-1}. It would be surprising if a perturbation strong enough to alter the intensities by large amounts did not also produce noticeable level shifts.

Another possibility arises from the recent observation that the singlet v=0 np1 series exhibits much more predissociation than had been expected [3]. The mechanism for dissociation of high Rydberg states is not clear. For the singlet states, repulsive curves calculated to cross near v=1 could be responsible. For triplet states, however, repulsive curves do not cross until near v=5 and are unlikely to be effective for v=0. It was suggested that perturbations from dissociating states of low principle quantum number and high vibration may be responsible. We offer the additional suggestion that predissociation of high Rydberg states may occur by direct coupling to the vibrational continuum of H_2 states dissociating to H(1s)+H(2s). Dissociation of n=4 levels via this route has been previously observed [5]. Although such couplings would be expected to be very weak, we note there is 1 μsec between

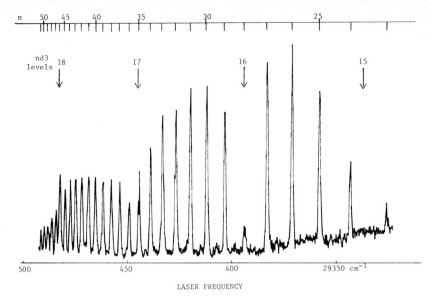

Fig. 3 Field ionization spectrum of the v=0 nd1 Rydberg levels of H_2.

excitation and detection for dissociation to occur. Since none of the peaks are entirely missing, an extremely weak predissociation rate ~10^5 sec^{-1} could produce the observed intensities. The inner walls of the potential curves come close together, so it is possible that the vibrational overlap integrals do not completely cancel. The degree of cancellation would, though, vary with principle quantum number.

Both of these explanations for the intensities will be checked in the near future. A narrower linewidth laser as well as direct probing of the nd3 series will measure the extent of perturbations. Also, a second laser to photo-ionize atomic H(2s) will detect any dissociation products.

This work is supported by National Science Foundation grant PHY-8503424.

References

1. R. Kachru and H. Helm, Phys. Rev. Lett. 55, 1575 (1985).
2. H. Rottke and K. Welge, J. Phys. (Paris) 46, C1-127 (1985).
3. W. L. Glab and J. P. Hessler, Phys. Rev. A 35, 2102 (1987).
4. R. D. Knight and Liang-guo Wang, Phys. Rev. Lett. 55, 1571 (1985).
5. E. E. Eyler and F. M. Pipkin, Phys. Rev. A 27, 2462 (1983).

Laser/Electric Field Dissociation Spectroscopy of Molecular Ions

N. Bjerre and S.R. Keiding

Institute of Physics, University of Aarhus, DK-8000 Aarhus C, Denmark

Detachment of weakly bound electrons in an external DC electric field is a well known phenomenon which has been observed and employed for detection in spectroscopy of Rydberg states and dipole bound states of negative ions. Instead of electron detachment, molecular ions can undergo field induced dissociation into an ion and a neutral atomic (or molecular) fragment. Unless extremely high electric fields are available, field dissociation can only occur for molecular ions in weakly bound states with a large vibrational amplitude. RIVIERE and SWEETMAN [1] observed electric field induced dissociation of H_2^+ in a wide distribution of vibrationally highly excited states.

Using a laser, one can selectively populate a specific level with large vibrational amplitude. For a single level, the field-induced dissociation sets in at a well defined threshold, which is directly related to the po-

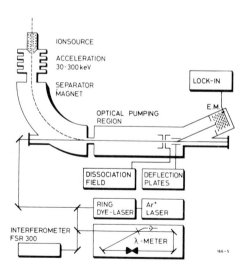

Fig. 1. Experimental setup for laser/electric field dissociation spectroscopy. O_2^+ ions formed in a radiofrequency discharge are accelerated to 120 keV. After mass selection, the ion beam is merged with the beam from a ring dye laser, which excites the molecular ions to states with a large vibrational amplitude. These states are dissociated in the high field gap between a pair of electrodes. The resulting O^+ ions are deflected out of the main beam and detected with an electron multiplier.

tential between the ionic and neutral fragments at distances of typically 10-20 Å. Thus one can directly measure this potential [2], which so far has only been available from calculations.

For the demonstration of the laser/electric field dissociation technique, we chose the ion O_2^+, which is well known from fast beam photofragment spectroscopy [3-4]. The experimental setup is shown in Fig. 1.

Keeping the electric field on a high value and scanning the dye laser, one obtains spectra like the one shown in Fig. 2, which illustrates two mechanisms of laser induced formation of O^+:

1. Excitation to levels above the dissociation limit. This leads to spontaneous dissociation of the molecule with $\tau < 1$ ns due to coupling to the $O + O^+$ continuum (predissociation). The lines from predissociation are homogeneously broadened.

2. Excitation to levels immediately below the dissociation limit. These have a long lifetime and the ions do not dissociate until they enter the high field region. Accordingly, the lines from electric field dissociation exhibit a Doppler limited profile which reflects the velocity spread in the fast beam.

Keeping the laser frequency fixed on resonance with a transition to a level below the dissociation limit and scanning the voltage on the dissociation field electrode, one observes the onset of field dissociation at a well defined threshold as illustrated in Fig. 3. The threshold occurs at higher fields for more strongly bound levels. This is quite useful in assigning

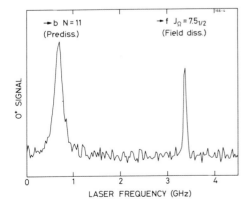

Fig. 2. Single frequency laser scan illustrating the formation of O^+ by the two types of dissociation processes: Predissociation and electric field dissociation. Note that the transition to the predissociated level is homogeneously broadened while the other transition is Doppler broadened.

203

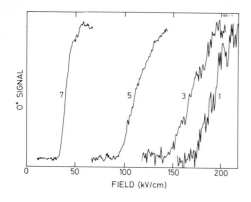

Fig. 3. Laser induced O^+ signal as a function of the electric field in the high field gap. The signal was recorded with the laser on resonance with different states with large vibrational amplitude. The onset of field dissociation occurs at a well-defined threshold field.

the field dissociation spectrum, but more important is that the threshold field provides an experimental determination of the long-range potential between O^+ and O.

Normally, a homonuclear molecule like O_2^+ has a symmetric charge distribution. Electric field induced dissociation can only occur if this symmetry is broken by the external field, so that the field can 'pull' the ionic end of the molecule away from the atom rather than just accelerating the molecule as a whole. The symmetry breaking is best visualized by considering an O_2^+ ion that performs a large amplitude vibration. At small internuclear distances, R, the two atoms will exchange electrons very fast compared to the timescale of the vibrational motion. This is the normal situation in a molecule. At large R, however, the exchange rate becomes very slow, so the molecule - on the timescale of a vibration - is indeed asymmetric and is best described as an O^+ ion bound to an O atom by the 'electrostatic' interaction of its charge with the quadrupole moment and the induced dipole moment of the atom. Such an asymmetric long-range molecular ion will immediately dissociate in an external electric field if the force of that field on the ion exceeds the force from the atom.

To derive an expression for the threshold field required to dissociate the molecule, we assume that the external electric field is parallel to the internuclear axis. Then the external field simply adds a term $-eER/2$ to the center of mass ion-atom potential, $V(R)$, so the total potential, $U(R) = V(R) - eER/2$, develops a maximum at an internuclear distance R_0, which is larger than the distance at the classical outer turning point in the zero field potential. At the threshold field, E_0, the maximum value of $U(R)$ equals the dissociation energy of the excited state, since tunneling of the heavy atomic fragments through the barrier is negligible. Figure 4 illu-

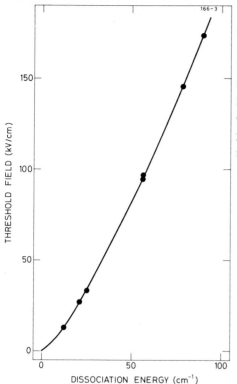

Fig. 4. Comparison of the experimental dissociation fields with those predicted classically for the theoretical potential from ref. [5]. There is only one free parameter in the plot: The energy of the dissociation limit. The experimental points have R_0 values ranging from 9.9 Å to 17.8 Å.

strates how well this simple classical model for the experimental threshold fields agrees with the theoretical long-range ion-atom potential calculated from the ab initio values for the polarizability and quadrupole moment of the oxygen atom [5].

References:

1. A.C. Riviere and D.R. Sweetman: Phys. Rev. Lett. 5, 560 (1960)
2. N. Bjerre and S.R. Keiding: Phys. Rev. Lett. 56, 1459 (1986)
3. J.C. Hansen, J.T. Moseley and P.C. Cosby:
   J. Mol. Spectrosc. 98, 48 (1983)
4. N. Bjerre, T. Andersen, M. Kaivola, and O. Poulsen:
   Phys. Rev. A 31, 167 (1985)
5. W.R. Gentry and C.F. Giese: J. Chem. Phys. 67, 2355 (1977).

Barrier Tunneling in He$_2$ c $^3\Sigma_g^+$

D.C. Lorents[1], S.R. Keiding[2], and N. Bjerre[2]

[1] Chemical Physics Laboratory, SRI International,
Menlo Park, CA 94025, USA

[2] Institute of Physics, University of Aarhus, DK-8000 Aarhus C, Denmark

The He$_2$ molecule exists only in electronically excited states. Many of these states have a potential barrier as illustrated in Fig. 1 for the two states of importance in this study: The metastable a$^3\Sigma_u^+$ state and the c$^3\Sigma_g^+$ state.

The lowest vibrational and rotational levels in the two states are well known from the emission studies of GINTER [1]. The highest level observed in fluorescence is v´ = 4 of the c-state. Semiempirical calculations [2] indicate that the level v´ = 5 is quasibound and decays by tunneling through the barrier, which leads to dissociation of the molecule into two helium atoms, one in the ground state and one in the lowest triplet state. We have studied this tunneling process by the fast beam laser photofragment technique introduced for hydrogen by HELM et al.[3].

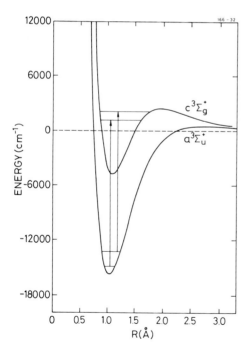

Fig. 1. Potential energy curves for the two states of He$_2$ relevant to the present study. The arrows indicate the transitions to the vibrational levels v´ = 4,5 that are induced by the dye laser. Tunneling through the barrier of the c-state leads to dissociation of the molecule into two He atoms, which are detected. The energies of the quasibound levels define the potential well, while the shape of the barrier can be derived from the tunneling rates.

A mass selected 1.7 keV beam of He_2^+ from a discharge ion source undergoes resonant charge transfer in Cs vapor to form a beam of He_2 in the metastable $a^3\Sigma_u^+$ state. This beam is merged with the beam from a CW dye laser with DCM dye. The dissociation releases a kinetic energy which is sufficient to deflect the atomic fragments out of the molecular beam so they can hit a channeltron detector placed 7 mm off the beam axis at the end of the 1 m long laser/beam interaction region. The typical beam current is 10^8 metastable He_2 molecules per second.

We observe tunneling for the rotational levels $N' = 8$-18 in $v' = 4$ and for $N' = 0$-10 in $v' = 5$ of the c-state. $N' = 8$ and 10 in $v' = 4$ appear with a Doppler limited linewidth of 100 MHz which allows partial resolution of the spin-spin fine structure. The higher rotational levels in $v' = 4$ and all levels in $v' = 5$ are homogeneously broadened because of the tunneling through the potential barrier. A scan recorded with a linear dye laser with 1 cm^{-1} resolution is shown in Fig. 2.

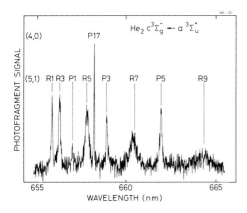

Fig. 2. Scan of the (5,1) c ← a bandhead region of the He_2 photofragment spectrum. The linewidth increases with the rotational quantum number of the upper state and becomes 23 cm^{-1} for the R9 line.

The observed tunneling decay times span four orders of magnitude, ranging from 0.23 ps for $v' = 5$, $N' = 10$ to more than 2 ns for $v' = 4$, $N' = 10$. At present we are analyzing the data to obtain an accurate determination of the He_2 $c^3\Sigma_g^+$ state potential near the top of the barrier.

References:

1. M.L. Ginter: J. Chem. Phys. 42, 561 (1965)
2. R.M. Jordan, H.R. Siddiqui, and P.E. Siska:
   J. Chem. Phys. 84, 6719 (1986)
3. H. Helm, D.P. de Bruijn, and J. Los: Phys. Rev. Lett. 53, 1642 (1984)

Radiative Lifetimes of Xe_2, Kr_2, and Ar_2 Excimers and Dependence on Internuclear Distance

B.P. Stoicheff and A.A. Madej

Department of Physics, University of Toronto,
Toronto, Ontario M5S1A7, Canada

1. EXPERIMENTAL METHOD AND RESULTS

Time-resolved fluorescence studies of the rare gas dimers Xe_2, Kr_2, and Ar_2 have been used to measure radiative lifetimes of the excimer states involved in the operation of electron-beam pumped excimer lasers. Dimers were formed in their weakly bound van der Waals ground states (at vibrational and rotational temperatures <10K) by expansion of high pressure gas through a pulsed supersonic nozzle into a region of 1-10 mTorr pressure. Excitation to selected vibronic levels of the Al_u excimer states (Fig. 1) was achieved by monochromatic and tunable vacuum ultraviolet radiation of ~3 ns pulse duration. This was generated by frequency mixing in Mg, Zn, or Hg vapor [1]. Fluorescence radiation was detected by a solar-blind photomultiplier with signal averaging and storage provided by a programmable transient digitizer. The time response of this detection system was ~2.5 ns.

The Al_u - $X0_g^+$ band systems of Ar_2, Kr_2, and Xe_2 occur at 107, 126, and 150 nm, respectively [2]. Because of the relative positions of the potential energy curves for the strongly bound excimer states and the shallow ground states, (Fig. 1), only high vibronic levels of the Al_u states are accessible from the ground states. Thus only the levels v' = 24-30 for Ar_2, v' = 32-38 for Kr_2, and v' = 36-43 for Xe_2 could be investigated. A typical fluorescence decay curve of the logarithm of intensity versus time is given in Fig. 2 for Ar_2. Such curves clearly revealed single-exponential decays with time for all three excimers, yielding the radiative lifetimes shown in Fig. 3 for each level investigated. Measured lifetimes were found to be independent of pressure over the small pressure range of this study (1-10 mTorr).

It is seen that for Ar_2 and Xe_2 the lifetimes are essentially constant for the high vibrational levels studied leading to average values of τ = 160 ±10 ns and 47±5 ns, respectively. For Kr_2 there is a small but noticeable increase in lifetime of τ = 50 ns at v' = 32 to τ ~68 ns at v' = 38, with an average of τ = 55±5 ns. These results for Ar_2, Kr_2, and Xe_2 differ significantly from values obtained earlier by use of charged particles [3] and synchroton radiation [4] for fluorescence excitation of these excimers formed at relatively high pressures (>100 Torr) in cell experiments. At these pressures, collisional frequencies are sufficiently high that rapid vibrational relaxation occurs, resulting in fluorescence emission from low vibrational levels (v' ~ 0) of the excited states. The values of τ for v' ~ 0 are 2.9(1) μs for Ar_2, 264(5) ns for Kr_2, and 99(2) ns for Xe_2. These results imply reductions in lifetimes by factors of ~20 for Ar_2, 5 for Kr_2, and 2 for Xe_2, in going from v' ~ 0 to v' ~ 20-40 in the Al_u states. While differences of a factor of two in radiative lifetimes for vibrational levels of the same electronic state are not uncommon in molecular spectroscopy, a factor of 20 is unique. Nevertheless such a large difference in lifetimes is not totally unexpected since SCHNEIDER and COHEN [5] had predicted a decrease by a factor

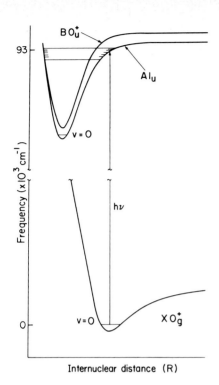

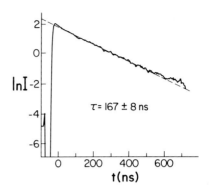

Fig. 1. Schematic diagram of potential energy curves for the lowest excimer states of Ar_2 showing transitions from v=0 of ground state to high vibrational levels of the Al_u state.

Fig. 2. Typical curve of logarithm of observed fluorescence intensity vs time for v'=26 of the Al_u state of Ar_2, illustrating single exponential decay.

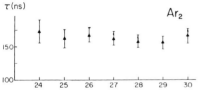

Fig. 3. Graphs of radiative lifetimes measured for high vibrational levels of the Al_u states of Ar_2, Kr_2, Xe_2.

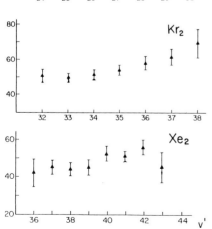

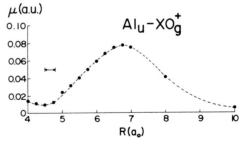

Fig. 4. Computed electronic transition moments $\mu(R)$ for the Al_u - XO_g^+ transition in Ar_2. The classical region of probability for the v'=0 level is indicated at $R\sim4.5a_0$.

209

>10 from v' = 0 to v' = 10 for the Al$_u$ state of Ne$_2$. The present results
for Kr$_2$ and particularly Ar$_2$ provide the only confirmation for such large
variations in lifetime with vibrational levels.

2. THEORY AND DISCUSSION

These large differences in radiative lifetimes arise from changes in the electronic transition moment μ(R) with internuclear distance R. The calculated
dependence of μ(R) on R for electric dipole transitions involving the Al$_u$
and ground (0$_g^+$) states of Ar$_2$ is shown in Fig. 4. In the united-atom and
separated-atom limits, transitions are forbidden, but at intermediate distances, spin-orbit coupling causes transitions to be weakly allowed. The
values of μ(R) given in Fig. 4 were computed on the basis of ab initio calculations by YATES, ERMLER, and WINTER [6] with the addition of the effect of
spin-orbit coupling. It is seen that the derived values of μ(R) increase
rapidly from ~1.0 x 10^{-2} a.u. at 4.5 a$_0$ (corresponding to v' ~ 0) to ~8 x 10^{-2}
a.u. at 6.8 a$_0$ (for v' ~ 30). These values were then used in calculations
of spontaneous emission probabilities and lifetimes from the well-known relation (in atomic units):

$$A_{v'} = \tau^{-1} = \frac{4}{3c^3} \sum_{v''} (\Delta E)^3 |<\phi_{2,v'}|\mu_{12}(R)|\phi_{1,v''}>|^2 .$$

Here ΔE is the energy between two vibronic states(1,v" and 2,v'), $\phi_{1,v''}$ and
$\phi_{2,v'}$ are vibrational wave functions for the ground and excited states, and
μ_{12}(R) is the electronic transition moment connecting the states 1 and 2.

The resulting values of radiative lifetimes for Ar$_2$ are compared with measured values in Fig. 5. There is excellent agreement for high vibrational
levels and an apparent discrepancy of a factor of 3 for v' = 0. However,
calculations have shown that this difference can arise from a 5% increase
in the equilibrium internuclear distance found for the Al$_u$ from our recent
spectroscopic data. This is well within the quoted error of ~10% in R$_e$ caused
by the necessarily long extrapolation of constants from v' = 24-30 to v' = 0
for Ar$_2$. Similar calculations of lifetimes for Kr$_2$ and Xe$_2$ have yielded good
agreement with measured values at high vibrational quantum numbers.

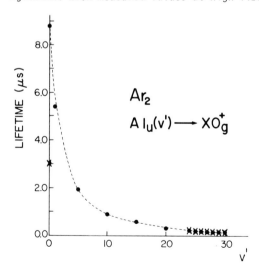

Fig. 5. Comparison of experimental (X) and calculated (●)
values of radiative lifetimes
for vibronic levels of the
lowest Al$_u$ state of Ar$_2$.

In summary, this investigation has revealed surprisingly large changes in radiative lifetimes between low (v' = 0) and high (v' ~ 30) levels of the Al_u states in Ar_2, Kr_2, and Xe_2. Calculations have confirmed the respective life-time ratios of ~20, 5, and 2, and shown that they arise from large variations of the electronic transition moments with internuclear distance. The development of tunable coherent sources in the vacuum ultraviolet has yielded new and precise data on lifetimes and spectroscopic constants of Ar_2, Kr_2, and Xe_2, which in turn may lead to a better understanding of excimer lasers.

This research was supported by the Natural Sciences and Engineering Research Council of Canada, and the University of Toronto.

3. REFERENCES

1. P.R. Herman, P.E. LaRocque, R.H. Lipson, W. Jamroz, and B.P. Stoicheff: Can. J. Phys. 63, 1581 (1985)
2. R.H. Lipson, P.E. LaRocque, and B.P. Stoicheff: J. Chem. Phys. 82, 4470 (1985); P.E. LaRocque, R.H. Lipson, P.R. Herman, and B.P. Stoicheff: J. Chem. Phys. 84, 6627 (1986); B.P. Stoicheff, P.R. Herman, P.E. LaRocque, and R.H. Lipson: In Laser Spectroscopy VII ed. by T.W. Hänsch and Y.R. Shen, (Springer, Berlin, Heidelberg 1985) p.174
3. L. Colli: Phys. Rev. 95, 892 (1954); J.W. Keto, R.E. Gleason, Jr., T.D. Bonifield, G.K. Walters, and F.K. Soley: Chem. Phys. Lett. 42, 125 (1976); P. Millet, A. Birot, H. Brunet, H. Dijols, J. Galy, and Y. Salamero: J. Phys. B15, 2935 (1982)
4. T.D. Bonifield, F.H.K. Rambow, G.K. Walters, M.V. McCusker, D.C. Lorents, and R.A. Gutcheck: Chem. Phys. Lett. 69, 290 (1980); J. Chem. Phys. 72, 2914 (1980)
5. B. Schneider and J.S. Cohen: J. Chem. Phys. 61, 3240 (1974)
6. J.H. Yates, W.C. Ermler, and N.W. Winter, Lawrence Livermore National Laboratory: Informal Report No. UCID-2-224 (1984), unpublished

Laser-Induced Fluorescence of Hg$_2$ and Hg$_3$ Excimers

J.B. Atkinson, L. Krause, R.J. Niefer, and J. Supronowicz

Department of Physics, University of Windsor,
Windsor, Ontario N9B 3P4, Canada

The continuum fluorescence bands of mercury at 335 and 485 nm have been known for over 50 years and it was recently shown that the 335 nm band is due to the D1$_u$ state of Hg$_2$ [1] and the 485 nm band to the A0$_u^\pm$ state of Hg$_3$ [2]. There have also been several more recent reports of absorption and emission bands involving more highly excited states of Hg$_2$ and Hg$_3$ [2,3], which are now being extensively studied in this laboratory using LIF techniques combined with 'pump and probe' methods [4]. The fourth harmonic of a Q-switched Nd:YAG laser (266 nm) is focused into a quartz cell containing pure Hg vapor at densities 10^{17}-10^{19} cm^{-3}, to populate the metastable A0$_g^\pm$ Hg$_2$ states. These were probed after a delay of 0.1 - 1 µs with a second collinear pulse from a dye laser to produce more highly excited Hg$_2$ states; a 5-10 µs delay was required to produce similar Hg$_3$ states. The fluorescence emitted in the decay of the various excited states was resolved with a monochromator and its time-evolution was registered with a transient digitizer. In this way we produced numerous fluorescence and excitation spectra whose analysis yielded accurate values of various molecular constants.

Figure 1 shows traces of H1$_u$ - X0$_g^+$ fluorescence bands (Condon diffraction patterns) emitted in the decay of selectively populated v' states [4]. These patterns enabled us to label unambiguously the various v' states. Figure 2 shows the superposition of two excitation spectra arising from: (a) I0$_u^+$ ← A0$_g^-$ and (b) J1$_u$ ← A0$_g^\pm$ transitions, which were resolved in fluorescence detection. The splitting between the J1$_u$ and I0$_u^+$ states is almost identical with the vibrational spacing in the A0$_g^+$ state; the doubling of each component in the J1$_u$ band gives the A0$_g^+$ - A0$_g^-$ splitting which is absent in the I0$_u^+$ band. The pump and probe experiments also revealed a broad fluorescence continuum at 217 nm, with a corresponding 415-510 nm unstructured absorption band [5]. The assignment of the absorption band to Hg$_3$ was based on the comparison between the time-evolutions of this band and the known Hg$_3$ fluorescence at 485 nm as shown in Fig. 3(d) and (a). The Hg$_2$ absorption and emission bands shown in Fig. 3(b) and (c) clearly exhibit a different time-dependence.

1. R.E. Drullinger, M.M. Hessel, E.W. Smith: J. Chem. Phys. <u>66</u>, 5656 (1977).
2. A.B. Callear, K.L. Lai: Chem. Phys. <u>69</u>, 1 (1982).
3. D.J. Ehrlich, R.M. Osgood: Chem. Phys. Lett. <u>61</u>, 150 (1979).
4. R.J. Niefer, J. Supronowicz, J.B. Atkinson, L. Krause: Phys. Rev. A. <u>34</u>, 1137 (1986).
5. R.J. Niefer, J. Supronowicz, J.B. Atkinson, L. Krause: Phys. Rev. A. <u>34</u>, 2483 (1986).

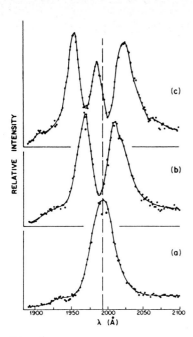

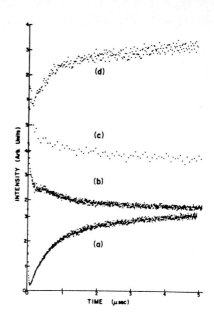

Fig. 1.
$H1_u \rightarrow XO_g^+$ fluorescence bands.
(a) $v' = 0$; (b) $v' = 1$;
(c) $v' = 2$

Fig. 3.
(a) and (b) show time evolutions of the Hg_3 485 nm and Hg_2 335 nm bands, respectively. (c) and (d) show the LIF intensities of the Hg_2 513 nm and Hg_3 217 nm bands, respectively, plotted against the pump-probe delay

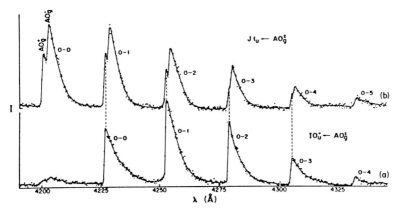

Fig. 2. (a) $IO_u^+ \leftarrow AO_g^+$ excitation spectrum; (b) $J1_u \leftarrow AO_g^{\pm}$ excitation spectrum. The $v' \leftarrow v''$ assignments are shown

Observation of Ba₂ Excimer Structure and Investigation of Ionization Processes of Rydberg Atoms and Molecules

Wu Dong-hong[1], *Yang Yu-fen*[1], *and K.T. Lu*[2]

[1]Institute of Atomic and Molecular Physics, Jilin University,
Changchun, People's Republic of China
[2]Atomic and Plasma Radiation Division, NBS,
Gaithersburg, MD 20899, USA

The thermionic diode is an excellent tool for studying the levels of Rydberg atoms and molecules /1/. In this paper we present not only a new structure of a Ba₂ excimer, but also a method of investigating ionization processes of Rydberg atoms and molecules with the thermionic diode.

Spectra of Ba₂ molecules /2/ and excimers /3/ have been reported during the past two years. In this experiment the band of the Ba₂ excimer was covered, via two-photon excitation, by the wavelength range of the R6G dye laser. We observed, in addition, five lines from a two-photon hybrid resonance and three lines from a two-photon resonance of Ba atoms, as shown in Fig. 1. The band was not complete because the laser energy dropped rapidly at the edges of the laser tuning curve. In the hybrid resonance transition, the population of atoms in the $6s6p$ $^1P_1^{\circ}$ level resulted from one-photon excitation of the ground state of Ba₂ and predissociation in the excited state. A greater line strength of the hybrid resonance implied a larger population in the ground state of the Ba₂ excimer, which was formed instantaneously upon collision.

The experiment was carried out in a thermionic diode which consisted of a cylindrical anode and an axially mounted cathode. The diode was operated at 800°C. Argon, at a pressure of 15 torr, was used as a buffer gas. The peak power, linewidth and pulse width of the laser were, 4 kW, 0.3 cm⁻¹ and 5 ns, respectively.

We have noticed that the peak of the ionization signal from atoms is earlier than that of molecules. The sampling aperture of the boxcar integrator was set at each peak in turn and the different spectral structures were observed by scanning the laser wavelength (Fig. 2). The differences were believed to result from the different ionization processes of Rydberg atoms and molecules /4/. According to studies on the amplification mechanism of a thermionic diode /5/, the production of ions

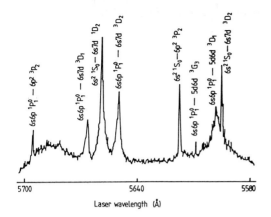

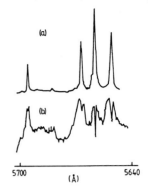

Fig. 1 Spectrum of Ba$_2$ and Ba I
via two-photon excitation.

Fig. 2 Part of spectra obtained
with different sampling
positions of the boxcar.
(a) at the atom signal peak,
(b) at the molecular signal peak.

causes an immediate increase in the diode current. However, the
amplification gain can only be maximal if most of the ions reach the nearest
part of the cathode. The time from the initial creation of ions to the
realization of maximum gain corresponds to the rise time of the ionization
signal. Among the many possible ionization processes of Rydberg atoms and
molecules, photoionization takes place only within the laser pulse width,
while other ionization processes usually take place within the lifetime of
the Rydberg atom or molecule. On the basis of the analyses of the spectrum
under the given experimental conditions (temperature, pressure of the
foreign gas, intensity of the laser, binding energy of the excited electron,
etc.) we concluded that the main ionization process of the Rydberg atoms of
Ba I was photoionization and that of the Rydberg molecules of Ba$_2$ was pre-
ionization.

References
1. K. Niemax: Appl. Phys. B38, 147 (1985).
2. R.M. Clements and R.F. Barrow: J. Chem. Soc. Faraday Trans. 81, 625
   (1985).
3. Wu Dong-hong (T.H. Wu) and K.T. Lu: ICAP-X Abstracts, I-83, ed. by H.
   Narumi and I. Shimamura, (Tokyo, Japan, 1986).
4. Wu Dong-hong, Yang Yu-fen and K.T. Lu: to be published.
5. Wu Dong-hong and Zhang Zai-xuan: to be published.

High Resolution UV Studies of Free Radicals

J.J. ter Meulen, G. Meijer, W. Ubachs, and A. Dymanus

Fysisch Laboratorium, Katholieke Universiteit Nijmegen,
Toernooiveld, NL-6525 ED Nijmegen, The Netherlands

The production of intense beams of free radicals has enabled us to perform high resolution spectroscopy on excited electronic states of these molecules by means of LIF detection. With narrow band UV radiation fine and hyperfine structures can be resolved yielding direct information about the electronic distribution inside the molecule. The resolution may be limited by a short lifetime of the excited state. From linewidth measurements rotationally dependent values for the natural lifetime can then be obtained.

The free radicals are produced either in a reaction (e.g. $H + NO_2 \rightarrow OH$, $C_2H_2 + O + H \rightarrow CH$) or in a microwave discharge (e.g. $NH_3 \rightarrow NH$) in front of the molecular beam source. For the LIF detection at about 15 cm from the source single frequency UV radiation between 296 and 337 nm is generated by frequency doubling in a $LiIO_3$ crystal inside the cavity of a ring dye laser.

In previous studies the diatomic hydrides $OH (A^2\Sigma^+)$ [1], CH $(A^2\Delta, C^2\Sigma^+)$ [2,3], NH $(X^3\Sigma^-, A^3\Pi, a^1\Delta, c^1\Pi)$ [4,5] and SH $(A^2\Sigma^+)$ [6] have been investigated. Large hyperfine splittings (100-1000 MHz) have been observed in the excited states, typically an order of magnitude larger than in the ground electronic states. This is mainly due to the Fermi contact interaction caused by the spin density of the open shell electrons at the site of the proton. When also the second nucleus has a non-zero spin complicated hyperfine splittings are observed as illustrated in Fig. 1 for the ^{17}OH radical. Recently Kristiansen and Veseth [7] have calculated magnetic hyperfine constants for OH, CH and NH in the excited states and obtained excellent agreement with our results.

In addition to the hydrides we now have studied the SiCl radical produced by a discharge in a mixture of Ar and $SiCl_4$. Measurements were performed on the $B^2\Sigma^+, v'=0 \leftarrow X^2\Pi, v''=1$ transition at 296-299 nm, yielding the rotational and fine structure in the ground and excited electronic state [8]. Contrary to the hydrides the hyperfine structure in the excited $B^2\Sigma^+$ state is much smaller than in the ground state which indicates that the open shell electron spin distribution is centered mainly on the silicon nucleus.

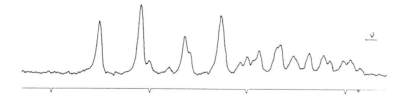

Fig. 1: Hyperfine structure in the $A^2\Sigma^+ \leftarrow X^2\Pi_{1/2}$, $^QP_{21}(2)$ transition of ^{17}OH. Distance between the markers is 598.8 MHz.

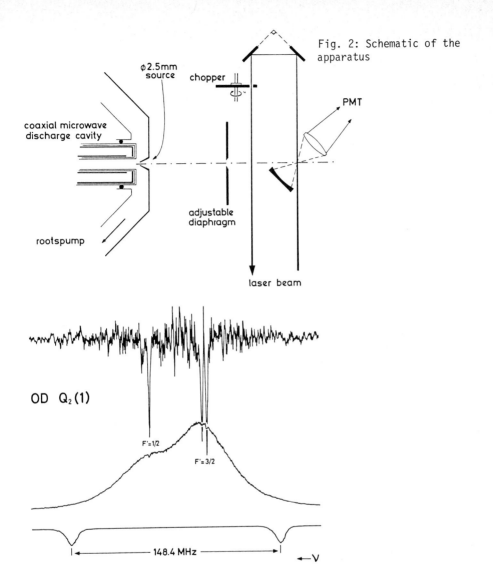

Fig. 2: Schematic of the apparatus

φ2.5mm source

chopper

coaxial microwave discharge cavity

PMT

rootspump

adjustable diaphragm

laser beam

OD Q₂(1)

F'=1/2

F'=3/2

148.4 MHz

V

Fig. 3: Lamb-dips of the $Q_2(1)$ transition of OD

The resolution in molecular beam LIF experiments is normally limited by the divergence of the molecular beam. Reduction of the linewidth by narrowing the beam can be achieved at the cost of sensitivity. Typical minimal linewidths obtainable are 20 MHz. In order to overcome this problem we have applied a detection technique in which the beam molecules are pumped and probed in separate regions by two counter propagating laser beams perpendicular to the molecular beam axis [9]. The set-up is shown schematically in Fig. 2. By modulating the pump beam a Lamb-dip against a flat zero line is observed in case of resonance as the result of absorption by molecules with velocities parallel to the axis. The method is tested on the $Q_2(1)$ transition ($A^2\Sigma^+_{1/2}$, $J'=1/2 \leftarrow X^2\Pi_{1/2}$, $J''=1/2$) of OD at 306 nm. The result is shown in Fig. 3. The Lamb-dips are visible in the directly observed LIF signal (middle trace in Fig. 3) but are much more pronounced with phase sensitive detection (upper trace). The two transitions to the $F'=3/2$ state are separated by 3.4 MHz, the hyperfine splitting in the $^2\Pi_{1/2}$, $J''=1/2$ state. The residual linewidth of 1.5 MHz is mainly determined by the divergence and frequency jitter of the laser.

In previous studies we have determined natural lifetimes τ of SH ($A^2\Sigma^+$) [6] and CH ($C^2\Sigma^+$) [2] from linewidth measurements. With the present Lamb-dip detection technique we can determine longer lifetimes; for SiCl ($B^2\Sigma^+$) we have measured $\tau = 9 \pm 2$ ns. We conclude that this experimental method can be used for the determination of lifetimes between $\tau = 0.1$ and 20 ns.

References

1. J.J. ter Meulen, W. Ubachs and A. Dymanus, Chem.Phys.Lett. **129** (1986) 533
2. W. Ubachs, W.M. v. Herpen, J.J. ter Meulen and A. Dymanus, J.Chem.Phys. **84** (1986) 6575
3. W. Ubachs, G. Meijer, J.J. ter Meulen and A. Dymanus, J.Chem.Phys. **84** (1986) 3032
4. W. Ubachs, J.J. ter Meulen and A. Dymanus, Can.J.Phys. **62** (1984) 1374
5. W. Ubachs, G. Meijer, J.J. ter Meulen and A. Dymanus, J.Mol.Spectrosc. **115** (1986) 88
6. W. Ubachs, J.J. ter Meulen and A. Dymanus, Chem.Phys.Lett. **101** (1983) 1
7. P. Kristiansen and L. Veseth, J.Chem.Phys. **84** (1986) 6336
8. G. Meijer, B. Jansen, J.J. ter Meulen and A. Dymanus, Chem.Phys.Lett. **136** (1987) 519
9. G. Meijer, W. Ubachs, J.J. ter Meulen and A. Dymanus, submitted to Chem. Phys.Lett.

High Resolution Diode Laser Spectroscopy of Transient Species

C.B. Dane, D.R. Lander, R.F. Curl, and F.K. Tittel

Rice Quantum Institute and Departments of Chemistry and
Electrical Engineering, Rice University, Houston, TX 77251, USA

An infrared diode laser spectrometer consisting of a tunable diode laser source (Spectra Physics Model SP-5150), an LSI-11/23 microcomputer, a 1 m multipass absorption cell and diagnostic instrumentation to provide frequency markers and reference gas spectra has been constructed in order to study the spectroscopy and kinetics of chemically reactive free radical species. For spectroscopy, these species are produced in the absorption cell from a suitable flowing precursor either by a DC discharge or by excimer laser flash photolysis, while for kinetics studies only the latter method is employed.

The apparatus, which is depicted in Fig. 1, uses a ZnSe beamsplitter to direct 30% of the laser radiation to diagnostic instrumentation consisting of a 3 GHz plane parallel etalon, a 500 MHz semi-confocal etalon, and a reference gas absorption cell. The computer regulates the diode current through a D/A convertor and adjusts the monochromator during frequency scans by means of a stepping motor. Signal outputs from the three diagnostic channels are processed through lock-in amplifiers and acquired through A/D convertors.

Extended continuous high-resolution frequency scans are obtained by making a series of short current scans between each of which the diode laser temperature is manually adjusted by viewing a display of 3 GHz etalon features on an oscilloscope during rapid computer-generated current ramping. The temperature steps are made such that small overlapping regions exist between sequential scans. Software has been developed to linearize each current spectrum with respect to frequency using the 500 MHz markers and to remove scan overlaps using matching of repeated etalon features. The resulting scan is calibrated using reference gas absorption lines by interpolating between the 500 MHz etalon peaks. Linear continuous scans of up to 7 cm^{-1} [1] have been collected in this manner, although the effec-

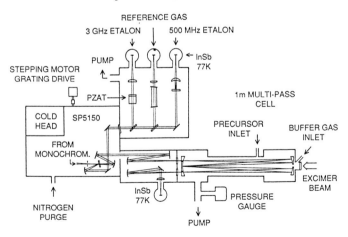

Fig. 1. Schematic representation of the diode laser spectrometer.

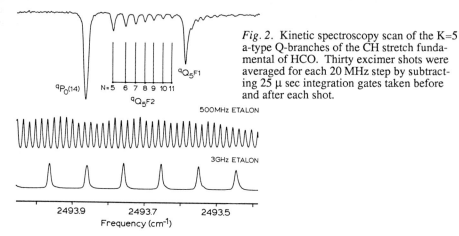

Fig. 2. Kinetic spectroscopy scan of the K=5 a-type Q-branches of the CH stretch fundamental of HCO. Thirty excimer shots were averaged for each 20 MHz step by subtracting 25 μ sec integration gates taken before and after each shot.

$^qP_0(14)$ N=5 6 7 8 9 10 11 $^qQ_5F1$
$^qQ_5F2$

500 MHz ETALON

3 GHz ETALON

2493.9   2493.7   2493.5

Frequency (cm^{-1})

tiveness of the technique is limited as usual by the temperature-current tuning characteristics of the diode laser.

When the DC discharge radical preparation scheme is used, magnetic rotation sensitivity enhancement [2] is employed. The AC magnetic rotation signal from the absorption cell detector-preamplifier is detected using a lock-in and acquired through an A/D convertor. When flash photolysis is used, the signals generated by IR absorptions of short-lived free radicals are acquired with a transient digitizer interfaced to the computer. The spectrum of a transient species is generated by converting the transient digitizer into a dual gated integrator by averaging points for a specified gate length before and after the excimer flash. Considerable reduction in noise arising from the probe laser results upon subtracting the before and after gate averages. The CH stretch fundamental vibration of HCO has been observed using this technique with 308 nm flash photolysis of acetaldehyde, and the spectrum has been fitted to rotational, spin-rotational, and centrifugal distortion constants [3].

Alternatively with flash photolysis, the time dependence of free radical decay can be measured by monitoring an absorption feature at a fixed probe laser frequency. Such decay rates in the presence of varying pressures of added reactant gases yield reaction rate constants, and the growth of absorptions due to product molecules provides information about reaction channels and mechanisms. Such measurements are complicated by the frequency instability inherent to the diode laser's operation. The solution to this problem developed in these studies is to modulate the 3 GHz etalon through a PZT. The laser can thereby be locked to the cavity and tuned into coincidence with the desired absorption line by applying a DC offset to the PZT. This method has a tremendous advantage for flash photolysis studies over methods which frequency modulate the laser. Dither of the laser frequencies is hard to synchronize with the excimer flash and, if not synchronized, has the effect that often the transient trace is started on the side of the absorption feature. When this happens, the frequency and thus the absorption signal is being systematically and undesirably varied during acquisition of a single transient trace. This leads to a very large apparent noise when many traces are averaged to produce a time scan. Using this technique, the chemical kinetics and mechanisms of various reactions involving HCO and C$_2$H are being investigated.

References

1. C.B. Dane, R. Brüggemann, R.F. Curl, J.V.V. Kasper, and F.K. Tittel: *Applied Optics* **26**, 95-98 (1987).
2. G. Litfin, C.R. Pollock, R.F. Curl, and F.K. Tittel: *J. Chem. Phys.* **72**, 6602-6605 (1980).
3. C.B. Dane, M.I.F. Ochsner, Y. Guo, C.B. Moore, D.R. Lander, R.F. Curl, and F.K. Tittel (to be submitted).

High-Resolution Laser Photofragment Spectroscopy of Near-Threshold Resonances in SiH$^+$

P.J. Sarre, J.M. Walmsley, and C.J. Whitham

Department of Chemistry, University of Nottingham,
University Park, Nottingham NG7 2RD, United Kingdom

1. Introduction

There is considerable current interest in the near-threshold dissociation of
diatomic molecules especially when one or both of the atomic fragments possess
electronic angular momentum [1,2]. In this situation nonadiabatic effects
including spin-orbit and Coriolis interactions are expected to play a
prominent role in determining the near-threshold energy level (resonance)
structure and in the dissociation dynamics. It has been predicted that photodissociation spectra of molecules such as CH$^+$ [1,2] should show evidence for
these nonadiabatic interactions, including the appearance of 'multichannel
resonances'. We report here the first experimental evidence for multichannel
resonances in photofragment spectra of SiH$^+$ [3]. The spectra recorded arise
almost totally from transitions to Feshbach resonances, which may be described
alternatively as predissociated energy levels which lie between the two fine
structure dissociation limits, Si$^+$($^2P_{3/2}$) + H and Si$^+$($^2P_{1/2}$) + H.

2. Experimental

A fast beam of SiH$^+$ ions is generated by 70 eV electron impact ionisation and
fragmentation of silane in a conventional mass spectrometer ion source, acceleration of the ions to an energy of 3.5 keV and mass separation by a small
electromagnet. The beam is irradiated coaxially over a 50 cm path length with
a single mode dye laser which induces electronic transitions from the X$^1\Sigma^+$
state to near-threshold predissociated levels (resonances). Predissociation
results in the formation of Si$^+$ ions which are separated from the parent beam
with an electrostatic analyser and are detected with an electron multiplier. A
laser photofragment spectrum is recorded by scanning the laser wavelength and
recording fragment ions. The translational energy released in the dissociation
can be measured by setting the laser wavelength to a molecular absorption line
and recording an energy scan of the photoproducts with the electrostatic
energy analyser. The energy release profiles carry information on the change
in rotational quantum number on excitation; for a Q line the profile has a
doublet shape (D) and for P and R lines a single peak (S) is recorded.
Further experimental details are given in refs. [3,4,5].

3. Results and Discussion

The photofragment spectrum of SiH$^+$ has been recorded between 15 600 cm^{-1} and
18 750 cm^{-1} and over seventy rotational lines were observed. The transitions
originate in v" = 3 and 4 of the X$^1\Sigma^+$ state and involve excitation to levels
which can be identified principally with the A$^1\Pi$ state. As the spectrum contains only a subgroup of all possible A - X transitions, involving just those
levels lying immediately above the dissociation threshold, the normal branch
structure of a $^1\Pi$ - $^1\Sigma^+$ transition does not appear. In order to assign the

Table 1   Linewidths, L [cm^{-1}], proton hyperfine splittings, Hfs [MHz], energy releases W [meV] and angular distributions (S/D) for some of the near-threshold resonances in SiH$^+$

v'	J'	Parity	L	Hfs	W(expt)	W(calc)	|ΔJ|	S/D
1	17	−	0.018	−	11.8(1)	11.4	1	S
1	17	+	0.002	150(10)	12.4(1)	11.4	0	D
T	17	−	0.0015	270(10)	6.1(1.5)	5.7	1	S
1	18	+	0.5	−	23.2(2.0)	23.7	1	S
1	18	−	3.0	−	26.7(3.0)	23.7	0	−
1	19	−	2.5	−	38.7(17.0)	36.0	1	−
1	19	+	4.5	−	−	−	0	−

spectrum it has been necessary to combine measurements of the transition frequencies, linewidths, hyperfine splittings, translational energy releases and fragment angular distributions with predictions of the line frequencies based on numerical solutions of the one-dimensional Schrödinger equation and intensities calculated using an ab inito r-dependent transition moment.

The resonances of table 1 are labelled according to the vibrational and rotational quantum numbers for the A$^1\Pi$ state (v', J') with the exception of the resonance 'T' which has triplet state parentage. The data show a wide variation in widths due to lifetime broadening and some lines exhibit proton nuclear hyperfine splittings. Extra lines involving excitation to 'multi-channel resonances' [1,2] appear, of which the resonance 'T' is one example. The extra lines arise because spin-orbit and Coriolis interactions couple the electronic states strongly in the region near to dissociation. It is notable that the '+' parity component of v' = 1, J' = 17 has a nuclear hyperfine splitting of 150 MHz even though the level is nominally derived from the A$^1\Pi$ state. This is further evidence that states are substantially mixed near to the dissociation asymptotes. Given the erratic couplings as illustrated by the hyperfine splittings and linewidths listed in table 1, it is unlikely that a quantitative account for these data can be provided by an approach based on perturbation theory. Consequently, a fully coupled calculation of the resonance positions and widths is needed and this is currently being undertaken.

References

1. S.J. Singer, K.F. Freed and Y.B. Band: Adv. Chem. Phys. LXI, 1, (1985)
2. C.J. Williams and K.F. Freed: J. Chem. Phys. 85, 2699 (1986),
   Chem. Phys. Lett. 127, 360 (1986)
3. P.J. Sarre, J.M. Walmsley and C.J. Whitham: Phil. Trans. Roy. Soc. (in the press)
4. C.P. Edwards, C.S. Maclean and P.J. Sarre: Molec. Phys. 52, 1453, (1984)
5. P.J. Sarre, J.M. Walmsley and C.J. Whitham: J. Chem. Soc. Farad. Discuss. (in the press)

High-Resolution Zero Kinetic Energy Photoelectron Spectroscopy of Nitric Oxide

W. Habenicht, R. Baumann, and K. Müller-Dethlefs

Institut für Physikalische und Theoretische Chemie der
Technischen Universität München, Lichtenbergstraße 4,
D-8046 Garching, Fed. Rep. of Germany

Conventional photoelectron spectroscopy (PES) which is based on a kinetic energy measurement of the ejected photoelectron is, compared to molecular spectroscopy, a rather low resolution method. In spite of considerable effort, all PES analysers seem to approach an almost invariant resolution limit of some 10 meV ($80 \ cm^{-1}$). This, in nearly all cases, is not sufficient to resolve rotational levels of the ion and hence has precluded the use of PES as a general tool to study rotationally selective photoionization dynamics.

To this end, we have developed the novel method of "Zero Kinetic Energy Photoelectron Spectroscopy (ZEKE-PES)" [1] which allows for a photoelectron energy resolution of better than 1 cm^{-1}.
The method is based on the principle of photoionization under field free conditions with application of a delayed pulsed electric field for the detection of zero kinetic energy photoelectrons only and can be used with laser (two-colour resonant MPI) or VUV synchroton radiation sources. Full details of the method have already been published [1,3]. Recently we have successfully used ZEKE-PES to study rotationally selective ionizing transitions in NO [2] and benzene [3].

In the previously reported 1+1 photon ZEKE-PE experiment of NO [3] we used P_1, $A^2 \Sigma^+$ ($v = 0$, N_A) $\leftarrow X^2 \pi(v=0,J)$ transitions to populate different rovibronic levels of the NO A state. The ZEKE-PE spectra from these levels, with $N_A=0,1,2,3$ ($J_A=N_A+1/2$) and definite parity $(-1)^N$, were obtained by tuning the second laser around the ionization threshold. We observed fully separated ionizing transitions into the lowest rotational levels of the $NO^+ \ X \ ^1\Sigma^+$ ($v^+=0,N^+$) state. The ZEKE-PE signal dependence of the observed $\Delta N^+=N^+-N_A \neq 0$ transitions on initial quantum number N_A was explained by the interaction between the ejected electron and the ion's quadrupole and dipole moment [2].

In order to further test this model and to investigate the dependence of the ZEKE-PE spectra on the alignment of the intermediate A state, we carried out further ZEKE-PE measurements by using Q_1, R_1 and S_{21} A←X transitions. Some results are shown in figs. 1 to 4. We observe that the ZEKE-PE spectra for $N_A=2$ (figs. 1 to 3) do not depend strongly on the A←X transition used to populate $N_A=2$. However, for $N_A=4$, populated by the A←X S_{21} transition, we observe stronger $\Delta N=N^+-N_A \neq 0$ transitions, compared to the $\Delta N^+=0$ transition, as would be expected

by our model /2/. This might be an indication that the alignment of the intermediate state is important for the interpretation of ZEKE-PE spectra.

ZEKE-PE measurements of several other molecules, also using
synchrotron radiation are on the way and will be reported in
the near future.

This research was funded by BMFT grant No. 05366 FAI.

/1/   K.Müller-Dethlefs, M.Sander and E.W.Schlag, Z.Naturf. _39a_,
       1089 (1984)
/2/   M.Sander, L.A.Chewter, K.Müller-Dethlefs, subm. to Phys.
       Rev.Lett.
/3/   L.A.Chewter, M.Sander, K.Müller-Dethlefs and E.W.Schlag,
       J.Chem.Phys. _86_, 4737 (1987)

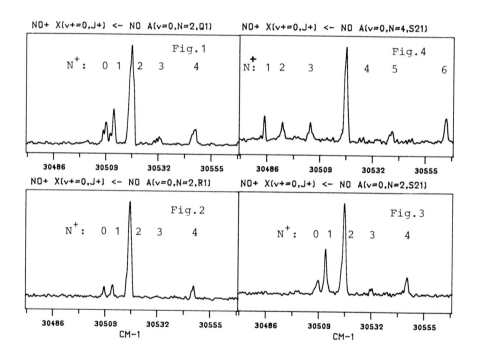

Coherent Laser Spectroscopy

A.M. Prokhorov

General Physics Institute of the USSR Academy of Sciences,
Vavilova St., 38, SU-117942 Moscow, USSR

The recent achievments in the field of quantum electronics and nonlinear optics radically changed the level of our knowledge about interactions of light with substance. They open the ways to a number of new spectroscopic methods based on the physical principles of coherent nonlinear-optical phenomena which apply the tunable frequency lasers as a source of excitation. These methods are more sensitive due to the rapid growth of nonlinear response of a medium with increase of laser radiation intensity. They also have high spectral resolution, limited only by spectral widths of the lasers applied.

The present report deals with a short review of the results of molecular gas investigations based on coherent anti-Stokes Raman Scattering (CARS) spectroscopy which were performed at the General Physics Institute. The investigations of molecular gases with CARS spectroscopy methods were carried out in the following directions :
1. Development of CARS-spectrometers.
2. High-resolution spectroscopy of molecules.
3. Spectroscopy of vibrational and rotational states in nonequilibrium excited molecules.
4. Applications : local nonperturbing diagnostics of gas density and temperature in flows.

Several generations of the continuous and pulsed CARS spectrometers were developed at the Institute. Here a scheme of one spectrometer which combines such merits as high sensitivity and spectral resolution, pulse operation and simultaneous recording of CARS and IR absorption spectra will be presented. This spectrometer incorporates two single-frequency CW dye lasers with line width 10^{-4} cm^{-1} pumped by argon-ion lasers. One of dye lasers is tuned and the frequency of another is fixed. Using two dye lasers makes it possible to study the Raman shifts from 0.01 to 4000 cm^{-1}. To increase the spectrometer sensitivity and to reach the temporal resolution two pulsed dye amplifiers with the peak gain coefficient 10^7 were employed. They are pumped by the second harmonic Nd:YAG laser with pulse duration 10 nsec and repetition rate 10 Hz. The line width of amplified radiation 10^{-3} cm^{-1} is limited by the pulse duration and determines spectral resolution of the spectrometer.

In this spectrometer there was realized a possibility of obtaining the CARS and IR-absorption spectra simultaneously during one frequency scan with the same spectral resolution. For this aim the infrared tunable radiation was obtained on the base of difference frequency generation in the non-linear optical $LiIO_3$ crystal excited by the same pumping lasers. The main advantage of this scheme is a common frequency scale for the CARS and IR-spectra. The operation range for IR-absorption is 5000-1800 cm^{-1}.

For the absolute frequency calibration of IR and CARS spectra we have developed a laser radiation wavemeter-spectrum-analyzer. It is based on four Fiseau type interferometers, and interference pictures are read by a 1024-channel CCD detector. The operation spectral range of the wavemeter was 0.4-1 μm, accuracy of the wavenumber measurement - 0.001 cm^{-1}, the pulsed and CW radiation energy threshold - I μJ. Experimental control, data acquisition and processing as well as wavenumber calculations were performed with the help of a microcomputer.

The capabilities of the CARS channel of this spectrometer are demonstrated by CARS spectra with resolved rotational structure of the totally symmetrical ν_1 band of $^{34}SF_6$ molecules (Fig.1). In these spectra the accuracy of vibrational frequency measurements is 10^{-3} cm^{-1}, and of the relative line shifts measurements - $5*10^{-4}$ cm^{-1}. This makes it possible to measure the value of the isotopic shift of the ν_1 mode of SF_6, $\Delta\nu_1 = 0.0574(5)$ cm^{-1}.

Figure 2 illustrates simultaneous recording of fully resolved CARS spectra of the Q-branch of ν_1 band (upper spectrum) and IR absorbtion spectra of the Q-branch of ν_3 band (lower spectrum) of $^{74}GeH_4$ obtained with the help of IR-CARS spectrometer. From a detailed quantitative analysis of such spectra, taking into account centrifugal distortions and Coriolis interaction, the parameters of effective rotational hamiltonian allowing to describe the frequency positions in CARS an d IR-spectra with accuracy of 0.004 cm^{-1} were calculated.

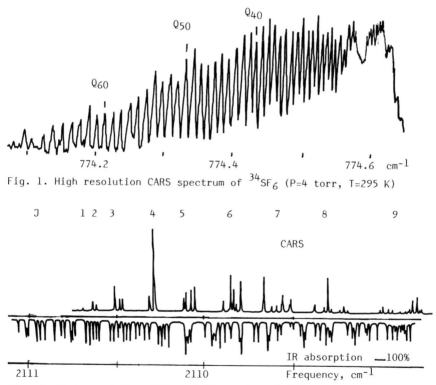

Fig. 1. High resolution CARS spectrum of $^{34}SF_6$ (P=4 torr, T=295 K)

Fig. 2. High resolution CARS and IR spectra of GeH_4.
CARS: P=5 torr, T=295 K.   IR: P=0.4 torr, T=295 K.

226

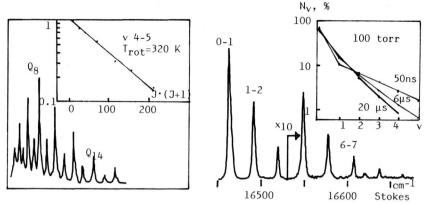

Fig. 3. CARS spectra of N_2 excited in pulsed electric discharge.

One more important application of CARS-spectroscopy is the study of molecular energy distribution over internal degrees of freedom as well as kinetics of relaxation processes. In this report the results of investigation of vibrational relaxation processes in N_2 and SF_6 excited molecules are presented. In the experiment the N_2 molecules were excited either by pulsed electric discharge with duration τ = 40 ns, or by the Raman process using biharmonic laser pumping. Strong vibrational excitation of up to 30% of the molecules on th first vibrational state was created in 10 ns. The rate of the collisional v-v single quantum vibrational population redistribution process was obtained by measuring the density population of the vibrational levels versus delay time between excitation and probing pulses. The CARS spectra of vibrational transitions of nitrogen excited in discharge are presented in Fig.3. Figure 3a displays the spectrum of the Q-branch of V=4 -> V=5 with a resolved rotational structure. From these spectra among rotational (a) and vibrational (b) molecular levels the distribution functions as well as the temperatures inside of these degrees of freedom was obtained. This method makes it possible to obtain the spectra during the discharge pulse and to observe transformation of the distribution functions when the pumping is switched off. Analysing a behaviour of the vibrational level populations in time and comparing with the solution of kinetic equations, derived within the framework of single-quantum resonance exchange, a rate constant of this process K_{01} was calculated to be $(1.0\pm0.4)*10^{-13}$ cm^3/s.

The method of CARS probing was applied also in investigations of vibrational population distribution functions of SF_6 molecules excited in the pulsed IR laser field. The essence of the method is in probing with a pulsed CARS spectrometer of the Raman active ν_1 mode of SF_6 molecules which is excited by a pulsed CO_2 laser with the frequency tuned to the IR-active ν_3 mode. As a result of excitation of molecular vibrational states the additional lines appear in CARS spectra. The frequency shift is determined by the constant of intermode anharmonism and the number of excited level. The amplitudes of lines are proportional to the squares of level populations. Thus, from CARS spectra one can determine both the populations and the mode type of excited levels. The sensitivity of CARS spectrometer makes it possible to record the SF_6 excited molecular spectra at pressures as small as 10^{-2} torr. Using CO_2 pulse laser duration of 30 nsec and minimum delays of probing relative to excitation of ~ 50 nsec makes it possible to detect the relaxational processes with the rate constant as short as 1 nsec*torr.

227

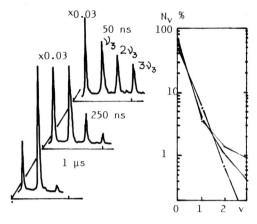

Fig. 4. CARS spectra of IR
laser excited SF_6 and ν_3 mode
vibrational distribution
function. P=0.05 torr, T=150 K.

Analysing the CARS spectra derived under minimum delays of probing we could obtain detailed information about the distribution function of the molecules excited in IR-field and to measure relative number of molecules on the discrete levels of the resonantly excited ν_3 mode at different IR field intensities. We could also determine the mean energy of molecules excited above the vibrational quasicontinuum boundary as well as the position of energetic quasicontinuum boundary of the molecules ,$E_k \approx 6000$ cm^{-1}. Study of the temporal behaviour of CARS-spectrum (Fig. 4a) allowed to investigate distribution kinetics of the population density between vibrational states of the excited mode which is caused by a single-quantum quasiresonance V-V exchange (Fig.4b). Modelling of temporal behaviour of populations allows to determine the constants of ν_3 mode (V-V) exchange process $K_{01}^{10} = 5*10^{-11}$ cm^3/s. The rate constant of interisotopic vibrational exchange between $^{32}SF_6$ and $^{34}SF_6$ molecules which turned out to be equal to $K_{01}^{10}(^{32},^{34}SF_6) = 2*10^{-11}$ cm^3/s. was also measured by this method.

We have also studied the collisional vibrational energy redistribution between the different SF_6 molecular modes and the rate of the process as a result of which the common vibrational temperature is reached in the set of vibrationally-excited molecules. From the viewpoint of explanation of the observed high collisional rates of this process (the rate constants of energy transfer into unexcited modes ν_4 and ν_6 in SF_6 molecules are ~ 100 nsec*torr) the mechanism of highly effective energy exchange between the molecules excited to the quasicontinuum with the unexcited molecules was experimentally established. In this process highly excited molecules with energies larger than quasicontinuum boundary where mode selectivity is lost are able to perform a resonance V-V exchange of quanta of any mode; the unexcited molecules receive these quanta during collision.

Thus, N_2 and SF_6 measurements of V-V exchange constant rates (the slowest for N_2 and the quickest for SF_6 as known from the published literature for V-V exchange processes) have demonstrated that CARS spectroscopy makes it possible to obtain not only qualitative but also quantitative information on the elementary processes of the collisional vibrational relaxation.

CARS Investigation of Relaxation Processes in Highly Excited Molecules

S.S. Alimpiev, A.A. Mokhnatyuk, S.M. Nikiforov, A.M. Prokhorov, B.G. Sartakov, V.V. Smirnov, and V.I. Fabelinskii

General Physics Institute of the USSR Academy of Sciences, Vavilova St., 38, SU-117942 Moscow, USSR

In our earlier works [1] an effective method of investigation of relaxation processes in vibrationally excited molecules has been developed. The method is based on CARS probing of vibrationally excited states, populated by intense IR CO_2 laser field. The measurements of vibrational distribution function (VDF) and of the intramode v-v energy exchange rate in SF_6 molecules have been successfully performed.

The aim of this study was to measure intermode v-v' energy exchange rate in the case when the molecules are excited to high lying vibrational quasicontinuum states. The measurements were performed for SF_6 and SiF_4 molecules. Detailed description of experimental technique may be found in [1].

Figure 1 illustrates temporal transformation of SF_6 (Fig. 1a) and SiF_4 (Fig.1b) CARS spectra due to collisions after the gas has been pumped up to the levels, when more than 30% of SF_6 or 15% of SiF_4 molecules populate quasicontinuum states. The analysis of these spectra and of similar ones, recorded at different IR CO_2 laser energy fluxes, shows that both molecules exhibit similar behaviour, that is:

1. Within 50 ns after the excitation, both the lines, corresponding to transitions from ν_3 mode levels, and the broad red-shifted stuctureless band, attributed to quasicontinuum states, are observed.

2. When the pump-probe delay time is increased to about 1 μs at 0.3 torr pressure, new lines which may be assigned to transitions starting from ν_4 mode levels, appear in the spectra. The VDF of the ν_4 ladder, derived from the spectra appears to be nearly Boltzmann with the temperature dependent upon IR energy flux and delay time.

3. Simultaneously, the bandwidth of the $0 -- \nu_1$ line slightly increases which is due to population of ν_6 (SF_6) and ν_2 (SiF_4) modes' levels. (The anharmonicity constants of these modes are small).

4. The broad maximum, most evident in SF_6 spectra, shifts to higher Raman frequencies when the pump-probe delay is increased.

5. The rate of ν_4 mode temperature change increases when the molecular excitation level is increased, as is illustrated in Fig.1c. The characteristic time necessary for the temperature of the ν_4 mode to reach its quasistationary value changes from 200 ± 50 ns torr at flux F=0.4 J/cm² to 90 ± 40 ns torr at F=1.2 J/cm² in case of SiF_4 and from 700 ± 100 ns torr at F=0.05 J/cm² to 100 ± 30 ns torr at F=0.4 J/cm² in case of SF_6.

The results obtained allow to conclude that the main process, determining the observed high rates of intermode vibrational relaxation, is the resonant collisional energy exchange between the molecules excited to quasicontinuum states and the molecules in the ground and low-lying discrete vibrational states. These cold molecules, can "extract" from the quasicontinuum the quanta of any mode practically without the defect of resonance.

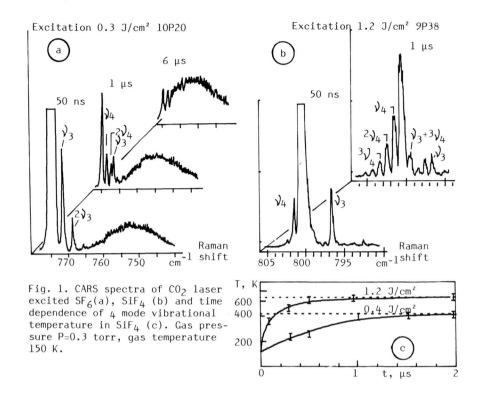

Fig. 1. CARS spectra of CO_2 laser excited SF_6(a), SiF_4 (b) and time dependence of ν_4 mode vibrational temperature in SiF_4 (c). Gas pressure P=0.3 torr, gas temperature 150 K.

In order to confirm this conclusion, the specially arranged experiment has been performed. The 1:10 mixture of $^{32}SF_6$ and $^{34}SF_6$ at 0.1 torr total pressure and 150 K temperature has been illuminated by 10P20 line of the CO_2 laser at flux higher than 1 J/cm². In this case more than 80% of $^{32}SF_6$ molecules were excited to high lying quasicontinuum states, while practically no $^{34}SF_6$ molecules were excited due to red 17 cm^{-1} isotopic shift of the ν_3 band of $^{34}SF_6$, thus forming the cold ensemble residing on ground vibrational state. The population of discrete (low-lying) states of the ν_3 mode was negligible and CARS signal from these states within 50 ns after the excitation lay beyond the sensitivity of our spectrometer. But after 1 μs delay the spectra distinctly exhibit the features corresponding to transitions from low-lying excited vibrational states of the ν_3 and ν_4 modes of SF_6. This experiment unambiguously shows the efficient (about 50 ns torr) vibrational energy exchange between highly excited molecular ensemble of $^{32}SF_6$ molecules and cold ensemble of $^{34}SF_6$ molecules.

It should be also noted that most probably the described mechanism is responsible for the complete redistribution of the absorbed energy among all vibrational modes of highly excited polyatomic molecules, establishing quasistationary vibrational temperature. The rate of this process depends on mode type, but the difference is not drastic. In the case of SF_6, the characteristic time of vibrational energy redistribution derived from the shift of broad maximum (see Fig.1a) is 600 ns torr.

1. Trudy IOFAN. Ed. A.M.Prokhorov, v.2, Moscow, Nauka, 1986.

Four-Wave Processes in Spectroscopy
of Vibrational Molecular States

S.S. Alimpiev, V.S. Nersisjan, S.M. Nikiforov, A.M. Prokhorov,
and B.G. Sartakov

General Physics Institute of the USSR Academy of Sciences,
Vavilova St., 38, SU-117942 Moscow, USSR

Phase conjugation reflection(PCR) of CO_2 laser radiation via degenerate four-wave mixing in the resonance gases has been investigated. This phenomenon is important from the point of view both of spectroscopy and applications. As a method of spectroscopy PCR process is based on the spectral dependence of non-linear medium susceptibility near the frequency of resonance transition. In this process both one- and two-photon resonances contribute to the nonlinear susceptibility [1].

The experiments were performed with the pulsed CO_2 laser radiation in the gases SF_6, CF_3Br, CD_4, which have the resonance absorption bands in 10 μm spectral range. The most essential feature compared with the works [2-4] was the use of high pressure CO_2 laser with continuously tunable frequency in region of 9.2-10.9 μm.The CO_2 laser linewidth was 0.005 cm^{-1}, a step of frequency tuning 0.05 cm^{-1}, which just determined the value of spectral resolution. The laser beam was split into two pumping beams illuminating the cell from opposite directions and a probing beam directed at gas cell with a small angle. The intensity of conjugate wave reflected from the gas was measured by the detector as a function of laser frequency. Simultaneously multiphoton absorption (MPA) at the gas was measured by the microphone installed at the cell.

Spectral dependences of PCR coefficient of the probing wave and MPA of the pumping wave in SF_6 and CF_3Br gases are presented in Figs.1 ,2. Analysis of these dependences shows that both one- and two-photon resonances are clearly distinguished in PCR spectra. The spectra well correlate with MPA spectra, but are considerably more contrast in the two-photon absorption range (peaks 1,2,4 in SF_6 and 1,2 in CF_3Br).Moreover, contrast of two-photon peaks considerably increases when the pump and probe waves polarizations are perpendicular.In the center of Q-branches of one-photon transitions (3 - SF_6, 4,5 - CF_3Br) one can observe the deeps arising from one-photon absorption saturation in the pumping field.

Figure 3 illustrates the PCR spectrum of CD_4 gas.In this gas one can resolve a rotational transition structure. All peaks of this spectrum were identified as one-photon transitions. The dependences of PC reflection coefficient on the pumping intensity at the frequencies of the most characteristic peaks are presented in Fig. 4.The most prominent feature of these curves is the increasing of SF_6 reflection at one-photon transition with decreasing pumping intensity down to 1 kW/cm^2 as well as the growth of the reflection at two-photon transitions in the range of high intensities.The dependence of the reflection efficiency on the gas pressure is square in the low pressure range that allows to increase the PCR coefficient. We observed 50% reflection in SF_6 at the pressure 2 Torr.At higher pressures the self-absorption of the probe and reflected waves decreases reflection efficiency.

Thus, along with rich spectroscopic information these results reveal a possibility of achieving high IR radiation PCR efficiency in molecular gases, while preserving a resonance character of reflection.

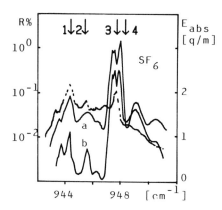

Fig.1 Spectral dependences of PCR coefficient R% and MPA (point line) in SF$_6$ for parallel (a) and perpendicular (b) polarizations of pump and probe waves.Frequencies of the peaks 1 - 944.44, 2 - 945.65, 3 - 947.90, 4 - 948.54 cm^{-1}.Pump intensity 3 MW/cm^2, P=0.5 Torr.T=150 K.

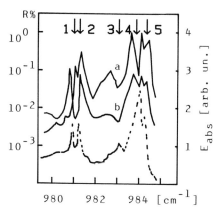

Fig.2 Spectral dependences of PCR coefficient R% and MPA (point line) in CF$_3$Br for parallel (a) and perpendicular (b) polarizations of pump and probe waves.Frequencies of the peaks 1 - 1081.00,2 - 1081.26, 3 - 1083.24, 4 - 1084.32, 5 - 1084.64. Pump intensity 3 MW/cm^2, P=0.5 Torr. T=150 K.

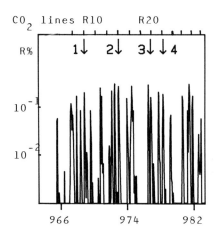

Fig.3 Spectral dependence of PCR coefficient R% at CD$_4$ for parallel polarization of pump and probe waves.Frequencies of the peaks 1 - 969.19,2 - 973.29,3 - 977.21, 4 - 978.53. Pump intensity 3 MW/cm^2, P=10 Torr.T=300 K.

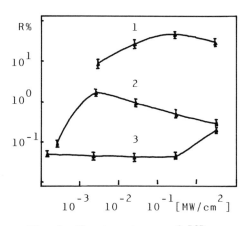

Fig. 4. The dependences of PCR coefficient R% on pump intensity at the frequencies of one-photon (1,2) and two-photon (3) resonances in SF$_6$. (2,3) - P=0.5 Torr (1) - P=2 Torr.

232

REFERENCES

1.Optical Phase Conjugation, ed. by R.Fisher, N.Y. Acad. Press (1983)
2.L.T.Bolotskikh, A.K.Popov: Appl. Phys. B31, 191 (1983)
3.N.G.Basov, V.I.Kovalev, F.S.Faizulov: Sov. Quant. El. 10, 1276 (1983)
4.V.N.Varakin, V.M.Gordienko: JETP Lett. 37 108 (1983)

Dynamic and Static Properties of Molecules in Highly Excited Vibrational States

T. Shimizu, Y. Matsuo, K. Nakagawa, and T. Kuga

Department of Physics, Faculty of Science, University of Tokyo, 7-3-1 Hongo, Bunkyo-ku, Tokyo 113, Japan

Molecules in highly excited vibrational states have been studied with great interest in recent years. Their energy level structure and dynamic behavior are directly concerned with mechanisms of multiphoton dissociation, chemical reaction, relaxation in the lasing medium, and so on. A localization of vibration energy into a particular bond may take place before the processes proceed. For detailed investigation of high vibrational states it is required that rotational constants and rotational relaxation rate constants in the individual vibrational states are measured.

We have observed the high overtone and combination tone band transitions with $\Delta v=5$ and 6 of NH_3 using a highly sensitive laser spectrometer in the visible region[1]. It consists of a ring dye laser and a multi-pass cell with 190 m absorption length. The Stark modulation method is employed. The ultimate sensitivity of this spectrometer is 3×10^{-9}/cm in the absorption coefficient, while relaxation rate constants can be determined for transitions with the absorption coefficient as weak as 5×10^{-8}/cm.

More than 1800 lines are observed in the frequency range of 14780 – 16200 cm^{-1} and 18000 – 18250 cm^{-1}. Most of the observed strong lines are concentrated in three frequency ranges of 15260 – 15590 cm^{-1} (the 645 nm band), 15900 – 16100 cm^{-1} (the 625 nm band), and 18000 – 18250 cm^{-1} (the 555 nm band), although the normal mode analysis predicts that several tens of bands appear in this range in roughly every 100 cm^{-1}. The rotational energy level structures associated with the 645 nm band are almost fully analyzed. The transitions belong to only two vibrational bands (parallel and perpendicular bands). Two vibrational states (A- and E-symmetry) are located close to each other and the rotational constants in these states are nearly equivalent. In the v=5 vibrational state the antisymmetric inversion level is located below the symmetric inversion level in the opposite order to the levels in the ground vibrational state.

All these characteristics of vibration-rotation energy level structure have been successfully explained by a local mode model in which three N-H stretching vibrations are described with the Morse potential and are combined with each other in consideration of the D_{3h} symmetry to form the base set[2,3]. The vibrational energy is calculated as functions of the anharmonicity (ωx) of the potential and the coupling constants (λ) among the stretching vibrations. The observed state of NH_3 which is well explained by the calculation with $\omega x/\lambda=2.0$ is much closer to the local mode limit ($\omega x/\lambda \to \infty$) than the normal mode limit ($\omega x/\lambda \to 0$).

We have systematically measured the pressure broadening parameters of vibration-rotation transitions in the 645 nm band of NH_3 to study dynamic properties of the highly excited vibrational states[4]. More than 40 transitions with various rotational quantum numbers J and K have been observed. The obtained parameters in the high vibrational states are to be compared with those in the ground state.

The obtained collision-induced relaxation rate constants of the excited states show strong dependence on J and K. This dependence is quite similar to that of the ground vibrational state[5,6]. Absolute values of rate constants of the excited states are only about 10 % larger than those of the ground state. It is well known that the rotational dependence of the relaxation rate constants in the ground state is due to the nearly resonant collision-induced transitions between inversion doubling levels (separations 0.6 - 0.9 cm^{-1}). In the v=5 vibrational state absolute values of the inversion splittings are 2.0 - 3.0 cm^{-1} which are considered to give sufficiently resonant collision with perturber molecules in the ground state. Even in the high vibrational states, vibration and rotation motions are well separated and the relaxation rate constants depend on only the local rotational energy level structure near the probed levels.

Since the high vibrational states have more relaxation channels than the ground state, larger relaxation rates are expected to be obtained. The 10 % difference in the absolute value of the relaxation parameters is rather smaller than expected. In the 645 nm band two vibrational levels (A and E) in the present local mode representation are very close. Collision-induced transitions between these levels are considered to give appreciable contributions to the relaxation rate constants in the high vibrational states, although these contributions do not show significant dependence on rotational states. Relaxation rate constants are almost predicted by the local structure of energy levels including nearby vibrational states.

References

1. T. Kuga, T. Shimizu, and Y. Ueda, Jpn. J. Appl. Phys. 24, L147 (1985)
2. L. Halonen and M.S. Child, J. Chem. Phys. 79, 4355 (1983)
3. T. Kuga, K. Nakagawa, Y. Matsuo, and T. Shimizu, XIV International Quantum Electronics Conference, San Francisco (1986)
4. Y. Matsuo, K. Nakagawa, T. Kuga, and T. Shimizu, J. Chem. Phys. 86, 1878 (1987)
5. W.E. Hoke, D.R. Bauer, J. Ekkers, and W.H. Flygare, J. Chem. Phys, 64, 5276 (1976)
6. T. Amano and R.H. Schwendeman, J. Chem. Phys. 65, 5133 (1976)

Highly Nonthermal Intramolecular Energy Distribution in Isolated Infrared Multiphoton Excited CF_2Cl_2 Molecules

E. Mazur, Kuei-Hsien Chen, and Jyhpyng Wang

Department of Physics and Division of Applied Sciences,
Harvard University, Cambridge, MA 02138, USA

When a polyatomic molecule with a strong vibrational absorption band is irradiated with an intense resonant infrared laser pulse it can absorb many (10 to 40) infrared photons.[1] If some initial energy deposition is 'localized'—preferably in one vibrational mode or in a subset of modes—it may become possible to induce 'mode-selective' reactions by infrared multiphoton excitation. The intramolecular dynamics of infrared multiphoton excited molecules has been studied by a variety of spectroscopic techniques.[2] One of these techniques is spontaneous Raman spectroscopy. In the past five years this technique has been successfully applied to monitor the vibrational energy in infrared multiphoton excited molecules.[3,4]

In this work we present experimental results of recent time-resolved spontaneous Raman experiments on collisionless infrared multiphoton excited CF_2Cl_2 molecules. The experiments show that the intramolecular energy distribution is highly nonthermal, and that a large part of the vibrational energy remains localized in the pump mode for a period of time long compared to the mean free time of the molecules.

The experimental procedure is described in detail in previous papers.[5] Briefly, a 15 ns CO_2-laser pulse excites the 919 cm^{-1} band of the CF_2Cl_2 molecules. After a short time delay a second 20 ns laser pulse from a frequency-doubled ruby laser probes the excited molecules. Raman scattered light is analyzed with a double monochromator and a high-gain photomultiplier. The time delay between the two laser pulses can be varied from 10 ns to 10 μs. The present measurements were carried out at a pressure of 400 Pa.

Fig. 1 shows the anti-Stokes spectrum of the multiphoton excited CF_2Cl_2. Signals from three Raman active modes, at 664, 919 and 1082 cm^{-1} are visible. The room temperature Stokes side of the Raman spectrum is shown in the same graph. At room temperature the intensity of the anti-Stokes peaks is too small to be measured at a pressure of 400 Pa. The intensity of the Raman peaks is a measure of the vibrational energy in each of the Raman-active modes.[4] Therefore, by measuring the time dependence of the anti-Stokes intensity, one can study the evolution of the vibrational energy distribution in multiphoton excited molecules.

The time dependence of the anti-Stokes Raman signals is shown in Fig. 2. The signals are normalized with their corresponding room temperature Stokes counterparts to correct for the different Raman cross sections. The rise time of the signals corresponds to the pulse duration of the laser pulses. The decay of the signals is due to a combination of collisional vibrational relaxation and diffusion of the excited molecules out of the excitation region. By comparing the intensity of the signals one can determine the distribution of energy in the vibrational modes. Fig. 3 shows the normalized intensity distribution 100 ns after excitation.

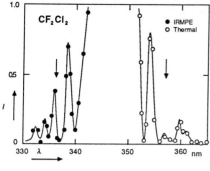

Fig. 1. Raman spectrum of infrared multi-photon excited CF_2Cl_2 at 400 Pa. The arrows show the position of the pump line.

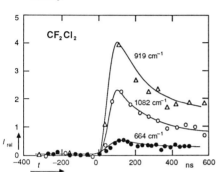

Fig. 2. Intensity of the normalized anti-Stokes signals as a function of the time delay between pump and probe pulse for CF_2Cl_2.

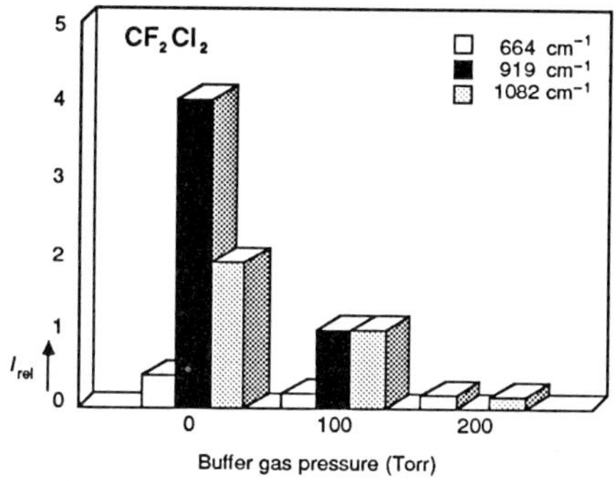

Fig. 3. Normalized anti-Stokes intensites for 3 modes of CF_2Cl_2 at 2 J/cm^2 for different N_2 buffer gas pressures.

From this graph it is clear that most of the excitation energy remains in the pump mode at 919 cm^{-1} for collisionless excitation without buffer gas. As an increasing amount of N_2 buffer gas is added the distribution tends toward thermal equilibrium, but the overall signal strengths decrease because of vibration-translation relaxation.

Summarizing, we present here time-resolved spontaneous Raman scattering measurements of infrared multiphoton excited CF_2Cl_2 at low pressure. The results show a highly nonthermal energy distribution among different modes, which persists even on time scales long compared to the mean free time of the molecules. A more detailed discussion of these results will be published elsewhere.[6] This work was supported by the Army Research Office and the Joint Services Electronics Program under contracts with Harvard University.[7]

References

1. See for instance the following publications and references therein: W. Fuss and K. L. Kompa, Prog. Quant. Electr. **7**, 117 (1981); D.S. King, *Dynamics of the Excited State*, Ed. K. P. Lawley (Wiley, New York, 1982)
2. V.N. Bagratashvili, V.S. Letokhov, A.A. Makarov, E.A. Ryabov, *Multiple Photon Infrared Laser Photophysics and Photochemistry* (Harwood, New York, 1985)
3. V.N. Bagratashvili, Yu.G. Vainer, V.S. Doljikov, S.F. Koliakov, A.A. Makarov, L.P. Malyavkin, E.A. Ryabov, E.G. Silkis, And V.D. Titov, Appl. Phys. **22**, 101 (1980)

4. Jyhpyng Wang, Kuei-Hsien Chen and Eric Mazur, Phys. Rev. A **34**, 3892 (1986)
5. Eric Mazur, Rev. Sci. Instrum. **57**, 2507 (1986)
6. Kuei-Hsien Chen, Jyhpyng Wang and Eric Mazur, submitted to Phys. Rev. Lett.
7. Contract numbers: DAAG29-85-K-0600 and N00014-84-K-0465, respectively

Stimulated Emission from Normally Non-fluorescent T.I.C.T. States of 7-DAMC

V. Chandrasekhar[1], B.M. Sivaram[1], B. Sivasankar[2], and S. Natarajan[3]

[1]Department of Physics, IIT Madras 600036, India
[2]Department of Chemical Technology, Anna University, Madras 600025, India
[3]Department of Physics, Anna University, Madras 600025, India

The 7 amino coumarins are an important class of laser dyes in the blue green spectral region. The dye 7 DAMC is an example of this class of dyes which has an amine moiety at the seventh position unrestricted by any substituent linkage. On optical excitation this dye can exist in the excited state in two different molecular conformations. The first excited conformation B* is a moderately polar, planar, intramolecular charge transfer (ICT) state and the other excited conformation A* obtained by rotation of the amine moiety at the seventh position leads to a non planar, highly polar, twisted intra molecular charge transfer (TICT) state. Both these ICT and TICT states are stabilized by substituents which are electron donating at the nitrogen site and on the aniline ring and electron withdrawing on the lactone ring, and also by solvent dipoles including specific interactions such as bonding.

The ICT-TICT model was first proposed by Rotkiewicz and co-workers for p dimethylaminobenzonitrile and has been extended to molecules of the coumarin series by Jones et al [1]. The intrinsic quantum yield for the TICT state is generally very low because it is a forbidden transition, a typical value for Q being 0.03 for DMABN in n butylchloride [2]. The TICT states of coumarins is apparently at best extremely weakly radiative and the TICT emission from 7 DAMC has not been noticed so far. But if high populations and population inversions could be achieved in the TICT states, then because of gain on the transition the TICT emission gets amplified strongly and the dye can work 'superradiantly' on this transition. We used a pulsed N_2 laser and transversely excited in a quartz cuvette, 5mM concentration solutions of 7 DAMC in n butyl acetate, ethyl acetate, p dioxane, chloroform, dichloroethane, trichloroethane, cyclo hexanone, ether etc. The dye solutions worked as lasers and produced A.S.E. generally at two wave lengths at 425 n.m. and 450 n.m. and the 425 n.m radiation is from the ICT state and the 450 n.m. radiation is from the TICT state as per convention and as deduced by us. The absorption and fluorescence for these dye solutions were also recorded at 0.01 mM and 5 mM concentrations at various temperatures from 5°C to 65°C. One absorption band with a peak at 365 n.m and one fluorescence band with a peak at 420 n.m, both without any shoulder structure was found for all the solutions, for all concentrations for this temperature range. The relative intensities and gains of the two laser transitions however varied with temperature as in Fig.1. For 7 DAMC in butyl acetate at low temperatures (5°C), the 450 n.m laser transition predominates and the 425 n.m laser transition is weaker. At 5°C, 7 DAMC in n butyl acetate has gains of 0.40×10^{-18} cm^2/mol/KW at 450 n.m and 0.20×10^{-18} cm^2/mol/KW at 425 n.m. As the temperature increases to 65°C the 450 n.m. laser transition weakens and the 425 n.m laser transition intensifies. At 65°C the gains of these transitions were measured as 0.21×10^{-18} cm^2/mol/KW at 450 n.m and 0.36×10^{-18} cm^2/mol/KW at 425 n.m.

The explanation offered for this observation of ours is as follows. The TICT formation rates K_1 in coumarins is of the order of 10×10^9/s as measured by Ref.[3]. The reverse back reaction of the TICT state reverting to the ICT state occurs at rates K_2 and K_1, K_2 increase with temperature. At temperatures below an equilibrating temperature T_{max} the forward reaction rate K_1 predominates over the back reaction rate and K_2 is small compared to the other deactivation processes of A* [2]. If short time duration intense pulses of excitation are used to pump 7 DAMC solutions then large transient populations and hence transient population inversions can occur in the A* state for times of the order of 1/deactivation rate of A*. If the gain on the A* transition is adequate then it produces amplified spontaneous emission at 450 n.m. As the temperature is reduced K_2 decreases [2] and hence the higher gain and intensities of the A* laser transition at lower temperatures. As temperature increases, because K_2 increases, the population in the B* state is enhanced and hence its laser intensity and gain increases with increasing temperature in the range studied. Pico second time domain spctroscopic studies are needed to understand the dependence of K_1 and K_2 on temperature and on solvents and their effects on the laser emission from these TICT states.

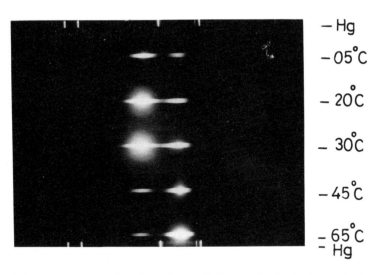

Fig.1 Laser emission from 7 DAMC in butyl acetate at various temperatures

REFERENCES

1. G.Jones, W.R.Jackson and A.M.Halpern, Chem. Phys. Lett. $\underline{72}$, 391, (1980).
2. W.Rettig and G.Wermuth, J. of Photochemistry $\underline{28}$, 351, (1985)
3. A.Declemy, C.Rulliere and Ph.Kottis, Chem. Phys. Lett. $\underline{101}$, 401, (1983).

Picosecond Spectroscopy of Molecular Dynamics of Proteins and Enzymes

R. Rigler[1], *A. MacKerell*[1], *H. Vogel*[2], *and L. Nilsson*[1]

[1]Department of Medical Biophysics, Karolinska Institutet,
  Box 60400, S-10401 Stockholm, Sweden
[2]Biocenter, University of Basel, CH-4056 Basel, Switzerland

Development of time correlated single photon detection techniques with picosecond time resolution based on laser (1) and synchrotron excitation (2) has allowed the study of motions of individual residues in protein structures and ennzymes in the picosecond time domain. We have started a project to compare experimentally determined rotational motions of aromatic amino acid residues as well as the motion of larger structural domains with molecular dynamics simulations (3) of the atomic trajectories in time intervals up to 200 ps. As an example the motion of the tryptophan residue 59 (Trp 59) in the nucleic acid specific enzyme Ribonuclease T_1 (Fig. 1) will be given. Its rotational relaxation time has been determined from the anisotropic decay of the polarized tryptophan emission r(t) and was found to lie in the range 100-200 ps depending on temperature and solvent viscosity (4).

Comparison of results from molecular dynamics trajectories with the experimental results requires calculation of the reorientational correlation between absorption $\hat{\mu}_a$ and emission $\hat{\mu}_e$ moments of Trp 59 and is given by the second order Legendre polynomial P_2 $r(t)=0.4<P_2[\hat{\mu}_a(0)\cdot\hat{\mu}_e(t)]>$. It was found that the calculated reorientational correlation time for Trp 59 (Fig. 2) agrees with the experimental one in this case where the tryptophan residue is embedded in a hydrophobic pocket devoid of solvent.

Calculation of the root mean square fluctuations from molecular dynamics trajectories for the enzyme without substrate shows that the domains with peaks in RMS fluctuations are constituted by the loop regions in the enzyme which interact normally with the substrate (Fig. 3).

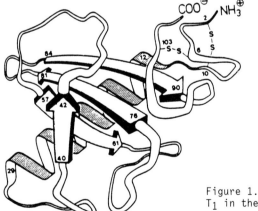

Figure 1. Structure of ribonuclease T_1 in the presence of 2'-GMP

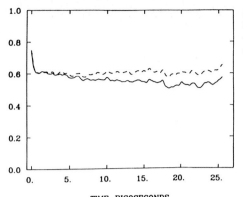

Figure 2. Correlation function P_2 using a $\hat{\mu}_a(0)$ of -38° and $\hat{\mu}_e(t)$ of -15° (---) and -61° (——) from free enzyme simulation.

TIME, PICOSECONDS

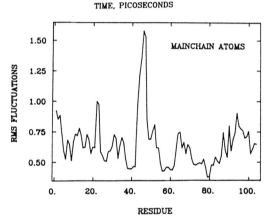

MAINCHAIN ATOMS

RMS FLUCTUATIONS

RESIDUE

Figure 3. Root mean square fluctuations (in Å) for the main chain (N, Cα, C, O) atoms of ribonuclease T_1 from the molecular dynamics simulation of the free enzyme.

A similar study has been performed with the peptide alamethicin which consists of an α-helix of 20 residues (5) similar to a previous study of melittin (6). In this peptide a tryptophan residue facing the solvent was investigated in various positions within the α-helix and in various solvents including phospholipids and methanol. In this case it was found that the reorientational correlation times are strongly dependent on the interaction of the tryptophan residues with the solvent.

The availability of time resolved multiwavelength spectroscopy with synchrotrons such as the storage ring MAX in Lund (2) will alllow a study of the motion of several aromatic residues simultaneously and will provide a detailed picture of the structural dynamics in biological structures.

References:

(1) Rigler,R.,Claesens,F. and Lomakka,G. (1984). In "Ultrafast Phenomena IV", D.H.Auston and K.B.Eisenthal, eds., Springer-Verlag, Berlin-Heidelberg, p. 472
(2) Rigler,R.,Kristensen,O.,Roslund,J.,Thyberg,P.,Oba,K. and Eriksson,M. (1987). Molecular structures and dynamics: Beamline for time resolved spectroscopy at the MAX synchrotron in Lund. Physica Scripta in press

(3) McCammon,A. and Karplus,M. (1983). Acc. Chem. Res. $\underline{16}$, 187
(4) MacKerell,A.D.Jr.,Rigler,R.,Nilsson,L.,Hahn,U. and Saenger,W. (1987) Protein dynamics. A time-resolved fluorescence energetic and molecular study of Ribonuclease T1. Biophysical Chemistry $\underline{26}$, 247
(5) Eisenberg,M.,Hall,J.E. and Mead,C.A. (1973). J. Membr. Biol. $\underline{14}$, 143
(6) Vogel,H. and Rigler,R. (1987). Orientational fluctuations of melittin in lipid membranes as detected by time-resolved fluorescence anisotropy measurements. In "Structure, Dynamics and Function of Biomolecules", Springer Series in Biophysics, Vol. 1, A.Ehrenberg et al., eds., Springer-Verlag, p. 289

244

Part VII

Clusters, Surfaces and Solids

CARS Spectroscopy of NH_3 Clusters in Supersonic Jets

H.D. Barth and F. Huisken

Max-Planck-Institut für Strömungsforschung,
D-3400 Göttingen, Fed. Rep. of Germany

During the past few years there has been increasing interest in the investigation of loosely bound van der Waals (vdW) complexes or clusters. Coherent anti-Stokes Raman spectroscopy (CARS) in the expansion region of supersonic jets, where these species are formed, provides an ideal means for the study of their formation and spectroscopy.

We have investigated the formation of clusters in supersonic jets of NH_3 seeded in various rare gases using the one-dimensional CARS geometry. The set-up consists of a Nd:YAG-dye laser system (Quantel) and a molecular beam machine equipped with a pulsed nozzle (Bosch automobile fuel injector). Bandwidths are 0.12 cm^{-1} for the pump beam at 532 nm and 0.11 cm^{-1} for the Stokes beam, respectively.

We have started our study by investigating the complete v_3 band of the NH_3 monomer between 3350 and 3600 cm^{-1}. Because of its widely spaced rotational lines this band is ideally suited to determine the rotational temperature in the beam [1,2]. At strong expansion conditions the formation of a broad structure, centered around 3370 cm^{-1}, was observed. Since we were tempted to associate this new structure with the formation of NH_3 clusters, we have investigated the spectral region between 3100 and 3500 cm^{-1} more systematically. We have measured CARS spectra for pure NH_3 and various NH_3/rare gas mixtures as a function of stagnation pressure and position in the jet. A selection of the measured spectra is shown in Fig.1. They have been obtained 2 mm downstream from the nozzle at a total stagnation pressure of 6 bar.

The upper spectrum shows the result for pure NH_3. The two prominent peaks are assigned to the v_1 Q branch at 3336 cm^{-1} and the $2v_4$ overtone at 3220 cm^{-1}. The spectrum in the middle is obtained with a mixture of 5 % NH_3 in He. As a striking feature now, the appearance of broad structures at 3204 and 3370 cm^{-1} is observed. These broad structures are even more pronounced if Ne instead of He is used as carrier gas. In the bottom spectrum, obtained with 5 % NH_3 in Ne, the peak at 3204 cm^{-1} has evolved to the dominant features.

In order to get information about the composition of our pulsed beams, we have analyzed them with the quadrupole mass spectrometer in our crossed molecular beam machine [3]. The results are shown in Fig. 2. In contrast to pure NH_3 where cluster formation is not very pronounced, in NH_3/rare gas mixtures strong clustering is observed. Further, it is noticed that in NH_3/Ne mixtures substantially larger clusters are formed than in NH_3/He.

This result suggests that the broad structures in our CARS spectra are primarily due to large $(NH_3)_n$ clusters with n = 5-50. The strongest support for this interpretation, however, is supplied by Raman studies of liquid

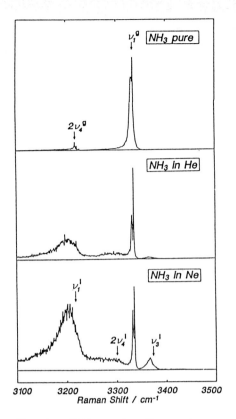

Fig. 1: CARS spectra for various NH$_3$/rare gas mixtures measured 2 mm downstream from the nozzle

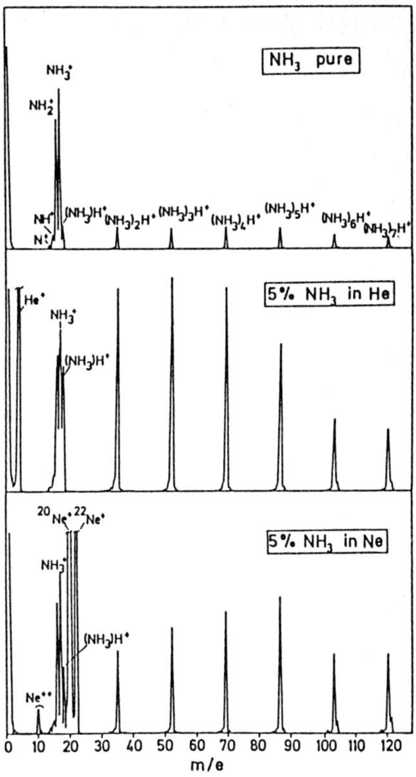

Fig. 2: Mass spectra showing the composition of the employed beams

NH$_3$ [4]. The three peaks observed in these studies are marked in the bottom spectrum of Fig. 1 by arrows. It can be seen that they coincide almost exactly with the positions that we observe.

Aside from the broad structures, also a narrow-peaked structure, red-shifted by 3.5 cm^{-1} from the ν_1 Q branch, can be observed in Fig. 1. It is attributed to small clusters (i.e. dimers and trimers) or to NH$_3$ molecules residing on the outside of the cluster whose H-atoms are free and not bound to the complex.

For a more complete account of this work the reader is referred to a paper presently in press [2].

1. F. Huisken and T. Pertsch, Chem.Phys.Lett. 123, 99 (1986)
2. H.D. Barth and F. Huisken, J.Chem.Phys. in press
3. F. Huisken and T. Pertsch, J.Chem.Phys. 86, 106 (1987)
4. M. Schwartz and C.H. Wang, J.Chem.Phys. 59, 5258 (1973)

Doppler-free Laser Spectroscopy of Na$_3$

H.-J. Foth and W. Demtröder

Fachbereich Physik, Universität Kaiserslautern,
Erwin-Schrödinger-Straße, D-6750 Kaiserslautern, Fed. Rep. of Germany

A knowledge of small metal clusters is important for the understanding of the fundamental mechanism of catalysis and surface chemistry. Molecular spectroscopy with lasers is also here a helpful tool to study the electronic potentials and molecular constants.

The optical spectrum of Na$_3$, first recorded by Delacrétaz and Wöste [1], shows four prominent band systems which correspond to transitions into different electronic states. The observed structure of these bands was analyzed and, perhaps between 600 and 625 nm, finally completely described by pseudo-rotation [2], which is a superposition of vibrational modes. Information about the rotational structure could not be obtained for these spectra because of the limited resolution due to the band width of 10 GHz of the pulsed lasers used by Wöste et al.

In our experiment the generation of Na$_3$ in a collimated molecular beam seeded in argon is similar to that of Wöste et al. Here, also resonant two photon ionisation (TPI photoions) is combined with detection of the photoions via a quadrupole mass spectrometer (tuned to M=69) to distinguish excitation lines of Na$_3$ against those of Na$_2$, which are in the same wavelength region. Therefore the trimers are excited by a cw single mode dye laser with a band width less than 1 MHz and ionized by an Ar$^+$-laser (20 W, multicolor).

The sub-Doppler spectrum of the system around 671 nm shows line widths of 30 MHz (Fig. 1), which means the lifetime of this excited state being longer than 5 nsec. Clearly this state is not dissociating or predissociating which is in good agreement with lifetime measurements performed recently by Wöste et al. [3]. Furthermore, vibrational spectroscopy by these authors gave nearly identical vibrational constants of the ground and this excited state. Since also the molecular dimensions, like bonding lengths and angles, should be nearly equal in both states, the whole rotational spectrum is compressed in a very small spectral range (~ 0.6 cm^{-1} as seen in Fig. 1). The disappointing result is: Even with this high resolution the huge line density in this spectrum gives only barely a chance to realize the rotational structure.

By developing the previously described experiment, optical-optical double resonance was set up as shown schematically in Fig. 2. The intensity of dye laser 1 is chopped with

248

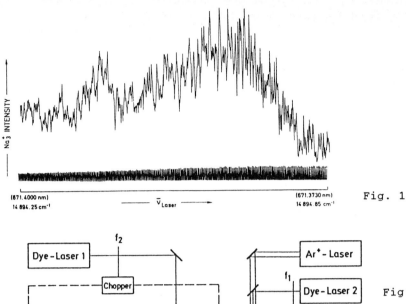

Fig. 1

(671.4000 nm)
14 894.25 cm⁻¹

$\bar{\nu}_{Laser}$

(671.3730 nm)
14 894.85 cm⁻¹

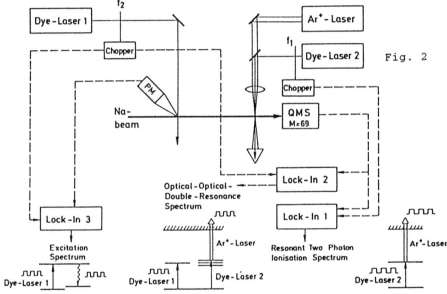

Fig. 2

frequency f_2, which modulates the population of a selected rotational level in the electronic ground state. As done previously, the photoions are produced by dye laser 2 via TPI. When dye laser 2 is tuned, the Lock-In 2 marks all those transitions which start from a level correlated to the selected ground state level.

As a typical result of an optical-optical double resonance spectrum, in Fig. 3 is seen that additionally to the main peak, where both laser wavelengths are identical, several lines appear on both sides indicating a symmetric structure. The line splittings are less than 100 MHz and are expected to present the hyperfine structure of this molecule.

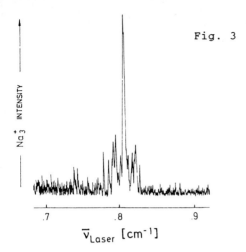

Fig. 3

The authors like to thank J. Gress for his contributions.

[1] G. Delacrétaz and L. Wöste, Surface Science 156, 770 (1985)
[2] G. Delacrétaz, E.R. Grant, R.L. Whetten, L. Wöste, and J.W. Zwanziger, Phys. Rev. Lett. 56, 2598 (1986)
[3] M. Broyer, G. Delacrétaz, P. Labastie, J.P. Wolf, and L. Wöste, Phys. Rev. Lett. 57, 1851 (1986)

Optical Spectrum of the Icosahedral C_{60} – "Follene-60": A Challenge for Laser Spectroscopy

A. Rosén, S. Larsson, and A. Volosov

Chalmers University of Technology and University of Göteborg, S-41296 Göteborg, Sweden

A molecule C_{60} has recently been discovered in vaporization of carbon species from graphite into a high density He flow using focused laser radiation /1/. The existence of C_{60} with unusual stability in an icosahedral structure was anticipated from Hückel theory /2/ and recent theoretical studies support this prediction. The bond lengths are alternating, with the pentagonal edge ones 0.07Å longer than the bonds shared with hexagonal rings /3/. Here we will use the name "follene-60" for C_{60}, which is derived from the latin word (follis) for football.

In a recent calculation using the discrete variational (DVM) Xα-method Hale obtained ionization energies and optical transition energies for the equal bond length case /4/. Internal configuration interaction (CI), however, usually plays an important role in aromatic molecules. It is therefore of interest to compare with results obtained within the CNDO/S method /5/ where configuration interaction is carried out between singly excited states of a singlet configuration. 250 configurations with the lowest energies were included. Calculations for C_{60} were done for the equal bond length case (1.421Å) and for the alternating bond length case (1.474Å for pentagonal edge and 1.400Å for bonds shared by hexagonal rings) /6/. Energy eigenvalues for the highest occupied molecular orbitals (HOMO) and the lowest unoccupied molecular orbitals (LUMO) for the case with unequal bond lengths are given in Fig. 1. Quantitatively similar results are obtained for the case with equal bond lengths.

The HOMO energy is -7.55 eV, which should be compared with a value of -7.35 eV for the equal bond length case and the calculated ionization energy of 6.4 eV by Hale. Experimentally it has been found /7/ that the C_{60} ion signal was two orders of magnitude larger when ionizing with the 7.87 eV photons (F_2-laser) than with the 6.42 eV photons (ArF-laser). This may indicate a near resonant ionization with the 7.87 eV photons while multiphoton ionization may take place with the 6.42 eV photons. Our calculation of the ionization energy seems therefore to be in good agreement with the present experimental data. An indication of the accuracy of our calculations was achieved by a calculation for naphthalene which gave a deviation of 0.4 eV to the experimental ionization energy. Calculated optical transitions were also found to be in good agreement with experimental absorption spectra for naphthalene.

In figure 1 the electronic transitions between molecular orbitals i and a are denoted by an arrow. Their energies above the ground state are given by

$$E_1 - E_0 = \varepsilon_a - \varepsilon_i - J_{ia} + 2K_{ia}$$

and represented along the horizontal axis. Optical transition energies resulting from CI calculations are given at the bottom of Fig. 1 disregarding the

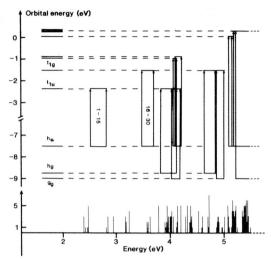

Fig. 1. Energy eigenvalues and optical transitions.

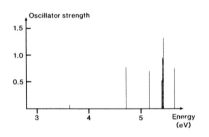

Fig. 2. Calculated oscillator strengths of allowed transitions.

oscillator strengths. In Fig. 2 the oscillator strengths have been included. There are 15 symmetry forbidden HOMO→LUMO transitions in the range 2.6-2.8 eV. The allowed transitions from HOMO→LUMO + 1, HOMO→LUMO + 2, HOMO - 1 → LUMO interact considerably by CI which in general shifts the intensity from lower energy at 3.6-4.2 eV to higher energies at 4.7, 5.15, 5.4 and 5.6 eV. The first triply degenerate allowed transitions are at 3.6 eV with a total oscillator strength of 0.08 which is very small compared to the oscillator strengths in the range of 0.25 - 0.90 for the transitions at higher energies. At 5.4 eV there are a number of close lying transitions which may enhance the signals. To get a test of our prediction of the optical spectrum two photon ionization or two colour experiments should be done which may be a challenge for laser spectroscopy.

1. H.W. Kroto, J.R. Heath, S.C.O. Brien, R.F. Curl and R.E. Smalley, Nature 318, 162 (1985).
2. R. Davidson, Theor. Chim. Acta (Berlin) 58, 193 (1981).
3. M.D. Newton and R.E. Stanton, J. Am. Chem. Soc. 108, 2469 (1986).
4. P.D. Hale, J. Am. Chem. Soc. 108, 6087 (1986).
5. J. Del Bene and H.H. Jaffé, J. Chem. Phys, 48, 1807, 4050 (1968).
6. S. Larsson, A. Volosov and A. Rosén, Chem. Phys. Lett. to appear.
7. D.M. Cox, D.J. Trevor, K.C. Reichmann and A. Kaldor, J. Am. Chem. Soc. 108, 2457 (1986).

Second-Harmonic and Sum-Frequency Generation for Surface Studies

J.H. Hunt, P. Guyot-Sionnest, and Y.R. Shen

Department of Physics, University of California,
Center for Advanced Materials, Lawrence Berkeley Laboratory,
Berkeley, CA 94720, USA

Second harmonic generation (SHG) has now been well established as a versatile surface-sensitive probe [1]. It has been used to study electrochemical processes at electrode surfaces, molecular adsorption and desorption at metal and semiconductor surfaces, orientational phase transition of molecular monolayers on water, surface reconstruction and epitaxial growth, and so on. More recently, it has been employed as a tool to monitor monolayer polymerization and other surface reactions [2], to probe polar order of molecules at interfaces [3], and to measure molecular nonlinearity [4]. While most surface techniques are restricted to the solid/vacuum environment, SHG is applicable to nearly all interfaces as long as the interfaces are accessible by light. In addition, SHG has the advantages of being capable of in-situ measurements with high temporal, spatial, and spectral resolutions.

The source responsible for the SHG is the nonlinear polarization $\vec{P}^{(2)}(2\omega)$ in the medium:

$$\vec{P}^{(2)}(2\omega) = \overleftrightarrow{\chi}^{(2)}(2\omega):\vec{E}(\omega)\vec{E}(\omega). \tag{1}$$

If the medium has an inversion symmetry, then the nonlinear susceptibility $\chi^{(2)}$ vanishes in the electric-dipole approximation. Since the inversion symmetry is necessarily broken at an interface, this makes SHG an effective surface probe. For a monolayer of molecules at a surface, the surface nonlinear susceptibility is typically $\sim 10^{-15}$ esu, which should yield an SH signal of $\sim 10^3$ photons/pulse with a laser pulse of 10 mJ energy and 10 nsec duration impinging on a surface area of 0.2 cm^2.

The resonant behavior of $\overleftrightarrow{\chi}^{(2)}(2\omega)$ can provide spectroscopic information about a surface or molecules adsorbed at the surface. This has been demonstrated in a number of cases [5]. However, since the optical wavelengths involved in the surface SHG experiments are generally in the 0.2-1 µm range, only electronic transitions of the molecules or surface structure can be probed. They usually have relatively broad bandwidths, making SHG not particularly useful for identification or selective monitoring of surface molecular species. Vibrational spectroscopy is more suitable for selective studies of adsorbed molecules and their interaction with the substrate. Unfortunately, vibrational modes appear in the IR range, and SHG in this part of the spectrum is not practical because of the poor sensitivity of the photodetectors. This problem can be solved by using instead of SHG, the IR-visible sum-frequency generation (SFG) [6].

In the SFG process, the IR input beam is tuned through the vibrational resonances, and the visible input beam up-converts the excitation to a sum-frequency output also in the visible, which can then be detected by photomultipliers. As a second-order process, SFG has nearly all the

advantages of SHG for surface probing, but in addition, it allows the
studies of surface resonant excitations in the IR. The nonlinear
susceptibility responsible for the SFG can be written as

$$\chi^{(2)}(\omega_S = \omega_{vis} + \omega_{IR}) = N_s \langle \alpha^{(2)}(\omega_S) \rangle, \tag{2}$$

assuming that the local-field correction is negligible, where N_S is the
surface density of molecular adsorbates, $\alpha^{(2)}$ is the nonlinear
polarizability, and $\langle \ \rangle$ denotes an average over the molecular
orientational distribution. In general, $\alpha^{(2)}$ consists of a resonant and a
nonresonant part.

$$\alpha^{(2)} = \alpha_R^{(2)} + \alpha_{NR}^{(2)} \tag{3}$$

with $\alpha_R^{(2)} = \sum_{\nu} \alpha_{R\nu}^{(2)}$, where ν indicates the νth mode near resonance. It can
be shown from the microscopic expressions that

$$[\alpha_{R\nu}^{(2)}(\omega_S)]_{ijk} = [\alpha_{R\nu}^{(1)}(\omega_{IR})]_{kk}[\alpha_{R\nu}^{(3)}(\omega_S = \omega_{vis} + \omega_S - \omega_{vis})]_{ijij} , \tag{4}$$

where $\alpha_{R\nu}^{(1)}$ and $\alpha_{R\nu}^{(3)}$ are resonant IR and Raman polarizabilities,
respectively. This points to the fact that only modes which are both IR
and Raman active can contribute to the SFG spectrum [6].

As a demonstration of SFG as a surface vibrational spectroscopic tool,
we have measured the C-H vibrational spectra of alcohol, methanol, and
isopropanol evaporated on glass in the 3.3 to 3.5 μm wavelength range. A
mode-locked Nd:YAG laser was used to generate a visible beam at 0.532 μm
by SHG in KDP and a tunable IR beam around 3.4 μm by parametric
amplification in $LiNbO_3$ [6]. The beam energies were 1.5 mJ and 0.2 mJ per
pulse, respectively. With the two beams simultaneously impinging on the
glass surface, SFG could be detected after proper spatial and spectral
filtering. Figure 1 displays the SFG spectra of the three adsorbed
molecular species together with their Raman spectra from the bulk liquid
[7]. The close agreement between the Raman and the SFG spectra allows us
to assign unambiguously the observed peaks in the SFG spectra.

In Fig. 1(a) for methanol CH_3OH on glass, two peaks at 2840 cm^{-1} and at
2960 cm^{-1} correspond to the CH_3 symmetric and asymmetric stretch
vibrations, respectively. In Fig. 1(b) for ethylene glycol, $C_2H_4(OH)_2$,
the symmetric and asymmetric CH_2 stretches appear at 2875 and 2935 cm^{-1},
respectively. Finally, for isopropyl alcohol, $(CH_3)_2CHOH$, we assign the
peak at 2885 cm^{-1} as the CH_3 symmetric stretch, the two peaks at 2950 and
2980 cm^{-1} as the degeneracy-lifted CH_3 asymmetric stretches, and the
little bump at 2920 cm^{-1} as the CH stretch. All the spectra were taken
with the visible light s-polarized and the infrared light unpolarized. It
is seen that the spectra for the three alcohols are distinctly different,
and various CH stretches are clearly distinguishable.

The methanol spectrum appeared to have changed in a few hours after the
sample was prepared, as shown in Fig. 2. This suggests that the methanol
might have been transformed into another molecular species, presumably
methoxy, CH_3O. The result is a manifestation of the capability of SFG to
monitor surface molecular reactions. We have also found that after
leaving the alcohol-covered samples in air for a while, the isopropyl
alcohol spectrum would disappear, but the other two would not. This
indicates that only isopropyl alcohol is desorbed completely from the
glass surface and explains why it is often used as the solvent for
cleaning glass.

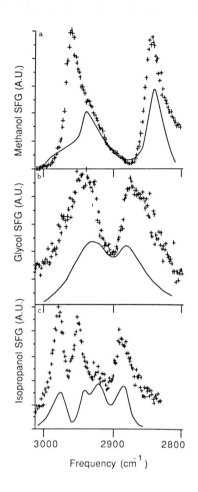

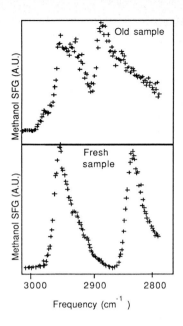

Fig. 2 SFG spectrum of methanol on glass of a fresh sample and an old sample. The latter was taken several hours later.

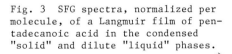

Fig. 1 Sum-frequency generation spectra as a function of infra-red input frequency. The bold lines are the bulk liquid Raman spectra.

Fig. 3 SFG spectra, normalized per molecule, of a Langmuir film of pentadecanoic acid in the condensed "solid" and dilute "liquid" phases. →

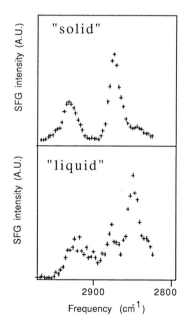

255

We have also used SFG to study monolayers of pentadecanoic acid on a water surface. Figure 3 shows two spectra of the CH stretches in the 2900 cm^{-1} region, one for the dilute liquid phase and the other for the more condensed "solid" phase. The "solid" spectrum is dominated by the CH_3 group at the end of the long chain molecule; the two peaks in the spectrum correspond to the symmetric and asymmetric CH_3 stretches. The liquid spectrum shows a new peak at 2850 cm^{-1} and a broad hump in the 2880-2930 cm^{-1} region, presumably arising from the CH_2 groups along the chain. A careful analysis of the polarization dependence of the spectra suggests that in the "solid" phase, the molecular chains are straight and normal to the surface (with perhaps an orientational distribution about the surface normal), but in the "liquid" phase, the chains may have coiled up more and orient more randomly on the water surface.

In conclusion, we have shown that while SHG has been demonstrated to be a successful surface analytical tool, SFG can be expected to be even more powerful because of its capability for IR spectroscopy measurements.

P.G-S. acknowledges support from the delegation generale de l'armement (France). This work was supported by the U.S. Department of Energy under Contract No. DE-AC03-76SF00098.

References

1. See, for example, Y. R. Shen, Ann. Rev. Mat. Sci. 16, 69 (1986); in Chemistry and Structure at Interfaces: New Laser and Optical Techniques, eds. R. B. Hall and A. B. Ellis (Verlag-Chemie, Berlin, 1986), p.151, and references therein.
2. G. Berkovic, Th. Rasing, and Y. R. Shen, J. Chem. Phys. 85, 7374 (1986); R. B. Hall, A. M. DeSantolo, and S. G. Grubb, J. Vac. Sci. Technol, to be published.
3. P. Guyot-Sionnest, H. Hsiung, and Y. R. Shen, Phys. Rev. Lett. 57, 2963 (1986).
4. Th. Rasing, G. Berkovic, and Y. R. Shen, Chem. Phys. Lett. 130, 1 (1986).
5. T. F. Heinz, C. K. Chen, D. Ricard, and Y. R. Shen, Phys. Rev. Lett. 48, 478 (1982); T. L. Mazely and W. M. Hetherington, J. Chem. Phys. 86, 3640 (1987).
6. X. D. Zhu, H. Suhr, and Y. R. Shen, Phys. Rev. B 35, 3047 (1987); J. H. Hunt, P. Guyot-Sionnest, and Y. R. Shen, Chem. Phys. Lett. 133, 189 (1987).
7. Selected Raman Spectral Data, ed. L. B. Beach (Thermodynamic Research Center Hydrocarbon Project Publications, College Station, Texas, 1983), Ser. No. 807,653,55.

Second Harmonic Generation on the (111) Surface of BaF$_2$

J. Reif[1], *P. Tepper*[1], *E. Matthias*[1], *E. Westin*[2], and *A. Rosén*[2]

[1]Fachbereich Physik, Freie Universität Berlin, D-1000 Berlin 33, Germany
[2]Institute of Physics, Chalmers University of Technology,
 S-41296 Göteborg, Sweden

In this contribution we present experimental results on the angular, polarization, and wavelength dependence of second harmonic generation (SHG)
on polished BaF$_2$ (111) surfaces in air. These measurements are part of a
program to investigate the electronic surface structure of alkaline-earth
fluoride crystals under various conditions. The first results on optical
SHG on surfaces of BaF$_2$ were reported as early as 1968 by Wang and
Duminski /1/. Meanwhile, guided by the pioneering work of Shen and col-
laborators /2/, the technique has matured into a powerful analytical tool
for the investigation of surface structures, and we find it worthwhile to
apply it further to surfaces of transparent ionic materials.

Before turning to surface structure investigations some methodical
studies were carried out. Most important is the determination of the
threshold for dielectric breakdown to assure that the UV light in the
expected SH range is not generated by broadband emission from a plasma.
Using the frequency-doubled 532 nm radiation of a Q-switched YAG-laser we
recorded the intensity dependence of the 266 nm radiation shown in Fig.1.
The two slopes clearly distinguish the SHG region from the one of plasma
radiation. By keeping the intensity of the fundamental below 10 MW/cm^2,
only SH light will be produced and there is no need for a monochromator.
To discriminate against the fundamental light we use a combination of
filters, prism dispersion, and a solar-blind photomultiplier.

Another methodical study is the SHG dependence on the polarization of
both fundamental and SH light. In agreement with theoretical predictions
/3/, we find among others that the intensity of s-polarized SH light is
strongest for a polarization angle $(2n+1)\pi/4$ (n=0,1,2...) of the fundamental with regard to the plane of incidence, and almost vanishes for s-
or p-polarized ($n\pi/2$) fundamental light. The combination of s-polarized
fundamental and s-polarized SH light represents the simplest physical
situation where both E(ω) and E(2ω) oscillate in the surface plane. This
situation was realized when taking the data shown in Fig. 2 in form of a
polar diagram. Here the SH yield was measured in reflection at 45° while
turning the crystal around its normal axis centered at the laser spot. The
geometric structure of the (111) plane is indicated. The SH yield reflects
the large electron density centered at the fluorine ions and the allowed
direction of oscillation under the influence of the electric field. This
conclusion is confirmed by cluster calculations of the valence electron
density distribution in the surface layer. Although we suspect that
higher multipole contributions from the bulk are strongly suppressed
/3,4/, it is at this stage not certain that the pattern in Fig. 2 represents surface SHG only, and more experiments need to be carried out to
establish this.

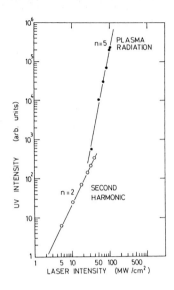

Fig.1 Optical SHG and plasma generation as a function of laser intensity at 532 nm.

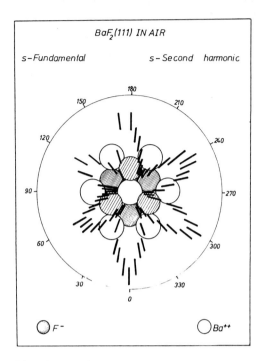

Fig. 2. Angular dependence of SHG for s-polarized fundamental and s-polarized SH light when turning the crystal around its normal. Angle of incidence 45°, fundamental wavelength 532 nm.

A strong argument that the SH intensity displayed in Fig. 2 indeed reflects the electronic density at the surface is obtained from the wavelength dependence of SHG. It was shown by Heinz et al. /5/ that the SHG can be resonantly enhanced by suitable excited states of adsorbates. It is conceivable that polished fluoride crystals exposed to air also might lead to resonant SHG because of occupied and empty states in the bandgap, similar to those predicted and observed for a non-stoichiometric surface structure of BaF_2 (111) in vacuum /6/. We have observed the wavelength dependence displayed in Fig. 4 for the experimental conditions indicated in Fig. 3 by a, b, c, i.e., three different angles ranging from maximum to minimum of the rotational angular dependence. For intensity reasons, these measurements were carried out with 45° polarization of both fundamental and SH light. We clearly observe two bands of enhanced SHG. Measurements with s-polarized fundamental and SH light reveal an inverted intensity of these two bands. Note from Fig. 3 that at position c there is still a large fraction (about 60 %) of SH yield which shows no spectral dependence. Therefore, we tentatively conclude that the SH fraction exhibiting a spectral dependence represents the local dipolar contribution while the "white" part may be due to nonlocal contributions /4/. The demonstrated combination of crystal rotation and wavelength dependence may be one way to separate the various contributions in surface SHG.

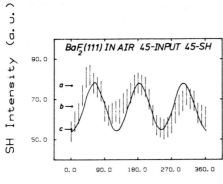

Fig. 3. Angular dependence of SHG when turning the crystal around its normal for 45° polarization with regard to the plane of incidence for both fundamental and SH light. Angle of incidence 45°, fundamental wavelength 532 nm.

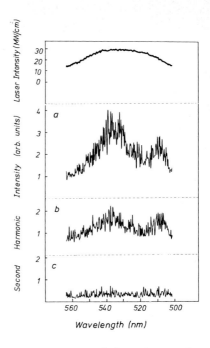

Fig. 4. Spectral dependence of SHG at three crystal azimuths mutually differing by 30°, as indicated by the corresponding letters in Fig. 3. The upper part of the figure shows the intensity variation of the fundamental light.

    This work was supported by Deutsche Forschungsgemeinschaft, Sfb 6, the Swedish Natural Science Research Council and the Swedish Board of Technical Development.

References

/1/ C.C. Wang and A.N. Duminski, Phys. Rev. Letters 20, 668 (1968)
/2/ Y.R. Shen, Annu. Rev. Mater. Sci. 16, 69 (1986), and references therein.
/3/ N Bloembergen et al., Phys. Rev. 174, 813 (1968)
/4/ P. Guyot-Sionnest and Y.R. Shen, Phys. Rev. B35, 4420 (1987)
/5/ T.F. Heinz et al., Phys. Rev. Letters 48, 478 (1982)
/6/ H.B. Nielsen et al., contributed paper EICOLS'87

Study of Adsorbates on a Silver Surface by Sum-Frequency Generation with Surface Plasmon Waves

Z. Chen, Y.J. Liu, J.B. Zheng, and Z.M. Zhang

Laboratory of Laser Physics and Optics, Fudan University, Shanghai, People's Republic of China

1. Introduction

Enhanced optical phenomena produced by the excitation of surface plasmon waves (SPWs) for surface studies have been employed extensively in the past years. Most of the non-linear optical effects on surfaces have been focused on the optical second-harmonic generation technique (SHG) and not on the more general case of sum-frequency generation (SFG), perhaps owing to the sophisticated techniques involved in keeping the two SPWs simultaneously excited during the experiment. However, the SFG seems to have a greater potential for surface studies, as discussed elsewhere /1/. Here, we will report on our investigations on the effects of the enhanced SFG due to adsorbates on a silver surface. Arachidic acid molecules are used as the adsorbate, and transmitted SFG signals from the bare silver surface and from the same surface covered with a mono-layer of the adsorbates are compared and fitted with theoretical calculations. Through such comparisons, we have determined the non-linear polarizability coefficient for a single arachidic acid molecule along the long-chain axis, d_{33}, to be $0.2 \cdot 10^{-29}$ esu. This value is consistent with the value for the same kind of molecule found using the SHG technique /2/.

2. Experimental

A pulsed YAG laser beam of $\lambda=10640$ Å and a beam from a dye laser, pumped by the same YAG laser tuned at $\lambda=6448$ Å, are used to excite the two counter-propagating SPWs through the inner hypotenuse side of a dense flint glass prism. The silver film, of thickness ~500 Å, is evaporated onto a flat glass plate made of the same material as the prism. Half of the silver surface is covered with the arachidic acid molecules, using the conventional Langmuir-Blodgett mono-molecular layer technique. The silvered glass plate is then attached to the hypotenuse side of the prism with a matching liquid. The dielectric constants of the silver film for the two frequencies of the laser light $\varepsilon_{Ag}(\omega_1)$ and $\varepsilon_{Ag}(\omega_2)$, as well as its thickness can be determined from the Attenuated Total Reflection Spectra (ATR) through the successive excitation of the two SPWs. The measurements were carried out with the optical set-up shown in Fig. 1, with one laser beam impinging on the prism only. Using the same procedure and the values obtained for $\varepsilon_{Ag}(\omega_1)$ and $\varepsilon_{Ag}(\omega_2)$, we can also measure the ATR spectra for the silver mono-layer system and calculate the corresponding dielectric constants $\varepsilon_m(\omega_1)$ and $\varepsilon_m(\omega_2)$, of the mono-molecular layer. This can be achieved by shifting the silvered glass plate along the prism so that the laser spot illuminates the portion of the plate covered by the mono-layer. For the simultaneous excition of the two SPWs at the silver-air and the silver-mono-layer interfaces, we first fixed the prism at the optimized position during the excitation of the SPW for $\lambda=10640$ Å then carefully adjusted the mirrors M and M_1 for the excitation of the SPW for $\lambda=6448$ Å, at the same point of

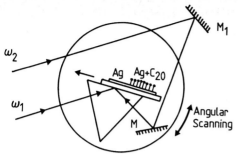

Fig. 1 Optical layout

incidence. As M is mounted on the same rotatable stage as the prism during the angular scanning (by rotating the stage with a stepping motor) the incident light beam of $\lambda=6448$ Å moves along the mirror M and hence the incident angle is scanned. In our configuration, the incident angles of the two laser beams change synchronously around the optimized position for the two SPWs to be excited at the same time and the same point. Transmitted SFG signals are measured with boxcar electronics.

3. Results

Figure 2 shows the experimental curves for the SFG for the bare silver surface and the mono-layer-covered part. These curves are least-squares fits to the experimental data points. Theoretical curves are calculated from the hydrodynamic model for non-linear effects on metal surfaces /3/. Both of the curves are normalized to the theorectical calculation for the bare silver surface at its maximum. The experimental curves are in agreement with the theoretical curves except for the broadening of the half-width, which is due to the "miss-overlapping" of the two SPWs and the unequal step lengths of the angular scanning of the two laser beams. The peak positions and the relative strengths of the SPW signals are in good agreement. By comparison and fitting of these experimental results to the theoretical analyses, the non-linear polarizability coefficient along the long-chain direction of the arachidic acid molecule was evaluated and found to be $d_{33} = 0.2 \cdot 10^{-29}$ esu.

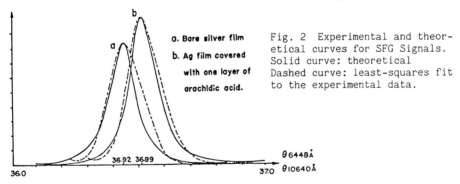

a. Bare silver film

b. Ag film covered with one layer of arachidic acid.

Fig. 2 Experimental and theoretical curves for SFG Signals. Solid curve: theoretical Dashed curve: least-squares fit to the experimental data.

References

1. X.D. Zhu, H. Suhr and Y.R. Shen: IQEC '86 Postdeadline Paper PD-28 (1986)
2. Z. Chen, W. Chen, J.B. Zhen, W.C. Wang and Z.M. Zhang: Opt. Commu. 54, 305, (1986).
3. J.E. Sipe, V.C.Y. So, M. Fukui and G.I. Stegeman: Phys. Rev. B21, 4389, (1980).

Adsorbate Resonance Enhancement in Optical Surface Second-Harmonic Generation

W. Heuer, L. Schröter, and H. Zacharias

Fakultät für Physik, Universität Bielefeld,
D-4800 Bielefeld, Fed. Rep. of Germany

Optical second-harmonic generation on interfaces between two media has been very actively investigated in recent years for it promises to be an effective tool for time resolved surface studies [1]. The influence of adsorbates on the intensity of the second harmonic (SH) generated has been observed during adsorption of O, CO, and alkali metals on Rh(111) surfaces [2]. In these cases the presence of the adsorbates was detected indirectly through their influence on the density of free electrons in the metal substrate. In this contribution we report an enhancement of the SH signal through an electronic resonance of the adsorbate itself, showing the possible adsorbate specifity of the harmonic signal. The second-harmonic was generated by sulfur atoms adsorbed on a Pd(100) single crystal surface in ultrahigh vacuum (2×10^{-10} Torr). As p polarized fundamental radiation, incident on the sample at an angle of 45°, served the output of a Q-switched Nd:YAG laser at $\lambda = 1064.2$ nm (pulse duration 8 ns, pulse energy 0.65 mJ, bandwidth 0.1 cm^{-1}). The laser beam diameter of about 0.1 mm resulted in an intensity of about 950 MW/cm^2 on the sample. The p polarized SH was detected after proper filtering by a photomiltiplier. The sulfur atoms were supplied by segregation from the bulk, and the coverage was monitored by Auger spectroscopy [3].

Figure 1 shows the second-harmonic intensity from the clean and sulfur covered Pd(100) crystal for increasing sulfur coverage. The signal intensity is normalized to that of clean Pd(100), which corresponds to about 100 photons generated per pulse at $\lambda = 532.1$ nm. A monotonic increase of the SH signal upon S adsorption is observed. No decrease of the signal level below the clean metal value occurred. Generally, adsorbates like O and S tend to localize free electrons of the metal substrate, which results in a decrease of the SH intensity induced by these electrons, as has been observed for O and CO adsorption on Rh(111) [2].

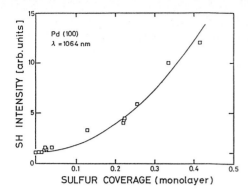

Fig. 1: Second-harmonic intensity from S/Pd(100) as a function of the adsorbate coverage. The full line represents a quadratic least-squares
fit to the data.

The signal shown in Fig.1 originates from the sulfur-adsorbate layer.
The SH intensity can be expressed by

$$I^{2\omega} = C* |e^{2\omega} \chi_{S,eff}^{(2)} e^{\omega} e^{\omega}|^2 (I^{\omega})^2, \tag{1}$$

where C denotes a constant, e^{ω}, $e^{2\omega}$ polarization vectors for the fundamental
and the SH wave, and I^{ω} the intensity of the fundamental radiation. In the
present case the second order surface susceptibility $\chi_S^{(2)}$ contains contribution from the metal electrons and the adsorbate.

$$\chi_S^{(2)} = \chi_{S,Pd}^{(2)} + e^{i\phi} \chi_{S,Ad}^{(2)} . \tag{2}$$

From a least squares fit to the data we conclude that $|\chi_{S,Ad}^{(2)}| \approx 8.1 *$
$|\chi_{S,Pd}^{(2)}|$ and a phase angle of $\phi \approx 60^{\circ}$. At the wavelength of the Nd:YAG laser
the susceptibility of the adsorbate is enhanced by an electronic resonance.
Free S atoms show a $3p^4$ $^1D_2$ resonance at 9239.0 cm^{-1}, only 157.7 cm^{-1} or
19.6 meV off the fundamental photon energy.

[1] Y.R. Shen, J. Vac. Sci. Techn. B 3, 1461 (1985); Ann. Rev. Mat.
Sci. 16, 69 (1986)
[2] H.W.K. Tom et al., Phys. Rev. Lett. 52, 348 (1984)
[3] G. Comsa, R. David, B.-J. Schumacher, Surf. Sci. 95, L210 (1980)

Narrow Resonances in CdSSe Doped Glasses:
An Application of Frequency Domain 4-Wave Mixing

J.T. Remillard and D.G. Steel

Departments of EECS and Physics, University of Michigan,
Ann Arbor, MI 48109, USA

Nonlinear laser spectroscopy enables the measurement of various relaxation rates in materials associated with states and superposition states and is based on the study of the temporal and spectral properties of the third order nonlinear optical susceptibility. The method enables the measurement of relaxation rates which cannot necessarily be measured by linear optical spectroscopy. In this work, we use cw frequency domain backward nearly degenerate 4-wave mixing (NDFWM) [1] to study relaxation in glasses doped with microcrystallites of CdSSe.

Earlier measurements of the nonlinear response of CdSSe doped glasses have been made using sub-picosecond dye lasers in transient four-wave mixing and have shown fast relaxation times [2]. However, using cw NDFWM at low powers ($\cong100$mW/cm^2), we show a new contribution to the third order optical susceptibility resulting in an ultranarrow resonance and a long, strongly temperature dependent relaxation time. The laser is tuned near the bandedge and the forward pump and probe beams interact in the material to produce a spatially modulated excitation similar to a grating. A backward pump beam scatters off of this excitation to produce a coherent signal beam.

The excitation relaxation rate is determined by measuring the nonlinear susceptibility as a function of the forward pump-probe detuning, δ. Physically, the detuned probe results in a moving spatial excitation or grating. As δ becomes larger than the excitation decay rate, the grating "washes out", resulting in a reduced signal. Fig. 1 shows a NDFWM spectrum recorded at 298K using the method of correlated optical fields to eliminate the effects of laser jitter [1]. The FWHM is 4500Hz, corresponding to a decay time of 70μsec. The linewidth is plotted as a function of inverse temperature in Fig. 2.

To assist in interpreting this data, we note that such long lifetimes are consistent with the assumption that the excitation results in filling a trap. Figure 3 shows a possible energy level diagram. For the experiment described above, the generation of the electron-hole pair is proportional to ($\mathscr{E}_f \mathscr{E}_p^*$ + c.c.) resulting in a spatial modulation associated with the pair ($\mathscr{E}_i$ is an optical field.) The forward pump and probe are resonant with the valence band to conduction band transition with energy difference $E_c - E_v$ as shown in the figure. As a result of the decay to the traps, the filled traps are also spatially modulated.

The signal results from a scattering of the backward pump off the resultant absorption and dispersion grating produced by the spatially modulated traps. The backward pump beam

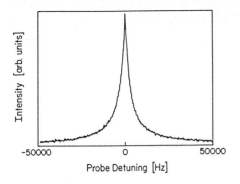

Fig. 1 NDFWM spectrum

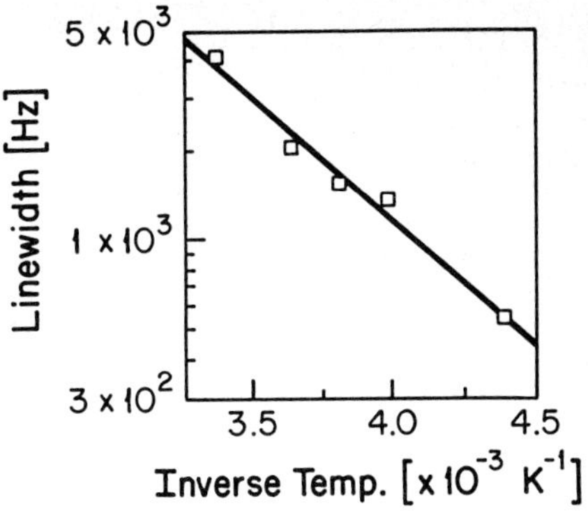

Fig. 2 Linewidth vs 1/T

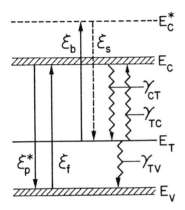

Fig. 3 Energy level diagram

interacts with the traps via a dipole between the traps and a state in the continuum with energy E_c^*.

Based on the above model and a rate equation approximation, the third order nonlinear polarization is given by:

$$P_{NDFWM}^{(3)} = \frac{c}{16\pi} \frac{\alpha_o(\omega)}{\hbar\omega} \tilde{\chi}_T(\omega) \mathcal{E}_f \mathcal{E}_b \mathcal{E}_p^* e^{-i\vec{k}_p \cdot \vec{x} - i(\omega-\delta)t}$$

$$\frac{1}{\left[i\delta + \gamma_{cv} + \gamma_{CT}\left[1 - \frac{\gamma_{TC}}{i\delta + \gamma_{TV} + \gamma_{TC}}\right]\right]} \frac{\gamma_{CT}}{(i\delta + \gamma_{TC} + \gamma_{TV})} + c.c.$$

where α_o is the absorption coefficient for a transition between the valence band (v) and the conduction band (c). $\tilde{\chi}_T$ is the linear susceptibility associated with a filled trap, γ_{ij} is the decay rate from level i to level j. In the limit that γ_{cv} is large compared to the other decay rates, we see that the decay rate of the nonlinear response is dominated by the inverse trap lifetime, $\gamma_{TC} + \gamma_{TV}$. We see that the NDFWM spectral response is a

265

Lorentzian (the signal is proportional to $|P^{(3)}|^2$) with a width determined by the inverse trap lifetime. If this lifetime is determined by phonon deexcitation, and we assume the usual exponential dependence of the lifetime on temperature, then we estimate a trap activation barrier of 150meV.

This work was supported in part by AFOSR and ARO.

1. D.G. Steel and S.C. Rand, Phys. Rev. Lett., 55, 2285 (1985).
2. P. Roussignol, et al., J. Opt. Soc. Am. B 4, 5 (1987). G.R. Olbright, et al., submitted to Phys. Rev. Lett., (1987).

Nonlinear Doppler-free Spectroscopy
of Gas-Phase Atoms at Glass-Vapor Interfaces

S. Le Boiteux, P. Simoneau, D. Bloch, and M. Ducloy

Laboratoire de Physique des Lasers, Université Paris-Nord,
av. J.-B. Clément, F-93430 Villetaneuse, France

Up to now, Doppler-free spectroscopy has been applied to atoms inside a gas sample ("*volume*" spectroscopy). We present here recent extensions of these methods to atoms in the vicinity of a *surface*. The basic principle consists of monitoring *reflectivity changes* produced when a laser beam ("probe"), incident on a glass-vapor interface, is brought into resonance with an atomic transition, and an auxiliary "pump" beam saturates both absorption and index of the atomic vapor. By comparison with ordinary transmission spectroscopy, the expected signal is $\approx \lambda/L$ smaller (L : cell length), which corresponds to the fact that the vapor is explored on a depth of the order of the wavelength only, and thus needs high-sensitivity (quantum-limited) detection methods [1]. In this article, we present saturated absorption and dispersion experiments performed either in evanescent wave spectroscopy, or in normal selective reflection, and discuss several extensions, now in progress.

A first series of experiments [2] has been performed in a Na cell, ended by a glass prism (index $n_1 \approx 1.45$), which is irradiated by two counter-propagating beams (issued from a c.w. single-mode dye laser), with incidence angle (θ_1) near the critical angle for total internal reflection $\left[\theta_c = \sin^{-1}(1/n_1) \right]$ (Fig. 1). One beam (pump) is amplitude-modulated at high frequency (≈ 1 MHz). The second beam (probe) is directed, after reflection on the interface, to a fast photodiode, which monitors the reflection modulation induced when the laser frequency is scanned across the Na (D_1 or D_2) resonance lines. As seen in Fig. 2, recorded D_2 spectra exhibit *Doppler-free absorption lineshapes* for $\theta_1 > \theta_c$, which originate in a saturated absorption (SA) process between the two evanescent waves (EW), for which the real components of the wave vectors are opposite. The EW pump modulates both absorption (δn_I) and dispersion (δn_R) indices of the vapor, as viewed by the counter-propagating EW probe (complex index change, $\delta n_2 = \delta n_R + i\delta n_I$). Like in ATR methods, due to energy balance between EW and reflected waves, the absorption process alters the probe reflection. For incidences far from the critical angle $(\Psi = \theta_1 - \theta_c \gg |\delta n_R|)$, the real index change δn_R does not modify the

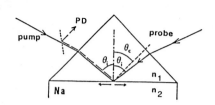

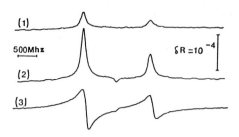

Fig.1 : Principle of EW Doppler-
free spectroscopy

Fig.2 : Probe reflectivity change
around the D2 line for
ψ = 1.4 mrd (1), 0.4 mrd (2) and
- 0.6 mrd (3). The hyperfine
structure of the 3S state is
well resolved (cf.[2]).

critical angle sufficiently to alter the total reflection process. The
Doppler-free nature of the resonances results from a velocity-selection
process along the surface (and in the incidence plane), and is analogous
to velocity selection in volume SA with freely-propagating waves. When
the incidence angle comes closer to the critical angle, the probe
reflectivity change increases sharply, to reach about 10^{-3}. At the
critical angle, the absorption lineshapes are dramatically altered, and
suddenly turn into *saturated dispersion* (SD) lineshapes, for $\theta_i < \theta_c$
(Fig. 2). This remarkable change (which appears in less than 1 mrd) may
be understood if one realizes that, in the *transmission mode* ($\theta_i < \theta_c$),
the probe reflection is no longer total, and undergoes very fast varia-
tions with the angular detuning, $\Psi = \theta_i - \theta_c$. Since θ_c is sensitive to
the *real* part of the refractive index, δn_2, any change in δn_R (i.e., Na
vapor dispersion) will affect the probe transmission, and thus its
reflection. This viewpoint is substantiated by means of Fresnel relations, which show that the probe reflectivity change is proportional to
$\delta n_I/\sqrt{\Psi}$ (for $\Psi > 0$) and $\delta n_R/\sqrt{|\Psi|}$ (for $\Psi < 0$) [with
$|\delta n_2| \ll |\Psi| \ll 1$]. The sharp increase of the signal for $\theta_i \to \theta_c$ is
correlated to the simultaneous increase of the EW amplitude. The reso-
nance linewidth is minimum for $\theta_i = \theta_c$, and increases on both sides. In
total reflection ($\theta_i > \theta_c$), the main broadening mechanism is caused by
atomic transit time in the evanescent wave. In the transmission mode
($\theta_i < \theta_c$), the two transmitted beams are no longer exactly counter-propagating, which produces a residual Doppler broadening. When the two incident beams are carefully collimated, the resonance linewidth (FWHM) has
been lowered to 40 MHz for the main SA resonances, and 20 MHz for some
fluorescence-induced cross-over resonances [2]. Figure 3 shows the
$3S_{1/2}$ (F=2) $\to$ $3P_{1/2}$ (F = 1, F = 2) D_1 resonances, and their (3S - induced)
cross-over for various incident polarizations. As expected from Fresnel

268

relations, the signal amplitude is generally larger for all
polarizations in the incidence plane. However, the amplitude of the
cross-over strongly depends on the beams' relative polarization. A cell
experiment ("volume" SA) has shown us that its peculiar behavior should
be related to its high sensitivity to the polarization-dependent optical
pumping time. We have also compared the position of all these EW resonances with the corresponding (volume) SA resonances : no systematic
frequency shift has been observed, which indicates a wall-induced shift
in EW spectroscopy smaller than 5 MHz.

Two extensions of the above work can be considered :
(i) the first one utilizes *fiber-optic evanescent waves.*If an optical
fiber without cladding is immersed in an absorbing gas,the propagating
light beam is attenuated due to EW resonant absorption in the gas [3].
In the case of a standing wave, a nonlinear coupling between the two
counter-propagating beams may be induced via SA processes in the EW.
This should allow one to perform SA with very long interaction paths
(e.g. very weak transitions).

(ii) The second extension relies on the inhomogeneous wave created via
surface plasmons. If, in the geometry of Fig. 1, a metallic (silver)
film is introduced between the prism and the gas, it is well known that
a surface propagating e.m. mode may be excited at the interface between
metal and gas, for an adequate beam incidence inside the prism. This phenomenon, known as surface plasmon-polariton, which is characterized by
an important e.m. field enhancement, has been used in a variety of nonlinear optics experiments [4]. With a standing-wave excitation, SA processes in the gas should couple the two counter-propagating inhomogeneous waves. A sensitive way of monitoring these couplings uses Attenuated Total Reflection (ATR) methods [4]. Fig. 4 shows the predicted
reflectivity modulation, dR/dn_2, on the interface, for small changes of
the complex gas index ($n_2 = 1 + \delta n_2$) as a function of the beam incidence
angle [5]. θ_p is the phase-matched incidence for the surface plasmon [ap-
proximately given by $(1 + \varepsilon_m) k_p^2 = \varepsilon_m \omega^2/c^2$ i.e. $n_1^2 \sin^2 \theta_p = \varepsilon_m /(1 + \varepsilon_m)$
where ε_m is the - negative - dielectric constant of the metal]. For
$\theta = \theta_p$, the reflectivity modulation originates only in the imaginary
part *(absorption)* of the index change ($dR/dn_R = 0$)while, for $\theta \neq \theta_p$, one
predicts a combination of nonlinear *absorption* and *dispersion* contributions. This comes from the fact that a change in the real index of the
gas simply shifts the plasmon reflection curve, while an imaginary index
change broadens this curve without shift [5]. An experimental check of
these predictions in Na is under way. Note that the important increase
in the e.m. field intensity (a factor 100 at $\theta = \theta_p$) should allow the
observation of higher-order nonlinear processes (two photon absorption
in the inhomogenenous wave, etc ...).

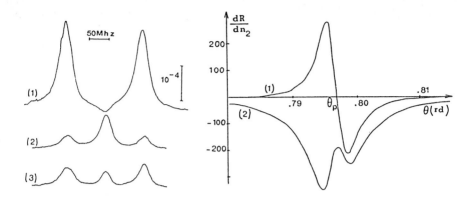

Fig.3 : Partial D1 reflectivity spectrum. Pump and probe polarization are parallel (1), or perpendicular (3) to the incidence plane. (2) : cross-polarized beams.

Fig.4 : Plasmon reflectivity modulation for real (1) or imaginary (2) changes of the gas index, n_2 (λ = 589 nm ; ϵ_m = - 15 + 0.4i ; Ag film thickness : d = 50 nm ; θ_p = .797 rd ; cf. [5]).

An alternate method to perform Doppler-free spectroscopy at an interface is to use a conventional SA geometry (two beams counter-propagating in a cell) and monitor the probe reflection on the entrance window (instead of its transmission). The modulation signal reflects the vapor *dispersion* as saturated by the counter-propagating pump beam. Contrary to the above methods, velocity selection is now operated perpendicularly to the surface. This has been experimentally demonstrated on the 3S-3P-4D cascade-up three-level system of Na [6]. The pump beam is tunable across the upper transition ($3P_{3/2}$-4D ; λ = 568 nm), while the probe beam is resonant for the $3S_{1/2}$ (F = 2) → $3P_{3/2}$ (F = 3) D_2 transition. Saturated selective reflection signals exhibit a resonance doublet associated to the $4D_{3/2}$, $4D_{5/2}$ fine structure (Fig. 5). Note that, as the 3P-4D transition is transparent to the pump, the signal mainly originates in *two-photon* dispersion [6]. By means of a proper choice of the beam frequencies, it becomes possible to select an arbitrary velocity group. This should allow one to monitor atoms either arriving onto, or leaving the surface, and, in particular, explore the transient behavior experienced by the latter ones.

In conclusion we have observed nonlinear Doppler-free resonances in selective reflection from a glass-vapor interface both near the critical angle and at normal incidence.EW spectroscopy makes possible Doppler-free diagnostics of optically thick media, and high-resolution

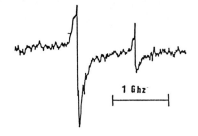

Fig.5 : Saturated selective
reflexion spectrum of sodium at
normal incidence (cf. [6]).

spectroscopy without light propagation in the sample.Since the probe reflectivity monitors the refractive index close to the interface, these
spectroscopic techniques yield a powerful tool to analyse the interactions between surface and gas-phase atoms at intermediate range
(0.1 - 1 μm), with both high sensitivity and resolution, and versatile
velocity selection. This could be applied to atoms-wall collisions
(transient atomic response, velocity re-distribution, dynamical behavior
of metastable species, ...) and radiative behavior of isolated atoms
near a dielectric or metallic surface : partial inhibition or enhancement of spontaneous emission [7] (experiments under way with surface
plasmons), etc. This work could be extended to the various nonlinear
techniques developed in ""volume" spectroscopy (EW two-photon fluorescence, four-wave mixing ...), as well as to normally forbidden transitions (possibly allowed in the vicinity of surfaces).
The authors would like to thank J.R. Leite, C.B. de Araujo and M. Fichet
for stimulating discussions and contributions to parts of this work

[1]   R.K. Raj, D. Bloch, J.J. Snyder, G. Camy and M. Ducloy,
        Phys. Rev. Lett. 44 (1980) 1251.
[2]   P. Simoneau, S. Le Boîteux, C.B. de Araujo, D. Bloch,
        J.R.R. Leite and M. Ducloy, Opt. Commun. 59 (1986) 103.
[3]   H. Tai, H. Tanaka and T. Yoshino, Opt. Lett. 12 (1987)
        437.
[4]   See, e.g. Y.R. Shen, Nonlinear Optics (Wiley, New York,
        1984) and references therein.
[5]   M. Fichet and M. Ducloy, to be published.
[6]   S. Le Boiteux, P. Simoneau, D. Bloch and M. Ducloy,J. Phys. B 20
        (1987) L 149.
[7]   W. Jhe, A. Anderson, E.A. Hinds, D. Meschede, L. Moi and
        S. Haroche, Phys. Rev. Lett. 58 (1987) 666.

Laser Photothermal Surface Spectroscopy

C. Karner, Q. Kong, G. Schmidt, and F. Träger

Physikalisches Institut der Universität Heidelberg,
Philosophenweg 12, D-6900 Heidelberg, Fed. Rep. of Germany

Recently, photothermal and photoacoustic techniques have been used in a number of experiments as versatile tools for surface studies [1-3]. Particularly in combination with laser excitation, these methods offer advantages like high sensitivity, high spectral and high temporal resolution. This makes it possible to investigate adsorbate-substrate as well as adsorbate-adsorbate interactions even for coverages as low as a few thousandths of a molecular monolayer [4]. In addition, photothermal laser spectroscopy has been used e.g. to study phase transitions at surfaces [5] or to investigate desorption phenomena [6]. Experiments were performed with continuous-wave as well as pulsed lasers and in combination with contact and non-contact surface probing.

This paper presents new results obtained by nanosecond time-resolved photothermal displacement spectroscopy with lasers [7,8]. In these experiments a thermoelastic surface deformation is induced with laser light and observed by measuring the change of the reflection angle of a low intensity probe laser beam. Measurements were performed with two goals in mind. First, a thermoelastic surface deformation reflects the temperature of the surface and its variation as a function of time. The measurement of surface temperatures is of vital significance for studies of desorption phenomena and laser-induced etching processes, or more generally speaking, for the understanding of surface reaction dynamics. Secondly, a time-resolved photothermal displacement signal reflects directly *both* the optical properties which can be derived from the amplitude of the signal and the thermal properties of the surface which follow from an analysis of the decay of the signal. One would therefore anticipate that the method can provide an interesting alternative for the existing techniques of thin film analysis.

In our experiments the beam of an excimer laser pumped dye laser with a pulse duration of 15 ns impinges on the sample surface. Part of the incident photon energy is absorbed and converted into heat by radiationless deexcitation. As a consequence, the temperature rises and the surface expands locally. The resulting displacement is probed with a He-Ne laser, whose beam is reflected from the shoulder of the thermoelastic surface deformation. The spatial deflection of the probe beam is measured with a position sensitive detector. The signal is processed by a transient digitizer with a sampling rate of 100 MHz. Actually, a *single* laser pulse is sufficient for registration of the time-resolved signal. The sample is attached to a manipulator in an ultrahigh vacuum system (base pressure 10^{-10} mbar). In order to facilitate the optical alignment of the laser beams, a precisely machined stainless steel block was installed inside the vacuum chamber. It serves for two purposes: Firstly, it carries the lenses for focussing of the laser beams. Secondly, the block defines the position of the surface under investigation. This allows for spatial translations or rotations of the sample with the manipulator in the vacuum system, e.g. for thin film deposition or surface analysis with ESCA, and simultaneously guarantees a reproducible realignment with respect to the foci of the light beams. There is also provision for heating and cooling of the sample and for coverage measurements with a quartz crystal microbalance.

Different samples such as metal surfaces and silicon wafers have been investigated under ultrahigh vacuum conditions. Also, thin films were studied. The signals generally exhibit a rapid increase when the laser pulse hits the surface followed by a slower decrease. The amplitude reaches its maximum value within about 40 ns. The decay typically lasts several microseconds and reflects the heat diffusion from the surface layer to the bulk of the material. The shape of the surface deformation and its change as a function of time have also been measured. Initially, the deformation is Gaussian reflecting the Gaussian laser beam profile. As the heat flows to the bulk of the material, however, the shape changes and can be described only by a complicated, non-Gaussian function. A comparison of the time-dependent signal with calculations of laser induced surface temperature changes has been made also. For this purpose one has to take into account that the surface deformation results from heating of a layer which is typically a few hundred Ångstroms thick whereas for most surface studies the temperature in the first few layers is of relevance. We conclude from our measurements that the calculations of Bechtel [9] provide a good description of transient surface temperatures.

Thin film studies have been performed e.g. with copper and stainless steel surfaces on which silver was deposited in situ. The shape of the time-resolved photothermal displacement signal was monitored as a function of film thickness. Even in the submicrometer range dramatic changes as compared to clean surfaces were observed. For the stainless steel substrate, for example, the amplitude decreases as a function of film thickness and the signal decays more and more rapidly. This reflects the larger reflectivity and the higher thermal diffusivity of Ag as compared to the substrate.

In addition to the measurements outlined above, theoretical calculations of the spatial and temporal temperature distribution at the surface have been performed. The goal is to provide an absolute calibration of the displacement effect, so that the technique can be used as an ultrafast thermometer for the determination of surface temperatures, e.g. during laser induced chemical reactions or in connection with laser induced desorption experiments.

In conclusion, we have described a very versatile surface probe with pulsed lasers. Advantages are instrumental simplicity, real time probing of transient heating processes at interfaces and the possibility to determine optical and thermal properties with a single laser pulse. In addition, the technique is non-destructive and does not require mechanical contact with the sample. It can be applied to clean and adsorbate covered surfaces as well.

References:
1. F. Träger, H. Coufal, T.J. Chuang, Phys. Rev. Lett. **49**, 1720 (1982)
2. F. Träger, H. Coufal, T.J. Chuang in "Surface Studies with Lasers", Springer Series in Chemical Physics Vol. 33, p. 110, Springer Verlag, Berlin, Heidelberg, New York, Tokyo, F.R. Aussenegg, A. Leitner, M.E. Lippitsch, eds. (1983)
3. P.E. Nordal, S.O. Kanstadt, Physica Scripta **20**, 659 (1979)
4. H. Coufal, F. Träger, T.J. Chuang, A.C. Tam, Surface Science **145**, L504 (1984)
5. H. Coufal, W. Lee in "Laser Processing and Diagnostics", Springer Series in Chemical Physics Vol. 39, p. 25, Springer Verlag, Berlin, Heidelberg, New York, Tokyo, D.Bäuerle ed. (1984)
6. I. Hussla, H. Coufal, F. Träger, T.J. Chuang, Ber. Bunsenges. Phys. Chem. **90**, 240 (1986)
7. M.A. Olmstead, N.M. Amer, S. Kohn, D. Fournier, A.C. Boccara, Appl. Phys. **A 32**, 141 (1983)
8. C. Karner, A. Mandel, F. Träger, Appl. Phys. **A 38**, 19 (1985)
9. J. Bechtel, J. Appl. Phys. **46**, 1975 (1975)

Dynamics and Topography
in Molecular Beam Scattering on Surfaces:
A Comparative Study of NO
on Diamond and Graphite

J. Häger, C. Flytzanis*, and H. Walther

Max-Planck-Institut für Quantenoptik,
D-8046 Garching, Fed. Rep. of Germany

The study of molecular beam scattering from solid surfaces /1/ combined with state-selective laser techniques for the diagnostics of the scattered molecules allows us to address some fundamental problems concerning the dynamics of the molecule-surface interaction. This problem is particularly intriguing compared to the one occurring in molecule-molecule collisions since the degrees of freedom of the two interacting partners are of fundamentally different character, individual and collective, respectively. The way the different degrees of freedom adapt to each other during the scattering process, how they exchange energy or how they equilibrate is of paramount fundamental and technological interest. Accordingly, in order to extract the relevant information from these investigations it is essential to maintain the specificity of each part, of the molecule as well as of the surface, in the description of the interaction and in the interpretation of the results.

Although the technique by now has been applied /1/ to a number of surfaces their apparent differences, metallic versus ionic, hardly outweigh their similarities and the features they share in common: They can all be visualized as close-packed plane, regular, isotropic arrays of spherical "atoms". This is the experimental situation. On the theory side /2/, with a few exceptions /3/, the point of view that prevails emphasises the atomic features at the expense of the surface-specific ones. Each surface "atom" is assumed to individually "collide" with the incident molecule and to experience its potential simultaneously as it experiences the potential of each of the other surface atoms. A rough agreement with experimental results is obtained by this approach. In more complicated cases, in particular with surfaces with covalent bonding, this type of calculation has not yet been applied.

Diamond is such a case and, as will be shown below, for the first time brings into focus some unexpected surface – specific features particularly when compared with graphite. As a matter of fact, diamond and graphite allow one to conduct a unique comparative study of such features as energy transfer and accomodation. The results for graphite have been reported /3/; here we concentrate on diamond.

Diamond and graphite structures are schematically depicted in Fig. 1. Along with the notable differences in the character of the chemical bonding, there are some striking differences in the topology of the two surfaces. Thus the uppermost plane of the graphite surface is a regular, planar

---

* Permanent Address: Laboratoire d'Optique Quantique, CNRS, Ecole Polytechnique, 91128-Palaiseau, Cedex, France

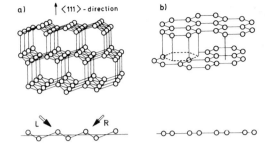

a) ↑ ⟨111⟩ - direction   b)

Fig.1 Diamond (a) and graphite (b) structures. The
diamond structure is so
oriented that one sees the
sawtooth profile of the surface as well as the channels
(in a different but equivalent surface).

array of strictly planar carbon hexagons and possesses the inversion symmetry; the plane below it has exactly the same structure but is shifted so
that its carbon atoms (hexagon corners) are located vertically beneath the
hexagon center of the uppermost plane and similarly for the subsequent
planes so that the surface shows no reconstruction (horizontal motion) or
hardly any relaxation (vertical motion). The same seems /4/ to be the case
for the (111) diamond surface but for drastically different reasons; this
is the surface of easiest cleavage as it requires the breaking of the smallest number of bonds per unit area and shows no reconstruction when kept
below 1000 °C. When viewed from the top it also appears as a regular array of
hexagons but these are tilted and not planar; indeed because of the rigid
tetrahedral bonding in diamond the atoms alternate around each ring in an
upper and a lower position, each group of three equivalent atoms forming an
isosceles triangle; the surface profile along any of the three equivalent
directions on the surface has a sawtooth appearance with asymmetric triangular "grooves", schematically depicted in Fig. 1. This implies that the
diamond surface is non centrosymmetric and thus differentiates between
"left" (L) and "right" (R) incoming molecules.

In addition to this asymmetric geometrical corrugation, another feature
distinguishes diamond from graphite and also from other metallic and ionic
surfaces: in the directions perpendicular to the large face of each sawtooth groove the crystal is pierced with channels of infinite extent that
end up in the surface and may act as traps with pronounced spatial asymmetry. Because diamond possesses an exceptionally high Debye temperature,
$\theta_D \approx 2000$ K for the bulk and ≈ 1800 K for the surface, a unique case in solid
state physics, all other materials having θ_D close to room temperature, the
sharpness of these features is not smeared out by thermal motion. One may
say that diamond possesses a "hard" surface par excellence compared to
other surfaces which may be termed as more or less "soft" and we anticipate
this to have important implications in the energy transfer and accomodation. In particular we expect the scattering to be dominated by geometrical
features.

The details of the experimental setup have already been described /3/ in
connection with the investigation on graphite. It suffices to repeat that
the scattering and the diagnostic take place in an ultrahigh vacuum chamber. A molecular beam of NO molecules, which can be seeded by Helium, if
necessary, is scattered from a polished diamond surface in the center of
the UHV chamber held at a background gas pressure of 10^{-10} mbar. The surface (6 mm x 4 mm) was obtained by polishing type II diamond along the
(111) plane and subsequently cleaning it: Throughout the preparation stage
and the experiments the surface temperature never exceeded 1000 °C, so that
we worked with an unreconstructed surface. The experimental setup allows

the measurement of the angular, rotational, vibrational and velocity distributions of the scattered NO molecules for different incident angles and surface temperatures ranging from $T_s=200$ K up to $T_s=700$ K; the measurements were repeated for two orientations of the sample obtained by turning it by 180°.

The influence of the left and right surface asymmetry is strikingly exhibited in the angular distribution of the scattered molecules recorded with the mass spectrometer and is depicted in Fig. 2: the single narrow lobe obtained in one direction contrasts with the broader and probably double lobe recorded in the opposite direction; this can be qualitatively traced back to the specific geometrical features of the diamond surface outlined above.

The same influence is also reflected by the two traces, Fig. 3, of the flux profiles of the scattered molecules obtained by the time-of-flight technique. For a given rotational level this is composed of two components, a sharp one and a broader one with a substantially longer flight time relative to the first one. The relative importance of the two components differs for the left and right directions and, in addition, depends on the surface temperature, the rotational state and the exit angle. The sharp component may be attributed to "fast" molecules leaving the surface right after experiencing a single (or possibly double) hard collision on the corrugated surface. The broad component corresponds to "slower" molecules showing a very long residence time on the surface ($\approx$1ms) and an asymmetric angular distribution. We therefore anticipate that these molecules were geometrically trapped inside the channels and reemerge, after "multiple" collisions on the hard walls of the channel in a prefered direction that reflects the channel direction rather than kinematic features. One expects then for the second component that the translation and rotation distributions are in equilibrium to each other characterized by roughly a single temperature; this was corroborated by our measurements on the rotational energy distribution.

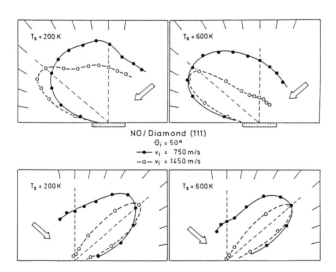

NO / Diamond (111)
$\Theta_i = 50°$
—•— $v_i = 750$ m/s
—o— $v_i = 1450$ m/s

Fig.2 Right (above) and left (below) scattering asymmetry for two diamond surface temperatures (T_s) and two incident NO beam velocities (v_i); incidence angle $\theta_i = 50°$.

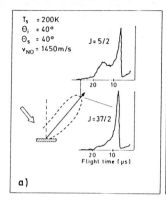
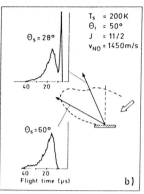

Fig.3 Time-of-flight profiles of scattered NO molecules in incident from left (a) and right (b) on a diamond surface.

The determination of the rotational state distribution also brought up some additional features which are strikingly different from those observed in graphite or other surfaces, metallic or ionic. As can be seen in Fig. 4a the scattered molecules of both velocity components emerge after the scattering with rotational state distributions corresponding to a Boltzmann distribution but with different rotational temperatures. It could be shown that the rotational energy almost entirely comes from the translation energy of the molecules: the diamond surface provides a very efficient mechanism for this energy conversion but does not get involved energetically in

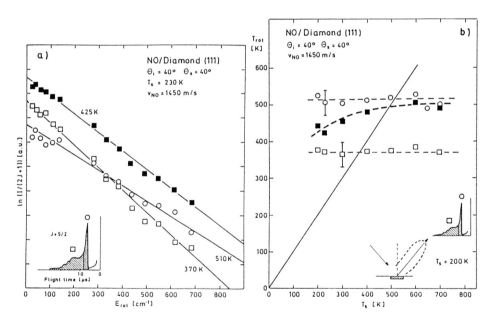

Fig.4 (a) Rotational population distribution of "fast" and "slow" NO molecules after scattering from a diamond surface; the straight lines are best fit Boltzmann distributions. (b) Rotational temperatures of scattered NO molecules versus diamond surface temperature for "left" beam incidence.

the process. A convincing indication for this is provided by the absence of any relation between the surface temperature and the rotational temperature of the scattered molecule for both velocity components demonstrated for "left" beam incidence in Fig. 4. Evaluating T_{rot} without distinguishing between fast and slow molecules leads to a variation of T_{rot} with T_s due to the temperature-dependent velocity components in the time-of-flight spectra (black squares in Fig. 4).

The above observations indicate that the attractive part of the diamond surface potential is negligible and the incident molecules either experience a single (or possibly double) collision with a hard core potential which strongly couples its translation and rotation without bringing them to equilibrium to each other or they plunge and get geometrically trapped into the channels where they experience many collisions before they emerge again in preferred directions, translation and rotation in equilibrium.

In conclusion, the study of diamond has shown that the topography of the surface can play a crucial role in the dynamics of molecule surface interaction. In this respect the left and right asymmetry is particularly striking and has some important implications in the energy transfer and accommodation.

Acknowledgement

One of the authors (C.F.) thanks the Alexander von Humboldt Foundation for a senior Humboldt award.

References

1. see for instance J.A. Barker and D.J. Auerbach, Surf. Sci. Repts. 4, 1 (1985)
2. see for instance references in J.W. Gadzuk, J. Chem. Phys. 86, 5196 (1987) or in Ref. 1
3. H. Vach, J. Häger, B. Simon, C. Flytzanis, and H. Walther, MRS Symposia Proceedings 51, 25 (1985, Materials Research Society)
4. P.G. Lurie and J.M. Wilson, Surf. Sc. 65, 453 (1977), or B.B. Pate, ibid 165, 83 (1986)

Multiphoton-Stimulated Emission of Electrons and Ions from the (111) Surface of BaF$_2$

H.B. Nielsen[1], *J. Reif*[1], *E. Matthias*[1], *E. Westin*[2], *and A. Rosén*

[1]Fachbereich Physik, Freie Universität Berlin,
D-1000 Berlin 33, Germany
[2]Institute of Physics, Chalmers University of Technology,
S-412 96 Göteborg, Sweden

The interaction of laser radiation and surfaces of optically transparent ionic materials is an important problem in surface science as well as in applied optics. For the latter, the question of how optical damage occurs, is of fundamental interest. Since the bandgap of ionic crystals is of the order of 10 eV, for visible light only multiphoton absorption can result in energy deposition into the surface. Structural defects, impurities, and adsorbates can generate surface states which are located somewhere in the bandgap, and can resonantly enhance multiphoton photoemission. This, in turn, can lead to desorption of ions and neutrals at comparatively low intensities. In this contribution we present results of photoemission and desorption studies carried out in ultra-high vacuum with cleaved (111) surfaces of BaF$_2$, at laser intensities in the MW/cm^2 range.

In order to obtain qualitative information about the electronic structure of the BaF$_2$ surface, calculations have been carried out which model the crystal surface as a planar cluster in an embedding crystal potential. If we assume a non-stoichiometric surface where the upper layer of fluorine ions is missing - which is described by a Ba$_1$Ba$_6$F$_3$ cluster in C$_{3v}$ symmetry - the calculations predict occupied states near the middle of the bandgap, and in its upper half strong resonances in the density of states, as shown in Fig. 1. The solid arrows indicate a possible resonantly enhanced two-photon photoionization. It is also conceivable that a five-photon absorption starting from the 2p valence band can proceed resonantly enhanced in its last two stages (dotted plus solid arrows).

In our experiments we measured wavelength and intensity dependencies of the yields for photoelectrons and positive ions, in order to identify effects suggested by the predictions in Fig. 1. From the almost iden-

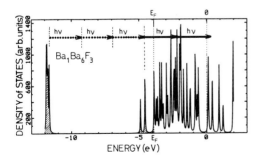

Figure 1. Calculated density of states for a non-stoichiometric cluster (shaded: occupied states)

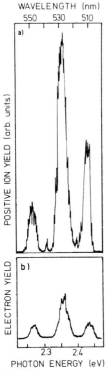

WAVELENGTH (nm)

Figure 2. Wavelength dependence

Figure 3. Intensity dependence

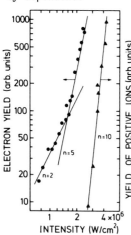

tical excitation spectra in Fig. 2 it is obvious that there is a strong relation between the emission of positive ions (in our case Ba^+, $(BaF)^+$, and F^+) and electrons. Hence we conclude that photoemission is, by creating holes in or near the surface layer, the initial cause for desorption. The strong wavelength dependence in Fig. 2 seems to confirm that resonantly enhanced multiphoton processes, as indicated in Fig. 1, do indeed take place. We also observed a strong dependence of the spectral shapes on the azimuthal orientation of the surface with respect to the laser polarization, which might give information on the spatial form of the surface wavefunctions.

Since the shape of the excitation spectra did not depend on the laser intensity, it was possible to measure the intensity dependence of electron and ion yields at fixed wavelengths. The result is shown in Fig. 3 for electrons (dots) and ions (triangles). At low intensities a two-photon ionization takes place which energetically must start from the middle of the bandgap, thus confirming the theoretical predictions of occupied states in that range. At only slightly higher intensities the slope changes and indicates a five-photon photoemission. Energetically, this could originate from the 2p valence band. Since the spectra remain unchanged, it must proceed, however, through the same intermediate resonances as the two-photon absorption.

The ion yield in Fig. 3 varies with the tenth power of the laser intensity. No such power dependence is observed for electrons. A true ten-photon absorption seems unlikely since it would amount to an initial state at about -24 eV (i.e. two photons below the highest barium band), with only few intermediate resonances near the final stages. Remembering in addition the similar wavelength dependencies for electrons and ions, we interpret the slope of ten as evidence that two holes in the 2p valence band are required for generating positive ion emission.

Although more measurements in other spectral ranges and for different crystals are needed to confirm our observations, the data show that multiphoton spectroscopy will be a valuable tool to investigate electronic surface structures of ionic materials.

This work was supported by the Deutsche Forschungsgemeinschaft, Sfb 6, the Swedish Natural Science Research Council and the Swedish Board of Technical Development.

Four-Wave Mixing Spectroscopy
of Metastable Defect States in Diamond

S.C. Rand

Hughes Research Laboratories, 3011 Malibu Canyon Road,
Malibu, CA 90265, USA

1. Introduction

Efficient, cw four-wave mixing can be observed in solids
whenever resonant excitation of defects is accompanied by
optical saturation at low intensities. Saturation of ground
state absorption generates a large third order nonlinearity and
occurs quite generally when intersystem crossing places defects
(excited on allowed transitions) in metastable excited states.

Recently it was shown[1] that by tuning the frequency of the
probe wave in a nearly-degenerate four-wave mixing (NDFWM)
experiment with stable radio-frequency techniques, spectroscopy
could be performed to study relaxation processes of such
metastable states. Very often, states of the metastable
manifold are inaccessible to direct study by conventional
techniques because transitions to the ground state are
forbidden. The four-wave mixing technique furnishes direct
measurements of metastable decay rates however, even for states
which relax completely by non-radiative means.

Here we present the results of cw NDFWM spectroscopy of the
N3 resonance of the N_3V° color center in diamond, for which the
known energy levels are shown as solid lines in Fig.1. Density
matrix calculations permit a full spectral analysis of the
ultranarrow, nonlinear optical resonance[1] in the ultraviolet
spectral region which reveals the existence of a new deep level
of the defect and furnishes its decay time.

2. Theory

In the low intensity limit, the result for four-wave mixing
signal intensity versus pump-probe detuning can be calculated
using perturbation theory. For a homogeneously-broadened,
four-level system[2] (Fig.1),

$$I_s \propto |\Omega_f \Omega_b \Omega_p{}^* \exp[i(\omega-\delta)t+ik_p x]|^2 \frac{(\gamma_{10} + \gamma_{21})^2 + \delta^2}{(\gamma_{10}{}^2 + \delta^2)(\gamma_{21}{}^2 + \delta^2)} . \qquad (1)$$

Here $\Omega_i = \mu \cdot E_i/2h$ are the three incident fields and γ_{ij} is the
decay rate from level i to j ($\gamma_{31}=\gamma_{20}=0$). Spectral width is
governed by the decay rate of <u>ground</u> state spatial hole-
burning, measuring a relaxation process quite different from
that of conventional excited state decay studies. If only

state $|2\rangle$ is metastable, the spectrum is dominated by a single Lorentzian of width γ_{21}. However, if two metastable states ($|1\rangle$ and $|2\rangle$) exist between $|0\rangle$ and $|3\rangle$, and $\gamma_{10} < \gamma_{21}$, the spectrum may be much narrower.

3. Experiment

A single-mode Kr^+ laser was used at 406 nm together with a 1.75 mm thick sample of natural diamond (OD=0.23). The probe beam was chopped at 10 Hertz for phase-sensitive detection and phase conjugate signal intensity reached values as high as 0.1% for equal pump intensities of only 5 W/cm^2.

Signal intensity versus pump-probe detuning is shown in Fig. 2. A simple Lorentzian expression with a full width γ_{21}/π=436 s^{-1} corresponding to the previously measured[3] population decay rate of state $|2\rangle$ alone (0.73±.03 ms) gives much too broad a spectrum. However a good fit is obtained with γ_{21} in (1) when an additional state (γ_{10}/π=272 s^{-1}) is introduced. The observation that the NDFWM spectrum is narrower than the contribution from $|2\rangle$ is inconsistent with $|2\rangle$ being the lowest excited state of the N_3V° center. It forces us to conclude that there is another metastable state with a decay time of $\tau = (272\pi)^{-1}$=1.2 ms lying deeper in the gap than the N2 electronic origin. This state may account for the unassigned N1 resonance (1.5 eV) in diamond. Supported by AFOSR Contract F49620-85-C-0058.

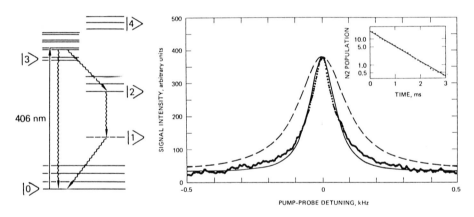

Fig.1.Energy levels of the N_3V° color center, showing the new state $|1\rangle$ (dashed).

Fig.2.N3 NDFWM spectrum. Theory without state $|1\rangle$(dashed); best fit including state $|1\rangle$(solid).

References

1. D.G. Steel and S.C. Rand, Phys. Rev. Lett. 55, 2285(1985).
2. S.C. Rand, Four-Wave Mixing Spectroscopy of Metastable Defect States in Solids, in Lasers, Spectroscopy and Ideas-A Tribute to A.L. Schawlow, Springer-Verlag, to be published.
3. S.C. Rand and L.G. DeShazer, Opt. Lett. 10, 481(1985).

Part VIII

Miscellaneous Laser Spectroscopy Experiments

Four-Wave Mixing and Stimulated Emission Processes in Strongly Driven Systems

Y. Shevy and M. Rosenbluh

Department of Physics, Bar-Ilan University, Ramat Gan, Israel

The propagation of a nearly resonant, intense laser pulse through an atomic vapor leads to the generation of waves at new frequencies which are emitted both parallel to the laser beam and in fully or partially formed concentric cones surrounding it. In spite of the complexities introduced by self-focusing, filament formation and transient effects, much has been learned about the fundamental processes involved in such interactions through measurements of the frequency spectrum of the emitted radiation as a function of laser intensity, detuning, vapor density and buffer gas pressure.

A schematic representation of those nonlinear interactions which lead to stimulated emission for the case of Na are shown in Fig. 1. Diagram (a) shows three photon scattering[1] (TPS)

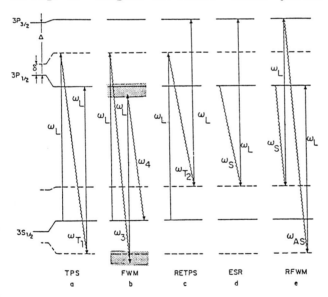

Fig. 1. Nonlinear interactions of an intense laser with Na: (a) three photon scattering, (b) parametric four-wave mixing over a broad region of gain, (c) resonance enhanced three photon scattering, (d) Raman scattering from the excited $P_{1/2}$ state, (e) parametric four-wave mixing involving the Raman scattered Stokes and anti-Stokes photons.

284

which leads to population transfer to the $P_{1/2}$ state and the emission of radiation at $\omega_{T1} = \omega_1 + \Omega'$, where Ω' is the generalized Rabi frequency. In (b) we show the parametric four-wave mixing[2,3] (FWM) process which has significant gain[4] over a range of the frequencies ω_3 and ω_4, provided $\omega_3 + \omega_4 = 2\omega_1$, the new waves have frequencies near ω_{T1} and the atomic D_1 resonance frequency, and the phase mismatch is small. The processes shown in (c)-(e) are possible due to the three-level nature of sodium. Population transfer via TPS to the $P_{3/2}$ level can be resonantly enhanced as shown in (c) and results in an emission at ω_{T2}. For high laser intensities parametric FWM near the D_2 resonance and ω_{T2} can also take place but is not shown. A population difference between the excited states leads to excited state Raman scattering (ESR) at either the Stokes frequency $\omega_s = \omega_1 - \Delta$, as shown in (d), or at the anti-Stokes frequency $\omega_{AS} = \omega_1 + \Delta$ (not shown), depending on the sign of the population difference. Raman type FWM, as shown in (e), is possible in a parametric process[3].

Not all of these interactions occur simultaneously under all the experimental conditions. We have shown[5] that for quite general conditions stimulated TPS and parametric FWM interfere destructively and suppress the emission at ω_{T1}. Thus FWM dominates the spectrum (and TPS is absent) for experimental conditions, such as self-trapped filaments, in which the phase mismatch is small. TPS is observed when the phase mismatch is large and in particular at high laser intensities since the saturation of FWM with laser intensity occurs well before the onset of TPS saturation. The Raman processes require a population in the excited state and are thus observed for high laser intensities and close to resonance where TPS populates the excited state.

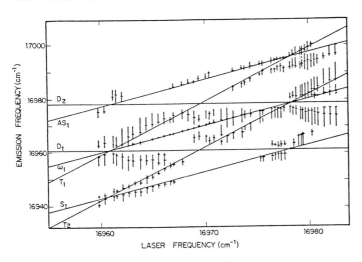

Fig. 2. The experimentally observed stimulated emisssion frequencies as a function of laser frequency. Vertical lines represent emission over a broad frequency range with the peak indicated by the cross. The continuous solid lines show the calculated emission frequencies for an unperturbed Na atom.

In Fig. 2 we show the frequencies of the observed emissions as a function of laser frequency, for experimental conditions under which both parametric FWM and stimulated TPS are present. ($I = 10$ MW/cm^2, Na density - 10^{15} atoms/cm^3 and He pressure of 1 torr.) The vertical bars in the figure represent broad emission over a range of frequencies with the emission peak(s) indicated. The solid lines represent the various processes shown in Fig. 1 assuming negligible light shifts. The broad emissions near to, but displaced from, D_1 and T_1 (and D_2 and T_2) correspond to parametric FWM such as shown in Fig. 1b. For very high laser intensities higher order Raman processes are also observed (not shown in the Figure) in the vicinity of the D lines, where strong RFWM is present.[3]

1) Y. Shevy, M. Rosenbluh and H. Friedmann, Opt. Lett. 11,85(1986).
2) D. J. Harter and R. W. Boyd, Phys. Rev. A 29, 739(1984).
3) Y. Shevy and M. Rosenbluh, Opt. Lett. 12, 257(1987).
4) R. W. Boyd, M. G. Raymer, P. Narum and D. J. Harter, Phys. Rev. A 24, 411(1981).
5) Y. Shevy and M. Rosenbluh, to be published.

Self-Oscillation Due to Four-Wave Mixing and to Pressure-Induced Two-Wave Mixing in Sodium

D. Grandclément, G. Grynberg, and M. Pinard

Laboratoire de Spectroscopie Hertzienne de l'Ecole Normale Supérieure, Tour 12, EO1–4, place Jussieu, F-75252 Paris Cedex 05, France

When a set of two-level atoms interacts with an intense pump wave, several gain processes are possible for a weak probe wave. We study here two of these processes (the four-wave mixing and the pressure-induced two-wave mixing), and we present the characteristics of the oscillating beam obtained when the active medium is enclosed in a ring cavity. The experiments are done with a low-density sodium vapor cell interacting with a c.w. beam whose frequency is nearly resonant with the D_2 transition of sodium. We show that the experimental results depend on the standing or travelling wave character of the pump beam and on the foreign gas pressure. In particular, results similar to those obtained with photorefractive materials |1| are only obtained when a buffer gas is introduced in the sodium cell.

Our experiments are done with a 5 cm quartz cell filled with sodium and enclosed in a ring cavity (Figure 1). The temperature of sodium is 160°C. The pump beam comes from a single-mode c.w. dye laser (P ∿ 300 mW). Its frequency is detuned by ± 2 GHz from the center of the absorption line. Depending on the presence of the totally reflective mirror M_0, the pump beam can be either a travelling wave (Figure 1a) or a standing wave (Figure 1b).

In the case of a travelling pump beam (Figure 1a), no oscillation in the cavity is observed with pure sodium. On the other hand, an oscillation is observed when a small amount of helium (p_{He}∿3 Torrs) is added in the sodium cell. The frequency ω_+ of the oscillating beam differs from the frequency ω_p of the pump beam. The difference $|\omega_+ - \omega_p|$ depends on the length of the cavity and varies between 10 MHz and 50 MHz. (Figure 2). This type of oscillation is very similar to the one already studied in photorefractive materials |1| but for the value of $|\omega_+ - \omega_p|$ which is much larger in our case. In fact, the order of magnitude of $|\omega_+ - \omega_p|$ is given by the inverse of the lifetime of the grating created by the pump beam and the oscillating beam. In our case, this time coïncides with the excited sodium atom lifetime. Collisions are necessary because they provide the extra energy required to really populate the excited state |2|. In fact,

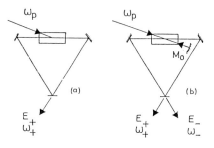

Figure 1 – Scheme of the experimental set-up. Two cases are considered : travelling pump beam (a) and standing pump beam (b).

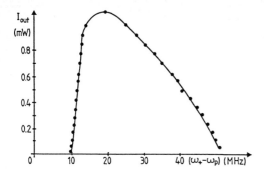

Figure 2 - The intensity I_{out} and the frequency ω_+ of the oscillating beam vary with the length of the cavity. We present here the variation of I_{out} versus $(\omega_+ - \omega_p)$.

the present effect has very close ties with the PIER 4 resonances |3| and can be predicted by studying the pressure-induced contribution to the non-linear susceptibilities |2|.

    In the case of a standing pump beam (Figure 1b), a strong oscillation is observed with sodium cells containing no buffer gas. The gain mechanism here is four-wave mixing. The oscillation now consists of two counterpropagating beams E_+ and E_- of same intensity ($|E_+|^2 = |E_-|^2$) and same frequency ($\omega_+ = \omega_- = \omega_p$). More precisely, we can assert from the study of the interference between two of these beams that $|\omega_+ - \omega_p|$ and $|\omega_- - \omega_p|$ is smaller than 0.1 Hz. The oscillation is thus perfectly degenerate. When this experiment is repeted with a cell containing 3 Torr of Helium, a very different behaviour is observed. The oscillation is now non-degenerate, $\omega_+ - \omega_-$ being of the order of 8 MHz. This new type of oscillation comes from the combined effect of four-wave mixing and pressure-induced non-linearities.

    The study of these types of oscillator can be extended by looking for applications in gyros and by analysing the statistics of emitted photons. In particular, in the case of degenerate four-wave mixing oscillation, it has been shown theoretically |4| that the number of photons in each counterpropagating beam must be equal ($N_+ = N_-$). The observation of such a property would be a new manifestation of squeezing.

1. K.R. MacDonald and J. Feinberg, Phys. Rev. Lett. 55, 821 (1985)
   J. Rajbenbach and J.P. Huignard, Opt. Lett. 10, 137 (1985)
2. G. Grynberg, E. Le Bihan and M. Pinard, J. Physique 47, 1321 (1986)
3. N. Bloembergen, Ann. Phys. Fr. 10, 681 (1985) and references therein
4. R. Horowicz, M. Pinard and S. Reynaud, Opt. Comm. 61, 142 (1987).

Two-Photon-Excited Parametric and Wave-Mixing Processes in Atomic Sodium

P.-L. Zhang and S.-Y. Zhao

Tsinghua University, Beijing, People's Republic of China

A pulsed dye laser beam (ω_L) pumped by a frequency-doubled Nd:YAG laser was focused into a heat-pipe oven containing Na vapor. Six- and four-wave mixing processes and parametric generation were investigated using two-photon excitation of sodium atoms.

1. Coherent Lines Due to Six-Wave Processes

The 7s $^2S_{1,2}$ level in sodium can be populated by two-photon excitation, and a series of cascade-stimulated emissions may occur. The photons of the stimulated emission will mix with the laser photons. Within the 254-349 nm region ten coherent radiation lines, resulting from six-wave mixing, are observed. Their wavelengths are 254.36, 298.93, 307.78, 309.30, 309.46, 309.75, 313.70, 318.81, 322.91 and 348.76 nm. For example, the coupling scheme of the first line is $\omega_{UV} = 2\omega_L - \omega(7S-6P) - \omega(6P-6S) + \omega(5S-4P)$ with 3-fold resonance and 1-fold near-resonance enhancement in the 5th-order susceptibility of the sodium atom. The other lines can be similarly explained. The line intensities, and their dependence on oven temperature, laser energy and wavelength have been experimentally investigated. After calculating the electric polarization using the density matrix method and relating the electric field and polarization by Maxwell's equations, a formula for the generated line intensity has been deduced. The line intensities are then explained qualitatively by the gain coefficients of the stimulated emissions, resonance enhancement in susceptibility and phase-matching conditions.

2. Coherent Lines Due to Four-Wave Mixing Processes

Six lines due to four-wave mixing processes were observed using two-photon excitation of the Na 7s $^2S_{1,2}$ level. Three of them, 254.10, 258.16 and 259.65 nm, are produced by sum-frequency mixing processes, $\omega_{UV} = 2\omega_L + \omega_{SE}$. The other three lines, 298.62, 341.85 and 342.04 nm are produced by difference-frequency mixing processes. Similarly, by using the Na 4d ^2D_J level, five lines are observed from sum-frequency mixing processes (255.78, 257.51, 257.54, 280.47 and 280.51 nm) and four lines from difference-frequency mixing processes (298.83, 298.87, 333.04 and 333.11 nm). It should be noted that for sum-frequency mixing processes the phase-matching condition cannot be fulfilled, therefore their intensities are weak and coherence lengths are calculated instead of phase-matching angles.

A special four-wave mixing process has been found when the laser wavelength is tuned to 574.43 nm, which is not resonant with any real energy level of sodium. The wavelength of the resulting coherent emission is 330.222 nm. Similarly, when the laser wavelength is 574.55 nm the output wavelength is 330.298 nm. After studying the excitation spectrum and the output wavelength

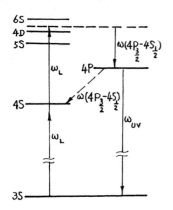

Fig.1 A special 4-wave
mixing process

versus oven temperature, laser wavelength etc., it was determined that the process was a collinear, phase-matched four-wave mixing process, $\omega_{UV}=2\omega_L-\omega(4P_{3/2}-4S_{1/2})$, as shown in the figure on the left. The infrared wave $\omega(4P_{3/2}-4S_{1/2})$ is generated by stimulated emission involving molecular processes of the sodium dimer, and is not generated by stimulated electronic hyper-Raman scattering. According to this mechanism, the calculated wavelengths are 330.212 and 330.291 nm, in agreement with the experimental results.

3. Coherent Lines Due to Parametric Generation

The parametric process $\omega_S+\omega_I=2\omega_L$ was investigated using the Na $4d\,^2D_J$ level. In the parametric processes the signal wave and idler wave grow together, following two-photon excitation. At lower oven temperatures, 280-405°C, two-pair components of the coherent radiation line (near 4D-3P and 3P-3S atomic transitions) are observed. Their wavelengths (568.225, 568.770, 589.072 and 589.650 nm at 381°C) are shifted, compared with the atomic transitions (4D-3P, 568.263 and 568.820 nm; 3P-3S, 589.995 and 589.592 nm) and the wavelength shifts of approximately a few hundreths of a nm increase with increasing oven temperture. These components correspond to the noncollinear phase-matching condition. At oven temperatures of 450°C or higher, the foregoing two-pair components gradually disappear and a new pair of coherent lines (568.443 and 589.412 nm) is generated instead. Their wavelengths do not change, while their intensities increase rapidly as the oven temperature increases. This pair corresponds to the collinear phase-matching condition. Near the 4P-3S transition, four components of parametric generation (330.229, 330.253, 330.299 and 330.315 nm, at 452°C) are observed. The first and third components correspond to the collinear phase-matching condition and the other two correspond to the noncollinear phase-matching condition. By two-photon excitation of the 7s $^2S_{1/2}$ and 8s $^2S_{1/2}$ levels in Na, parametric generation lines at 259.39, 268.04 and 285.29 nm (near 7P-, 6P- and 5P-3S transitions, respectively) are also observed.

The wavelength shift and intensity of the parametric generation lines and their dependence on oven temperature are explained by a theoretical analysis based on the coupled wave equation, atomic polarization and phase-matching condition.

References

1. W. Hartig: Appl. Phys. 15, 427 (1978).
2. A.V. Smith and J.F. Ward: IEEE J. Quantum Electron. 17, 525 (1981).
3. P-L. Zhang and A.L. Schawlow: Can. J. Phys. 62, 1187 (1984).

Narrow Resonances in Four-Wave Mixing Due to Radiative Decay

Jing Liu[1], *G. Khitrova*[2], *D. Steel*[1], *and P. Berman*[2]

[1]Departments of Physics and EECS, University of Michigan,
 Ann Arbor, MI 48109, USA
[2]Department of Physics, New York University, New York, NY 10003, USA

The early work of Bloembergen and co-workers on pressure induced extra resonances (PIER-4) verified earlier predictions that resonances in nonlinear spectroscopy can result from incomplete cancellation of quantum mechanical amplitudes in the presence of collisions[1]. This work was followed by a demonstration of collision induced narrow resonances in backward nearly degenerate four-wave mixing (NDFWM)[2] which have a similar physical origin.

In the current work, we demonstrate that such resonances can be observed in collisionless systems due to spontaneous emission. In a two level Doppler broadened system, the NDFWM response is characterized by two resonances as a function of the pump-probe detuning, δ. The first resonance, at $\delta = 0$, has a linewidth determined by the relaxation of the ground state and excited state. The second resonance, at δ equal to twice the pump-resonance detuning, has a width determined by the decay rate of the dipole coherence. In this work, we are concerned with the first resonance only.

If the system is closed (i.e., the excited state can decay only to the ground state) then the population difference decays at the spontaneous emission rate, and the linewidth of the $\delta = 0$ resonance is given by γ_{sp}. However, if the system is "open"

(for example, the upper state can radiatively decay to a long lived state other than the initial state such as a different

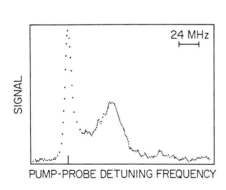

24 MHz

SIGNAL

PUMP-PROBE DETUNING FREQUENCY

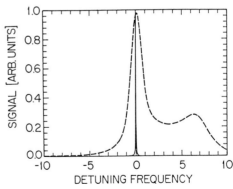

SIGNAL [ARB. UNITS]

1.0
0.8
0.6
0.4
0.2
0.0

-10   -5   0   5   10
DETUNING FREQUENCY

Fig. 1a NDFWM spectra on the $S_{1/2}(F=2)$ to $P_{1/2}(F=2)$ transition.

Fig. 1b Calculation normalized to γ_{sp}. The dashed line accounts for Doppler broadening.

hyperfine or Zeeman sublevel of the ground state), then the population difference decays with two components that have decay rates determined by the inverse upper and lower state lifetimes.

These effects are observed in NDFWM measurements of atomic sodium. Fig. 1a shows the NDFWM spectrum recorded on the $S_{1/2}(F=2)$ to $P_{1/2}(F=2)$ transition. In the absence of the $P_{1/2}(F=2)$ decay to the F=1 ground state, the system would be closed and both of the resonances would have the same width, 20 MHz. However, the observed linewidth of the $\delta=0$ resonance is 7 MHz. This width is in excellent agreement with the analysis (Fig. 1b) based on an open system which gives a nonvanishing amplitude for the narrow resonance (solid curve). The dashed line includes Doppler broadening. The results show a long-lived component in the population difference with a width determined by the transit time and residual Doppler effects. (Note, *unlike a coherent Raman effect, all three input beams are nearly resonant with the* $S_{1/2}(F=2)$ *to* $P_{1/2}(F=2)$ *transition.*)

Fig. 2a shows the NDFWM spectrum recorded on the $S_{1/2}(F=2)-P_{3/2}(F=3)$ transition. The data shows a new feature which is a dip in the first resonance. To understand this, we note that earlier descriptions have ignored the effects of alignment and orientation which contribute to the response in a system with magnetic substate degeneracies. For the F=2 to F=3 transition, the population is conserved and the system is closed for population. However, since the field can transfer net orientation or alignment to the atom, one can show the system is not closed. The resulting ground state alignment and orientation lead to a nonvanishing amplitude for the narrow component which manifests itself as a dip in the resonant structure centered at $\delta=0$. The width of this dip is again determined by the inverse transit time and residual Doppler width. For this transition, the sign of the ground state component is opposite to the sign of the upper state component, producing the dip. Agreement with theory is excellent, as shown (for the first resonance) in Fig. 2b.

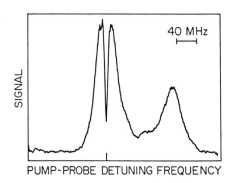

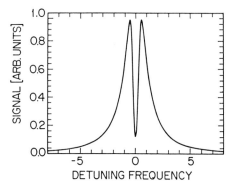

Fig. 2a NDFWM spectrum of the $S_{1/2}(F=2)$ to $P_{3/2}(F=3)$ transition.

Fig. 2b Calculation of the first resonance, normalized to γ_{sp}.

In summary, these data and analysis show the effects of decay of the excited state to a state other than the initial state of excitation. We anticipate that this understanding will be important in understanding frequency or time domain measurements in other materials such as the spectroscopy of resonant systems in solids.

This work was supported by ARO, ONR, and NSF (PHY8415781).

1. N. Bloembergen, in Laser Spectroscopy, edited by H. Walther and K.W. Rothe (Springer, Berlin, 1979), p340.   Yehiam Prior, etal, Phys. Rev. Lett. 46, 111 (1981).
2. J.F. Lam, D.G. Steel, and R.A. McFarlane, Phys. Rev. Lett. 49, 1628 (1982), also 56, 1679 (1986).

Experimental Observation of Nonlinear Birefringence in a J=1/2 to J=1/2 System

D.E. McClelland, R. Holzner, D.M. Warrington, R.J. Ballagh, and W.J. Sandle

Department of Physics, University of Otago,
P.O. Box 56, Dunedin, New Zealand

1. Introduction

There has recently been considerable interest in the non-linear propagation of laser light through birefringent media (e.g.[1-3]). The features of the propagation, whether using a medium with inherent[1], or electrically induced[2], birefringence are basically similar: the principal axes of the ellipse describing the polarisation state of the light are found to either oscillate or rotate with constant or varying handedness depending on initial conditions, particularly on the intensity of the laser field.

We wish to report experimental observation of oscillatory behaviour in sodium vapour for which a propagating laser beam (near D1 resonant) modifies the birefringence induced by a large transversely applied magnetic field. This field serves an important subsidiary purpose of decoupling the hyperfine structure via the Back-Goudsmit effect, thus allowing, for the first time, nonlinear polarisation dependent behaviour near the D1 transition to be quantitatively treated in terms of a J=1/2 to J=1/2 atomic model.

2. Experiment

Ring dye laser light, linearly polarised at 45^0 to a 3.3kG magnetic field (B: oriented perpendicular to the propagation direction), is incident upon a sodium vapour cell. The cell is trapezoidally shaped to allow the dependence of emergent polarisation on propagation distance to be explored. This polarisation was analysed in terms of circular and linear components; measurements were carried out as a function of the laser frequency, scanned over 19 GHz.

Illustrative data for left and right circular components are displayed in Fig.1(a) The basic interpretation of the behaviour is simple. In the large transverse B field, the J=1/2 to J=1/2 linear σ and π resonances are fully resolved (positions are as shown), and the differing refractive indexes and absorptions for these components lead to an emergent polarisation state which is a strong function of laser frequency. Furthermore this 'structure' becomes more pronounced for increasing optical thickness, owing to the phase difference between linear π and σ increasing monotonically with penetration distance. Fig. 2 shows, in an ellipticity (e) versus major-axis angle (Φ) diagram, the way the polarisation progresses with penetration distance at fixed frequency, f. The 'spiraling' of the diagram towards e=0, Φ =-90^0 is a result of linear dichroism (stronger absorption of the π component at this particular frequency).

3. Theoretical Comparison

To obtain the complete saturation behaviour for a J=1/2 to J=1/2 system in a general transverse magnetic field, the density matrix has been analytically solved within the semiclassical approximation and the RWA. With the magnetic field used in the experiment sodium atoms (D1 transition) are well modelled as independent J=1/2 to J=1/2 systems with inhomogeneously distributed component transition frequencies distinguished by the value of nuclear spin com-

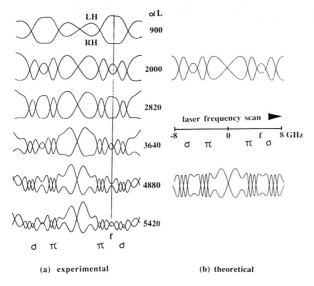

LH — RH — αL 900

2000

2820

laser frequency scan ▶

3640

-8 0   f   8 GHz
σ   π   π   σ

4880

5420

σ   π   π   σ
f

(a) experimental   (b) theoretical

Fig. 1. Right-hand (RH) and left-hand (LH) circularly polarised transmitted signals. Graphs are labelled according to optical thickness αL. Parameter values: B, 3.3kG; laser power, 6mW; beam waist, 90μm; relaxation rates - free atom values except for Γ_{lower}, 1MHz (~inverse transit time).

ponent (-3/2 to +3/2) and by the atomic velocity. The laser field is assumed plane wave and with an initial gaussian intensity envelope; propagation is treated within the slowly varying envelope approximation.

Fig.1(b) compares theoretical predictions with the experimental data for two representative propagation lengths. The homogeneous weak field absorption coefficient is the only free fitted parameter. Excellent structural agreement has been obtained indicating the basic correctness of the model. However, in detail, disagreement in total absorption is apparent particularly in the longer path length scan and the e-Φ diagram. Further experiments are underway to elucidate this.

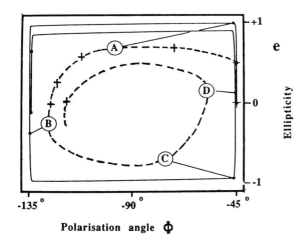

Ellipticity

+1

A

D

0

B

C

-1

-135°   -90°   -45°

Polarisation angle Φ

e

Fig 2. Plot of polarisation ellipticity vs. major-axis angle, parametrised by optical thickness, at frequency f (see Fig.1.) Dashed curve indicates the progression of experimental points (+). Comparison with theory (solid line) is explicitly indicated at A (αL=900), B (2000), C (2820) and D (3640).

295

References

1. H.G. Winful: Opt. Lett.11, 33 (1986).
2. K.L. Sala: Phys. Rev. A29, 1944 (1984).
3. E. Wielgolaska-Górecka and M. Roman: J. Opt. Soc. Am. B 3, 1305 (1986)

Laser Modified Birefringence in a Doppler Broadened Medium

P.J. Manson, H.-A. Bachor, and R.J. Sandeman

Laser Physics Centre, Department of Physics and Theoretical Physics, Australian National University, P.O. Box 4, Canberra 2601, Australia

The response of an atomic system to a strong laser field has been extensively studied, both experimentally and theoretically. Most of this work has concentrated on two or three level, stationary atoms because the theoretical treatment is simpler and physical processes are more obvious. However, this simplicity is generally offset by the increased complexity of experiments used for comparison with theory.

We have attempted to develop a theoretical description of a more typical experiment, and to do a detailed comparison between the theory and an experiment using a low density sodium vapour. Because of the hyperfine and Zeeman structure of sodium (Fig. 1), the theoretical treatment had to account for 16 levels, some of them degenerate. Doppler broadening also had to be included.

To probe the behaviour of the strongly pumped atoms, variations on standard saturated absorption and polarisation spectroscopy were used. The pump and probe beams were produced by different lasers and copropagated through the cell, with the pump laser frequency held constant during a probe laser frequency scan. The sodium vapour was created in a continuously pumped heated vacuum cell (without buffer gas), with Helmholtz coils to cancel stray magnetic fields.

The theoretical description of this system involved a solution of the equation of motion for the density matrix using an extension of the method of Ben-Reuven and Klein [1]. This technique gives a simple matrix expression for the signal which can be evaluated by computer. The signal is derived from the time dependent part of the density matrix, integrated from the start of the interaction until the steady state is reached. It therefore includes contributions from times before steady state, so systems with losses to external levels can be modelled (see for example [2]). The steady state density matrix for such a system is zero (i.e. all population is pumped out of the

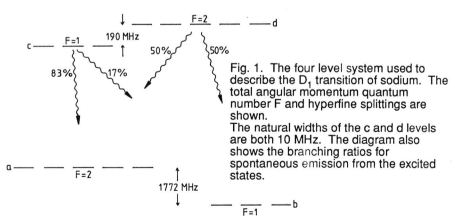

Fig. 1. The four level system used to describe the D_1 transition of sodium. The total angular momentum quantum number F and hyperfine splittings are shown.
The natural widths of the c and d levels are both 10 MHz. The diagram also shows the branching ratios for spontaneous emission from the excited states.

297

system). However, experiments using c.w. excitation record a non-zero signal, so it is important to include the transient effects in the calculation. This technique allows these effects to be included without solving the Bloch equations for the time dependence of the density matrix.

Figure 2 shows typical saturated absorption profiles for a range of pump laser detunings, and the theoretical results for corresponding conditions. The origins of the features in these profiles are quite different from standard saturated absorption because the two lasers were co-propagating. The peak which appears when the pump and probe laser frequencies are equal is due to a combination of all possible "I" resonances, i.e. with both lasers in resonance with the same transition. The other features are due to the various combinations of Λ, V and N systems which create the "crossover resonances" in saturated absorption spectroscopy.

The four level system used in the theoretical calculation is a complete description of the sodium D1 transition and there are no losses via spontaneous emission to external levels. However, propagation of the atoms through the laser beams introduces a loss mechanism to the system. While this atom loss rate can be ignored relative to the spontaneous decay rates of the excited levels, it must be included as a relaxation rate for the ground states and has a strong effect on relative peak

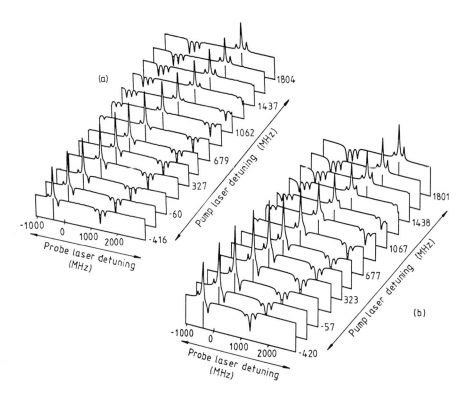

Fig. 2. Experimental (a) and theoretical (b) saturated absorption profiles for a pump laser Rabi frequency of 10 MHz and a Doppler width of 2 GHz. Detunings are relative to the |a>-|c> transition. The vertical lines indicate equal pump and probe detunings.

intensities. The good agreement between theory and experiment achieved here shows that the technique used is a good compromise between the simple steady state solution and the full integration of the Bloch equations.

References
1. A. Ben-Reuven and L. Klein Phys. Rev. A $\underline{4}$ 753 (1971)
    L. Klein, M. Giraud and A. Ben-Reuven Phys. Rev. A $\underline{10}$ 682 (1974)
2. P.T.H. Fisk, H.-A. Bachor and R.J. Sandeman Phys. Rev. A $\underline{33}$ 2418 & 2424 (1986)

Nonlinear Magneto-Optic Effects in a J=1 to J'=0 Transition

W. Lange, K.-H. Drake, and J. Mlynek

Institut für Quantenoptik der Universität Hannover,
Welfengarten 1, D-3000 Hannover 1, Fed. Rep. of Germany

Magneto-optic effects have proven a useful tool in linear spectroscopy for many years. Systematic studies of the magneto-optic effects in intense coherent light fields, however, do not seem to have been performed up to now. In this paper we give evidence for the existence of nonlinear versions of the Faraday effect and the Voigt effect. They can occur in very low magnetic fields, corresponding to Larmor frequencies in the kHz range, at modest laser powers.

The experimental set-up is very simple. A linearly polarized cw single mode dye laser beam passes a dilute atomic sample and the corresponding rotation of polarization is monitored by means of an analyzer. The magnetic field is either longitudinal ("Faraday geometry"; this case has recently been discussed in a theoretical paper [1]) or transverse ("Voigt geometry"). The experiments have been performed in the 570.7 nm line of Sm 152, which corresponds to the $4f^6 6s^2 \; ^7F_1 - 4f^6 6s6p \; ^7F_0$ transition. An experimental result obtained in the transverse case is shown in Fig. 1. The narrow structure around B = 0 represents the nonlinear effect to be discussed here. The situation is similar in the longitudinal case.

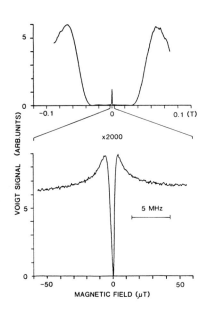

Fig. 1: Signal in the transverse case. The lower curve shows the central region on an enlarged scale

In the interpretation of the experimental results of both cases we use a perturbative treatment, perform the Doppler integration by means of the plasma dispersion function and take propagation effects into account. In Fig. 2 experimental and theoretical results are compared. We have performed systematic studies of the dependence of the signals on the laser intensity, the angle between polarizer and analyzer, the buffer gas pressure and the laser frequency and obtain very fair agreement between calculated and experimental results, when using realistic values of the atomic constants.

The narrow structures discussed here are due to the creation of a long-lived alignment in the J=1-level of the Sm ground state multiplet. The sensitivity of the alignment to external fields produces the dramatic field dependence observed in the experiment. Obviously the nonlinear magneto-optic effects are closely related to the phenomenon of "stimulated level crossing". It should be emphasized that even spurious magnetic fields may be sufficient to induce signal distortion in experimental schemes sensitive to the direction of polarization. Thus these effects are an example of the importance of the often neglected magnetic substructure in nonlinear spectroscopy and nonlinear optics.

This work was supported by the Deutsche Forschungsgemeinschaft.

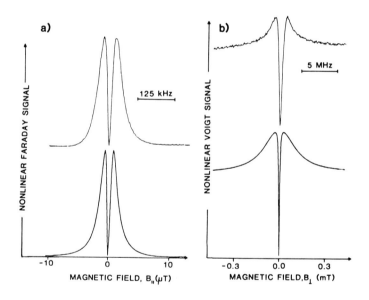

Fig. 2: Experimental (upper curves) and calculated (lower curves) signals in the (a) longitudinal and (b) transverse case

1. S. Giraud-Cotton, V. P. Kaftandjian, and L. Klein: Phys. Rev. A32, 2211 (1985) and A32, 2223 (1985)

Present addresses: [1] Physikalisch-Technische Bundesanstalt, Braunschweig; [2] Eidgenössische Technische Hochschule Zürich

Collision-Induced Ramsey Resonances in Sm Vapor

J. Mlynek[1], E. Buhr[2], and W. Lange[2]

[1]Institut für Quantenelektronik, ETH Zürich,
 CH-8093 Zürich, Switzerland
[2]Institut für Quantenoptik, Universität Hannover,
 Welfengarten 1, D-3000 Hannover 1, Fed. Rep. of Germany

Ramsey's method of separated fields for the observation of narrow radiofrequency (rf) resonances is well known from atomic and molecular beam experiments /1/. In this contribution, we demonstrate the occurrence of similar Ramsey resonances in an atomic vapor due to collisional velocity diffusion of sublevel coherence within an optical Doppler distribution; here the velocity diffusion of sublevel coherence between the excitation and detection processes plays a role similar to the spatial motion of the active atoms in conventional Ramsey experiments. The new phenomenon is observed using coherent resonance Raman processes to optically induce and detect Zeeman coherence in the Sm $\lambda=570.68$ nm $J=1-J'=0$ transition.

In the experimental configuration shown in Fig. 1 , counterpropagating laser fields are used for the coherent excitation and phase-sensitive detection of oscillating Zeeman coherence. For a nonzero laser detuning with respect to the Doppler-broadened optical transition ($\Delta_E \neq 0$), both fields interact with different velocity subgroups whose widths are determined by the homogeneous optical linewidth 2Γ (see Fig. 2d). In a Doppler-free optical-optical double resonance process (Fig. 2a) that uses carrier and sidebands of a modulated excitation field, the sublevel coherence is driven with frequency ω_M for atomic velocities $-v_0$. In the presence of velocity-changing collisions (VCC), the active atoms are redistributed from $-v_0$ to other velocity subgroups. During the time τ required for velocity diffusion, the sublevel coherence further evolves at its eigenfrequency Ω (Fig. 2b). In the velocity interval centered around $+v_0$, the oscillating coherence is monitored by coherent resonance Raman scattering of the probe field (Fig. 1c). The steady state Raman sidebands of frequency $\omega_E \pm \omega_M$ and the probe field (ω_E) then yield a heterodyne beat signal of frequency ω_M at the photodetector (Fig. 1); the phase sensitive detection of this rf beat signal allows to measure magnitude and phase of the sublevel coherence with high sensitivity /2/. As a result of the resonance conditions for the excitation process shown in Fig. 2a, rf resonance signals will be observed if the sublevel detuning $\Delta_M = \omega_M - \Omega$ varies through zero. For a certain velocity diffusion time τ, however, the rf beat signal at the photodetector is phase-shifted by $\Delta_M \tau$, thereby introducing

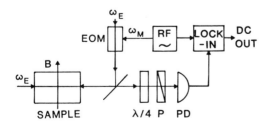

Fig. 1. Experimental scheme:
EOM, electrooptic modulator;
$\lambda/4$, retardation plate;
P, analyzer; PD, detector

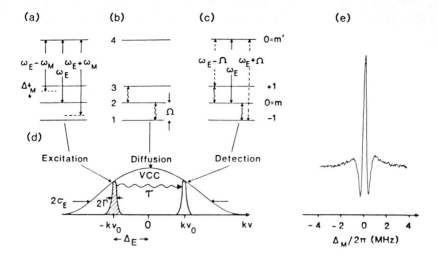

Fig. 2. Excitation (a), diffusion (b) and detection (c) process of Zeeman coherence; (d) Schematic of the Doppler distribution; (e) Typical Raman heterodyne signal measured at large detuning (see text)

Ramsey-type interference patterns proportional to $\cos(\Delta_M \tau)$ in the detected line shapes at the lock-in amplifier. As a consequence, narrow resonance structures should occur for large velocity diffusion times τ, i.e. for large laser detunings Δ_E within the Doppler-broadened transition.

A typical experimental result obtained with He collision partners (p=240 Pa) for a Doppler detuning $\Delta_E/2\pi=550$ MHz is shown in Fig. 2e; the measured lineshape clearly demonstrates the presence of a Ramsey-type interference pattern. This interpretation of our experimental findings is in good agreement with calculations based on a Fokker-Planck equation /3/; the theoretical fits of the measured sublevel resonance curves directly yield the time constant for collisional thermalization and the cross section for depolarizing collisions.

References
/1/ N.F. Ramsey, Molecular beams (Oxford Univ. Press, London, 1963)
/2/ J. Mlynek, K.H. Drake, G. Kersten, D. Frölich, and W. Lange,
      Opt. Lett. 6 (1981), 87
/3/ E. Buhr and J. Mlynek, Phys. Rev. Lett. 57 (1986), 1300

Optical Resonances with Subnatural Linewidths Resulting from Nonstationary Optical Pumping

W. Gawlik

Instytut Fizyki, Uniwersytet Jagiellónski,
PL-30-059 Kraków, ul. Reymonta 4, Poland

A reduction of the width of optical resonances below the limit imposed by the natural lifetimes has long been a subject of considerable interest (see e.g. /1/ and refs therein). Recently, we have reported on new signals with subnatural linewidths in a polarization spectrsocopy (PS) experiment on sodium vapor /2,3/. The most characteristic feature of the signals is an additional narrow structure superimposed on the Dopper-free power broadened resonances. These structures (dips) can be as narrow as 2.6 MHz for the Na D1 line, whereas the natural linewidth of this transition is 10 MHz. The dips appear only for those transitions where optical pumping (either Zeeman, or hyperfine) is possible, and even if the light beams are too weak to include saturation effects. The appearance of the dips depends substantially on the interaction time of atoms with the light fields. The apparatus for observation of the new subnatural structures is a typical PS arrangement with a narrow-band dye laser, circular pump beam and linear probe beam polarizations, supplemented with magnetic shields and with provision for changing the diameter of the light beams, i.e. the interaction times.

The described subnatural structures are explained by the theory of nonstationary velocity-selective optical pumping in a four-level system /4/. Nonstationarity arises here from the fact that the atoms move across the light beams and interact with them only for a limited period of time. If the probe beam is strong enough it alters atomic populations as much as the pump. For an appropriately long (but finite) interaction time the two-beam nonstationary pumping results in a dip in the center of a Doppler-free PS signal (see Fig. 1).

Since the dips are much narrower than the homogeneous width Γ, in ref. 4 we have considered their applicability to high-resolution spectroscopy. It turns out that the dips can be, in principle, infinitesimally narrow. However, the constraints that are due to the finite depth of the dip and finite signal-to-noise ratio limit the practically attainable resolution. Moreover, for closely spaced lines, contributions from different transitions interfere, which makes the resolution of two lines that are closer than Γ impossible. Nevertheless, for isolated transitions (i.e. those whose homogeneous profiles do not overlap) the estimated practical resolution offered by the described dips could be more then two times better than the theoretical resolution limit of a standard Doppler-free PS /5/. By the practical resolution we mean here the precision with which the line center may be determined at a given signal-to-noise ratio. We hope, therefore, that the narrow dips which originate from the nonstationary optical pumping may find applications in, e.g. precision measurements of not too close level structures, subtle line shifts in a low-pressure vapor phase, laser-frequency stabilization, etc.

304

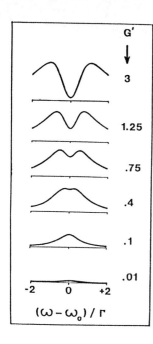

Fig.1 Doppler-free PS resonances calculated for an isolated transition of frequency ω_0 and for different values of the saturation parameter G' of the probe beam. The saturation parameter, G, of the pump beam was kept constant at G=0.5. For G'$\geqslant$0.4 a dip appears at the center of the resonance. Its width and depth increase with increasing values of G'. We define $G=4(\beta^2/\Gamma)T_{tr}$ and $G'=4(\beta'/^2\Gamma)T_{tr}$ where β and β' denote the Rabi frequencies associated with the pump and probe beams, and T_{tr} is related to the atomic transit time across the light beams.

This work was supported by the Polish Research Program CPBP 1.06. The participation of J. Kowalski, F. Träger amd M. Vollmer in the experimental part of this work (performed at the University of Heidelberg) is gratefully acknowledged.

References

1. D.P. O'Brien, P. Meystre and H. Walther: Adv. Atom. Molec. Phys. ed. by D. Bates and B. Benderson, vol. 21, p.1 (Acad. Press, New York, 1985).
2. W. Gawlik, J. Kowalski, F. Träger and M. Vollmer: Phys. Rev. Lett. 48, 871 (1982).
3. W. Gawlik, J. Kowalski, F. Träger and M. Vollmer: J. Phys. B 20, 997 (1987).
4. W. Gawlik: Phys. Rev. A34, 3760 (1986).
5. C. Wieman and T.W. Hänsch: Phys. Rev. Lett. 36, 1170 (1976).

A New SO(4)-Based Scheme for Producing Atoms in Circular Rydberg States

D. Delande and J.C. Gay

Laboratoire de Spectroscopie Hertzienne de l'Ecole Normale Supérieure, Tour 12, EO1-4, place Jussieu, F-75252 Paris Cedex 05, France

We present a new scheme for exciting atoms in circular Rydberg states with guaranteed high efficiency whatever the atomic species. The method only requires the use of a weak magnetic field crossed to a slowly varying electric field.

Compared to the Hulet-Kepler method based on conservation of the angular momentum in the interaction of atoms with microwave fields, we fully exploit the O(4) symmetry of the Rydberg system |1||2|. One single global process allows the transfer of low m-laser excited atoms into circular states with an efficiency nearly 100%. This as well applies to alkali Rydberg atoms.

Fig. 1 is an SO(4) vectorial model of the dynamics of the n Rydberg shell in crossed fields. Here we made use of an SO(3) ⊗ SO(3) type description of this symmetry introducing the pair of 3-dim. angular momenta, $\vec{j}_1$ and $\vec{j}_2$, such that :
$$\vec{j}_{1,2} = 1/2 \ (\vec{L} \pm \vec{A}).$$
$\vec{L}$ and $\vec{A}$ are, respectively, the angular momentum in real space and the Runge-Lenz vector.

Actually, $\vec{L}$ and $\vec{A}$ build the six components of the 4-dim. angular momentum $\mathscr{L}(\vec{L},\vec{A})$, the conservation of which expresses the rotational invariance of the n Coulomb shell in a 4-dimensional space |3|.

A consequence of this dynamical symmetry is the tunability of Rydberg atoms, the structure of which can be controlled with external fields. Another consequence is the structure of the energy diagram (Fig. 1) reproducing itself identically from the Zeeman to the Stark limit. These are the basic elements for producing, in particular, circular Rydberg states |3|.

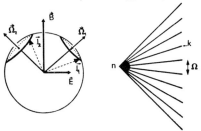

Figure 1 - The low field crossed $(\vec{E},\vec{B})$ fields situation as seen from the SO(4) vectorial model. The two quasi-spin $(\vec{j}_1,\vec{j}_2)$ are precessing around the axis $\vec{\Omega}_{1,2} = \vec{\omega}_L \pm \vec{\omega}_S$ with an angular velocity $\Omega = (\omega_S^2 + \omega_L^2)^{1/2}$. This is the spacing of the energy levels.

One of the simplest "universal" schemes for producing atoms in circular states is actually as follows.

1- Start with a crossed field situation such that $\vec{\omega}_s \gg \vec{\omega}_L$ ($\vec{\omega}_s$ the linear Stark pulsation ; ω_L the Larmor one). The n manifold is thus quantized in nearly the Stark regime. Efficient laser excitation of states with $L_x=M=0$ and maximum values of the Lenz vector ($|A_x|=n-1$) can then be performed. These states are those at both ends of the Stark manifold associated with electronic densities concentrated up-field or down-field. In conditions as in Fig. 1, this means $j_{1x} = - j_{2x} = \pm(n-1)/2$.

2 - Second and last step is to switch off the electric field adiabatically to zero. The adiabaticity condition is $d\omega_s/dt \ll \Omega^3/\omega_L$. The rotation of the axis $\vec{\Omega}_{1,2} = (\vec{\omega}_L \pm \vec{\omega}_s)$ is slow enough compared to the velocity of precession Ω of $\vec{j}_1$ and $\vec{j}_2$ around them. The final state such that $j_{1z} = j_{2z} = \pm(n-1)/2$ is thus a circular state freely evolving in the $\vec{B}$ field alone as $L_z = n - 1$ and $A_z = 0$.

From the 4-dimensional point of view, the transformation so achieved is a 4-dim. rotation in which the components of $\mathcal{L}$ are transformed into each other :

$$\vec{\mathcal{L}}(0,0,0,n-1,0,0) \to \vec{\mathcal{L}}(0,0,n-1,0,0,0).$$

From a classical point of view, the degenerate classical Kepler ellipse (a straight line) along $\vec{E}$ is transformed into a circular trajectory in the plane perpendicular to $\vec{B}$.

Minor modifications of this scheme allow it to be applied as well to alkali Rydberg atoms. Actually, this is not a surprising feature as the scheme is based on the SO(4) symmetry which any Rydberg system realizes more or less perfectly well.

1. F. Penent, D. Delande, F. Biraben and J.C. Gay - Optics Comm. 49 (1984) 184
2. J.C. Gay in "Atoms in Unusual Situations", J.P. Briand ed. (Plenum) 1987 D. Delande and J.C. Gay - submitted for publication
3. F. Biraben, C. Chardonnet, D. Delande, J.C. Gay, F. Penent - in Le Courrier du CNRS - Images de la Physique - Supplément n° 59, CNRS (Paris, 1985)

Transmission Zeeman Beat Spectroscopy for Determination of Depolarisation Rates in Atomic Ground States

P. Hannaford, R.M. Lowe, and R.J. McLean

CSIRO Division of Materials Science and Technology,
Locked Bag 33, Clayton, Victoria 3168, Australia

We report an accurate technique, based on the detection of Zeeman quantum beats in transmission [1-3], for the direct determination of relaxation rates for both orientation ($\Gamma_1(\ell)$) and alignment ($\Gamma_2(\ell)$) in atomic ground states. A short intense pulse of laser radiation generates coherence between non-degenerate Zeeman states in the ground level of an atomic sample in the presence of a transverse magnetic field $\underline{B}$ and the transmission of a weak cw probe beam through nearly-crossed polarisers monitors the time evolution of the coherence through its coupling to the complex susceptibility of the sample. By adjusting a Babinet Soleil compensator in the pump beam to either half-wave or quarter-wave retardation, the atoms are pumped respectively with circularly polarised light to induce an orientation coherence in the ground state $\rho^1_{\pm1}(\ell)$ (quantisation axis along $\underline{B}$) or with light linearly polarised at 45° to $\underline{B}$ to induce a $|\Delta m|=1$ alignment coherence $\rho^2_{\pm1}(\ell)$. For a weak probe laser polarised at right angles to $\underline{B}$, the evolution of the induced coherence results in a transmission signal of the form

$$I_T \propto \theta^2 - \theta C e^{-\Gamma_k(\ell)t}\cos\omega_L t + [De^{-\Gamma_k(\ell)t}\cos\omega_L t]^2 \qquad (1)$$

where θ is the uncrossing angle between the polarisers and C,D include the strength and detuning of the probe laser and atomic relaxation rates. If θ is large enough for the final term in (1) to be neglected, then the beat signal is modulated at the Larmor frequency ω_L and decays at a rate given simply by $\Gamma_1(\ell)$ (circularly polarised pump) or $\Gamma_2(\ell)$ (45° linearly polarised pump). For the case of perfectly crossed polarisers ($\theta=0$), the beat signal is modulated at $2\omega_L$ and decays at the rate $2\Gamma_k(\ell)$. A polarisation-rotation technique similar to that described here but not involving quantum beats has previously been used to determine depolarisation rates in <u>excited</u> atomic states [4].

We have applied the transmission Zeeman beat technique to the determination of both $\Gamma_1(\ell)$ and $\Gamma_2(\ell)$ for the $4f^66s^2$ $^7F_1$ 293 cm^{-1} near-ground level of Sm I in the presence of various rare-gas perturbers. The $^7F_1$ is the lower level of the widely used 570.68 nm $^7F_1$-$^7F_0$ transition and direct information on collisional depolarisation rates for this level is of importance in the analysis of a number of recent laser spectroscopic investigations on Sm I [5-8]. Two of these investigations [7,8] have yielded widely different values (factor of 50) for the cross sections for depolarising collisions with the rare gases.

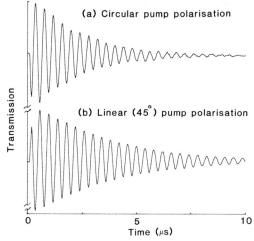

(a) Circular pump polarisation

(b) Linear (45°) pump polarisation

Transmission

0   5   10
Time (μs)

Figure 1. Experimental traces of transmission Zeeman beat signals for the $^7F_1$ 293 cm^{-1} near-ground level in Sm I. Samarium vapour produced by cathodic sputtering in 0.25 Torr Ar. Magnetic field: 1.2 Gauss. Nearly crossed polarisers

Figure 1 shows typical transmission Zeeman beat signals for the Sm $^7F_1$ level recorded with the (pulsed) pump laser tuned to the strong 478.31 nm $^7F_1$-$^7D_1$ transition and the (cw) probe laser tuned to the weak 570.68 nm $^7F_1$-$^7F_0$ transition. A significant difference in the two depolarisation rates for the $^7F_1$ level is evident $(\Gamma_1(\ell)/\Gamma_2(\ell)=1.56)$, indicating substantial anisotropy in the collisional relaxation in the level. This ratio is found to remain constant within 3% over a wide range of pressure for each of the rare gases. Such anisotropic behaviour has important implications in studies of, for example, polarisation switching [6] and collision-induced Hanle resonances in four-wave mixing in Sm [8]. The pressure dependence of $\Gamma_1(\ell)$ and $\Gamma_2(\ell)$ for the various rare gases yield disalignment and disorientation cross sections [9] with an accuracy of typically 5%. Comparison with the previous determinations indicates reasonable agreement between our disalignment cross sections and those of ref. [7].

References

1. W. Lange and J. Mlynek: Phys. Rev. Lett. <u>40</u>, 1373 (1978)
2. J. Mlynek and W. Lange: Opt. Comm. <u>30</u>, 337 (1979)
3. T. Dohnalik, M. Stankiewicz and J. Zakrzewski:
   In "Laser Spectroscopy VI", Eds. H.P. Weber and W. Lüthy
   (Springer, Berlin 1983), p. 58.
4. A.P. Gosh, C.D. Nabors, M.A. Attili, J.E. Thomas and
   M.S. Feld: Phys. Rev. Lett. <u>53</u>, 1333 (1984)
5. R.J. McLean, D.S. Gough and P. Hannaford:
   In "Laser Spectroscopy VII", Eds. T.W. Hänsch and Y.R.Shen
   (Springer, Berlin 1985), p. 220.
6. C. Parigger, P. Hannaford and W.J. Sandle: Phys. Rev.
   A <u>34</u>, 2058 (1986)
7. Chr. Tamm, E. Buhr and J. Mlynek: Phys. Rev. A <u>34</u>,1977 (1986)
8. Y.H. Zou and N. Bloembergen: Phys. Rev. A <u>34</u>, 2968 (1986)
9. R.M. Lowe, D.S. Gough, R.J. McLean and P. Hannaford
   (to be published)

Broadband Light Photon Echoes and Collisional Energy Transfers

A. Débarre, J.-C. Keller, J.-L. Le Gouët, and P. Tchénio

Laboratoire Aimé Cotton, Bât. 505, C.N.R.S. II, F-91405 Orsay, France

Broadband (incoherent) light pulses can be characterized in the time domain by the pulse duration τ_L and by the inverse of the spectral width i.e. the coherence time τ_c. For standard pump-probe experiments the time resolution is governed by τ_L. In a nonlinear experiment involving the autocorrelation of the light field, the ultimate time resolution is determined by the coherence time τ_c. Subpicosecond time resolution has been demonstrated in the measurements of transverse relaxation time T_2 using nanosecond light pulses [1]. We report on the measurement of the evolution of a longitudinal structure with time resolution $\tau_c \ll \tau_L$.

In our experiment, two laser pulses obtained by splitting of a single , resonantly excite the $^3P_1$ level in calcium vapor. The time delay t_{12} between these excitation pulses is responsible for the construction of a modulation (modulation period λ/t_{12}) in the velocity distribution of the transition upper level population. A probe laser pulse builds up a Stimulated Photon Echo from this modulation [2]. One observes the collisional relaxation of this modulation or its collisional transfer to other levels due to resonant (depolarizing) or non resonant (fine structure changing) processes. The time resolution, as measured with an urea crystal correlator, is 80 ps while the pulse duration is 5 ns.

The transfer $^3P_1 - >^3P_2$ ($\Delta E = 105cm^{-1}$), induced by collisions with helium, is illustrated on Fig.1 . The broken line curve represents the Stimulated Echo signal intensity that would be recorded in the absence of collisional velocity changes, by exciting level $^3P_1$ and probing level $^3P_2$. The full line curve is the actual experimental recording. The build up of the echo signal when t_{12} is increased from negative to positive values is a consequence of the angled beam geometry of the experiment. It occurs on a time scale that corresponds to the inverse Doppler width [3]. The decay of the echo signal for positive values of t_{12} directly reflects the velocity changes which are associated with the collisional transfer under investigation. The characteristic features in Fig.1 could not have been observed on a $\tau_L = 5ns$ time resolution basis .

We have also observed the decay of the echo signal due to power effects and have started in solving the corresponding Stochastic Bloch equations in the strong field regime. As they probe high order correlation functions of the optical fields , coherent transients produced by broadband sources give some opportunity for checking the source models.

310

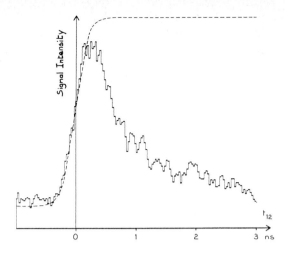

Figure 1. Evolution of the fine structure transfer echo signal as a function of t_{12} (helium pressure: 0.6 torr).

1 - M.Tomita and M.Matsuoka : J.Opt.Soc.Am.B. 3 (1986) 560

2 - J.-C. Keller and J.-L. Le Gouët : Phys.Rev.A. 32 (1985) 1624 and references therein.

3 - M.Defour , J.-C. Keller and J.-L. Le Gouët : J.Opt.Soc.Am.B. 3 (1986) 544

Competition Effects Among
Nonlinear Optical Processes

D.J. Gauthier, M.S. Malcuit, J.J. Maki, and R.W. Boyd

Institute of Optics, University of Rochester, Rochester, NY 14627, USA

Recently a number of examples have been reported in which two different nonlinear optical processes compete with one another [1-4]. In some cases, the occurrence of one nonlinear optical process can entirely prevent the occurrence of another process which would normally be expected to occur under the prevailing conditions. We recently observed a form of competition in which a nonlinear optical process was suppressed by another process that was nominally many times weaker. We observed that four-wave mixing (FWM) in the near forward direction pumped by two-photon excitation of the sodium 3d level could lead to the nearly complete suppression of amplified spontaneous emission (ASE) from the 3d to the 3p level, even though the gain for the ASE process calculated in the absence of competition effects was many times larger than the calculated gain of the four-wave mixing process [3]. We have also presented a theoretical model that shows why suppression of ASE occurs under these conditions [4]. Our calculation is based upon finding a spatially invariant solution of the coupled amplitude equations describing the the mutual interaction of the fields at frequencies ω_1, ω_2, and ω_3. The nonlinear susceptibilities used in this calculation are obtained from the perturbation solution to the density matrix equations of motion for the three-level atomic system shown in Fig. 1.

The origin of the competition can be understood in terms of the energy-level diagram shown in Fig. 1. Near the entrance window of the sodium cell, only the laser field at frequency ω_1 is present (Fig. 1a). This field can two-photon excite the 3d level, leading to a population inversion and subsequently ASE between the 3d and 3p levels. However, four-wave mixing also occurs within the sodium cell (Fig. 1b), which leads to the generation of two new fields at frequencies ω_2 and ω_3. These new fields lead to a second excitation pathway connecting the ground and excited states (Fig. 1c). Our

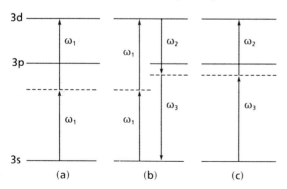

Fig. 1.   Energy level diagrams showing the origin of the competition.

calculation shows that for sufficiently long interaction lengths the fields generated by the FWM process always attain values such that an exact destructive interference between the pathways shown in (a) and (c) occurs. For regions of the cell in which the fields attain these values, no population is excited to the 3d level and no ASE can be generated.

In order to examine the nature of the competition process more fully and to determine its dynamical properties, we have recently measured the temporal evolution of the fields generated by the ASE and FWM processes. An incident laser beam was weakly focused to an intensity of 100 MW cm^{-2} into a 5-cm-long sodium cell containing 10^{16} atoms per cm^3. A streak camera system having a time resolution of 12 ps was used to monitor the time evolution of the fields generated by the ASE and FWM processes. The ASE process was monitored by measuring the field generated in the backward direction, whereas the FWM process was monitored by measuring the field generated near the 3d→3p transition frequency in the phase-matched near-forward direction. The results of this measurement are shown in Fig. 2. The ASE signal is seen to be present only for a brief time interval near the begining of the laser pulse. As the laser pulse approaches its maximum value, the ASE signal turns off as the FWM process simultaneously turns on. Under the experimental conditions, the gain coefficient for the ASE process is approximately 100 times larger than that for FWM. Even the weak leading edge of the laser pulse is able to excite the ASE process efficiently. However, near the peak of the laser pulse the FWM process has sufficient gain that the generated fields reach their spatially invariant values after propagating only a small fraction of the length of the interaction region, and ASE is completely suppressed once this occurs.

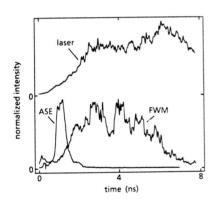

Fig. 2. Temporal evolution of the laser intensity and the intensity of the emission at the 3d→3p transition frequency.

This work was supported in part by the US Army Research Office.

1. J. C. Miller, R. N. Compton, M. G. Payne, and W. W. Garrett, Phys. Rev. Lett. 45, 114 (1980).
2. D. J. Jackson and J. J. Wynne, Phys. Rev. Lett., 49, 543 (1982).
3. M. S. Malcuit, D. J. Gauthier, and R. W. Boyd, Phys. Rev. Lett. 55, 1086, (1985).
4. R. W. Boyd, M. S. Malcuit, D. J. Gauthier, and K. Rzazewski, Phys. Rev. A 35, 1648, (1987).

Resonant Optical Suppression
of the Van der Waals Force

J.F. Lam, S.C. Rand, and R.A. McFarlane

Hughes Research Laboratories, 3011 Malibu Canyon Road,
Malibu, CA 90265, USA

1. Introduction

The behavior of a single two-level atom in the presence of
strong resonant light fields was first described by Mollow[1]
and Cohen-Tannoudji[2], and observed by Schuda[3]. The
interaction of light with two atoms coupled by Van der Waals
forces has been described previously only well off
resonance[4], however. Here we show that stationary pair
systems of this kind, driven by light fields resonant with
double-excitation pair transitions, exhibit unique features[5].
These include a seven-line resonance fluorescence spectrum and
resonant suppression of the Van der Waals force.

2. Theory

The system Hamiltonian is

$$H = H_o - \mu_a \cdot E(R,t) - \mu_b \cdot E(R,t) + H_{dd}, \qquad (1)$$

where μ_i is the electric dipole moment of atom i, $E(R,t)$ is the
light field and H_{dd} is the static electric dipole-dipole
interaction between atom a and atom b. H_o is the unperturbed
Hamiltonian of the two-atom system. Both atoms have s-type
ground states $|Ag\rangle$, $|Bg\rangle$ and one electric-dipole-allowed
transition. A single p-type excited state $|Ae\rangle$ was assumed for
atom a, but atom b was assigned one s and one p-type excited
state, $|Be\rangle$ and $|Bf\rangle$ respectively. This is the simplest model
with (normally forbidden) electric-dipole-allowed cooperative
transitions of the type $|AgBg\rangle \leftrightarrow |AeBe\rangle$.

Dressed atom results for stimulated pair emission versus
detuning of the near resonant light field are illustrated in
Fig. 1. Several features of the spectrum are worth noting.
First, the maximum of the emission process is broadened and
shifted away from the coupled pair transition by the AC Stark
effect. Second, two sharp dips appear, symmetric about zero
detuning. These unexpected features occur when

$$\omega_R = \pm 2|K|/\sqrt{3} \, h. \qquad (2)$$

That is, coherent nulls occur when the generalized Rabi
frequency $\omega_R = (\Delta^2 + [\mu E/h]^2)^{1/2}$ is comparable in magnitude to the
matrix element K of H_{dd} on the pair transition. Stimulated
pair emission drops sharply, despite near resonance excitation.

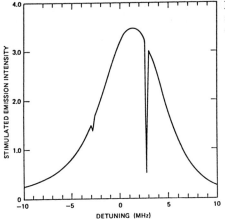

Fig.1.Stimulated pair emission showing predicted nulls at the electric "magic angle".

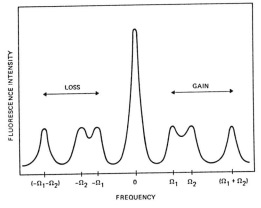

Fig.2.Resonance fluorescence spectrum of coupled atoms.

This behavior may be easily understood by realizing that when the two atoms are driven coherently, such that their interaction in the rotating frame vanishes, their cooperative emission must cease. This optical effect is analogous to the decoupling of magnetic spin interactions[6] by application of rotating magnetic fields at the "magic angle". Here, however, it is the rotating electric field of incident light which decouples the pseudospins of the electric dipole–dipole interaction. The separation of the decoupling features should be useful for precise determinations of H_{dd} in systems of very cold atoms in traps or stationary atoms in solids.

The resonance fluorescence spectrum calculated using the dressed atom approach of Ref.[2] is shown in Fig.2. Unlike the two-level problem in which three lines arise, a total of seven lines emerges in the two-body case. The coupled, dressed states[5] form an infinite ladder of triplet states of the form

$$|Ain> = a_i \sin\theta \left[\cos\theta |AgBg, n+1> + \sin\theta |AgBfn>\right] + \\ + b_i \cos\theta \left[\sin\theta |AgBg, n+1> + \cos\theta |AgBfn>\right] + c_i |AeBen>.$$

From these expressions the relative intensities $I(\omega_o):I(\omega_o\pm\Omega_1):I(\omega_o\pm\Omega_2):I(\omega_o\pm\Omega_1\pm\Omega_2)$ of the sidebands may be obtained.

$$[|a_1|^2|a_1+b_1|^2+|a_2|^2|a_2+b_2|^2+|a_3|^2|a_3+b_3|^2] \quad : \quad |a_1+b_1|^2|a_2+b_2|^2 \\ : \quad |a_2+b_2|^2|a_3+b_3|^2 \quad : \quad |a_1+b_1|^2|a_3+b_3|^2 \tag{3}$$

The pair resonance fluorescence spectrum should exhibit probe gain on one side and loss on the other side of center frequency, just as in two-level systems. This work was supported by AFOSR Contract F49620-85-C-0058.

References

1. B.R.Mollow, Physical Review 188, 1969(1969).
2. C.Cohen-Tannoudji and S.Reynaud, in Multiphoton Processes, eds. J.Eberly and P.Lambropolous, Wiley (NY), pp.103-118, 1978.
3. F.Schuda, C.Stroud Jr. and M.Hercher, J.Phys.B7, L198(1974).
4. D.Dexter,Phys.Rev.126,1962(1962);L.Gudzenko and S.Yakovlenko Zh. Eksp. Teor. Fiz.62,1686(1972)[Sov.Phys.-JETP35, 877(1972)].
5. J. Lam and S.C. Rand, Physical Review 35A, 2164(1987).
6. M. Lee and W.I. Goldburg, Phys. Rev. 140, A1261(1965).

Light-Induced Cherenkov Emission

R. Shuker[1] and I. Golub[2]

[1]Department of Physics, Ben-Gurion University of the Negev,
84105 Beer-Sheva, Israel
[2]LROL, Département de Physique, Université Lâval,
Quebec G1K 7P4, Canada

1. Introduction

The passage of a laser light detuned to the blue of a resonant atomic transition results in a conical emission around the laser beam [1, 2]. This conical emission shell has a half-angle of a few degrees around the laser axis, is red detuned from the transition and is spectrally broad (~ 10 cm^{-1}).

We propose a Cherenkov-type process for the production of the ring emission [2]. All the features of this emission are accounted for by this model. The laser light passing through the metal vapor induces moving polarization of the medium, which propagates at a velocity in excess of that of the red-shifted light, resulting in this Cherenkov-type emission.

2. The Model

A blue detuned laser propagates in a resonant medium at a velocity that exceeds the speed of light. It induces a medium polarization that results in the Cherenkov emission. This polarization is essentially the medium excitation moving at the laser radiation group velocity, v_{gr}. The medium emits at frequencies ω within the radiative transition linewidth that fulfill the Cherenkov condition, namely $v_{gr}(\omega L) > v_{ph}(\omega)$. Here $v_{ph}(\omega) = c/n(\omega)$ is the phase velocity of the emitted light. The Cherenkov condition is fulfilled for the light emitted to the red of the transition. The maximal spectral intensity of the Cherenkov emission occurs where $n(\omega)$, the index of refraction, is maximal. The maximum value of the saturated index of refraction which is correlated with the cone peak emission frequency ω is at $\omega - \omega_0 = \omega_0 - \omega_L$, making the laser and cone frequency symmetric about the transition frequency ω_0, as measured experimentally.

The angle of the ring emission is given by the Cherenkov relation, $\cos\theta = v_{ph}(\omega)/v_{gr}(\omega_L)$. We have calculated the dispersive response of the index of refraction in terms of the off-diagonal elements of the density matrix $\rho_{ba}(\omega)$ and have found that $dn/d\omega|_{\omega=\omega_L} = 0$. Thus, $\cos\theta = n_L/n_C$. n_L and n_C are the refractive indices at the laser and emission frequencies, respectively. For small angles, $\theta \simeq [2\Delta n_L - \Delta n_C]^{\frac{1}{2}}$, where $\Delta n = 1 - n$.

The Cherenkov radiation has the characteristics of a surface phenomenon which does not conserve the transverse component of the linear momentum. Thus, the ring emission should originate mainly at the surfaces of the self-trapped filaments, and not in their bulk.

3. RESULTS AND DISCUSSION

The measured forward light for a laser detuned to the blue of the D_2 transition contains a central beam and the ring emission. The central beam is composed of the laser radiation and a coherent peak to the blue of the D_1 transition [3]. The conical emission has a wide spectrum spanning the frequency region where the index of refraction n_c, is larger than 1. For sodium densities of $\sim 10^{16}$ cm^{-3} the emission has a component even to the red of the D_1 line, where the index of refraction is also larger than 1. The cone angle is $1°-3°$ and increases as the laser frequency approaches the atomic transition and with increasing sodium density. We find good agreement between measured cone angles and values calculated according to the Cherenkov emission model, as illustrated in Fig. 1. Here the vertical bars represent the measured range of conical shell angles. The Cherenkov emission angle θ_C are calculated for two saturation degrees Ω, namely 0.5Δ and 2Δ, which occur near the filament boundary. Here $\Delta = \omega_L - \omega_0$. The ring angle calculated from the phase matching condition of the four-wave mixing model [1], $\cos\theta = 1/n_c$, does not agree with the experimental results. A quantum model describing the emission as medium-laser excitation is being developed.

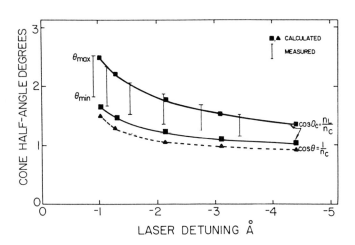

Fig. 1. Cone half-angle as a function of laser detuning. [Na] - 1.8·10^{15} cm^{-3}. ■ - Calculated according to the Cherenkov model and ▲ - according to the four-wave mixing model. Vertical bars represent the measured range of angles.

REFERENCES

1. D.J. Harter and R.W. Boyd: Phys. Rev. A29, 739 (1984).
2. I. Golub, G. Erez and R. Shuker: J. Phys. B19, L115 (1986).
3. I. Golub, R. Shuker and G. Erez: J. Phys. B, 20, L62 (1987).

Part IX

Laser Spectroscopic Diagnostics

Recent Developments
in CARS Combustion Spectroscopy

A.C. Eckbreth

United Technologies Research Center,
Silver Lane, East Hartford, CT 06108, USA

Coherent anti-Stokes Raman spectroscopy (CARS) has become a well-established
technique for the remote, spatially and temporally precise probing of instru-
mentally-hostile combustion processes for both fundamental research and practical development. A major limitation of CARS, in comparison with spontaneous
Raman scattering, is the inability to interrogate more than one molecular species at a time as normally implemented. In this paper, recent research into
multi-color CARS approaches, which permit the simultaneous measurement of temperature and two or more species, will be reviewed. These involve multiple
pump or Stokes lasers and synergistic combinations thereof. An ancillary benefit has been the development of a new method for generating pure rotational
CARS from pump and broadband Stokes lasers having arbitrary spectral separation.

1. CARS - RECENT RESEARCH EMPHASIS

For diagnostic applications, CARS is generally performed in a manner quite
dissimilar from the approaches used in spectroscopic investigations. Diagnos-
tically, high spatial and temporal resolution are required leading, respec
tively, to the utilization of crossed-beam phase matching and broadband gen-
eration/detection approaches. Much of the early work in the field was oriented toward achieving these aims and at demonstrating the utility of CARS to a
wide array of practical combustion devices [1-3]. Having been proven as a
broadly applicable and powerful tool for combustion research, much of the
activity of the past few years has been directed toward refining the accuracy
of CARS measurements and further extending its capabilities. Accuracy has
been enhanced by research investigations into in-situ normalization [4], new
laser bandwidth convolutions [5-7], Raman linewidth scaling models [8,9], improved collisional narrowing models [10,11] and laser field statistical effects [12]. Single pulse CARS generation has received much experimental [13-
16] and theoretical attention [17,18] and remains an intense research area
where improved accuracy and reproducibility are still desired. Much of the
capabilities research has been on electronic resonance enhancement [19,20] and
simultaneous multi-species detection [21] which will be the main focus of this
paper.

2. MULTI-COLOR CARS APPROACHES

Figure 1 summarizes the various approaches conceived to date for simultaneous,
multiple species CARS measurements. Dual Stokes approaches [22] are a
straightforward extension of CARS. A separate Stokes source is introduced for

320

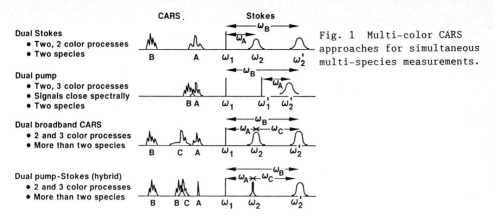

Fig. 1 Multi-color CARS approaches for simultaneous multi-species measurements.

each constituent and each is measured in an independent two-color process. Similarly, in dual pump approaches [23], two narrowband pump lasers are used in conjunction with a single broadband Stokes laser to monitor two species via two separate, three-color processes. Both of the foregoing monitor two constituents and can be extended to a greater number of species by adding more Stokes or pump lasers but, clearly, at the cost of increased complexity.

Approaches which permit many species, i.e. at least three, to be measured with no further increase in complexity involve the synergistic use of the laser frequencies involved. In dual broadband CARS [21,24], two broadband Stokes lasers are used in conjunction with a pump. In addition to the two, two-color sequences, there is the three-color wave mixing sequence for Raman resonances corresponding to the frequency difference range of the two Stokes sources assuming all processes have been phase matched. There is also an unresolved, diffuse three-color background due to Raman coherences between the pump and high-Raman-shift Stokes source from which the remaining Stokes source scatters. The difference frequency range spanned by two broadband sources is quite large blanketing Raman resonances over a several hundred wavenumber range. Spectral positioning of the broadband sources depends on the application and species of measurement interest. For hydrocarbon-fueled combustion, the spectral location of the major species resonances is quite fortuitous. Positioning the Stokes sources to generate two-color CARS from the major combustion products, CO_2 and H_2O, results in three-color generation from N_2 and possibly CO, the major intermediate, if sufficiently abundant. An example of simultaneous CARS generation from CO_2, N_2 and H_2O in the post reaction zone of a premixed flame is displayed in Fig. 2. It should be noted that the three-color dual broadband process is not as intense as normal two-color mixing due to the intensity difference between a pump and Stokes laser. It will not be as weak, however, as one might suspect due to the spectral integration aspect of dual broadband CARS, i.e. many frequency combinations drive each Raman resonance. For the same reason, a beneficial spectral averaging may occur resulting in improved single pulse spectral quality. Such dual broadband CARS approaches also lend themselves quite readily to conditional sampling strategies [25].

The last technique in Fig. 1 is a hybrid and termed dual pump-Stokes due to the fact that the low-Raman-shift source serves as both a pump and Stokes

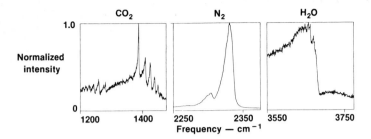

Fig. 2   Simultaneously generated dual broadband CARS signatures from the postflame zone of a premixed ethylene/air flame.

source. If the low frequency Raman resonance is narrow, as with H_2 transitions, the strength of the two-color process is greatly enhanced by contracting the ω_2 bandwidth. This then sharpens the underlying, normally-diffuse background in dual broadband CARS for resonances between the pump and high-Raman-shift Stokes source producing two well-defined signatures in close spectral proximity as in dual pump approaches. Dual pump-Stokes approaches arise in formulating multi-color strategies for H_2-air combustion [26]. One Stokes source is centered at 3657 cm^{-1} to track the appearance of H_2O product. Subtracting the N_2 Raman shift approximately, the low-Raman-shift source is positioned at 1246 cm^{-1} to generate H_2 CARS from its pure rotational S(4) transition. By making this laser narrow to enhance the two-color mixing, the H_2O signature appears along with N_2 in the three-color "dual broadband" process as discussed above.

3. NEW APPROACH TO PURE ROTATIONAL CARS

A variant of the dual broadband technique has led to a new and very simple approach to pure rotational CARS [27]. To perform dual broadband CARS for rotational Raman resonances in the 0 to 150 cm^{-1} range, the two Stokes sources have to be considerably overlapped. So overlapped in fact that one can simply employ a single broadband dye laser. Different frequency pairs within the broadband laser excite the rotational coherences from which the pump scatters. Quite importantly, the Stokes laser can be arbitrarily positioned in a spectral sense. The dye laser frequency can be selected on the basis of maximiz-

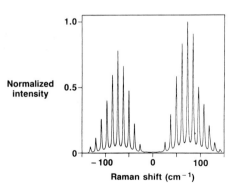

Fig. 3   Pure rotational CARS/CSRS spectrum of O_2 with arbitrary pump-Stokes spectral separation.

ing the output signal or exciting a selected vibrational mode in addition [28]. Figure 3 displays the pure rotational CARS spectrum of O_2 generated using this new approach. Note also that the CSRS spectrum is simultaneously generated as well. This new technique for pure rotational CARS may also lead to improved single shot spectral quality due to the dual integrative aspect of the process.

4. REFERENCES

1. R. J. Hall and A.C. Eckbreth: in Laser Applications, ed. by J. F. Ready and R. K. Erf, Vol. 5 (Academic, Orlando FL, 1984) p. 213.
2. S. A. J. Druet and J-P.E. Taran: Prog. Quant. Elec. 7, 1 (1981).
3. A. C. Eckbreth: Proc. SPIE 621, 116 (1986).
4. R. L. Farrow, R. P. Lucht, G. L. Clark and R. E. Palmer: Appl. Opt. 24, 2241 (1985).
5. H. Kataoka, S. Maeda and C. Hirose: Appl. Spect. 36, 565 (1982).
6. R. E. Teets: Opt. Lett. 9, 226 (1984).
7. R. L. Farrow and L. A. Rahn: J. Opt. Soc. Am. B 2, 903 (1985).
8. L. A. Rahn and R. E. Palmer: J. Opt. Soc. Am. B 3, 1164 (1986).
9. L. A. Rahn and D. A. Greenhalgh: J. Mol. Spect. 119, 11 (1986).
10. M. L. Koszykowski, R. L. Farrow and R. E. Palmer: Opt. Lett. 10, 478 (1985).
11. J. P. Sala, J. Bonamy, D. Robert, B. Lavorel, G. Millot and H. Berger: Chem. Phys. 106, 427 (1986).
12. L. A. Rahn, R. L. Farrow and R. P. Lucht: Opt. Letts. 9, 223 (1984).
13. D. R. Snelling, R. A. Sawchuk and R. E. Mueller: Appl. Opt. 24, 2771 (1985).
14. A. C. Eckbreth and J. H. Stufflebeam: Expt. Flds. 3, 301 (1985).
15. D. A. Greenhalgh and S. T. Whittley: Appl. Opt. 24, 907 (1985).
16. M. Pealat, P. Bouchardy, M. Lefebvre and J.-P. Taran: Appl. Opt. 24, 1012 (1985).
17. R. J. Hall and D. A. Greenhalgh: J. Opt. Soc. Am. B 3, 1637 (1986).
18. S. Kroll, M. Alden, T. Berglind and R. J. Hall: Appl. Opt. 26, 1068 (1987).
19. B. Attal, D. Debarre, K. Muller-Dethlefs and J.P. E. Taran: Rev. Phys. Appl. 18, 39 (1983).
20. B. Attal-Tretout and K. Muller-Dethlefs: Ber. Bun. Phys. Chem. 89, 318 (1985).
21. A. C. Eckbreth and T. J. Anderson: Appl. Opt. 24, 2731 (1985).
22. G. L. Switzer, D. D. Trump, L. P. Goss, W. M. Roquemore, R. P. Bradley, J. S. Stutrud and C. M. Reeves: AIAA Paper 83-1481 (1983).
23. R. P. Lucht: Opt. Lett. 12, 78 (1987).
24. A. C. Eckbreth and T. J. Anderson: Appl. Opt. 25, 1534 (1986).
25. A. C. Eckbreth, T. J. Anderson and G. M. Dobbs: Proc. 21st Symp. Combustion, The Combustion Institute, Pittsburgh PA (1987).
26. A. C. Eckbreth, T. J. Anderson and G. M. Dobbs: Proc. 23rd JANNAF Combustion Mtg., CPIA, Laurel MD (1987).
27. A. C. Eckbreth and T. J. Anderson: Opt. Lett. 11, 496 (1986).
28. M. Alden, P. E. Bengtsson and H. Edner: Appl. Opt. 25, 4493 (1986).

The Rotational Dual Broadband Approach and Noise Considerations in CARS Spectroscopy

M. Aldén[1], P.-E. Bengtsson[1], H. Edner[1], S. Kröll[1], D. Nilsson[1], and D. Sandell[2]

[1]Department of Atomic Physics, Lund Institute of Technology, Box 118, S-221 00 Lund, Sweden
[2]Department of Mathematical Statistics, Lund Institute of Technology, Box 118, S-221 00 Lund, Sweden

New experimental approaches to broadband CARS have recently been presented [1-3]. Rotational dual broadband CARS (RDBC) has shown to have several interesting properties for CARS thermometry [3]. By using an appropriate phase matching arrangement simultaneous recording of rotational and vibrational CARS spectra have been demonstrated. This is attractive for temperature determination in turbulent media with high temperature gradients. In rotational CARS there are no computational difficulties due to pressure induced narrowing as in vibrational Q-branch CARS where the Raman lines overlap at higher pressures. Non-Gaussian pump laser statistics [4,5] also does not have to be taken into account. Fitting to theoretical spectra for thermometry has been performed [6] in particular if instead of the complete spectral signature only the integrated intensities of the Raman resonances are fitted the computation time is reduced about two orders of magnitude. A quantitative analysis of the thermometry possibilities of RDBC is in progress [7]. Using RDBC a factor of two lower noise in non-Raman resonant CARS spectra than in a corresponding set-up using the conventional CARS approach has been observed [3]. There has also been a lively debate whether a single-mode or a multimode pump laser is the most appropriate one to use in CARS thermometry [8,9]. A qualitative model which could explain why the observed noise in Raman resonant CARS was much lower when a multimode pump laser was used has been presented [10]. That analysis failed, however, to explain why the opposite behaviour was observed in non Raman resonant CARS. A quantitative model have thus been developed [11]. The multimode laser fields E(t) are modelled as

$$E(t) = \sum_{k=-\infty}^{\infty} a_k \exp\left[i(\omega_o+k\Omega)t+\phi_k\right] \exp\left[-k^2\Omega^2/2\Gamma^2\right]$$

where Ω, Γ and ω_o are mode spacing, bandwidth and center frequency of the laser fields, respectively. The mode amplitudes and phases are all independent stochastic variables. The phases are assumed to be uniformly distributed in the interval $[0,2\pi]$ and the stochastic part (a_k) of the mode amplitudes to have an exponential probability distribution [12]. The lasers are assumed to operate only in their lowest order spatial mode and spatial effects are neglected. Given the assumptions above, the exact form (with no stochastic variables) for expectation value and variance for the the nonresonant CARS intensity can be obtained expressed as a summation over all possible mode combinations. The approach is in principle applicable to nonlinear optical phenomena in general in the limit of weak conversion efficiency and if the polarization can be expressed by a perturbation expansion in the electric fields. The analysis makes it possible to separate noise contributions from mode amplitude fluctuations, phase incoherence (both mode beating and non-mode beating terms) and also to identify the contribution from each laser source. The model agrees with the observed increase in noise in non-Raman resonant CARS when using a multimode laser

[9]. A multimode pump laser will add noise to the non-resonant spectra through mode beating with the dye laser. In principle the same thing should happen in a Raman resonant spectra. However, this may be outweighted by the strong reduction of noise from mode amplitude fluctuations in the dye laser since a multimode pump laser here significantly increases the number of dye laser modes contributing to the excitation of the Raman resonance [11]. The model, however, fails to explain such a strong decrease of the noise as has been observed in RDBC. There may be several reasons for this. In particular Eq. 1 may be too crude an approximation for the laser fields due to e.g. spontaneous mode-locking [13] or transverse mode structure [12,9] in the broadband dye laser or non-Gaussian pump laser statistics [4,5]. However, because of the possibility to separate and quantify different types of noise contributions the model may still provide a better understanding of the origin of noise in broadband CARS (and other nonlinear processes) and to optimise experimental approaches.

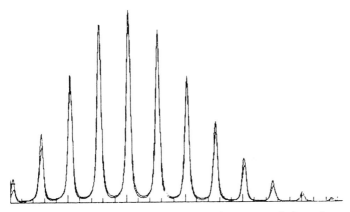

Figure 1. Room temperature oxygen spectrum recorded using rotational dual broadband CARS together with theoretical fit.

References

1. A.C. Eckbreth, T.J. Anderson: Appl. Opt. 24, 2731 (1985)
2. A.C. Eckbreth, T.J. Anderson: Opt. Lett. 11, 496 (1986)
3. M. Aldén, P.-E. Bengtsson, H. Edner: Appl. Opt. 25, 4493 (1986)
4. R.J. Hall: Opt. and Quant. Elect., 18, 319 (1986)
5. G.S. Agarwal, R.L. Farrow: J Opt. Soc. B3, 1596, (1986)
6. D. Nilsson: Lund Reports on Atomic Physics, LRAP-76 (1987)
7. M. Aldén, P.-E. Bengtsson, H. Edner, S. Kröll, D. Nilsson: in progress
8. R.J. Hall, D.A. Greenhalgh: J. Opt. Soc. B3, 1637 (1986)
9. D.R. Snelling, G.J. Smallwood, R.A. Sawchuk and T. Parameswaran: Appl. Opt. 26, 99 (1987)
10. S. Kröll, M. Alden, T. Berglind, R.J. Hall: Appl. Opt. 26, 1068 (1987)
11. S. Kröll, D. Sandell: to be submitted to J. Opt. Soc B
12. D.A. Greenhalgh, S.T. Whittley: Appl. Opt. 24, 907 (1985)
13. L.A. Westling, M.G. Raymer: J. Opt. Soc. B3, 911 (1986)

Resonance CARS Spectroscopy of the OH Radical in Flames and Discharge

B. Attal-Trétout, P. Berlemont, and J.P. Taran

Office National d'Etudes et de Recherches Aérospatiales,
BP 72, F-92322 Châtillon Cedex, France

Resonant CARS spectra of the OH radical have been obtained in a low pressure discharge under triple resonance conditions. A hydrogen-air flame has also been investigated under controlled conditions. We have achieved optimum sensitivity in our experiments by using three different exciting frequencies ω_1, ω_2 and ω_3.

A simplified expression of the CARS susceptibility has been derived from a quantum mechanical treatment of the nonlinear four-wave mixing process [1,2]. One obtains

$$\underline{\chi}^{(3)}(\omega_4) = \frac{N}{\hbar^3} \sum_{a,b,n,n'}{}' (A - B_1 + B_2 - B_3)\, \mu_{an},\, \mu_{n'b}\, \mu_{bn}\, \mu_{na} \tag{1}$$

with

$$A = \rho_{aa}^{(0)}\, [(\omega_{ba} - \omega_1 + \omega_2 - i\Gamma_{ba})(\omega_{na} - \omega_1 - i\Gamma_{na})(\omega_{n'a} - \omega_4 - i\Gamma_{n'a})]^{-1}$$

$$B_1 = \rho_{bb}^{(0)}\, [(\omega_{ba} - \omega_1 + \omega_2 - i\Gamma_{ba})(\omega_{nb} - \omega_2 + i\Gamma_{hn})(\omega_{n'a} - \omega_4 - i\Gamma_{u'a})]^{-1}$$

$$B_2 = \rho_{bb}^{(0)}\, [(\omega_{n'n} - \omega_3 + \omega_2 - i\Gamma_{n'n})(\omega_{n'b} - \omega_3 - i\Gamma_{n'b})(\omega_{n'a} - \omega_4 - i\Gamma_{n'a})]^{-1}$$

$$B_3 = \rho_{bb}^{(0)}\, [(\omega_{n'n} - \omega_3 + \omega_2 - i\Gamma_{n'n})(\omega_{nb} - \omega_2 + i\Gamma_{bn})(\omega_{n'a} - \omega_4 - i\Gamma_{n'a})]^{-1}$$

where μ_{an} is a matrix element of the dipole-moment operator, ω_{na} is an absorption frequency with damping Γ_{na}, $N\rho_{aa}^{(0)}$ and $N\rho_{bb}^{(0)}$ are the initial number densities of molecules in states $|a\rangle$ and $|b\rangle$ and Γ_{ba} is the Raman linewidth in the ground electronic state. The condition for a triple resonance to appear is easily seen to be [3]

$$\omega_1 - \omega_2 = \omega_{ba} \quad \text{with} \quad \omega_1 = \omega_{na} \quad \text{and} \quad \omega_3 = \omega_{n'b} . \tag{2}$$

The spectral properties of three-colour CARS have been studied as a function of $(\omega_1 - \omega_2)$, ω_1 and ω_3 [4].

Among several aspects, the particular character of the Raman resonances of the excited electronic states $(\omega_{n'n})$ can be pointed out. They split from the main Raman line (ω_{ba}) when the laser frequency ω_1 (or ω_3) is shifted out of resonance. Starting from a single line given by (2), the spectral evolution will result in a quadruplet if both ω_1 and ω_3 are detuned from exact resonance. The resonant CARS spectra may be contributed by a single rotational line only. Note that resonance CARS spectroscopy in OH is highly selective since the rotational splitting is large ($B_e \simeq 18$ cm^{-1}).

A computer program was written in order to fit the experimental spectra. This program calculates the four major susceptibility contributions with due account for the Doppler effect, the CARS rotational line strengths [5] and the laser linewidth function.

The experimental set-up is composed of three frequency-doubled dye lasers which produce pulses of spectral width between 0.05 and 0.1 cm^{-1} with 30-50 μJ energy in the red. Frequency calibration was carried out first with a commercial Fizeau wavemeter. The absolute calibration was then improved down to 0.01 cm^{-1} by recording excitation spectra of iodine.

The resonance conditions were selected using the spectroscopic data of COXON [6]. The frequencies ω_1 and ω_3 were tuned respectively to the $P_1(0-0)7.5$ and $R_1(1-1)5.5$ lines of the A $^2\Sigma^+$ - X $^2\Pi$ electronic system of OH. The ω_2 frequency was then tuned to scan the fundamental O(1-0)7.5 Raman transition which is located at $\omega_{ba} = 3065.303$ cm^{-1}. A triple resonance is thus achieved for the main CARS line $[O_1(1-0), P_1(0-0), P_1(1-0)]7.5$. Note that the rules and notations used for the labeling of resonance CARS lines were defined in a previous paper [7].

The OH signal was first detected by focusing the three laser beams into the afterglow of a microwave discharge. The latter was fed with a flowing mixture of Helium and water vapor at 9 mb total pressure. The CARS spectra of Fig. 1 were recorded with $\omega_3 = 32011.55$ cm^{-1} ($\approx \omega_{n'b}$). As ω_1 is down shifted (from (a) to (c)), the $\omega_{n'n}$ contributions separate from the main Raman line. This phenomenon was not observed in two-colour CARS spectroscopy [3]. The different contributions are identified on the spectra and their spectral positions are marked on the horizontal axis. The OH CARS line appears above a negligible background caused by the electronic noise. The detectivity in the discharge ranges between 10^{11} and 10^{12} cm^{-3} depending on the saturation threshold of the one-photon transitions.

The reaction zone of an atmospheric pressure flame was then investigated. A variety of premixed hydrogen-air and methane-air flames was studied with most of the measurements made in a stoechiometric mixture (equivalence ratio = 1, cold gas velocity = 49 cm/s). Beams were focused at about 0.5 mm above the burner porous. Two experimental spectra are shown in Fig. 2 along with a computer simulation for $\omega_1 = 32122.09$ cm^{-1} and $\omega_3 = 32012.08$ cm^{-1} both slightly detuned from exact resonance. The non-resonant background is due to air surrounding the flame. It was suppressed by crossing the beams in a "BOXCARS" arrangement [8]. The CARS multiplet is just resolved and the structure has been unambiguously identified. Saturation was not observed in the flame. We estimate that the OH density is about 10^{15} cm^{-3} in the reaction zone. Background interference from water vapor was not observed [9].

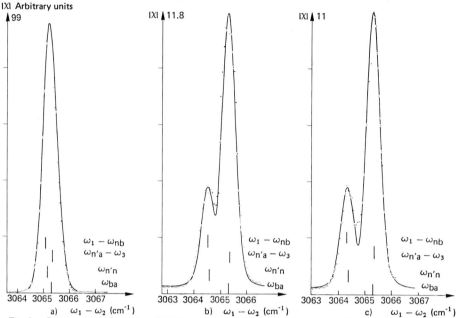

Fig. 1 – *Resonance CARS spectra of OH in a microwave discharge with computer simulations (full lines). The positions of the various single resonances are indicated with bars ; a) $\omega_1 = 32123.19\ cm^{-1}$, b) $\omega_1 = 32122.60\ cm^{-1}$, c) $\omega_1 = 32122.42\ cm^{-1}$*

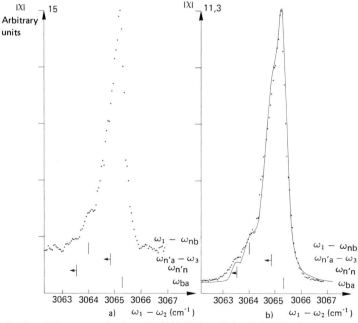

Fig. 2 – *OH spectra in the flame. At 2200 K, the collisional width is taken as $\Gamma = 0.085\ cm^{-1}$ for $J = 7.5$ in the simulation ; a) collinear beams, b) crossed-beams*

1. S.A.J. Druet, B. Attal, T.K. Gustafson, J.P. Taran: Phys. Rev. 18 A (1978) p.1529
2. N. Bloembergen, H. Lotem, R.T. Lynch Jr: Indian J. Pure Appl. Phys. 16 (1978) p.151
3. B. Attal, D. Débarre, K. Müller-Dethlefs, J.P. Taran: Rev. Phys. Appl. 18 (1983) p.39
4. B. Attal-Trétout, P. Berlemont, J.P. Taran (to be published)
5. B. Attal-Trétout, K. Muller-Dethlefs: Ber. Bun. fur Phys. Chem. 89 (1985) p.318
6. S.A. Coxon: Can. J. Phys. 58 (1980) p.933
7. B. Attal, O.O. Schnepp, J.P. Taran: Opt. Commun. 24 (1978)
8. A.C. Eckbreth: Appl. Phys. Lett. 32 (1978) p.421
9. J.F. Verdieck, R.J. Hall, A.C. Eckbreth: AIAA Paper n° 83-1477 (1983)

Electronically Resonant CARS Spectroscopy of C_2 in a Flame

C.G. Aminoff, M. Kaivola, and T. Virtanen

Department of Technical Physics, Helsinki University of Technology, SF-02150 Espoo, Finland

Coherent Anti-Stokes Raman Scattering (CARS) has proven to be a powerful laser technique for combustion diagnostics. However, the presence of a nonresonant background susceptibility sets the detection limit for CARS at a level of 1 % in practical applications. By tuning one or several of the laser frequencies involved into resonance with an electronic transition in the molecule studied (see Fig. 1), a strong enhancement of the resonant CARS susceptibility relative to the background is achieved. This yields a detection sensitivity superior to conventional CARS by several orders of magnitude. Electronically resonant CARS may thus provide a sensitive technique for the detection of trace species in flames such as radicals and combustion products [1,2]. On the other hand, the signal is easily affected by saturation of the electronic transitions, which may be difficult to avoid in practice when small molecular concentrations are to be measured.

The present work is an investigation of resonance enhanced CARS spectra of the C_2 radical in an acetylene/oxygen flame at atmospheric pressure. We have studied the CARS signal within the Swan band ($d^3\Pi_g - a^3\Pi_u$) at two different spectral positions where the pump laser field, the Stokes laser field and the anti-Stokes field are all in resonance or close to resonance with electronic transitions ($\omega_v \approx \omega_{v'}$). These near-triple resonances involve the triplet components $J = 18, 19, 20$ of the lowest vibrational level $v = 0$ and they occur at the pump laser wavelength 512.8 nm and the Raman shifts 1611 cm^{-1} (Q branch) and 1743 cm^{-1} (S branch), respectively [2].

For the experiment we used two narrowband pulsed dye lasers pumped by an excimer laser. A crossed-beam (BOXCARS) configuration was used for good spatial resolution.

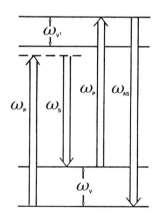

Fig. 1. Schematic of electronically resonant CARS in four-level system; ω_v, $\omega_{v'}$ denote rovibrational frequency separations and ω_P, ω_S, ω_{AS} are pump laser, Stokes laser and anti-Stokes frequencies, respectively

The burner nozzle had the diameter 0.7 mm. A two-dimensional mapping of the CARS spectra in the vicinity of the near-triple resonances in the Q and S branches was carried out by scanning the laser frequencies. Fig. 2a represents the CARS signal intensity measured at the Q branch resonance using the pulse energy 2 μJ (peak power 130 W) in each laser beam. The spectrum contains contributions from one-photon, Raman, and three-photon resonances and reveals the presence of saturation effects. The prominent double-peak feature is due to the partly resolved triplet structure.

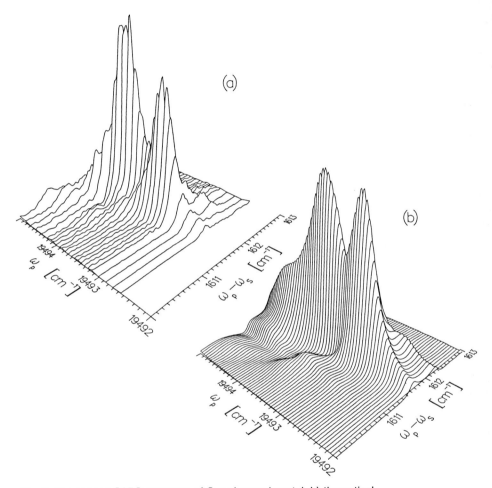

Fig. 2. Resonance CARS spectrum of C_2; a) experimental, b) theoretical

    The power dependence and the saturation behaviour of the peak signal were also studied. The resonance enhancement yielded detectable CARS signals from the flame for laser peak powers as low as 25 W. Based on our present equipment, the detection limit for C_2 molecules in a flame at atmospheric pressure was estimated to be $10^{11}/cm^3$, which represents an increase in sensitivity of the order of 10^6 compared with conventional CARS.

For the theoretical analysis we have used a model based on the density matrix description of a four-level system (cf. Fig. 1). Using given molecular parameters [2] the resonance CARS spectrum of the almost triply resonant Q branch multiplet of C_2 has been simulated. The third-order calculation of the CARS signal was found not to adequately reproduce the detailed structure of the spectrum in Fig. 2a. To allow for higher order effects in the spectrum the steady state density matrix equations were solved numerically. A good fit to experimental results was found by adjusting relaxation rates and temperature. The theoretical spectrum in Fig. 2b calculated for the temperature 2000 K includes saturation and is seen to reproduce the essential features of Fig. 2a.

This work has been financially supported by Neste Oy and the Finnish Academy of Sciences.

1.   S. Druet, B. Attal, T.K. Gustafson, and J.P. Taran: Phys. Rev. A **18**, 1529 (1978)
2.   B. Attal, D. Débarre, K. Müller-Dethlefs, and J.P. Taran: Revue Phys. Appl. **18**, 39 (1983)

CARS Thermometry of a H_2 Supersonic Jet

G. Marowsky and A. Slenczka

Max-Planck-Institut für Biophysikalische Chemie, Abteilung Laserphysik, D 3400 Göttingen, Fed. Rep. of Germany

In a recent study [1] both pure rotational and rotational-vibrational transitions in molecular hydrogen have been used for thermometry with coherent anti-Stokes Raman scattering (CARS). A dual dye-laser system, pumped with a powerful excimer-laser, allowed the observation of the $J = 5 \rightarrow J = 7$ rotational transition in the vibronic groundstate of H_2 under room temperature and 100 Torr pressure in a single-shot experiment. Although rotational CARS techniques are superior in sensitivity and accuracy at low temperature, vibrational Q-branch spectroscopy has been applied as a convenient experimental tool for the characterization of H_2 supersonic jet expansion under various temperature conditions.

Scanning-CARS with tuning of the dye-laser generating the pump-wavelength for improved spectral resolution was used for monitoring rotational temperatures by evaluation of the contributions from the Q_0 to Q_5 transitions in the temperature range 240 K to 260 K. H_2-expansion through a nozzle of 200 μm internal diameter under room temperature and 0.5 atm stagnation pressure results in a rotational temperature of 240 K as shown in Fig. 1. This figure also shows the temperature determination using several Q-branch intensity ratios and the concomitant scatter of experimental data, typical ±3 % after weighted averaging. Figure 2 summarizes the experimental results:

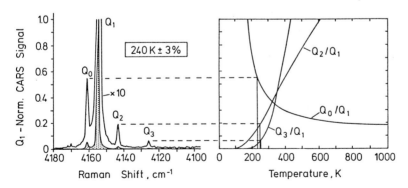

Fig. 1. Q-branch thermometry of molecular hydrogen: well-resolved CARS spectra, normalized to the peak intensity of the Q_1-transition (left), yield immediate temperature information upon comparison with calibration curves (right)

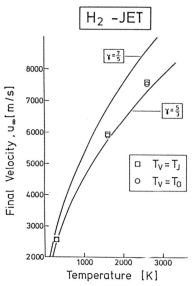

Fig. 2. Final velocity of H_2-jet versus nozzle temperature for specific heat ratios $\gamma = 7/5$ and $\gamma = 5/3$

Molecular hydrogen relaxes under supersonic jet expansion rather as a monoatomic gas than as a diatomic molecular gas. Figure 2 shows the final velocity for the expansion of a monoatomic gas, characterized by the specific ratio $\gamma = 5/3$, and a diatomic molecular gas ($\gamma = 7/5$) [2] in a supersonic jet versus nozzle temperature. In comparison with theoretical considerations outlined in Ref. [3] the experimentally determined rotational temperature shows that only 20 % of the internal energy of the H_2-molecules contributes to the final velocity.

References

1. G. Marowsky, A. Gierulski, B. Dick, U. Sowada, R. Vehrenkamp: Appl. Phys. B 39, 47 (1986)
2. P.P. Wegener (ed.): Molecular Beams and Low Density Gasdynamics (Marcel Dekker, New York 1974)
3. A. Slenczka, G. Marowsky, A. Vodegel: in preparation - to be submitted to Appl. Phys. B

Subharmonic Resonances in Higher-Order Collision-Enhanced Wave Mixing in a Sodium-Seeded Flame

R. Trebino and L.A. Rahn

Combustion Research Facility, Sandia National Laboratories, Livermore, CA 94550, USA

Collisional damping has been shown to give rise to new or "extra" resonances in four-wave-mixing processes[1-3]. We report the observation of collision-enhanced resonances in N-wave mixing in a sodium-seeded hydrogen-air flame using high-resolution (<60 MHz) pulsed lasers. Our spectra yield evidence for N as high as fourteen.

We have performed $2\omega_1-\omega_2$ four-wave-mixing experiments on ground-state Zeeman and hyperfine resonances of sodium. At high intensities, new spectral lines appear at the frequency differences $\pm\omega_{hfs}/2$, where ω_{hfs} is the hyperfine splitting 0.059 cm^{-1}[2]. Additional experiments, using higher-order beam geometries, verify that these resonances result from higher-order wave mixing. Specifically, we use geometries phase-matched for six- and eight-wave mixing (see Fig. 1). RAJ et al.[4] have used a similar method for higher-order grating effects. Our geometries are planar and have the k-vector equations $k_{sig}=2(k_1-k_2)+k_1'$ for N = 6, and $k_{sig}=3(k_1-k_2)+k_1'$ for N = 8. These geometries eliminate lower-order wave-mixing effects, but allow N-, (N+2)-, (N+4)-, ... -wave effects. In these experiments, two pulse-amplified cw dye lasers operate single-mode and are tuned 2 cm^{-1} below the D_1 line of sodium. One of these beams is split into two pulses of approximately equal energy. The resulting three beams cross, unfocused, in an approximately stoichiometric, hydrogen-air flame seeded with sodium (~200 ppm). The beams having k-vectors k_1 and k_1' have frequency ω_1 and are polarized vertically, and the beam with k-vector k_2 has frequency ω_2 and is polarized horizontally.

Figure 2a shows a typical experimental spectrum using a six-wave-mixing $(3\omega_1-2\omega_2)$ geometry for 7 kW/cm^2 in each of the three input beams. The subharmonic resonances are now clearly evident at $\delta = \pm\omega_{hfs}/2 = \pm0.03$ cm^{-1}. The hyperfine components at $\delta = \pm\omega_{hfs} = \pm0.06$ cm^{-1} are also present, but weak. In addition, careful observation at $\delta = \pm\omega_{hfs}/3$ reveals *dips* in the wings of the Zeeman resonance line, indicating the possible presence of eighth- or higher-order wave-mixing processes.

An eight-wave-mixing geometry reveals additional subharmonics as shown in the spectrum of Fig. 2b. The central Zeeman component and the $\delta = \pm\omega_{hfs}/2$ subharmonics remain relatively strong, but the hyperfine resonances at $\delta = \pm\omega_{hfs}$ are now extremely weak. The $\delta = \pm\omega_{hfs}/3 = 0.02$ cm^{-1} subharmonics now appear to be interfering with the $\omega_{hfs}/2$ wing, and new subharmonics now appear at $\delta \approx \pm\omega_{hfs}/4 = 0.015$ cm^{-1} and $\delta = \pm\omega_{hfs}/5 = 0.012$ cm^{-1}. While eight-wave-mixing processes can accomodate $\pm\omega_{hfs}/2$ and $\pm\omega_{hfs}/3$ subharmonics, at least ten-wave mixing is required for the $\pm\omega_{hfs}/4$ subharmonic. Fourteen-wave mixing is the minimum-wave process that can yield $\pm\omega_{hfs}/5$ subharmonics in our eight-wave-mixing geometry.

A theory of collision-enhanced higher-order wave mixing was recently developed by AGARWAL and NAYAK[5], who used a continued-fraction approach for a two-level system. This work predicted fractional resonances at the subharmonics of the Rabi frequency. AGARWAL[6] has recently extended this theory to a three-level system,

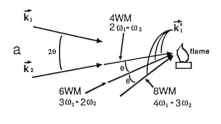

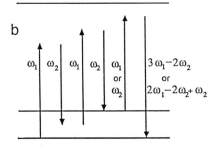

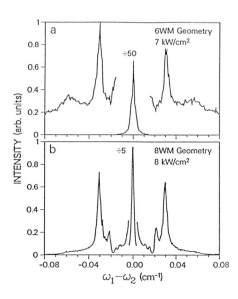

Fig. 1 a) Six- and eight-wave-mixing planar beam geometries using phase-matching that excludes lower-order wave-mixing signals. b) Energy level diagram for six-wave-mixing showing a four-photon $(2(\omega_1-\omega_2))$ subharmonic resonance.

Fig. 2 a) Spectrum of collision-enhanced resonances in Na in a flame obtained using a six-wave-mixing geometry $(3\omega_1-2\omega_2,$ in lowest order). b) Spectrum obtained using an eight-wave-mixing geometry $(4\omega_1-3\omega_2,$ in lowest order).

predicting subharmonic resonances between the two ground electronic states. Standard diagrammatic perturbation theory also predicts subharmonics in higher-order wave-mixing processes, and we have used a diagrammatic approach[7] to derive theoretical expressions for collision-enhanced six-wave mixing that predict subharmonic resonances. We are, however, at the limits of applicability of the perturbative expansion of χ in the applied electric fields. Consequently, a high-intensity theory, such as that of AGARWAL, will ultimately be required.

The authors would like to acknowledge the useful advice of G. S. Agarwal. This work was supported by the U.S. Department of Energy, Office of Basic Energy Sciences, Chemical Sciences Division.

References
1. L.J. Rothberg and N. Bloembergen, Phys. Rev. A **30**, 820 (1984), and references therein.
2. R. Trebino and L.A. Rahn, in *Advances in Laser Science II*, edited by R.C. Powell, W.C. Stwalley, M. Lapp, G.A. Kenney-Wallace, and R. Gross (AIP, 1987).
3. J.F. Lam, D.G. Steel, and R.A. McFarlane, Phys. Rev. Lett. 56, 1679 (1986).
4. R.K. Raj, Q.F. Gao, D. Bloch, and M. Ducloy, Opt. Commun. 51, 117 (1984).
5. G.S. Agarwal and N. Nayak, Phys. Rev. A 33, 391 (1986).
6. G.S. Agarwal, private communication.
7. R. Trebino, submitted to Opt. Lett., 1987.

Recent Advances in Flame Diagnostics Using Fluorescence and Ionization Techniques

J.E.M. Goldsmith

Combustion Research Facility, Sandia National Laboratories,
Livermore, CA 94550, USA

1. Introduction

Laser-based spectroscopic techniques are proving to be powerful methods for studying combustion environments. The needs for non-intrusive measurements of physical parameters (e.g., temperature and species concentrations) with high spatial and temporal resolution are well matched by the capabilities of laser spectroscopy. The techniques that have received the most attention to date are coherent anti-Stokes Raman spectroscopy (CARS), laser-induced fluorescence (LIF), and ionization detection. A recent review of applications of LIF can be found in [1]. An accompanying paper in this volume describes recent developments in CARS, a powerful technique that has proven especially useful for measuring temperature and the concentration of the major flame species (e.g., fuel, oxidant, and end gases, typically present in concentrations >1% mole fraction). This paper will highlight a few of the recent advances in flame diagnostics based on LIF and ionization techniques. These methods provide capabilities for studying the chemically reactive radical or minor species that are present in lower concentrations (typically <1%), but that play a dominant role in controlling the flame chemistry.

2. Imaging

One of the major advances in applications of fluorescence detection to flame diagnostics has been the rapid development from conventional point measurements to one (line), two (plane), and even three (volume) dimensional images of combustion properties. More important than the obvious advantage of acquiring measurements much more rapidly than by a series of point observations, these developments have been driven from the combustion viewpoint by the need to take "snapshots" in turbulent environments. Planar images recorded in a single laser shot (or volumes represented by a series of planar measurements made with closely spaced laser pulses) yield far more detail about turbulent structures than could possibly be provided using point-measurement techniques.

Imaging techniques are generally adaptations of methods such as LIF that had previously been used to make point measurements. Applications of these methods have largely been limited by the difficulties of providing sufficient illumination to the flowfield, and the ability to record, process, and read out the resulting image from the photosensitive device. Improvements in laser and camera technology have therefore been crucial to this field. Several flowfield parameters, including species concentrations, temperature, density, velocity, and pressure, have now been addressed with imaging techniques. Single-photon-excited LIF schemes have received the most attention, but multiphoton-excited LIF has been used to image H and O along lines, and CO in a plane. Further discussion of these methods is beyond the scope of this paper, but an excellent review of the field appears in [2].

3. Multiphoton Excitation

The suitability of a technique for measuring a given flame species is determined primarily by the spectroscopy of the atom or molecule in question. For example, the OH radical, one of the most important species in flame chemistry, has strong electronic transitions in the convenient spectroscopic range of frequency-doubled rhodamine dye lasers, and thus has had many methods applied to its detection. Several other species of interest in combustion research (e.g., H, O, and CO) require vacuum-ultraviolet radiation to excite their first electronic resonances. This radiation is not transmitted by flame gases, but resonant multiphoton excitation using wavelengths longer than 200 nm provides convenient spectroscopic access to these species. Following the excitation step, the atom or molecule can be observed in several ways, including monitoring the photons emitted from radiative decay to lower-lying levels (fluorescence detection), or by monitoring the ion/electron pairs produced by subsequent photoionization.

Atomic hydrogen has been the subject of a large number of multiphoton excitation studies. Seven excitation schemes have been used to detect hydrogen in flames, four using ionization detection and three using fluorescence detection [3]. While this may seem an excessive number, there are significant trade-offs among the methods, and the best choice depends to a large extent on the details of the environment being probed. In particular, as will be discussed in the next section of this paper, high-intensity, short-wavelength pulses can cause significant photochemical production of the species being monitored, so that comparisons of profiles measured using multiple excitation schemes are very desirable. Atomic oxygen and CO have also been measured in flames using both fluorescence and ionization detection. In a two-photon excitation study of the CO B $^1\Sigma^+ \leftarrow$ X $^1\Sigma^+$ transition using a few mJ/pulse of 230–nm radiation to excite the transition in a methane-air flame, we could see the blue fluorescence resulting from B $^1\Sigma^+ \rightarrow$ A $^1\Pi$ decay by eye over the blue chemiluminescence from the flame. In addition, we could detect the excitation by *ear* because of the optoacoustic effect.

One of the advantages of two-photon excitation is the relative ease with which Doppler-free spectra can be obtained, assuming that the combination of sufficient laser resolution and peak power is available. One excellent source that fills these needs, and that has been developed at several laboratories, is an injection-seeded, single-mode, frequency-doubled Nd:YAG laser used to pump the dye pulse amplifier for a single-mode cw dye laser system. We have used the 620-nm output from such a system for Doppler-free fluorescence spectroscopy of OH A $^2\Sigma^+ \leftarrow$ X $^2\Pi$ transitions in low-pressure hydrogen-oxygen flames, completely resolving the hyperfine structure of the A state [4]. We have also recorded Doppler-free fluorescence and ionization spectra of CO [5]. To generate narrowband, tunable 230-nm laser radiation to excite the CO transition, we first frequency-doubled the output of the pulse-amplified dye laser system operating at 587 nm. The resulting 294–nm radiation was in turn frequency mixed with the residual 1.06–μm radiation from the single-mode Nd:YAG laser used to pump the dye pulse amplifier, producing up to 0.5 mJ/pulse. We have recorded CO spectra in a cell to date, and hope to extend this method to flame studies as well.

4. Photochemical Effects

One of the frequently mentioned benefits of laser-based combustion diagnostics, as compared to probe sampling techniques, is the non-intrusive nature of the former. This is generally true, but as the wavelength of the laser beam is decreased, and its peak power increased, the probability of producing photochemical perturbations in the sample can become significant. In particular, the laser beam may create significant quantities of the species it is being used to detect, so that the resulting signal cannot be directly related to relative concentration, even in the absence of other difficulties (such as variations in quenching, discussed in the following section). We have previously found direct evidence for this type of perturbation using two-photon-excited fluorescence detection of both atomic hydrogen [6] (205-nm excitation) and atomic oxygen [7] (226-nm excitation) in a lean, atmospheric-pressure hydrogen-oxygen flame.

Of the seven demonstrated methods for detecting H in flames [3], we have had the most success using a two-step scheme (two-photon excitation of the 1S-2S transition with 243–nm radiation, followed by subsequent 2S-3P excitation at 656 nm, and detection of the resulting fluorescence at 656 nm). We have found evidence for photochemical production of H using this technique, though it was not as serious a perturbation as found in 205-nm excitation. In addition, the mechanism producing the photochemical effect was very different from that observed previously. We have studied these effects by separately monitoring with LIF the OH fragment produced by the far-UV laser pulses. In the 205-nm study, production of excess OH was linear in pulse energy, with no resonance observed as the 205-nm wavelength was scanned across the hydrogen transition; the excess OH (and H) was due simply to single-photon photolysis of water in the flame [6]. This contrasts our result using 243-nm radiation, in which we found that the OH photolysis signal was *nonlinear* in pulse energy, and that there was a *resonance* in the mechanism coincident with the atomic hydrogen two-photon 1S-2S transition.

5. Quantitative Measurements

One of the major difficulties encountered in using techniques such as those described above for making quantitative measurements in combustion environments is the large variation in quenching that can occur through a flame. For example, in a stoichiometric hydrogen-oxygen flame, the local environment varies from a room-temperature mix of 67% H_2 and 33% O_2, through the reaction zone with a complex mixture of species, to the post-flame gases consisting of nearly 100% H_2O at a much higher temperature (up to ~3000 K). Thus in the measurement of a relative concentration profile, it cannot necessarily be assumed that a measured signal is simply proportional to concentration throughout the flame.

Several approaches have been taken to this problem. The most straightforward is to calculate the local quenching rate, and use the calculated value to correct the quantum yield in the signal. Unfortunately, application of this method requires knowledge of the flame temperature and species concentrations at the appropriate point in the flame, and information on the temperature-dependent quenching rates of each of the collision partners. Of course if the former were completely known, there would be no need to make the measurement; in many cases, however, reasonable estimates (or measurements using other methods) of the major species concentrations can be made. Knowledge of the appropriate quenching parameters is still very limited, although significant advances are being made in the collection of this basic spectroscopic information. A recent assessment of the data base for the hydrides OH, NH, and CH has recently been made, with the conclusion that it should be possible to estimate the fluorescence quantum yield for OH in methane and ammonia flames burning in oxygen to within ~30%, but only within a factor of perhaps five for other molecules [8]. Measurements of quenching by several species of two-photon-excited H and O in a discharge-flow reactor have also been reported [9], but care is needed in extrapolating these measurements to flame conditions.

An alternative that has taken several forms is to operate under conditions where some other physical mechanism occurs at a rate faster than quenching, in which case variations in the quenching rate can be unimportant. Predissociation of levels in the O_2 B $^3\Sigma_u^-$ state typically occurs on the picosecond time scale, so fluorescence quantum yields from Schumann-Runge excitation, while small, do not vary substantially under flame conditions [10]. Most other molecules do not provide such an easy solution, however, requiring some ingenuity on the part of the experimentalist. The relative insensitivity of laser-saturated fluorescence (LSF) to quenching has received a great deal of attention, and quantitative applications for flame studies have been examined in detail [11,12].

Photoionization from the laser-excited level at a rate faster than quenching also minimizes the affect of variations in quenching rate. This mechanism can be exploited directly in ionization measurements. A more novel application, given the acronym PICLS (photoionization-controlled loss spectroscopy), has also been demonstrated [13]. In this method, two-photon-excited fluorescence of atomic hydrogen occurred in the presence of a separate laser beam that only interacted with the atom by photoionizing from the excited level. By adjusting the intensity of this beam such that the fluorescence signal was reduced to perhaps 10% of its original level, relative independence to variations in quenching rates could be assured, although at the cost of the reduction in the signal level. We have recently demonstrated an ionization-detection variation of this technique for atomic hydrogen. In this method, we excited the 1S-2S transition by a 243–nm laser beam, which also photoionized a small portion of the excited atoms. A separate 355–nm beam, overlapping the 243–nm beam but with a larger spot size, was varied in intensity, and the ionization signal detected by a probe wire in the flame. The resulting dependence of the signal on 355–nm pulse energy demonstrated nearly complete saturation of the ionization signal. In addition to providing (relatively) quenching-independent ionization measurements, careful characterization of the 355-nm laser pulses could provide direct information on collisional quenching rates of the laser-excited level.

6. Conclusion

This paper can present only a small sampling of the recent advances in flame diagnostics using fluorescence and ionization detection. In addition to advances in techniques, contributions to the spectroscopic database available for atoms and molecules continue to add new species to the list of those accessible in flame studies using laser-based methods. Advances in laser spectroscopy and technology will continue to have a significant impact on applications such as combustion diagnostics, so that the field will continue to grow, and methods that currently fall into the class of "new techniques" will mature into working tools for combustion measurements.

This work is supported by the U.S. Department of Energy, Office of Basic Energy Sciences, Division of Chemical Sciences.

1. R. P. Lucht, in *Laser Spectroscopy and Its Applications*, Leon J. Radziemski, Richard W. Solarz, and Jeffrey A. Paisner, eds. (Marcel Decker, New York, 1986), pp. 623-676.
2. R. K. Hanson, in *Twenty-First Symposium (International) on Combustion*, The Combustion Institute (in press).
3. J. E. M. Goldsmith, in *Advances in Laser Science - I, Proceedings of the First International Laser Science Conference*, William C. Stwalley and Marshall Lapp, eds. (American Institute of Physics, New York, 1986), pp. 279-282.
4. J. E. M. Goldsmith and L. A. Rahn, JOSA B (to be submitted).
5. J. E. M. Goldsmith and L. A. Rahn (in preparation).
6. J. E. M. Goldsmith, Opt. Lett. **11**, 416 (1986).
7. J. E. M. Goldsmith, Appl. Opt. (in press).
8. N. L. Garland and D. R. Crosley, in *Twenty-First Symposium (International) on Combustion*, The Combustion Institute (in press).
9. U. Meier, K. Kohse-Höinghaus, and Th. Just, Chem. Phys. Lett. **126**, 567 (1986).
10. M. P. Lee, P. H. Paul, and R. K. Hanson, Opt. Lett. **12**, 75 (1987).
11. J. T. Salmon and N. M. Laurendeau, Appl. Opt. **24**, 1313 (1985).
12. K. Kohse-Höinghaus, R. Heidenreich, and Th. Just, in *Twentieth Symposium (International) on Combustion*, The Combustion Institute, p. 1177 (1985).
13. J. T. Salmon and N. M. Laurendeau, Opt. Lett. **11**, 419 (1986).

LIF with Tunable Excimer Lasers as a Possible Method for Instantaneous Temperature Field Measurements at High Pressures

P. Andresen[1], A. Bath[1], H.W. Lülf[1], G. Meijer[2], and J.J. ter Meulen[2]

[1]MPI für Strömungsforschung, D-3400 Göttingen, Fed.Rep.of Germany
[2]Katholieke Universiteit, Toernooiveld,
   NL-6525 ED Nijmengen, The Netherlands

A new method for instantaneous temperature field measurements at high pressures is presented. This method is based on spectrally dispersed LIF studies of OH, O_2 and H_2O in an open atmospheric flame with a tunable KrF-excimer laser. The crucial problem of quenching of the fluorescence at high pressures is almost completely eliminated by excitation to a fast predissociating state. In this case the fluorescence is emitted only during the short predissociation lifetime τ of the excited electronic state. For a sufficiently short predissociation lifetime there will be no quenching within this lifetime, even at high pressures. The fact that the excited state molecules radiate only in the short time τ means on the other hand that only a very small part of the molecules does fluoresce. Although this causes a tremendous loss in LIF signal, this is partly compensated because the densities are much higher than in normal LIF experiments. The large power of the KrF-laser is therefore also of particular importance. This laser delivers 450 mJ per pulse in a bandwidth of 0.5 cm^{-1}. For two dimensional measurements a large sheet can be illuminated with this laser, still having efficient excitation of even weak transitions.
   Many different vibrational-rotational transitions of OH, O_2 and H_2O, all ending up in fast predissociating electronic states, were found to lie in the narrow tuning range (248.0-248.9 nm) of the KrF-laser. In the figure the excitation spectrum of the OH ($A^2\Sigma^+$, v'=3 $\leftarrow$ $X^2\Pi$, v"=0) transition is shown. As various rotational levels with a large energy separation in the ground state are probed, accurate temperature measurements are possible. Dispersion spectra of single laser prepared states show explicitly that rotational and vibrational quenching within the predissociation lifetime is small, even at atmospheric pressures.

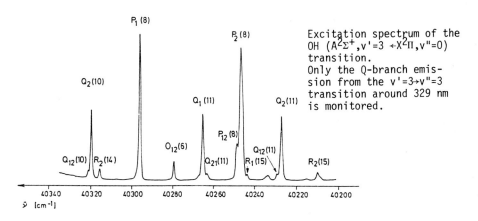

Excitation spectrum of the OH ($A^2\Sigma^+$,v'=3 $\leftarrow X^2\Pi$,v"=0) transition.
Only the Q-branch emission from the v'=3$\rightarrow$v"=3 transition around 329 nm is monitored.

Collision-Free Energy Distribution of OH Radicals After H_2O_2 Photolysis Using LIF in a Flow System

A. Jacobs, M. Wahl, R. Weller, and J. Wolfrum

Physikalisch-Chemisches Institut, Universität Heidelberg,
Im Neuenheimer Feld 253, D-6900 Heidelberg, Fed. Rep. of Germany

In order to investigate chemical elementary reactions, it is necessary to know the internal energy distribution of the species involved in such a reaction. In field of combustion research the OH radical is of great importance. Using the LIF-method we measured the collision-free internal energy distribution of ground state $OH(X\ ^2\Pi)$ radicals after photolysis of H_2O_2 at $\lambda = 193$ nm.

Hydrogen peroxide was flowed permanently through a quartz cell at a pressure of 5 mTorr, and photolyzed by an ArF-excimer laser at 193 nm. 50 nsec after photofragmentation the generated OH radicals were pumped to the first electronically excited state $A\ ^2\Sigma^+$ by a pulsed frequency doubled dye laser. In contrast to former measurements /1/ the experimental conditions realized here ensured collision-free determination of the OH energy distribution. The OH fluorescence was detected by a photomultiplier and a boxcar system. A microcomputer was used for data sampling, storing and analysing of the OH excitation spectra. Special care was taken to work in the linear regime with both photolysis and probe laser allowing shot-to-shot normalization of fluorescence signals to the laser energies. The computer additionally controlled the triggering of lasers and boxcar and performed the scanning of the probe laser. The system typically worked at repetition rates of 5 Hz. For further details see /2/.

To obtain the relative populations of an absorbing OH state from line intensities in the LIF-excitation spectra, it is necessary to know the Einstein coefficients of the absorbing transitions. We used the relative coefficients from /3/ to calculate the rotational distribution of the OH radicals. With a least squares fit procedure /4/ we were able to include overlapping transitions leading to a complete rotational distribution of the $OH(X\ ^2\Pi,\ v" = 0)$ state.

Although the available energy after photolysis of H_2O_2 at 193 nm would be high enough to leave one of the OH radicals in the lowest electronically excited $A\ ^2\Sigma^+$ state, no emission from this state was detectable directly after the photolysis pulse. Only weak signals were found for vibrationally excited $OH(X\ ^2\Pi,\ v" = 1)$, indicating that due to the angled structure of H_2O_2, OH acts as a spectator during photodissociation. We estimated that approximately 1% of the available energy appears as vibrational energy. The rotational distribution we obtained for OH($X\ ^2\pi$, v" = 0) is very similiar for the spin components ($^2\pi_{3/2}$, $^2\pi_{1/2}$) and the two lambda components ($^2\pi^+$, $^2\pi^-$) showing a clear peak at rotational quantum number K" = 12. Figure 1 represents the overall distribution of rotational levels in the OH ground state. We found 16% of the available energy to be converted into rotational energy leaving a maximum of 83% for translation. The measurements of the translational energy with pulsed high-resolution probe laser are still in progress.

The OH rotational distribution found here is similar to the CO distribution after photolysis of H_2CO /5/ and seems to be explainable by a semiclassical theory /6/.

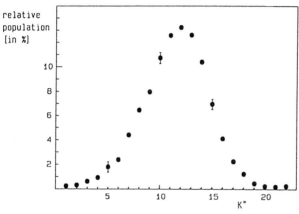

Fig. 1   Rotational distribution of OH($X\ ^2\pi$, v" = 0) normalized to unity

1. G. Ondrey, N. van Veen, R. Bersohn: J. Chem. Phys. 78, 3732-3737 (1983)
2. A. Jacobs, M. Wahl, R. Weller, J, Wolfrum: Appl. Phys. B 42, 173-179 (1987)
3. I.L. Chidsey, D.R. Crosley: J. Quant. Spectrosc. Radiat. Transfer 23, 187-199 (1980)
4. J. Stoer, R. Bulirsch: Introduction to Numerical Analysis (Springer, New York 1980) pp. 197-201
5. D.J. Bamford, S.V. Filseth, M.F. Foltz, J.W. Hepburn, C.B. Moore: J. Chem. Phys. 82, 3032-3041 (1985)
6. R. Schinke: J. Chem. Phys. 90, 1742-1751 (1986)

Laser Diagnostics of Radicals in Catalytic Reactions

A. Rosén, S. Ljungström, T. Wahnström, and B. Kasemo

Department of Physics, Chalmers University of Technology,
S-41296 Göteborg, Sweden

Oxidation of hydrogen to water in gas phase only takes place at rather high
temperatures due to the existence of a high activation barrier. However, in
the presence of a catalyst the hydrogen and oxygen molecules can dissociate
on the catalyst and recombine to form water. In both the gas and surface
reactions H, O and OH are important radicals participating in the reaction.
Identification and measurements of possible intermediate species and their
properties in chemical reactions (including catalytic reactions) is one of
the most direct ways to establish reaction routes and map out the reaction
mechanisms /1/. In the present work we use the laser induced fluorescence
LIF technique for detection of intermediate OH radicals in the oxidation of
hydrogen to water on a polycrystalline Pt foil. The production of water is
simultaneously measured by evaluation of the chemical energy released at the
catalyst.

The experimental setup, which has been described in detail in an earlier
work /2/, consists of a vacuum chamber. The partial pressure of hydrogen and
oxygen can be varied individually. The catalyst is a high purity Pt-foil
heated resistively in the reactant gases in the pressure range 0.01-0.20 torr.
The laser beam is parallel to the foil surface and its center axis lies about
4 mm from the foil. In the detection of thermally desorbed OH radicals from
the surface the laser is scanned through the absorption band of OH with
simultaneous detection of the emitted fluorescence light.

The most commonly adopted reaction scheme is dissociative H_2 and O_2 adsorption followed by recombination of O and H to a shortlived OH radical,
which in turn can combine with a second H atom to form a water molecule.
Alternatively two OH radicals may react to form water and chemisorbed oxygen.

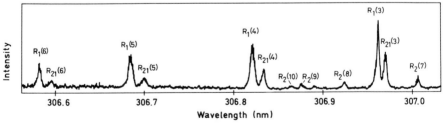

Fig. 1 A scan of the fluorescence intensity vs laser wavelength from a Pt
catalyst at 1100 K in 0.05 torr. The peaks in the spectrum reveal, by
comparison with known spectroscopic constants for OH, the presence of OH
molecules /3/ . Since the peaks disappear when the foil temperature is
decreased below 900 K it is ascertained that the detected OH originates
from the surface catalyzed reaction.

The H_2O molecule is thermally desorbed from the surface at temperatures well above 140 K. The reaction sequence can be summarized as follows:

$$H_2 + 2* \rightarrow 2H*$$ (dissociative hydrogen adsorption) (1)

$$O_2 + 2* \rightarrow 2O*$$ (dissociative oxygen adsorption) (2)

$$H* + O* \rightarrow OH*$$ (OH formation) (3)

$$OH* + H* \rightarrow H_2O + *$$ (water formation step) (4)

$$OH* \rightarrow OH + *$$ (desorption process) (5)

$$2OH* \rightarrow O* + H_2O* \rightarrow O* + H_2O$$ (water formation step) (6)

where * denotes a bonding site on the catalyst, i.e. one or a few Pt-atoms. Reaction step (5) which is utilized in this work is a minority route but sufficiently efficient to allow detection of OH by laser spectroscopic methods.

Figure 2 shows the emission intensity of OH radicals from the Pt-foil catalyst and the simultaneous production rate of H_2O, both as functions of the relative H_2 concentration in a $H_2 + O_2$ mixture. A maximum in the OH intensity is obtained as a function of H_2/O_2 ratio, as expected since the production of OH should vanish when the H_2/O_2 ratio goes to infinity or zero. The water production is strongly dependent on gas composition and its maximum does not coincide with the OH intensity maximum. This indicates that the production of H_2O proceeds via the $OH + H \rightarrow H_2O$ on the surface, i.e. reaction step (4) and not by reaction step (6).
Summarizing, LIF has shown to be a very helpful tool to establish reaction kinetic schemes by making it possible to study intermediate radical production.

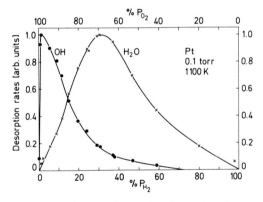

Fig. 2 OH desorption and H_2O production on Pt as functions of % H_2 in H_2+O_2 at 1100 K and 0.1 torr total pressure. The intensities of the curves are, respectively, normalized to unity

1. B. Kasemo and B.I. Lundqvist, Comments At. Mol. Phys. 14, 229 (1984)
2. A. Rosén, S. Ljungström, T. Wahnström and B. Kasemo, J. Electr. Spec. Rel. Phenomena 39, 15 (1985)
3. G.H. Dieke and Crosswhite, J. Quant. Spectrosc. Transfer. 2, 97 (1962)

Development of Element Determination Methods by Resonance Ionization Spectrometry

E.-L. Lakomaa[1], *I. Auterinen*[1], *J. Likonen*[1], *R. Zilliacus*[1], *and R. Salomaa*[2]

[1]Technical Research Centre of Finland, Reactor Laboratory,
  Otakaari 3 A, SF-02150 Espoo, Finland
[2]Helsinki University of Technology, Department of Technical Physics,
  Rakentajanaukio 2 C, SF-02150 Espoo, Finland

1. Introduction

Increasing needs in the semiconductor industry as well as research on environmental pollution and clinical chemistry call for more sensitive element analysis methods than have been available earlier. The realization of the laser resonance ionization spectroscopy (RIS) for analytical work /1,2/ has enabled stepping into the ultralow concentration region in element determinations. Working with ultralow concentrations, however, requires more attention to be paid towards the development of proper sample handling and quality control regimes than when working at higher concentration levels.

The demand for any analysis method is that accurate results would be obtained in different types of matrices in a reasonable time and at a reasonable cost. In order to reach these goals in RIS determinations the following principles, which we have found important when working with our RIS-device /3/, should be established into the determination procedures. The following is based on experiments with aluminium and gallium determinations.

2. Sample Handling

The avoidance of contamination is essential, for even one dust particle mixed with the sample can jeopardize the whole determination. In our work a dust-poor room of class 100 (< 4 particles/l) has been constructed. A dissolved or solid sample (1-40 mg) is introduced into a graphite crucible in this room. The crucible is transported in an air-tight box to the RIS-instrument and further into the analytical chamber. All devices coming into contact with the sample have been selected so that they do not contain measurable amounts of the element to be determined. In addition all devices are precleaned before use by acids or other cleaning agents. Aluminium, which is a common contaminant, could not be determined in concentrations below ng/g without the dust-poor environment.

3. Selection of the Parameters for the Determination

3.1. Excitation Schemes

When determining an element in a complicated matrice the excitation scheme has to be selected so that side reactions due to the elements or molecules in the matrice do not cause a considerable yield of ions or of excited atoms. Time of flight mass separation of the ions formed from the excited atoms will considerably decrease the effects of the side reactions.

3.2. Field Ionization

We use a pulsed electric field to ionize the excited atoms from a Rydberg-state. The strength of the electric field has to be optimized so that all Rydberg atoms get ionized, but not excited atoms on lower energy levels. For gallium the field was 7.5 kV/cm.

3.3. Atom Beam Formation

In our method the thermal atomization of the sample takes place from a pyrolytically coated graphite crucible in a vacuum better than 10^{-4} Pa .The crucible can be heated up to 3000 ^{0}C by an AC current by means of a selected program. The heating procedure needs optimization, which depends on the element and the matrice involved. The final temperature has to be high enough so that all atoms of the element to be determined come in the atomic form out of the sample. The heating rate must not be so fast that the ions formed will saturate the ion detector. The elements may be in the sample in the molecular form or at low temperatures volatile compounds may be formed during the heating procedure. These phenomena cause results which are too low and which can be avoided by using reagents which modify the matrix or change the atomization conditions by other means.

4. Standards

The analytical signal of a sample is compared with signals of known standards. The element atomized from the standard should behave similarly to that atomized from the sample for quantitative determination. Addition of reagents into the sample may change the atomization behaviour. Thus the standards should contain the same concentrations of the interfering or matrix modifying compounds as the samples.

5. Evaluation of the Precision and the Accuracy

For achievement of accurate results the precision has to be good and, in addition, systematic errors must be excluded from the determination. The precision is studied by determining several parallel samples at different concentration levels. In RIS the standardization has to be performed daily, for several instrument parameters have effect on the size of the signal. Our experience is that the fluctuation of the lasers and the electric noise due to the numerous instruments are the most prominent causes for poor precision.

Accuracy can be studied by using standard reference materials which have a similar matrix as the samples to be analyzed. Also determinations by other analytical methods should be used when available. The problem when working at concentrations lower than μg/g and especially below ng/g is that standard reference materials are scarce. Also the detection limits of RIS seem to be at such a level that only few alternative methods can detect so low amounts of elements. The future challenge of RIS will be the achievement of accuracy.

6. Quality Control of the Determinations

A quality control program has to be in use continuously. This includes the use of "null"-samples giving information about the contamination and electric noise in the runs and the use of reference standards resembling the real samples under determination. The reference standards are treated like other samples, and thus the accuracy of the determinations is checked. Intercalibrations with other laboratories are needed, especially when reference standards are not available.

7. Literature

1. G.S. Hurst, M.G. Payne, S.D. Kramer & J.P. Young: Rev Mod Phys <u>51</u> (1979) 767.
2. G.I. Bekov and V.S. Letokhov: Appl Phys B30,(1983) 161.
3. G.I. Bekov, Yu.A. Kudryavtsev, I. Auterinen & J. Likonen: In <u>Proc</u>. 3rd <u>Int. Symp. Res. Ion. Spectr. and Its Appl</u>.,ed. by G.S. <u>Hurst</u> & C.G. Morgan Inst. of Phys. Conf. Ser. <u>84</u>,97 (Inst. of Physics, Bristol 1987).

Laser Monitoring in the Atmosphere

G. Megie

Service d'Aéronomie du CNRS, BP 3,
F-91371 Verrières-le-Buisson Cedex, France

Remote measurements of trace constituents and physical parameters in the Earth's upper atmosphere using an active technique like the lidar have been made possible by the rapid development of powerful tunable laser sources which have opened a new field in atmospheric spectroscopy by providing sources which can be tuned to characteristic spectral features of atmospheric constituents. The interaction of a laser beam with the atmosphere is in principle dominated at all wavelengths by elastic processes i.e. Rayleigh scattering from atmospheric gas molecules and Mie scattering due to particles of different natures and shapes. By considering the simplified lidar equation written as

$$n_r = n_o \cdot \frac{A_o}{R^2} \, (\beta_R + \beta_M + \beta_i) \cdot \Delta R \cdot \exp \, (-2 \, (\tau_R + \tau_M + \tau_i))$$

(where n_o, n_r are the number of photons emitted and received per pulse from range R ; A_o is the receiver area ; β_R, β_M and β_i are the volume backscattering coefficients by Rayleigh scattering, Mie scattering and resonant scattering by a given species i ; τ_R, τ_M and τ_i being the corresponding optical thickness for the path length R ; ΔR is the range resolution), the determination of a given constituent number density N_i is then possible by two means :

- either β_i is much larger than $\beta_R + \beta_M$ allowing the measurement of N_i based on a scattering process ;
- or τ_i is much larger or of the same order of magnitude than $\tau_R + \tau_M$ allowing the measurement of N_i based on an absorption process.

These two methods have been used for the measurement of stratospheric and mesospheric trace constituents from the ground :

Resonance scattering has allowed the measurement of trace metals between 80 and 100 km with a very large sensitivity down to the 10^{-12} mixing ratio range (less than 1 atom.cm^{-3}). Historically these developments went back to 1967 and the first apparition of flashlamp pumped dye lasers followed by the measurement of the sodium atoms vertical distribution in 1969. They have been extended until recently by both large increases in spatial and temporal resolution and detection of other metallic species including potassium, lithium, neutral and ionized calcium, iron. This extended set of data has led to the solution of the questions raised for 50 years on the origin and behaviour of these constituents [1]. Furthermore and by using a narrowband (200 MHz) dye laser tunable through the Doppler broadened absorption profile of the sodium atoms, measurements of the atmospheric temperature between 80 and 100 km have been achieved recently [2].

The Differential Absorption Laser (DIAL) technique is the most widely used for the measurements of minor constituents in the troposphere and stratosphere. Its general principle as proposed by SCHOTLAND [3] is based on the emission of two laser lines at two different wavelengths which correspond to a difference in the absorption properties of the constituent under study. In the case of a cw laser source, such measurements will give access to the total content of a constituent between the emission point and a remote target, whereas the use of a pulsed laser source as the transmitter will lead, as in most lidar systems, to the direct retrieval of the spatial distribution of the measured parameters by a temporal analysis of the backscattered signals.

To select the most appropriate spectral region to be used for a specific measurement, reference has to be made to the constituent number density, to its altitude profile and also to potential interferences due to other absorbers in the same spectral range. In the UV and visible spectral range, the DIAL technique has been used for ozone, sulfur dioxide and nitrogen dioxide measurements. A large number of operational systems have been developed in many countries for both pollution type studies and free tropospheric and stratospheric measurements. These include ground based systems for ozone monitoring up to 50 km altitude [4/5], mobile systems for SO_2 and NO_2 measurements in polluted areas [6/7/8] so as airborne systems for the study of pollution episodes and long range transport [9]. In the near infrared (0.7 μm - 0.9 μm), the DIAL technique can be used to measure three very important meteorological parameters, i.e. water vapor concentration, atmospheric pressure and temperature. Taking into account the present day requirements for high resolution profiling of such parameters in the depth of the atmosphere, such potentialities have led to a very rapid development of both ground based [10/11/12] and airborne systems devoted to these applications [13]. In the infrared (9-11 μm), several systems using either cw or pulsed TEA CO_2 laser sources have been developed to monitor atmospheric pollutants, ozone and water vapor in the troposphere [14/15/16]. The use of an heterodyne detection technique further increases the sensitivity of these measurements [17].

The reliability and operational characteristics of a lidar system are greatly dependent on the laser source itself. In most of the measurements, tunable, high power sources are required and over the years, various laser systems have thus been used : dye lasers, solid state lasers (ruby, alexandrite), gas lasers (CO_2, exciplexe...). The large progresses in laser technology during the last decade and the advent of new types of lasers have led to a rapid evolution from quite complex systems involving complicated wavelength generation schemes towards much simpler and more reliable sources. As an example, the first ozone measurement in the UV wavelength range were made using a frequency doubled dye laser excited by the second harmonic of a solid state Nd:Yag laser [4]. Exciplexe laser are now used which directly, or using simple Raman shifter techniques, generate the appropriate UV wavelengths. Similarily, the Nd:Yag pumped dye lasers used in the near infrared for water vapor, pressure and temperature measurements are progressively replaced by solid state Alexandrite lasers which provide again directly the appropriate laser lines [18].

Following twenty years of rapid methodological and technological developments, the laser techniques for atmospheric sounding have now reached a very high degree of maturity. Ground based and airborne systems have already proven the ability of lidar to provide unique informations and to complement passive sensors informations in various fields of

350

atmospheric physics : weather prediction, climate and environmental studies. Progresses are yet tightly connected to high resolution spectroscopy measurements and to the development of reliable narrow line tunable laser sources in the UV, visible and infrared laser ranges.

References

1. C. Granier, J.P. Jégou, G. Mégie : Geophys. Res. Letters, 12, 10 (1985)
2. K.H. Fricke, U. Von Zahn : J. Atm. Terr. Phys., 47, 499 (1985)
3. R.M. Schotland : Proc. Third Symposium on Remote Sensing of Environment, Ann Arbor, U.S.A., (1964)
4. J. Pelon, G. Mégie : J. Geophys. Res., 87, C7, 4947 (1982)
5. J. Pelon, S. Godin, G. Mégie : J. Geophys. Res., 91,D8,8667-8671 (1986)
6. K. Fredriksson, B. Galle, K. Nystrom and S. Svanberg : Appl. Optics, 20, 4181-4189 (1981)
7. J.G. Hawley : Laser Focus, Mar, 60-62 (1981)
8. W. Staehr, W. Lahmann, C. Weitkamp : Appl. Opt., 24, 1950-1956 (1985)
9. E.V. Browell. : Optical Eng., 21, 128-132 (1982)
10. C. Cahen, P. Flamant and G. Mégie : J. Appl. Meteor., 21, 1506 (1982)
11. E.V. Browell, T.D. Wilkerson and T.J. McIlrath : Appl. Opt., 18, 3774-3483 (1978)
12. J.E. Kalshoven, C.L. Korb, G.K. Schwemmer and M. Dombroski : Appl. Opt., 20, 1967 (1981)
13. E.V. Browell, S. Ismail and S.T. Shipley : Appl. Opt.,24,2827-2836 (1985)
14. R.T. Menzies : Laser heterodyne detection techniques, Laser Monitoring of the Atmosphere (Ed. E.O. Hinkley), Springer Verlag (1976)
15. D.K. Killinger and N. Menyuk : IEEE J. Quant. Electr., QE-17, 1917-1929 (1981)
16. R.M. Hardesty : Appl. Opt., 23, 2545-2553 (1984)
17. R.T. Menzies, M.S. Schumate : J. Geophys. Res., 83, 4039 (1978)
18. J. Pelon, G. Mégie, C. Loth, P. Flamant : Opt. Comm., 59, 3, 213-218 (1986)

Applications of Laser and Lidar Spectroscopy to Meteorological Remote Sensing *

T.D. Wilkerson[1], G.K. Schwemmer[2], K.J. Ritter[3], U.N. Singh[1], and R. Mahon[1]

[1]Institute for Physical Science and Technology, University of Maryland, College Park, MD 20742, USA
[2]Laboratory for Atmospheres, NASA Goddard Space Flight Center, Greenbelt, MD 20771, USA
[3]Martin Marietta Laboratories, 1450 South Rolling Road, Baltimore, MD 21227, USA

The development of energetic, tunable, narrowline pulsed lasers has made it possible to utilize numerous optical radar techniques for remotely probing the atmosphere; i.e., to measure constituents (H_2O, O_3, NO_2, etc) and the thermal state. We describe progress on using the O_2 A-band (760-770 nm) for lidar profiling of atmospheric temperature, pressure, and density [1]. First we illustrate the impressive capabilities expected for lidar systems, given equipment performance within reach of current technology. Second we outline a new body of high resolution spectroscopic data on the O_2 A-band, data vitally needed for accurate interpetation of lidar observations that make use of this band. Third we indicate an ongoing problem of spectral purity for the narrowband lasers that are required to be tuned onto molecular absorption lines, for certain lidar methods to be successful.

With the differential absorption lidar (DIAL) technique [2], absolute humidity is measured using H_2O absorption lines in the near IR, and ozone concentration using continuum absorption in the UV [3]. By taking advantage of the uniform mixing ratio of oxygen, we may measure temperature by measuring the variation of absorption on thermally populated transitions of the A-band near 770 nm. Alternatively, the laser may be tuned to absorption troughs in the wings of two strongly absorbing A-band transitions having low temperature dependence to measure atmospheric pressure or density [1]. The great potential of DIAL for measuring atmospheric density vs altitude is illustrated in Fig. 1 where we have plotted the expected error as a function of altitude given realistic system parameters and measurement noise sources.

To make accurate measurements of the oxygen A-band absorption line parameters, we used both absorption and photoacoustic cells. For each set of measurements, a cw narrow linewidth tunable dye laser scanned over the absorption lines. Measurements included line strengths, widths and frequency shifts; widths and shifts were also measured as a function of temperature. To analyze these measurements, we fit standard line profiles to the observed profiles using a least-squares fitting routine. Our linewidth results, along with measurements by other investigators, are shown in Fig. 2. The details of these measurements and references to prior work are given elsewhere [4,5].

*Research supported by NASA,NOAA,ONR, and the University of Maryland

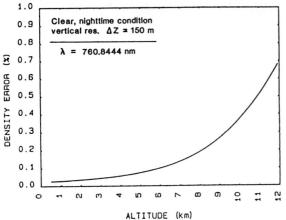

Fig. 1 Error analysis for atmospheric density measurements

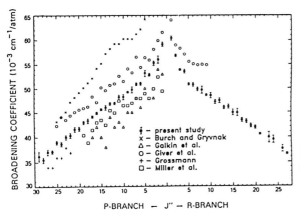

Fig. 2 Linewidth measurements in the oxygen A-band

A frequency-doubled Nd:YAG-pumped Quanta-Ray PDL-2 dye laser (modified) in combination with a 1 meter Raman cell was used to generate pulsed, tunable (760-770 nm), narrowband (0.02-0.03 cm^{-1}) laser radiation for oxygen absorption measurements [6]. With the experimental arrangement shown in Fig. 3, we continually monitored all the performance parameters of the laser system and measured the differential transmission at the center of all 29 lines (FWHM 0.06-0.1 cm^{-1} at 1 atm) in the P-branch of the O_2 A-band in air at NTP contained in a White cell of 60 m path length. Two sets of experimental data are shown in Fig. 4, corresponding in one case to the radiation being directly generated in a dye laser, and, in the second case, to the use of Raman shifted dye laser radiation. Also shown for comparison are predictions [4] based on a Voigt profile. Since the latter were obtained using a cw dye laser (10^{-4} cm^{-1} bandwidth), they have a higher inherent accuracy than the pulsed measurements. The discrepancies can possibly be attributed to the finite linewidth of the pulsed laser or to the presence of Amplified Spontaneous Emission (ASE). Efforts are underway to resolve the problems due to unabsorbed light, so that the cw and pulsed measurements can be reconciled. To realize the great potential of lidar for atmospheric monitoring, very accurate spectroscopic data are needed, and quality laser transmitters must be developed and utilized.

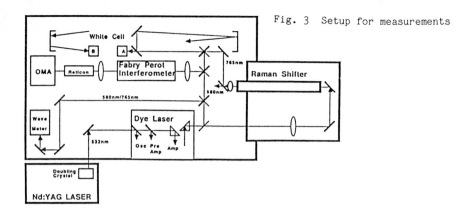

Fig. 3 Setup for measurements

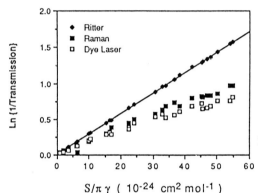

Fig. 4 O_2 absorption results

1. C.L. Korb and C.Y. Weng: J. Appl. Meteor. 21, 1346 (1982); Appl. Opt. 22, 3759 (1983).
2. R.M. Measures: Laser Remote Sensing: Fundamentals and Applications, John Wiley, New York (1984); D.K. Killinger and N. Menyuk: Science 235, 37 (1987).
3. E.V. Browell,A.F. Carter,S.T. Shipley,R.J. Allen,C.F. Butler, M.N. Mayo,J.H. Siviter, Jr., and W.M. Hall: Appl. Opt. 22, 552 (1983).
4. K.J. Ritter and T.D. Wilkerson: J. Molec. Spectrosc. 121, 1 (1987).
5. K.J. Ritter: Ph.D. Thesis, University of Maryland (1986).
6. U.N. Singh,R. Mahon and T.D. Wilkerson: Proc. Laser and Optical Remote Sensing, OSA Topical Meeting, Cape Cod, Massachusetts, Sept. 28-Oct. 1 (1987).

Laser Remote Sensing of the Atmosphere*

T.D. Wilkerson

Institute for Physical Science and Technology, University of Maryland, College Park, MD 20742, USA

The field of laser applications to atmospheric remote sensing (lidar, or optical radar) has seen major advances in recent years, owing to improvements in laser technology and the advent of inexpensive computer equipment for high speed data acquisition. Further technical advances and the realization of new lidar methods will make it possible to routinely use aircraft and satellite platforms for wide area surveys of:

- pollutants and natural constituents (gases and particles)
- thermal conditions (temperature, density, and stability)
- states of motion (three dimensional winds and turbulence)

This paper summarizes recent developments that depend on the atomic and molecular spectra of atmospheric species. An earlier review [1] treated the potential of lidar observations for studies of optical and microwave propagation. More general treatments [2-6] cover all aspects of lidar research through 1985. Here we will omit discussions of coherent and incoherent laser scattering by atmospheric particulates; i.e., interactions that are used to map aerosol distributions and to measure winds by means of the Doppler effect. Also the lidar applications of Rayleigh scattering are not discussed in this paper.

The optical interactions of interest here are:

- Raman scattering
- Absorption
- Fluorescence
- Nonlinear processes

A broad set of distinctions between these categories is the following:
(1) Raman scattering is particularly advantageous for "major-minor" species such as H_2O and CH_4 whose abundances are great enough to offset the small size of the scattering cross section; (2) fluorescence is the most useful for high altitudes (e.g., Na and OH) where the laser-induced excitation is not quenched by collisions; (3) differential absorption, using two lidar transmissions at wavelengths "on" and "off" a spectral absorption feature, applies to many species and often requires great care for the spectral quality of the lidar transmitter; (4) while nonlinear optical processes are widely used in generating lidar transmissions, their exploitation in the atmosphere itself for environmental measurements [7] awaits the development of more energetic lasers.

Raman scattering applications, particularly for H_2O, have been carried out for twenty years [8-18]. Typically the mixing ratio $n(H_2O)/n(N_2)$ is

*Research supported by ONR, NASA, NOAA and the University of Maryland

measured by means of the relative intensities of the first Stokes (vibrational) Raman lidar returns. Rotational Raman bands have been used for close-in temperature profiling [15,18]. Recent work has concentrated on high spatial resolution [16] and long-term, high accuracy profiling of H_2O [17].

Fluorescence measurements of stratospheric OH have been made with balloon-borne lidar instruments [19,20], and methods have been proposed [21] to use OH fluorescence lidar to measure mesospheric temperature and pressure. The possibility of lidar measurements of NO_2 in the upper atmosphere has been supported by detailed laboratory studies of laser-induced fluorescence of NO_2 [22]. Ground based lidars have probed the mesospheric sodium layer for many years [23,24] principally by means of resonance fluorescence on the Na D_2-line. This work has been extended to other elements (K, Li, Ca, Ca^+), and to monitoring the sodium layer as an indicator of mesospheric temperature, atmospheric waves, and sudden changes in the ionosphere [23-39]. Experiments have begun on using the lidar-induced sodium fluorescence beam spot in the upper atmosphere as a light source for adaptive optics corrections for astronomical imaging [40,41].

Differential absorption lidar ("DIAL") is closely analogous to dual beam absorption spectroscopy, in that the "off line" optical channel is affected by the selfsame processes (e.g. particulate scattering and absorption) that occur in the nearby "on line" channel, except for the molecular absorption which one uses quantitatively to derive the concentration of the absorbing species. Such concentration measurements can also be interpreted in terms of temperature or pressure, if the absorption line strength or width is strongly temperature or pressure dependent. Detailed accounts of DIAL techniques, results, and problems have been published [2,3,5,6,42-46]. Pollutant gases have been an important target for tropospheric DIAL measurements [5,42,45,50-57], principally SO_2, NO_2, O_3 and C_2H_4. The folding of naturally occurring stratospheric ozone into the troposphere [58,59] has been observed by means of an airborne ultraviolet lidar system [60]. Groundbased lidar monitoring of ozone from the troposphere through the lower stratosphere and up to altitudes as high as 50 km is becoming an important environmental application of the DIAL technique [61-70].

The other major application of differential absorption lidar is meteorology, or more precisely the remote measurement of atmospheric profiles of humidity, temperature, pressure and density. Tropospheric H_2O has been measured with a variety of DIAL laser sources spanning the wavelength range 0.7-10 μm, making use of the rich vibration-rotation band spectrum of the water molecule [71-81]. The so-called A- and B-bands of O_2 (~ 760 nm and 690 nm, respectively) have been proposed for lidar measurements of temperature [82-86], pressure [87] and density [88,89]. Confirmation of these techniques has been obtained in a few observations [90,91]. It is generally recognized [92] that these methods put great demands on the spectral stability and purity of the tunable laser transmitter, because the absorption lines are quite narrow (FWHM ~ 0.2 and 0.08 cm^{-1}, respectively, for H_2O and O_2 at sea level). Attention to this problem is reflected in two papers to be presented later at this meeting [93,94], which are also concerned with ongoing high resolution studies of the H_2O and O_2 absorption spectra in their own right. Compared to the measurements with other species whose absorption features are relatively broad, the meteorological DIAL work has proved to be more difficult and needs further improvement, at least as regards the laser sources employed. The quantitative spectroscopic data required for these near infrared transitions of H_2O and O_2 are being checked and updated in various experiments [93-100]. Improvements in

356

the requisite laser technology are expected to result from work on stimulated Raman scattering [101], the alexandrite laser [102], and the $Ti:Al_2O_3$ laser [103]. When the meteorological lidar capability is fully developed, it will be possible to obtain detailed weather data over large portions of the atmosphere, and thereby provide better initialization and verification of numerical weather prediction codes than at present.

The development of lidar concepts and measurement techniques over the past twenty years makes it possible to carry out unique atmospheric observations over the entire altitude range from the ground to the ionosphere. This capability, given improvements in laser and space technology, is likely to expand to global coverage and to a better understanding of our atmospheric environment. Lidar methods are already available for protecting populations from a variety of atmospheric pollutants, and will be used more widely with improvements in equipment cost and reliability.

References

1. T. D. Wilkerson: "Lidar Profiling of Atmospheric Properties that Influence Propagation", Proc. Internat. Conf. on Optical and Millimeter Wave Propagation and Scattering in the Atmosphere, Florence, Italy (May, 1986).
2. D. K. Killinger and N. Menyuk: Science 235, 37 (1987).
3. E. D. Hinkley (editor): Laser Monitoring of the Atmosphere, Springer-Verlag (1976).
4. A. I. Carswell: Canad. J. Phys. 61, 378 (1983).
5. D. K. Killinger and A. Mooradian (editors): Optical and Laser Remote Sensing, Springer-Verlag (1983).
6. R. M. Measures: Laser Remote Sensing: Fundamentals and Applications, John Wiley (1984).
7. T. J. McIlrath, R. Hudson, A. Aikin and T. D. Wilkerson: Appl. Optics 18, 316 (1979).
8. D. A. Leonard: Nature 216, 142 (1967).
9. J. A. Cooney: Appl. Phys. Lett. 12, 40 (1968).
10. S. H. Melfi, J. D. Lawrence, and M. P. McCormick: Appl. Phys. Lett. 15, 295 (1969).
11. R.G. Strauch, V.E.Derr, and R.E. Cupp: Appl. Optics 10, 2665 (1971).
12. S. H. Melfi: Appl.'Optics 11, 1605 (1972).
13. J. C. Pourny, D. Renaut, and A. Orszag: Appl. Optics 18, 1141 (1979).
14. D. Renaut, J. C. Pourny, and R. Capitini: Optics Lett. 5, 233 (1980).
15. J. A. Cooney: Opt. Eng. 22, 292 (1983).
16. J. Cooney, K. Petri, and A. Salik: Appl. Optics 24, 104 (1985).
17. S. H. Melfi and D. Whiteman: Bull. Am. Meteor. Soc. 66, 1288 (1985).
18. Y. F. Arshinov, S. M. Bobrovnikov, V. E. Zuev, and V. M. Mitev: Appl. Optics 22, 2984 (1983).
19. W. S. Heaps: Appl. Optics 19, 243 (1980).
20. W. S. Heaps and T. J. McGee: J. Geophys. Res. 90, 7913 (1985).
21. T. J. McGee and T. J. McIlrath: Appl. Optics 18, 1710 (1969).
22. C. S. Dulcey: "A Study of Laser-Induced Fluorescence Cross Sections in Nitrogen Dioxide", Ph.D. thesis, University of Maryland (1982).
23. A.J. Gibson and M.C.W. Sandford: J. Atmos. Terr. Phys. 33,1675 (1971).
24. J.E. Blamont, M.-L. Chanin, and G. Megie: Ann. Geophys. 28,833 (1972).
25. J. R. Rowlett, C. S. Gardner, E. S. Richter, and C. F. Sechrist: Geophys. Res. Lett. 5, 603 (1978).
26. G. Megie, F. Bos, J. E. Blamont, and M.-L. Chanin: Planet. Space Sci. 26, 27 (1978).
27. A.J. Gibson, L. Thomas, and S.K. Bhattacharyya: Nature 281,131 (1979).

28. L. Thomas and S. K. Bhattacharyya: Proc. 5th ESA-PAC Symp. on European Rocket and Balloon Programmes and Related Research, ESA-SP-152 (1980).

29. G. Megie, M.-L. Chanin, G. Tulinov, and Y. P. Doudoladov: Planet. Space Sci. 26, 509 (1978).

30. P. Juramy, M.-L. Chanin, G. Megie, G. Tulinov, and Y. P. Doudoladov: J. Atmos. Terr. Phys. 43, 209 (1981).

31. C. Granier, J. P. Jegou, and G. Megie: Geophys. Res. Lett. 12, 655 (1985).

32. C. S. Gardner and D. G. Voelz: Geophys. Res. Lett. 12, 765 (1985).

33. C. S. Gardner, D. G. Voelz, C. R. Philbrick, and D. P. Sipler: J. Geophys. Res. 91, 12131 (1986).

34. K. H. Fricke and U. von Zahn: J. Atmos. Terr. Phys. 47, 499 (1985).

35. U. von Zahn, K. M. Fricke, R. Gerndt, and T. Blix: J. Atmos. Terr. Phys. 48, (1986).

36. U. von Zahn, P. von der Gathen, and G. Hansen: Geophys. Res. Lett. 14, 76 (1987).

37. U. von Zahn and R. Neuber: "Thermal Structure of the High Latitude Mesopause Region in Winter", in Beitrage Phys. Atmos., in print (1987).

38. U. von Zahn and C. Tilgner: "The Sodium Layer at 69° N Latitude in Wintertime", Proc. 8th ESA Symposium on European Rocket and Balloon Programmes and Related Research, Sunne, Sweden (May 1987); ESA Report No. SP-270 (July, 1987).

39. U. von Zahn, K. H. Fricke, R. Gerndt, and T. Blix: "Mesospheric Temperatures and the OH Layer Height as derived from Ground-Based Lidar and OH* Spectrometry", J. Atmos. Terr. Phys., in print (1987).

40. R. Foy and A. Labeyrie: Astron. and Astrophys. 152, L29 (1985).

41. L. A. Thompson and C. S. Gardner: "Laser Guide Star Experiments at Mauna Kea Observatory for Adaptive Imaging in Astronomy", Nature, in print (1987).

42. E. V. Browell: Opt. Eng. 21, 128 (1982).

43. E.V. Browell, S. Ismail and S.T. Shipley: Appl. Optics 24, 2827 (1985).

44. W. C. Braun: Appl. Optics 26, 2123 (1987).

45. J. Altmann, W. Lahmann and C. Weitkamp: Appl. Optics 19, 3453 (1980).

46. W. Staehr, W. Lahmann and C. Weitkamp: Appl. Optics 24, 1950 (1985).

47. N. Menyuk and D.K. Killinger: Appl. Optics 22, 2690 (1983).

48. W. B. Grant and R. T. Menzies: J. Air Pollut. Control. Assoc. 33, 187 (1983).

49. D. K. Killinger, N. Menyuk, and W. E. DeFeo: Appl. Phys. Lett. 36, 402 (1980).

50. K. W. Rothe, U. Brinkman, and H. Walther: Appl. Phys. Lett. 4, 181 (1974).

51. K. W. Rothe: Radio Electron. Eng. 50, 567 (1980).

52. K. Asai, T. Itabe, and T. Igarishi: Appl. Phys. Lett. 35, 60 (1979).

53. G. Ancellet, G. Megie, J. Pelon, D. Renaut, and R. Capitini: "Lidar Measurements of Sulfur Dioxide and Ozone in the Boundary Layer during the 1983 Fos/Berre Campaign", to be published in Atmos. Environ.

54. A.-L. Egeback, K. A. Fredriksson, and H. M. Hertz: Appl. Optics 23, 722 (1984).

55. K. A. Frederiksson and H. M. Hertz: Appl. Optics 23, 1403 (1984).

56. H. Edner, S. Svanberg, L. Uneus, and W. Wendt: Optics Lett. 9, 493 (1984).

57. E. V. Browell, G. L. Gregory, R. C. Harriss, and V. W. J. H. Kirchhoff: "Tropospheric Ozone and Aerosol Distributions across the Amazon Basin", to be published in J. Geophys. Res. (1987).

58. M. A. Shapiro: J. Atmos. Sci. 12, 994 (1980).

59. E. V. Browell, E. F. Danielsen, S. Ismail, G. L. Gregory, and S. M. Beck: J. Geophys. Res. 92, 2112 (1987).

60. E. V. Browell, A. F. Carter, S. T. Shipley, R. J. Allen, C. J. Butler, M. N. Mayo, J. H. Siviter, and W. M. Hall: Appl. Optics 22, 522 (1983).

61. G. Megie, J. Y. Allain, M.-L. Chanin, and J. E. Blamont: Nature 270, 329 (1977).
62. O. Uchino, M. Maeda, T. Shibata, M. Hirono, and M. Fujiwara: Appl. Optics 19, 1475 (1980).
63. J. Pelon and G. Megie: J. Geophys. Res. 87, 4947 (1982).
64. J. Werner, K. W. Rothe, and H. Walther: Appl. Phys. B32, 113 (1983).
65. O. Uchino, M. Tokunaga, M. Maeda, and Y. Miyozoe: Opt. Lett. 8, 347 (1983).
66. O. Uchino, M. Maeda, H. Yamamura, and M. Hirono: J. Geophys. Res. 88, 5273 (1983).
67. G. Megie, G. Ancellet, and J. Pelon: Appl. Optics 24, 3454 (1985).
68. J. Pelon, S. Godin, and G. Megie: J. Geophys. Res. 91, 8667 (1986).
69. J. Werner, K. W. Rothe, and H. Walther: "Measurement of the Ozone Profile up to 50 km Altitude by DIAL", in Atmospheric Ozone: Proc. of the Quadrennial Ozone Symposium (editors Zerefors and Ghazi), Chalki-diki, Greece (September 1984); publ. by D. Reidel, Dordrecht (1985).
70. J. Werner, K. W. Rothe, and H. Walther: "Lidar Techniques for Long Term Measurements of Stratospheric Ozone Concentrations", Proc. NATO/CCMS Workshop on Application of Advanced Air Pollution Assessment Methods and Monitoring Techniques, Lindau, Germany (October 1985); publ. by NATO Committee on the Challenges of Modern Society, Report No. 153 (1985).
71. R. M. Schotland: "Some Observations of the Vertical Profile of Water Vapor by a Laser Optical Radar",. Proc. 4th Symp. on Remote Sensing of Environment, University of Michigan, Ann Arbor (1966).
72. R. M. Schotland: J. Appl. Meteor. 13, 71 (1974).
73. V. V. Zuev, V. E. Zuev, Y. S. Makushkin, V. N. Maricher, and A. A. Mitsel: Appl. Optics 22, 3742 (1983).
74. E. V. Browell, T. D. Wilkerson, and T. J. McIlrath: Appl. Optics 18, 3474 (1979).
75. E.V. Browell, A.F. Carter, and T.D. Wilkerson: Opt. Eng. 20, 84 (1981).
76. C. Cahen, G. Megie, and P. Flamant: J. Appl. Meteor. 21, 1506 (1982).
77. P. W. Baker: Appl. Optics 22, 2257 (1983).
78. R. M. Hardesty: Appl. Optics 23, 2545 (1984).
79. R. M. Hardesty: "Measurment of Range-Resolved Water Vapor Concentra-tion by Coherent CO_2 Differential Absorption Lidar", NOAA Tech. Memo. ERL/WPL-118 (March 1984).
80. T. D. Wilkerson and G. K. Schwemmer: Opt. Eng. 21, 1022 (1982).
81. E. V. Browell, A. K. Goroch, T. D. Wilkerson, S. Ismail, and R. Markson: "Airborne DIAL Water Vapor Measurements over the Gulf Stream", Proc. Twelfth International Laser Radar Conference, Aix-en-Provence, France (August 1984).
82. J. B. Mason: Appl. Optics 14, 76 (1975).
83. H. I. Heaton: J. Quant. Spectrosc. Radiat. Transfer 16, 801 (1976).
84. H. I. Heaton: Astrophys. J. 212, 936 (1977).
85. G. K. Schwemmer and T. D. Wilkerson: Appl. Optics 18, 3539 (1979).
86. C. L. Korb and C. Y. Weng: J. Appl. Meteor. 21, 1346 (1982).
87. C. L. Korb and C. Y. Weng: Appl. Optics 22, 3759 (1983).
88. C. L. Korb and C. Y. Weng: "A Two-Wavelength Lidar Technique for the Measurement of Atmospheric Density Profiles", Proc. CLEO, Phoenix, Arizona (April 1982).
89. T. D. Wilkerson, L. J. Cotnoir and G. K. Schwemmer: "Lidar Probing of Tropospheric Density, Temperature, Pressure, and Humidity for Ballis-tics Corrections", Report for Battle-Columbus Scientific Services Program to U.S. Army Research Office and Atmospheric Sciences Labora-tory (April 1986).
90. J. E. Kalshoven, C. L. Korb, G. K. Schwemmer, and M. Dombrowski: Appl. Optics 20, 1967 (1981).

91. C. L. Korb, G. K. Schwemmer, M. Dombrowski, and R. H. Kagann: "Remote Sensing with a Tunable Alexandrite Laser Transmitter", in Tunable Solid State Lasers for Remote Sensing (editors R. L. Byer, E. K. Gustafson and R. Trebino) Springer-Verlag (1985).

92. C. Cahen and G. Megie: J. Quant. Spectrosc. Radiat. Transfer 25, 151 (1981).

93. T. D. Wilkerson, G. K. Schwemmer, K. J. Ritter, U. N. Singh, and R. Mahon: "Applications of Laser and Lidar Spectroscopy to Meteorological Remote Sensing",. Proc. EICOLS '87, Are, Sweden (June 1987).

94. B. E. Grossmann and E. V. Browell: "High Resolution Water Vapor Spectroscopic Measurements in the 720-nm Region for Lidar Meteorological Applications", loc. cit.

95. K. J. Ritter and T. D. Wilkerson: J. Molec. Spectrosc. 121, 1 (1987).

96. J. Bosenberg: Appl. Optics 24, 3531 (1985).

97. B. E. Grossmann, C. Cahen, J. L. Lesne, J. Benard, and G. Leboudec: Appl. Optics 25, 4261 (1986).

98. C. Cahen, B. E. Grossmann, J. L. Lesne, J. Benard, and G. Leboudec: Appl. Optics 25, 4268 (1986).

99. T. D. Wilkerson, G. Schwemmer, B. Gentry and L. P. Giver: J. Quant. Spectrosc. Radiat. Transfer 22, 315 (1979).

100. L. P. Giver, B. Gentry, G. Schwemmer, and T. D. Wilkerson: J. Quant. Spectrosc. Radiat. Transfer 27, 423 (1982).

101. B. E. Grossmann, U. N. Singh, N. S. Higdon, L. J. Cotnoir, T. D. Wilkerson, and E. V. Browell: Appl. Optics 26, 1617 (1987).

102. J. Pelon, G. Megie, P. Flamant, and C. Loth: Optics Comm. 59, 213 (1986).

103. P. Brockman, C. H. Bair, J. C. Barnes, R. V. Hess and E. V. Browell: Optics Letters 11, 712 (1986).

High-Resolution Water Vapor Spectroscopic Measurements in the 720-nm Region for Lidar Meteorological Applications

B.E. Grossmann[1] *and E.V. Browell*[2]

[1]Old Dominion University, Research Foundation,
 Norfolk, VA 23508, USA
[2]Atmospheric Sciences Division, NASA Langley Research Center,
 Hampton, VA 23665, USA

The Differential Absorption Lidar (DIAL) technique allows one to remotely measure profiles of atmospheric gases such as water vapor. This method makes use of the difference observed in pulsed laser backscatter from the atmosphere when a pulsed laser transmitter is tuned on and off a water vapor absorption line. This technique requires accurate water vapor spectroscopic data on line strength, pressure broadening, and pressure shifts in the wavelength region of interest. As part of the atmospheric research program, NASA is currently involved in the Lidar Atmospheric Sensing Experiment (LASE)[1]. This DIAL system uses tunable Alexandrite lasers and operates from a high altitude ER-2 (extended range U-2) aircraft. To support this experiment, a high-resolution spectroscopy setup consisting of a CW ring dye laser and two White cells is being used to provide measurements of needed water vapor line parameters.

1. EXPERIMENTAL SETUP

The experimental setup consists of a CW ring dye laser (Spectra Physics, 380D) in conjunction with two long-path absorption cells. Each cell can be adjusted to have a path length of up to 120 meters. The dye laser linewidth is 500 kHz, and it can be scanned over a 30 GHz range. Details of the setup are given in Figure 1. For measurements of pressure shifts and pressure broadening, both cells are used in parallel to avoid any laser frequency drift during the measurement. A high-resolution Fabry-Perot with a free spectral range of 0.1 cm^{-1} is used in parallel with the absorption cells for relative frequency calibration. As many as four photodiodes are used simultaneously with one of them being used for the laser power reference. An A/D converter with a 12 bit resolution and 10 kHz sampling frequency is used for digitization. Data acquisition and the laser scan are simultaneously started by the computer. The water vapor density is measured by using a temperature controlled pressure transducer (MKS-220B) operating in the 0-100 torr range with an accuracy of 0.15%. A capacitive water vapor sensor (Vaisala, HMP 114Y) is also used to check that the humidity remains the same when buffer gases are introduced in the cells. This way, adsorption and desorption can be easily monitored during the course of the experiment. Also, to minimize adsorption, both cells are Teflon coated and temperature controlled within 0.3 K and can be heated up to 340 K. This allows us to work in high water vapor pressure conditions and, therefore, gives us the ability to investigate weak lines. Also, both cells can be used in serial to double the path length if necessary. When the buffer gas (nitrogen or oxygen) is introduced in the cell, the flow rate is adjusted as low as 20 torr/min to avoid any condensation and to minimize adsorption on the cell walls. Then the cells are allowed to reach equilibrium, that is when the capacitive sensor reading reaches the same value as it was before.

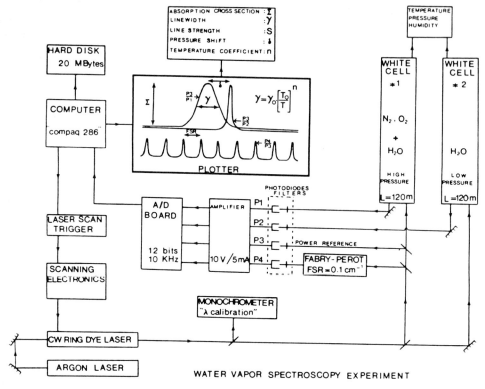

Fig. 1: Experimental setup

2. RESULTS

Measurements of water vapor line strengths, pressure shifts, and pressure broadening coefficients have been made for lines between 726.5 and 730 nm. Pressure shifts in air were measured by using both cells in parallel, one filled with pure water vapor and the other one with air and water vapor. Negative pressure shift coefficients averaging 0.015 cm^{-1}/atm have been observed for several lines. The average shift is almost as large as the Doppler linewidth (≈ 0.02 cm^{-1}). The same spectrum was also used to derive an air pressure broadening coefficient. For the low J lines in this region, the broadening coefficient average is about 0.1 cm^{-1}/atm. Self broadening was also measured. For this measurement, both cells were filled with two different water vapor pressures (10 and 20 torr). Figure 2 shows an example. On this spectrum, the self-broadening effect is easily observable by looking at the dip between these two H_2O lines. At low pressures, the dip is more pronounced. For this line, the self broadening coefficient is about 0.45 cm^{-1}/atm. We were also able to observe a negative pressure shift due to increasing the water vapor pressure; however, the shift was too small to make an accurate measurement. Use of higher water vapor pressures will enable us to conduct such measurements. Pressure shifts in nitrogen and oxygen, as well as the pressure broadening temperature dependence, will also be investigated.

362

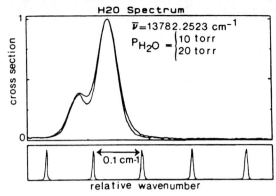

Fig. 2:
Recorded spectrum at two different water vapor pressures (10 and 20 torr). The lower trace represents the spectral throughput of the etalon.

Benoist Grossmann is working at the NASA Langley Research Center under NASA NCCI-32 while on leave from Electricité de France, Chatou, France.

[1] E. V. Browell et al.: In Proc. of the Thirteenth International Laser Radar Conference, Toronto, Canada, August 11-15, 1987, pp. 6-9.

Analysis of Surface Films on Liquids by Pulsed Laser Photoacoustic Spectroscopy

M.W. Sigrist, Z.H. Chen*, and D. Scherrer

Institute of Quantum Electronics, ETH, CH-8093 Zürich, Switzerland
*Permanent address: Institute of Physics, Chinese Academy of Sciences,
  Beijing, People's Republic of China

1. Introduction

The photoacoustic (PA) generation of plane acoustic waves in strongly absorbing or opaque liquids by pulsed laser radiation has been reviewed recently [1]. We have also applied the PA technique to spectroscopic studies of absorbing liquid surface films spread on liquids, i.e. of liquid/liquid interfaces, for the first time [2]. Here we only report on new results obtained for liquid and solidified or crystallized surface films of ≥ 0.5 μm thickness present on water (H_2O).

2. Experimental

Our experimental setup has been described in detail [1,2]. Briefly, the pulses from a tunable hybrid CO_2 laser system are directed unfocussed onto the H_2O/surface film compound where the acoustic transients are generated by the thermoelastic process. The acoustic signals are detected piezoelectrically in the bulk of H_2O. A new theoretical model which takes the altered boundary condition at the liquid/liquid interface into account has been developed on the basis of the previous model [1]. It yields qualitative agreement with experimental data on liquid surface films [3] yet does not cover our recent experiments on solidified surface films which are discussed in the following.

3. Results and Discussion

As shown in Figs. 1 and 2 for the case of an octadecyl methacrylate (ODMA, $C_{22}H_{42}O_2$) film on H_2O, the PA spectra differ drastically between a liquid and a solidified or crystallized surface layer of identical composition and thickness. The PA spectra have all been recorded within the 10R branch of the CO_2 laser at a temperature T = 23.5°C where the ODMA film is liquid or at T = 20°C where ODMA is crystallized. Figure 1a represents the case of H_2O with a free surface, Figs. 1b and 1c that of H_2O carrying a liquid or crystallized ODMA layer, respectively, with a thickness $d_\ell = 17.4$ μm $<$ $1/\alpha_\ell$, where α_ℓ represents the absorption coefficient of the layer medium at the laser wavelength. Figure 2a-c represents an identical measurement yet the thickness of the ODMA film is 29.0 μm, i.e. $d_\ell > 1/\alpha_\ell$. In all cases an enhancement of the PA amplitudes with respect to the case of H_2O with free surface is observed. The PA spectrum in Fig. 1b follows the IR spectrum of ODMA as expected for a liquid surface layer with $d_\ell < 1/\alpha_\ell$ [2,3]. The spectrum of Fig. 2b represents an intermediate case with $d_\ell \geq 1/\alpha_\ell$ and no dependence on α_ℓ is observed. Additional measurements for $d_\ell \gg 1/\alpha_\ell$ have shown that the typical $1/\alpha_\ell$ dependence observed in that case [2] also applies to this system.

The presence of a crystallized instead of a liquid surface film (Figs. 1c und 2c) gives rise to a flat spectrum, independent of d_ℓ. However, the signals are further enhanced with respect to the liquid films of identical d_ℓ. Also in this case the amplitudes increase with d_ℓ. It should be noted that the PA spectra of Figs. 1c and 2c are identical to those obtained for a rigid H_2O surface realized e.g. by placing a transparent AR/AR coated ZnSe plate in contact with H_2O. Yet the enhancement of the PA amplitude compared to H_2O with free surface is even larger in the latter case and reaches a factor of ≈ 8.8 [1] instead of ≈ 6.3 (Fig. 2).

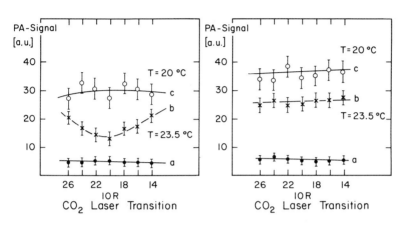

Fig. 1 PA spectra
a. H_2O with free surface
b. H_2O with liquid ODMA film
    (d_ℓ = 17.4 μm)
c. H_2O with crystallized ODMA
    film (d_ℓ = 17.4 μm)

Fig. 2 PA spectra
a. H_2O with free surface
b. H_2O with liquid ODMA film
    (d_ℓ = 29 μm)
c. H_2O with crystallized ODMA
    film (d_ℓ = 29 μm)

In conclusion, it is demonstrated that laser PA spectroscopy represents a simple, contactless and versatile method for the analysis of thin surface films on liquids.

References

1. M.W. Sigrist: J. Appl. Phys. 60, R83 (1986)
2. M.W. Sigrist and Z.H. Chen: Appl. Phys. B 43, 1 (1987)
3. M.W. Sigrist and D. Scherrer: In Photoacoustic and Photothermal Phenomena, ed. by P. Hess and J. Pelzl, Springer Series in Optical Sciences (Springer, Berlin, Heidelberg 1987), to be published

Spectroscopic Diagnosis for Control
of Laser Treatment of Atherosclerosis

R.R. Richards-Kortum[1], A. Mehta[1], T. Kolubayev[1], C. Hoyt[1],
R. Cothren[1], B. Sacks[1], C. Kittrell[1], M.S. Feld[1], N.B. Ratliff[2],
T. Kjellstrom[2], G. Bordagaray[2], M. Fitzmaurice[2], and J. Kramer[2]

[1]George R. Harrison Spectroscopy Laboratory,
Massachusetts Institute of Technology, Cambridge, MA 02139, USA
[2]Department of Cardiology, Cleveland Clinic Foundation,
Cleveland, OH 44106, USA

Healthy artery tissue is composed of three layers: intima, media, and adventitia.
With the onset of atherosclerosis, fatty and fibrous deposits build up in the intimal
layer, narrowing the lumen for blood flow, leading to stroke and heart attack. Laser
radiation of modest power can vaporize these deposits, restoring normal flow [1]. This
can be accomplished without open heart surgery, using a catheter containing optical
fibers to deliver the laser light to the obstruction. However, without appropriate
control, arterial perforation can occur. As part of our efforts in this area [2,3], we are
developing a spectroscopic guidance system to diagnose the presence and control the
removal of atherosclerotic blockages.

Laser induced fluorescence was obtained from freshly excised human cadaver aorta.
Normal and diseased artery exhibit distinct fluorescence spectra when excited by low
power 476nm argon-ion laser light. The normal artery fluorescence has more structure
and is three times as intense as the diseased artery fluorescence, as shown in Fig. 1.

Two processes combine to generate the observed spectra. Fluorescence is induced
by the light incident on a given region of the tissue. As this light returns to the
surface of the tissue, where the detector is located, it is modulated by the wavelength
dependent absorption of the tissue. Given this, we wish to understand the connection
between the observed spectra and the anatomical layers of the artery.

To study this, we sectioned aortic artery samples into the various anatomic layers
and recorded fluorescence and absorption spectra for each.

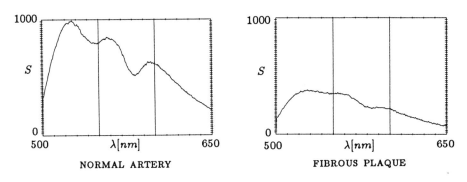

Figure 1: Normal artery wall and fibrous plaque fluorescence spectra

With normal artery tissue, as shown in Fig. 2, the primary source of the composite spectrum is the intima, with a small additional contribution made by the underlying media. The difference between the observed spectrum and the intimal fluorescence spectrum can be explained by the intimal absorption spectrum. The observed spectrum has deeper valleys, corresponding to the absorption peaks at 540 and 580nm. Also, the peak-to-peak intensity ratio ($S(600)/S(550)$) is larger for the composite spectrum than the intimal fluorescence, since the absorption drops around 600nm, allowing the penetration of relatively more medial fluorescence at this wavelength.

With fibrous plaque, the composite and intimal spectra are equivalent, indicating that the composite tissue spectrum is generated entirely in the diseased intima. The media and adventitia spectra are the same as those of healthy tissue, within sample to sample variations. This is to be expected, since the layers are not generally affected by the disease.

In order to more fully understand the composite tissue fluorescence spectrum, we have developed a two-layer model of tissue fluorescence. The geometry of this model is illustrated in Fig. 3. This model assumes that there is exponential attenuation of light in tissue and that the second tissue layer is opaque, so the composite spectrum is generated in the top two layers. The amount of fluorescence $dI(\lambda)$ contributed by each small layer of tissue, thickess dz, can be written as

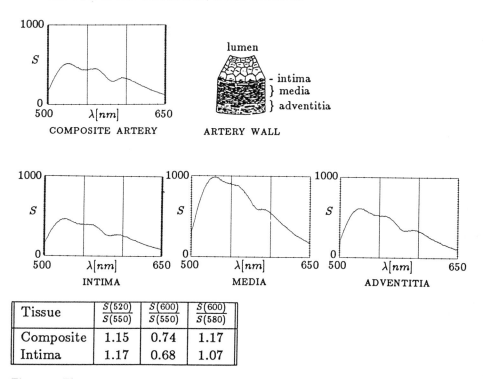

Tissue	$\dfrac{S(520)}{S(550)}$	$\dfrac{S(600)}{S(550)}$	$\dfrac{S(600)}{S(580)}$
Composite	1.15	0.74	1.17
Intima	1.17	0.68	1.07

Figure 2: Fluorescence spectra of normal artery wall and its component layers

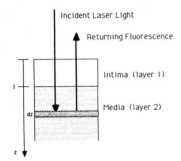

Incident Laser Light

Returning Fluorescence

Figure 3:
Model geometry

Intima (layer 1)

Media (layer 2)

$$dI(\lambda) = I_o e^{-\alpha_o z}\ \Phi(\lambda)\ e^{-\alpha(\lambda)z}dz, \tag{1}$$

where

I_o = incident intensity @476nm,

$\Phi(\lambda)$ = intrinsic fluorescence parameter, proportional to the density and absorption cross section of the fluorescing components,

$e^{-\alpha_o z}$ = attenuation of incident light,

$e^{-\alpha(\lambda)z}$ = attenuation of fluorescence emitted from dz.

Each layer ($i = 1, 2$) can be characterized by a separate fluorescence parameter, $\Phi_i(\lambda)$, and absorption parameter, $\gamma_i(\lambda) = \alpha_{io} + \alpha_i(\lambda)$. The observed signal, $S(\lambda)$, is

$$S(\lambda) = \frac{\Phi_1(\lambda)}{\gamma_1(\lambda)}[1 - e^{-\gamma_1(\lambda)l}] + \frac{\Phi_2(\lambda)}{\gamma_2(\lambda)}e^{-\gamma_1(\lambda)l}, \tag{2}$$

which can also be written as

$$S(\lambda) = L_1(\lambda) + L_2(\lambda)e^{-\gamma_1(\lambda)l}, \tag{3}$$

where $L_1(\lambda)$ and $L_2(\lambda)$ are the observed spectra for layers 1 and 2. Equation (3) can be solved to yield $\gamma_1(\lambda)$,

$$\gamma_1(\lambda) = -\frac{1}{l}\ln\left[\frac{S(\lambda) - L_1(\lambda)}{L_2(\lambda)}\right]. \tag{4}$$

In the sectioning experiments we measured $S(\lambda)$, the composite tissue fluorescence spectrum; $L_1(\lambda)$, the intimal spectrum; $L_2(\lambda)$, the medial spectrum; and l, the intimal thickness. From this we can extract $\gamma_1(\lambda)$. Figure 4 shows that this agrees with the measured absorption spectrum of intima, $\alpha_1(\lambda)$.

This agreement demonstrates that the model can provide an understanding of the way in which intima and media together generate the composite normal tissue fluorescence lineshapes.

The intimal absorption spectrum exhibits peaks at 350, 415, 540 and 580 nm. An absorption spectrum of whole blood also exhibits peaks at these four wavelengths. The porphyrin heme is responsible for the absorption in blood at these wavelengths,

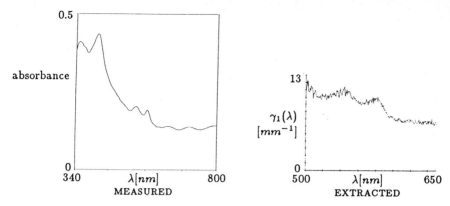

Figure 4: Intimal absorption spectrum of normal artery wall

and this suggests that a heme containing compound is responsible for much of the absorption in the arterial wall, although other metallo-porphyrin compounds could also contribute to absorption at these wavelengths.

The above experiments were conducted with cadaverous tissue. It is important to establish that equivalent spectra are generated in living tissue. As a first step, we are performing *in vivo* experiments on the carotid arteries of dogs. Preliminary results show that there are only subtle differences between the spectral lineshapes of living and cadaverous dog artery.

In conclusion, we have developed a framework for connecting the spectroscopic characteristics of arterial tissue with its anatomical structure. Equation (2) provides the basis for determining the intimal thickness, l, and the degree of disease from the composite fluorescence signal. Spectroscopic diagnosis of atherosclerosis for guidance and control of a laser catheter thus appears feasible.

References

1. J.M. Isner, R.H. Clarke: *IEEE J. Quantum Electronics*, **QE-20**, p. 1406 (1984).

2. R.M. Cothren, G.B. Hayes, J.R. Kramer, B. Sacks, C. Kittrell, M.S. Feld: *Lasers in the Life Sciences*, **1**, p. 1 (1986).

3. C. Kittrell, R.L. Willett, C. de Los Santos-Pancheo, N.B. Ratliff, J.R. Kramer, E.G. Malk, M.S. Feld: *Applied Optics*, **24**, p. 2280 (1985).

Laser Spectral Analysis
of Human Atherosclerotic Vessels

A.A. Oraevsky[1], *V.S. Letokhov*[1], *V.G. Omelyanenko*[2], *S.E. Ragimov*[2],
A.A. Belyaev[2], *and R.S. Akchurin*[2]

[1]Institute of Spectroscopy, Academy of Sciences USSR,
SU-142092 Troitzk, Moscow region, USSR
[2]Institute of Experimental Cardiology, Academy of Medical
Sciences USSR, SU-121552 Moskow, USSR

Laser angioplasty, i.e., clearing obstructed atherosclerotic blood vessels with laser radiation, is a new medical application of lasers. Studying the spectral properties of vessel walls is essential both to diagnose atherosclerotic lesions and to realize specific action by laser light on atherosclerotic plaques.

Lately, reports have appeared in the literature about some difference between the fluorescence spectra of normal and pathological zones (plaques) in human atherosclerotic vessels [1]. This difference is that the spectrum of normal vessel wall features three peaks instead of single wide fluorescence band typical of fibrous plaques. We also have measured fluorescence spectra of human cadaver thoracic and ventral aorta samples[2]. The results obtained enabled us to draw certain conclusions as to the causes for the difference in the spectral properties of fibrous plaques and normal wall areas of atherosclerotic vessels.

Figure 1 presents the fluorescence spectra for the normal zone and fibrous plaque of an atherosclerotic aorta under excitation at λ = 480 nm. The plaque spectrum is a wide band covering the range 490-650 nm with a maximum at λ = 515 nm. This fluorescence is due largely to the oxidized flavoproteins of the respiratory chain of the cells. At the same time, the spectrum of the normal vessel wall zone, as distinct from the plaque spectrum, has three characteristic peaks (λ =515, 560 and 595 nm) in the above region

The key to the understanding of the reason for the difference in these fluorescence spectra was provided by fluorescence measurements taken from the adventitia side (i.e., the exterior) of the vessel samples. This spectrum has more pronounced peaks as compared with its counterpart taken from the intima (lumen) side. It is well-known that the microcapillary blood network (vasa vasorum) supplying major vessels is located much closer to the adventitia than to the intima, and we hypothesized that the above difference between the fluorescence spectra is due to the presence of hemoglobin.

Actually, by subtracting the absorption spectrum of oxygenated hemoglobin from the fluorescence spectrum of fibrous

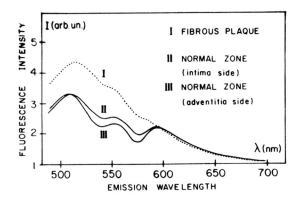

Fig. 1. Fluorescence spectra of human atherosclerotic
vessel wall taken under excitation at λ = 480 nm

plaques we obtained a spectrum very similar in shape to the fluorescence spectrum characteristic of normal wall zones of atherosclerotic vessels.

Thus, the difference between the fluorescence spectra of
Fig.1. is due to reabsorption in oxyhemoglobin present in the
vessel walls of radiation emitted by luminescent cells. And the
plaque tissue, thickened and infiltrated with lipoproteins (primarily cholesterol) as it is, strongly absorbs and scatters the
exciting light. So, the absorption of radiation by oxyhemoglobin has no effect on the emission spectrum of fibrous plaques.

The measurements performed using optical fibers to deliver
and gather radiation give reasons to believe that it will be
possible in future to realize local spectral diagnostics of
atherosclerotic lesions in vivo. The analysis should be made
on the basis of the ratio between the fluorescence intensities
at the wavelengths corresponding to the maxima and minima in
the spectra (I_{595} / I_{578} and I_{560} / I_{544}). During fluorescent
diagnostics one needs to replace temporarily the blood in the
vessel by the normal saline solution.

References

1. C.Kittrell, R.L.Willet, C.de los Santos-Pacheo, N.B.Ratliff,
   J.R.Kramer, E.G.Malk, M.S.Feld: Appl.Opt. <u>24</u>(15), 2280-2281,
   1985

2. A.A.Oraevsky, V.S.Letokhov, S.E.Ragimov, V.G.Omelyanenko,
   A.A.Belyaev, B.V.Shakhonin, R.S.Akchurin: Lasers in the
   Life Sciences, in press.

Diagnostics of Cancer Tumours and Atherosclerotic Plaque Using Laser-Induced Fluorescence

P.S. Andersson[1], J. Johansson[1], E. Kjellén[2], S. Montán[1], K. Svanberg[2;3], and S. Svanberg[1]

[1]Department of Physics, Lund Institute of Technology,
P.O. Box 118, S-221 00 Lund, Sweden
[2]Department of Oncology, Lund University Hospital,
S-221 85 Lund, Sweden
[3]Department of Internal Medicine, Lund University Hospital,
S-221 85 Lund, Sweden

In tissue diagnostics using laser-induced fluorescence it is important to utilize the full spectral information, e.g. for clearly demarcating tumours from surrounding healthy tissue. Both for point monitoring[1] and for imaging measurements[2,3] it has been found to be favourable to compress the spectral information into optimized contrast functions. These functions should be dimensionless to ensure immunity to distance variations, surface topography and variations in the excitation and detection efficiency. In cancer tumour detection, tissue auto-fluorescence and characteristic features of injected hematoporphyrin derivative (HPD) can be utilized. We have performed extensive studies of tissue fluorescence in rat tissue. Different photosensitizers have been used and the importance of the excitation wavelength for the achievable contrast between tumour and surrounding tissue has been investigated. Recently, our fluorescence diagnostics techniques have also been applied to humans undergoing photodynamic therapy.

An important aspect of tissue fluorescence characteristics is the possibility of guidance in assessing radicality in surgical tumour resection. Here the real-time capability of the fluorescence technique is of particular interest. We have studied a rat brain tumour system and found very encouraging results[4]. In Fig. 1 an example of a point monitor scan across the tumour is shown. In a characteristic tumour spectrum, induced by N_2 laser radiation, the relevant signal intensities are defined. It should be noted, that essentially no contrast is obtained if conventional HPD 630 nm monitoring (A') is used. Using background-free monitoring (A), the contrast is enhanced. Since the tumour is also characterized by a sharp decline in the blue fluorescence level (B), the dimensionless ratio A/B exhibits a very strong contrast. There seem to be good prospects for tumour diagnostics at substantially reduced ambient light sensitization levels if the techniques indicated here are utilized.

Recently, auto-fluorescence monitoring has been applied to the characterization of atherosclerotic plaque[5]. We have extended our medical diagnostic work into the field of vessel monitoring[6]. In Fig. 2 spectra from normal aortic arch and arch with plaque are shown for N_2 laser excitation. Several spectral features distinguish the samples. For plaque demarcation we have tested different contrast functions in point monitoring as well as in imaging measurements.

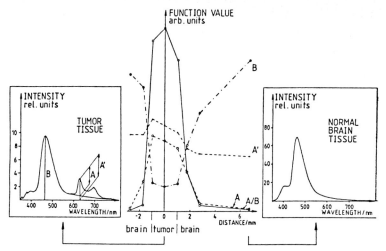

Fig. 1. Laser-induced fluorescence data from a scan along a line starting in normal rat brain, passing a tumour and extending again into normal brain tissue. Spectra with signal levels indicated are shown together with evaluated data with contrast enhancement illustrated.

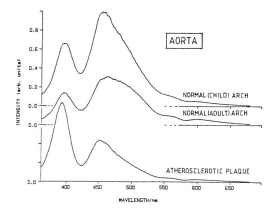

Fig. 2. Laser-induced fluorescence spectra from normal tissue and atherosclerotic plaque in human cadaver aortic arch.

1. K. Svanberg, E. Kjellén, J. Ankerst, S. Montán, E. Sjöblom, S. Svanberg: Cancer Res. 46, 3803 (1986)
2. S. Montán, K. Svanberg, S. Svanberg: Opt. Lett. 10, 56 (1985)
3. P.S. Andersson, S. Montán, S. Svanberg: IEEE J. Quant. Electr., October 1987 issue
4. P.S. Andersson, E. Kjellén, L.G. Salford, K. Svanberg, S. Svanberg: unpublished results
5. C. Kittrell, R.L. Willett, C. de los Santos-Pacheo, N.B. Ratliff, J.R. Kramer, E.G. Malk, M.S. Feld: Appl. Opt. 24, 2280 (1985)
6. P.S. Andersson, A. Gustafson, U. Stenram, K. Svanberg, S. Svanberg: Lasers in Med. Sci. , to appear

Part X

Spectroscopic Techniques

Towards the Ultimate Laser Resolution

J.L. Hall, D. Hils, C. Salomon†, and J.-M. Chartier*

Joint Institute for Laboratory Astrophysics, National Bureau of
Standards and University of Colorado, Boulder, CO 80309, USA

An early and productive theme in laser stabilization work was the use of
molecular absorbers such as iodine and methane to provide sharp absolute
frequency references for stabilization of HeNe lasers. Progress was rapid
and by 1975 the hyperfine splitting and even the small recoil structure of
the methane transition had been resolved/1/ using a cell of large transverse dimension so that the interaction time could be sufficient to allow
the desired resolution, $\sim 10^{11}$. Detailed theory/2/ led to an understanding
that the effective velocity distribution of the contributing absorbers was
non-thermal. One expected small frequency shifts with power and other
parameter changes, due to changes in the second-order Doppler redshift of
the resonant molecules. The presence of such a shift at the 10^{-12} level,
calculable to perhaps 10% accuracy, stood in disappointing contrast to the
stability attainable, 3×10^{-14} and later/3/ 5×10^{-15}. This made clear
the absolute necessity of developing laser or other techniques to control
the velocity distribution of gas-phase absorbers. Coincidentally, 1975
brought two landmark proposals/4,5/ for just such optical cooling. Now,
about one decade later, various groups are able to slow, stop, and even
reverse alkali atomic beams and, indeed, atom cooling and trapping work
has become a prime growth industry.

One problem affecting any laser stabilization scheme based on molecular
absorption is its specificity. There simply are lots of megaHertz in the
visible domain -- 80 GHz per Å in the red, for example -- and it is not
yet convenient to measure frequency intervals much beyond such a value.
This leads us to recognize the beauty of the comb of resonances offered by
a Fabry Perot resonator: one can use one of the resonances for controlling
a dye laser and quite another one for controlling the resonator itself
relative to a stabilized gas laser. Another advantage of the linear,
passive nature of the cavity resonator is that -- within limits -- one may
increase the available control signal by using more laser power for the
purpose, without concomitant power broadening. Significant progress has
been made toward constructing powerful and general spectrometers of this
general type, as effective and broadband frequency-shifting acousto-optic
and electro-optic modulators have recently become available.

Still the promise of interaction times on the order of a second to give
resolution in the $10^{-14} - 10^{-15}$ range, using ion trap confinement for
example, requires the laser spectroscopist to carefully consider his
strategy. The basic problem is that the I_2-stabilized laser might offer
$\sigma \simeq 10^{-11.5}/\sqrt{\tau}$ stability, with the averaging time τ expressed in seconds.
Although the newer method of modulation transfer/6/ has given a 100-fold

† Permanent address: Lab. de Spectroscopie Hertzienne, Paris
* Bureau International des Poid et Mesures; Sèvres, France

performance gain relative to the conventional I_2-stabilized lasers, even with this technique we need ~250 s to average the random noise down to 1 Hz. Somehow one cannot help but feel that a golden opportunity is passing by: if an interferometer of robust design were isolated from thermal, acoustic and mechanical perturbations, it seems reasonable that it could be extremely stable in the short term, just where the actively-stablized reference laser is weakest. In the long term, a day or a month or a year for example, one surely would not expect a mere bar of some low expansion material to exhibit much stability. Still, long term "creep" rates down toward 10^{-9} per year (or even less) are known for some materials such as ULE and fused silica/7/. This drift, now expressed as 10^{-15} for 1/2 a minute, begins to sound very attractive again. In any case it is only the change from a predictable drift rate that affects the cavity's utility as the frequency reference for our dreams of "fantastico-resolution" with cooled, trapped ions or with Ramsey's fringes in atoms of Zacharias' fountain. The idea then becomes clear: We lock the laser tightly to a high-finesse Fabry Perot cavity designed with a low expansion, stable spacer. We explore the quantum absorber's resonance by a suitable algorithm for a few seconds (up to perhaps several minutes) to develop our best knowledge of the frequency offset between the studied resonance and the laser reference cavity. One then can refine the offset value supplied initially and continue the process of interrogating the atomic resonance. Also, after some time improved information will be available for the cavity drift-rate correction. It seems likely that this concept will offer new quantum frequency standards with unprecedented accuracy, surely $1:10^{15}$ or better.

Another interesting potential application of such "flywheel" lasers based on cavity stabilization is as local oscillators in any experiment needing high stability in the domain ≈ 10 s. One extremely challenging such application is the proposed space-based laser heterodyne gravitational wave antenna/8/. In this concept, a pair of satellites separated by $\sim 10^6$ km from the central station would contain transponders phase-locked to the outgoing laser beams produced by the central station. The laser frequency fluctuations observed over one arm of the antenna would be used in a correction algorithm to remove the jitter in the measured difference in the length of the two arms. The minimum gravitational wave period detectable with full sensitivity is approximately the light travel time of 6 s. To reach the expected sensitivity level $\delta \ell / \ell \approx 10^{-20} / \sqrt{Hz}$, it would be convenient to have the phase jitter be less than 1 radian. So we are asking for a laser with phase stability of 1 radian after 3×10^{15} optical cycles, corresponding to a stability of $\sim 1 \times 10^{-16}$ at 6 s. Surprisingly enough, we have been able to demonstrate locking at this level and below. Evidently this kind of performance also gives entry into the delicious "garden of advanced tests of our cherished physical principles." We turn now first to a brief description of locking two lasers to a single cavity to facilitate sensitive investigation of the question, "How well can we do the locking?" This report concludes with a discussion of what we now know about the short and long term stability of our present generation of reference cavities.

In the theory of servo control systems it is easy to show that the errors within a servo loop can be driven below any small value one wishes, merely by increasing the closed loop gain. Of course more gain can be usefully employed, considering loop stability issues, only if we can also increase the bandwidth and reduce the time delay if it approaches more than $\sim 1/6$ of the inverse bandwidth. The system output is not noise free under high feedback conditions however, since there is no way to distinguish fluctua-

tions of the photodetector output from a correspondingly-scaled frequency
excursion. Indeed, the loop will drive the total input error signal to
zero. The bottom line is that a precise servo ideally will suppress
intrinsic noise essentially without limit, replacing it with measurement
noise. This argument was articulated in more detail recently/9/.

Several types of laser frequency discriminators have been considered. It
appears that modulation techniques will always be preferred if the limiting
performance is desired, since sources ordinarily exhibit appreciably more
noise below some characteristic frequency ($\sim$5-10 kHz for small HeNe
lasers, $\sim$3 MHz for jetstream dye lasers, and $\sim$1 GHz for cw GaAs diode
lasers). The base noise level will be set by photodetection shotnoise.
This amplitude noise must be interpreted as frequency noise by the system
when it is working at the fundamental limit, and so it is clearly advantageous to have a large discriminator slope. Then a few millivolts will
represent fewer kilohertz laser jitter. The most direct way to increase
the slope is to decrease the cavity resonance linewidth, which introduces
a potential for unwelcome time delay inside the control loop. DREVER et
al./10/ have discussed a topology based on sensing the cavity reflection
which has the happy property of reducing the static light level on the
photodetector (and thus reducing the shotnoise) while giving a crossover
to phase-detection above $\Delta \nu_c /2$. The resonant field inside the interferometer cannot change rapidly, and its leakage back through the input
coupler forms the necessary stable local oscillator for (optical) phase-
sensitive detection of the laser's fast phase error. (We have also been
able to use the forward "darkness wave" of resonant molecules for the same
purpose. Interestingly, the Doppler width provides us with fresh phase-
reference atoms if we wish to slowly tune the laser. Of course saturation
effects make this phenomenon richer than the cavity locking case.)

In view of the above remarks from servo theory that all systems can eventually be quiet enough so that the measurement noise can dominate, we have
chosen to perform our current set of experiments using HeNe lasers of a
compact design intended for supermarket scanner applications. The tube is
frequency-tuned by heating the glass to expand it, stretching it directly
a fringe or so with a surrounding PZT tube, or modifying the plasma's
index of refraction slightly via a small change in the discharge current
($\sim$3 kHz/μA). Since the reference cavity mirrors at present are not well
optimized in their transmission/loss ratio our resonance fringe height is
only $\sim$1% of the input power. This makes it attractive to work in
transmission since the residual AM produced by the AD*P phase modulator
($\sim$1:10^6) appears as a ppm-level effect, rather than a 10^{-5} effect in
reflection. The measured fringewidth is 72 kHz with a 500 MHz fsr (F $\simeq$
6500). The signal-to-shotnoise ratio in the transmitted beam is calculated
to be 40,000:1 in the 5 kHz control bandwidth. The resulting rms frequency
excursion then should be 1.8 Hz. It is important to note that this is not
the Lorentz linewidth of the laser due to the low phase modulation index
associated with the high frequency part of the shotnoise spectrum. For
our example, the noise spectral density Δ = 1.8 Hz/(5000 Hz)$^{1/2}$ and,
following ELLIOTT et al. /11/, the Lorentz laser linewidth will be $\Delta \nu_L = \pi \Delta^2$
$\simeq$ 2 mHz! Certainly we would be satisfied to measure such a value.

It is useful to note the extreme demands for optical isolation in the
experiment. Our cavity resonance width is 72 kHz and the phase will shift
by $\pm \pi /4$ over this range. On the other hand we potentially have a linewidth in the millihertz domain, almost 10^{-8}-fold smaller. A spurious
little optical phasor 10^{-8}x weaker than the main beam can thus pull the
cavity lock point as much as the noisewidth. To approximate this required

378

160 dB isolation we are presently using two Faraday isolators (~30 dB each) and two acousto-optic frequency shifters (>70 dB each) one before and one just after the resonant cavity. The indicated isolation/directivity values were measured directly using the elements singly. Scattering on surfaces will ultimately be the limiting effect since it operates to create weak low-finesse interferometers in series with the laser beam line, without the possibility to accumulate isolation in between. When the temperature or air pressure changes, these "fringes" are changed in phase. At present our best result is a setting precision ~1 Hz, estimated by alternately locking a pair of lasers on a number of axial orders and counting the resulting beat frequencies. A recent calculation/12/ indicates possibly beneficial results from the use of harmonic locking, although the lineshapes in this fast-modulation regime are not as simple as in the usual quasi-static limit.

Our experiments/13/ to test the possibility of such tight locking utilize two lasers locked to a single cavity to largely cancel the effects of cavity instability, per se. To avoid ambiguity and cross-locking, the lasers are locked on adjacent axial orders of the Fabry Perot, c/2L = 500 MHz apart. Preliminary measurements of the rf beat showed it to be narrow indeed, $\ll$ 10 Hz. For further analysis the 500 MHz optical difference frequency was heterodyned to ~10 Hz, using an extremely stable synthesizer. It was necessary to replace the internal low drift crystal reference oscillator with a "research grade" 5 MHz quartz oscillator to avoid conspicuous phase noise buildup after ~10 sec. The counter measurements were of millihertz resolution and 1 s gate time, followed by 130 ms deadtime for data transfer and storage. Figure 1 shows representative data for 9000 measurements, presented in the form of an Allan variance plot. At τ = 1 s the implied frequency uncertainty is 100 millihertz in the beat, ~50 mHz for each laser. This would give a phase coherence time of ~3 s for the laser's optical field. We emphasize that this value is

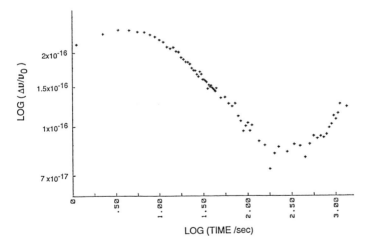

Fig. 1. <u>Allan variance of cavity-locked He-Ne laser</u>. To investigate the precision of locking, two lasers were independently locked to adjacent orders of a stable high finesse Fabry-Perot cavity. The 500 MHz optical beat was heterodyned with a stable rf source to a few Hz for analysis. Over this 10^4 s record, the 1/e correlation time of the frequency was about 4 s.

obtained as an average over a ~10⁷ s interval. Within restricted time
domains it is easy to find intervals of 10 s during which the beat phase
does not change by as much as 1 radian. One unexplained observation is
the presence of a large noise peak, with a phase modulation amplitude of
~1 radian occurring at a frequency of ~1/6 Hz. See Fig. 1. Without this
noise the coherence time before 1 radian phase noise occurs would be
rather satisfying. Time domain plots of the smoothed beat frequency give
ideas about the upturn of the variance in Fig. 1 at longer times. We find
there is a tendency for the beat frequency to jump rather abruptly between
two levels ~0.5 Hz apart, but have not yet isolated the cause. Looking at
records extending over several months, one sees a stability of the beat
frequency of ~2 Hz (4×10^{-15}), only somewhat worse than the ~1 Hz
apparent systematic offset of the lock point.

To summarize, it is indeed possible to lock to a reference cavity with a
reproducibility $\approx 10^{-15}$ fringe widths and a stability a decade or so
better, producing a laser linewidth of ~50 milliHertz. Heterodyne tests
against an I_2-stabilized reference show a frequency drift rate of +6100
Hz/hr stable to 10^{-3} per day and ~1% for a week/14/. Feed-forward
compensation for this smooth drift then offers a source for spectroscopy
with a frequency uncertainty < 100 milliHertz for about one minute, as
will be appropriate to scan over the "quantum telegraph" optical
resonances of single trapped ionswith long metastable lifetimes/15/.

REFERENCES

1. J. L. Hall, C. J. Bordé, and K. Uehara, Phys. Rev. Lett. 37, 1339
    (1976)
2. C. J. Bordé, J. L. Hall, C. V. Kunasz and D. G. Hummer, Phys. Rev. A
    14, 236 (1976)
3. S. N. Bagaev, L. S. Vasilenko, V. G. Gol'dort, A. K. Dmitriev, and A. S.
    Dychkov, Sov. J. Quant. Electr. 7, 665 (1977); S. N. Bagayev and V. P.
    Chebotayev, Appl. Phys. 13, 291 (1977)
4. T. W. Hänsch and A. L. Schawlow, Opt. Commun. 13, 68 (1975)
5. D. J. Wineland and H. G. Dehmelt, Bull. Am. Phys. Soc. 20, 637 (1975)
6. J. H. Shirley, Opt. Lett. 7, 537 (1982)
7. F. Bayer-Helms, H. Darnedde, and G. Exner, Metrologia 21, 49 (1985);
    B. Justice, J. Res. NBS 79A, 545 (1975); J. W. Berthold III, S. F.
    Jacobs, and M. A. Norton, Metrologia 13, 9 (1977); A. Sakuma, BIPM,
    Sèvres, France, private communication.
8. J. E. Faller, P. L. Bender, J. L. Hall, D. Hils, and M. A. Vincent, in
    Proc. Colloquium "Kilometric Optical Arrays in Space," Cargese
    (Corsica), 23-25 Oct. 1984.
9. J. L. Hall, in Quantum Optics IV, J. D. Harvey and D. F. Walls, eds.
    (Springer, 1986) p. 273
10. R. W. P. Drever, J. L. Hall, F. V. Kowalski, J. Hough, G. M. Ford, A.
    J. Munley, and H. Ward, Appl. Phys. B 31, 97 (1983)
11. D. S. Elliott, R. Roy, and S. J. Smith, Phys. Rev. A 26, 12 (1982)
12. D. Hils and J. L. Hall, Rev. Sci. Instr. in press (Aug. 1987)
13. C. Salomon, D. Hils and J. L. Hall, J. Opt. Soc. Am. B, to be
    published
14. D. Hils and J. L. Hall, in XV International Conference on Quantum
    Electronics, Baltimore MD, April 1987, paper WDD3
15. See contributions in this volume by colleagues in Boulder, Seattle
    and Hamburg

Effects of Curvature in Laser Spectroscopy with Strong Fields

Ch.J. Bordé[1], Ch. Chardonnet[2], and D. Mayou[3]

[1]Laboratoire de Gravitation et Cosmologie Relativistes, Paris and
  Laboratoire de Physique des Lasers, F-93430 Villetaneuse, France
[2]Laboratoire de Physique des Lasers, F-93430 Villetaneus, France
[3]L.E.P.E.S., C.N.R.S., F-38042 Grenoble, France

This paper deals with the interaction of two-level systems either with a single traveling wave or with two counterpropagating beams, for systems with lifetimes longer than the transit time ($\tau=w/v_x$) and hence for which, the curvature of wavefronts plays a dominant role. In the case of a single traveling wave, one of us has demonstrated, first theoretically and then experimentally [1], that the Rabi nutation [2] was replaced by adiabatic fast passage (Fig. 1). A detailed theoretical presentation of this phenomenon can be found in reference [3] from which Fig. 1 and 2 are reproduced. In the case of two counterpropagating beams new consequences of curvature have been recently discovered through the catastrophic distortions of the saturated absorption lineshape reported here for the first time. But we shall first outline some of the results obtained for the single wave case. We shall focus on the final angle (Fig. 2) of the pseudo-spin as a function of three parameters respectively proportional to the Rabi pulsation $|A|=\Omega_{ba}w_0/v_x$, to the distance to the waist B=4z/b (confocal parameter b) and to the detuning $C=(\omega_0-\omega+kv_z)w/v_x$. The equation for the two-level spinor is written with Pauli matrices as

$$id/dt\begin{pmatrix}b\\a\end{pmatrix}=-(1/2)\vec{\Omega}.\vec{\sigma}\begin{pmatrix}b\\a\end{pmatrix}, \text{ with } \Omega_1\tau=2A\exp-(t/\tau)^2, \ \Omega_2=0, \ \Omega_3\tau=-B(t/\tau)-C$$

in the suitably rotated frame and equivalent equations in other rotating frames. Different theoretical approaches have been used in the low, intermediate and high field domains.

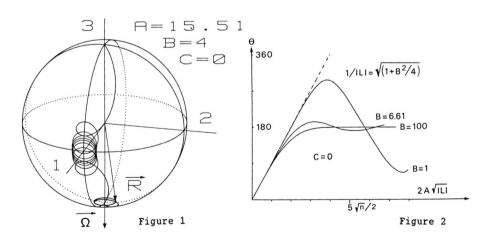

Figure 1

Figure 2

381

In the low field limit, very general perturbative calculations including relaxation can be performed with density matrix diagrams (Fig. 3) and associated rules given in [4]. These diagrammatic rules provide a way to include most of the physics relevant for moving atoms : first and second-order Doppler effects, recoil shift, transverse and longitudinal transit effects, space-time dependent light shifts, acceleration of the atoms in an external field [5] and even small angle elastic collisions. For simplicity we shall give here only the 2x2 first-order S matrix element in the absence of relaxation : $S_{ba}^{(1)}=iA(\pi L)^{1/2}\exp(-LC^2/4)$ with $L=1/(1-iB/2)$, which explains the common tangent at zero field for all the curves of Fig. 2 when plotted as functions of $A/(1+B^2/4)^{1/4}$. The first condition for adiabatic reversal is that this parameter should be larger than 1. The second order Magnus expansion is also easily calculated with the error function of complex arguments.

In the intermediate field region, apart from specific time dependences of the field, we know only numerical methods solving coupled differential equations by predictor-corrector methods and the choice is again between density matrix calculations including relaxation and a very simple system which couples two of the spinor Euler angles [3] :

$$d\varphi/dt=\Omega_1\cos\varphi/tg\theta-\Omega_3 \quad , \quad d\theta/dt=\Omega_1\sin\varphi \ .$$

An infinite number of pairs of functions for Ω_1 and Ω_3 which give analytic solutions satisfying $d\varphi/dt=0$, has been derived in [3], thus generalizing the well-known sech/tanh pulse solution given in [6]. The Gaussian laser structure does not satisfy the criteria for these functions but still retain the property of a fall in a $\varphi=\varphi_0$ plane with $tg\varphi_0=2/B$ in the early part of the spinor motion, which sets $B>2$ as the second criterion for adiabatic evolution.

In the strong field regime, we have derived analytic formulas for the S matrix in two cases for which the evolution from 0 to $+\infty$ can be divided in two parts :
1) In the case of arbitrary B, an intermediate time can be found before which the adiabatic approximation can be applied and after which the purely exponential pulse solution [7] can be used. The evolution operator U(+∞,0) is

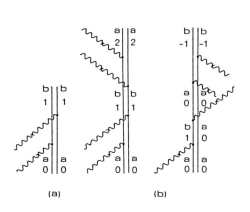

Figure 3   Figure 4

obtained as a product of 2x2 matrices from which the S matrix is obtained through

$$S(A,B,C,+\infty,-\infty) = U(A,B,C,+\infty,0)\sigma_3 U^{-1}(A,B,-C,+\infty,0)\sigma_3 .$$

The interesting result is that the final population of the initial state (e.g. b=1 at t=-∞) is : bb*=sech²(πB/4)cos²ψ where ψ is the (complicated) phase angle responsible for the oscillations of Fig. 2 whose amplitude has therefore a very simple expression, consistent with the previous condition for adiabatic reversal.

2)In the case of strong B, a similar approach is used : before some critical time t_0 the analytic solution for a quadratic phase pulse of constant amplitude [8] is used followed by the adiabatic approximation from t_0 to +∞. In this case again, a very simple result is obtained :

$$bb^*=exp[-(2\pi A^2/B)exp(-2C^2/B^2)]$$ which corresponds to the non-oscillating limit of the curves of Fig. 2 (θ=Arccos(2bb*-1)).

In the case of two counterpropagating waves, the problem is considerably complicated by the possibility for the molecules to exchange many quanta of momentum h/λ through successive interactions with both waves (see Fig. 4). The density matrix elements acquire Fourier components at expi(m_α-m_β)kz but this series is truncated in a resolution-dependent way if the recoil splitting is comparable to the linewidth. The only analytic solutions known to us, in the presence of curvature, are either perturbative [9] or apply to the special case of zero detuning and negligible recoil splitting [3]. In the latter case, one can show that for matched beams in the absence of relaxation, the final probability for the upper state is given by

$$bb^*=(1+cos\Theta)/2 \text{ with } \Theta=4A(\pi|L|)^{1/2}exp(-kbv_z^2/4v_x^2)cos(kz+\Phi)$$

and is easily expanded in a series of Bessel functions. Since the leading term for cosΘ is J_0, the adiabatic reversal property is lost in this case,

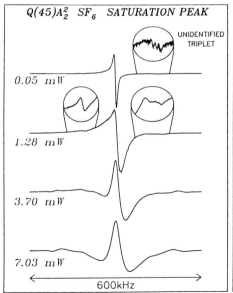

Figure 5   Figure 6

but because of curvature, a new phenomenon appears off-resonance which originates from the symmetry breakdown between both wave directions and which manifests itself in saturation spectroscopy. It is already known from perturbation theory [9] that, in this case, a lineshape asymmetry of opposite sign appears for either choice of probe/saturation wave owing to an interference resulting from the phase shifts induced by the opposite curvature of the saturation wave and by the detuning (the diagrams of Fig. 3b correspond to each recoil peak for the (-) probe wave). With strong fields this asymmetry forces the populations to build up preferentially in one direction or in the other on the diagram of Fig. 4 and hence favours absorption from one wave followed by emission into the other (one wave is amplified at the expense of the other). With strong fields this results in a narrow dispersion feature which appears on top of a broad asymmetric pedestal so that the lineshape derivative obtained with a constant modulation index evolves from the usual first derivative of the absorption peak toward a "second derivative" symmetric shape. This behaviour was discovered as we were measuring weak resonances such as the crossover resonances in SF_6 (Fig. 5) and for which it was necessary to use much stronger fields than usual. Figure 7 is a striking example in which a weak OsO_4 line keeps its normal shape while the strong line in close coincidence with the CO_2 10R(10) line is severely distorted. The evolution of the unmodulated lineshape has also been carefully studied (Fig. 6) to confirm these conclusions and demonstrate the sign change of the asymmetry with wavefront curvature conditions in our large cell. These conditions turn out to be very critical, since the reversal of the bottom signal of Fig. 6 or the sign change of the derivative illustrated in Fig. 8, occur as the beam waist is moved by less than ±70m from the entrance to the exit of our cell (for comparison the confocal parameter b≈1km).

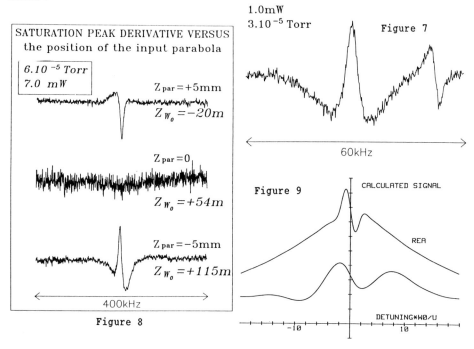

SATURATION PEAK DERIVATIVE VERSUS the position of the input parabola

6.10^{-5} Torr
7.0 mW

$Z_{par} = +5mm$
$Z_{W_0} = -20m$

$Z_{par} = 0$
$Z_{W_0} = +54m$

$Z_{par} = -5mm$
$Z_{W_0} = +115m$

400kHz

Figure 8

$1.0mW$
3.10^{-5} Torr
Figure 7

60kHz

Figure 9

CALCULATED SIGNAL

REA

DETUNING*W0/U
-10   10

The **theoretical approach** that we have used is based on a set of
coupled equations for the Fourier components of the density matrix which is
equivalent to equations (16) of reference [9] after proper rotation to a
frame where the field has a constant phase. Two differential equations are
added to take care of the energy exchange with each wave. As explained in
[9], these equations are solved numerically by a predictor-corrector method
followed by velocity integration. In a first step, we have used the
so-called [10] rate equation approximation (REA) in which density matrix
elements harmonics at exp(inkz) with $|n|>1$ are neglected. The resulting
lineshape for $\gamma w_0 / v_x = 1$ and A=15 for both waves, is given in Fig. 9 in the
case of matched beams at B=1 . A narrow central dispersion feature has grown
up from the low-field curvature asymmetry with its upper part on the
blue/red side for a diverging/converging probe beam and is absent if both
beams have opposite curvature. If the system is truncated to only three
energy-momentum states (e.g. $m_a = 0$, $m_b = \pm 1$), which is the case of a well-
resolved recoil peak, this dispersion feature disappears. When the exp±2ikz
terms are restored, it is replaced by a reversed asymmetry and a splitting
which increases with power. This corresponds to a dynamic Stark effect for
plane waves which gets asymmetric for curved waves. Without the $|n|=2$ terms,
the dispersion feature slowly reappears if the number of coupled energy-
momentum states is increased (over 50 for A=15) and the REA result is reco-
vered. This proves the direct link between this resonance and the motion in
momentum space. If the full Fourier series is kept (up to 45 coupled equations for the A=15 case of the Fig. 9) the lineshape is a combination of the
central dispersion resonance with an asymmetric pedestal dominated by the
$|n|>1$ harmonics and which is in good qualitative agreement with the experimental observations. This critical asymmetry of the lineshape, which appears
in a very strong field, is in fact a sensitive way to achieve very symmetric
lineshapes for weak fields. Its coupling to self-focusing/defocusing gas
lens effects remains to be investigated. Finally, let us emphasize that
these phenomena discovered in the case of beam curvature, are formally equivalent to what would be obtained with molecules falling on a standing plane
wave in a gravitational field.

[1] S. AVRILLIER, J.-M. RAIMOND, Ch.J. BORDÉ, D. BASSI and G. SCOLES
    Opt. Commun.,39, 311-315 (1981).
[2] Ch.J. BORDÉ, S. AVRILLIER, A. VAN LERBERGHE, Ch. SALOMON, Ch. BRÉANT,
    D. BASSI and G. SCOLES, Applied Physics B, 28, 82 (1982).
[3] Ch.J. BORDÉ, Revue du CETHEDEC, Ondes et Signal, NS83-1, 1-118 (1983).
[4] Ch.J. BORDÉ, Density matrix equations and diagrams for high resolution
    non-linear laser spectroscopy, in Advances in Laser Spectroscopy,
    p. 1-70, NATO ASI Series, Plenum (1983).
[5] Ch.J. BORDÉ, J. SHARMA, Ph. TOURRENC and Th. DAMOUR, J. Physique
    Lettres, 44, L-983-L-990 (1983).
[6] L. ALLEN and J.H. EBERLY, Optical resonance and two level atoms,
    J. Wiley, p. 102 (1975).
[7] D.S.F. CROTHERS, J. Phys. B 11, 1025 (1978).
[8] P.HORWITZ, Appl. Phys. Lett. 26, 306 (1975).
[9] Ch.J. BORDÉ, J.L. HALL C.V. KUNASZ and D.G. HUMMER, Phys. Rev.,
    14, 236-263 (1976).
[10]B.J. FELDMAN and M.S. FELD, Phys. Rev. A1, 1375 (1970).

Single Recoil Component Optical Ramsey Fringes at 514.5 nm in an I_2 Supersonic Beam *

G. Camy, N. Courtier, and J. Helmcke**

Laboratoire de Physique des Lasers, Université Paris-Nord,
Avenue J.B. Clément, F-93430 Villetaneuse, France

This article reports on the first observation of optical Ramsey fringes in molecular iodine at 514.5 nm wavelength on the 43-0 P(13) transition (a_2).

In conventional saturated absorption experiments utilizing I_2 absorption cells, the acccuracy in determining the line center is ultimately limited by : unresolved recoil structure /1/ ; residual transit-time broadening ; second-order Doppler broadening and shift ; and collisions (nonlinear pressure broadening and shift). These limitations can be greatly reduced by applying the method of separated field excitation /2/ in a supersonic I_2 beam.

The experiments were carried out using a high-resolution I_2-stabilized Ar^+-laser spectrometer at 514.5 nm and a collimated supersonic beam. The spectral width of the Ar^+-laser was estimated to be in the range of 10 kHz. From the high density and the narrow velocity distribution of the supersonic beam we expect to achieve good signal-to-noise ratio, reduced second order Doppler broadening and improved knowledge of the second order Doppler shift.

The experimental setup is shown in Fig. 1. The I_2 molecules of the supersonic beam are excited sequentially by two pairs of travelling waves propagating parallel but in opposite directions. A coherence between the initial and the excited states is built up in the first interaction zone /4/. It then precesses freely through the field-free region. In the second zone, two "population gratings" of the initial state and the excited state are created in the v_z-space, depending on the phase difference between the coherence and the exciting field. The two gratings are slightly shifted relatively to each other due to the recoil splitting. The distance a between the first two interaction zones is limited to about 3 mm to avoid excessive decay of the upper state of the selected absorption line. During the free flight between the second and third interaction zones, separated by a distance d = 35 mm, the upper state population decays completely into many ro-vibrational levels of the electronic ground state, leaving the population grating of the initial state practically unchanged. Correspondingly, only the initial state population grating will contribute to the Ramsey signal generated after the final two interactions and only the recoil component associated with the initial state will contribute to the Ramsey fringes /1,3,4/.

---

* Work partially supported by B.N.M.
**Permanent address : P.T.B, Bundesallee 100, D-3300 Braunschweig, FRG

The pair of two parallel travelling waves is generated by means of a carefully collimated laser beam passing through a double slit near the molecular beam. Each slit has a width of 0.2 mm and a height of 20 mm, respectively. The Ramsey signal is observed by monitoring the fluorescence. In order to select only the nonlinear part of the signal the first two interaction zones are chopped and the output of the photomultiplier is phase-sensitively detected. Figure 2 shows three typical recordings of optical Ramsey fringes observed in I_2 at separations between neighbouring zones of a = 1.2 mm, 2.2 mm, and 3.2 mm. The corresponding widths of the central fringes are 85 kHz, 50 kHz, and 30 kHz, respectively. We note, that the latter fringe width corresponds to about one half of the natural line-width of the investigated absorption line.

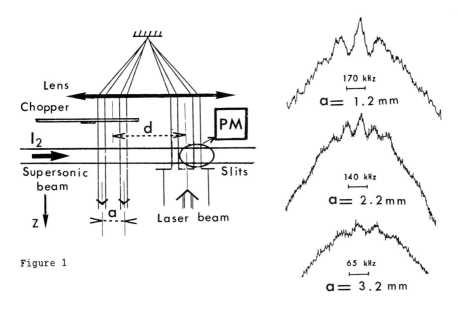

Figure 1

Figure 2

References

/1/ Ch.J. Bordé, G. Camy, B. Decomps : Phys. Rev. A 20, 254 (1979)
/2/ Ye.V.Baklanov, B.Ya. Dubetsky, V.P. Chebotayev : Appl. Phys. 9, 171
     (1976) ; see also Ch.J. Bordé et al. : Appl. Phys. B 28, 82 (1982),
     and J. Helmcke et al. : Appl. Phys. B 28, 83 (1982)
/3/ A.N. Goncharov, M.N. Skvortsov, and V.P. Chebotayev :
     Sov. J. Quantum Electron. 13, 1429 (1983).
/4/ Ch. J. Bordé, et al. : Phys. Rev. A 30, 1836 (1984)

High Resolution Optical Multiplex Spectroscopy

K.P. Dinse[1], M.P. Winters[2], and J.L. Hall[2]

[1]Institut für Physik, Universität Dortmund,
  D-4600 Dortmund 50, Fed. Rep. of Germany
[2]Joint Institute for Laboratory Astrophysics, University of Colorado
  and National Bureau of Standards, Boulder, CO 80309, USA

Cross-correlation techniques have been applied in a variety of fields to obtain the impulse response function of a system. The Fourier transform of the impulse response function is the transfer function (or spectrum) of the system under investigation. The most widely used correlation technique is Fourier transform infrared (FTIR) spectroscopy, where a Michelson interferometer is used as an analog correlator. If the sample is placed in one arm of the interferometer the resulting cross correlation spectrum results in the unknown spectrum directly after Fourier transformation.

The correlator output is therefore equivalent to a Free Induction Decay (FID) signal, obtained after a single pulse stimulation of the sample. This equivalence was investigated in detail in magnetic resonance experiments /1/. The essence of the method is the stimulation of the system with broad-band noise, being characterized by a correlation time τ. In a true multiplex fashion the sample is excited within a spectrum intervall $\Delta\Omega \simeq 2/\tau_C$ and the response is unfolded by cross-correlation with a varying delay. Owing to the continous excitation in contrast to the single pulse excitation, the power requirements to excitation are greatly reduced. In addition a signal/noise advantage of appr. $\{\Delta\Omega/\Delta\omega\}^{1/2}$ ($\Delta\omega$ denoting the single line spectral width) can be obtained when comparing the noise excitation experiment with a conventional scan experiment.

As in FTIR the spectral resolution of the spectrometer is limited by the maximum path length difference $\Delta l \cong c \cdot \Delta t$. Aiming for a spectral resolution of appr. 10 kHz, path length variations > 10 km would have been required. Therefore, instead of using Gaussian noise, a pseudo-random binary sequence (PRBS) was generated, which could be dublicated and electronically delayed by arbitrary Δt. Utilizing standard ECL IC's, a correlation time of 5 ns corresponding to an usable excitation width of 200 MHz could be realized. The PRBS sequence was used to phase-modulate a frequency-stabilized Ar^+-laser. Details of the spectrometer designed for saturation spectroscopy are given elsewhere /2/.

Figure 1 shows the cross-correlation signal when tuning the unmodulated laser frequence close to the b8/b9 region of the R(15) 43-0 transition of I_2. The signal has the appearance of an OFID with zero dead-time and after spectral analysis with a Maximum Entropie algorithm (MEM) /3/ results in a spectrum depicted in Fig. 2. This algorithm, utilizing the time correlations in the OFID signal is used because of its ability to separate close-by spectral lines.

1. R.R. Ernst, J. Magn. Reson. 3, 10 (1970)
2. K.P. Dinse, M.P. Winters, and J.L. Hall, J. Opt. Soc. Am., submitted
3. E.D. Lane, J. Skilling, J. Staunton, S. Sibisi, and R.G. Brereton,
   J. Magn. Reson. 62, 437 (1985)

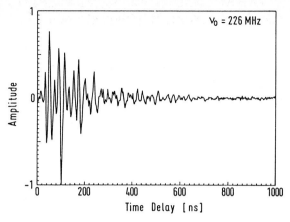

Fig. 1   Cross-correlation signal obtained by exciting part of the R(15) 43-0 transition of $^{127}I_2$. Zero time delay is at 35 ns

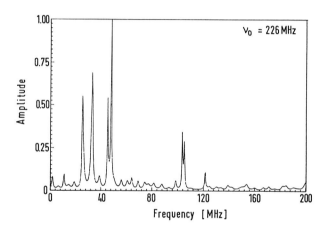

Fig. 2   Doppler-free I_2-spectrum obtained by analyzing the time-domain data with a MEM routine for optimum spectral resolution

Fourier Transform Heterodyne Spectroscopy: A Simple Novel Technique with Ultrahigh (150 mHz) Resolution

E. Mazur

Department of Physics and Division of Applied Sciences,
Harvard University, Cambridge, MA 02138, USA

Light beating spectroscopy has been used from the early days of the laser to study light scattering.[1] By detecting the beating signal between the scattered light and a 'local oscillator' field derived from the same laser, resolving powers of 10^{14} have been achieved. The Fourier transform heterodyne spectroscopy presented here is simpler and more direct than the conventional heterodyne techniques using autocorrelators or spectrum analyzers.

The techniques to measure spectra of scattered light fall into two categories as illustrated in Fig. 1: frequency-domain and time-domain spectroscopy. In the frequency domain (Fig. 1a) one first spectrally filters the incoming light and then detects the transmitted signal. The spectral resolution of this technique is limited by either the resolution of the filter (monochromator, interferometer, etc.), or by the bandwidth of the laser. In the time domain (Fig. 1b) one detects the beating of the scattered light—with itself or with part of the incident light—and then analyzes the spectrum of the detector signal with a spectrum analyzer or autocorrelator. The main advantage of this scheme is that fluctuations in the phase of the incident laser field, which limit the resolution of frequency-domain spectroscopy, cancel—provided the two fields reaching the detector are coherent. Also, the *total* signal I_s is detected, rather than the filtered intensity I_ω, yielding a higher signal-to-noise ratio.

In our present detection scheme an acousto-optically shifted local oscillator (Fig. 1c) allows us to study spectra of scattered light near the incident laser frequency. The detector signal is sampled for a certain length of time, stored in a microcomputer, and then a fast Fourier transform is applied to the sampled waveform. It can readily be shown that the resulting datapoints correspond precisely to the (shifted) spectrum of the scattered light $S(\omega)$. This scheme bypasses the need to filter or process the detector signal with an analog spectrum analyzer or autocorrelator.

The local oscillator can also be frequency shifted by reflecting it from a moving mirror (Fig. 1d): a speed of 1 mm/s results in a shift of 3 kHz on a He-Ne beam. Stability in mirror motion then limits the resolution. With a servo mechanism, a stability better than 0.02% can be achieved, so that a resolution of 1Hz is possible with a shift of 5 kHz. The data shown below were all obtained by acousto-optic frequency shifting.

We are currently applying this technique to the study of hydrodynamic fluctuations in liquid-vapor interfaces.[2] Fig. 2 shows the spectrum of light scattered from the liquid-vapor interface of water at room temperature. The full Rayleigh-Brillouin triplet, centered around the 4.9 kHz frequency shift of the local oscillator, is visible. The central quasi-elastic Rayleigh scattering is due to nonpropagating fluctuations in the interface, whereas the Brillouin peaks at a shift of 1.2 kHz result from propagating fluctuations (capillary waves). Without frequency shifting the two Brillouin peaks merge into one single peak at 1.2 kHz (see position marked

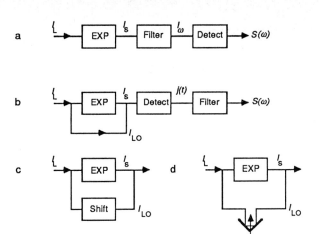

Fig. 1. Spectroscopy configurations: (a) frequency and (b-d) time domain detection techniques.

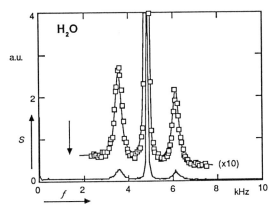

Fig. 2. Fully resolved Rayleigh-Brillouin triplet of light scattered from the liquid-vapor interface of water at room temperature.

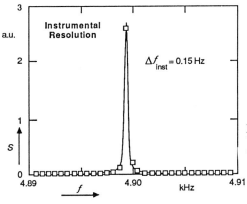

Fig. 3. Instrumental resolution obtained by replacing the sample with a mirror. The halfwidth of the Lorentzian fit is 150 mHz.

with arrow). A full discussion of these results will appear elsewhere.[3] Fig. 3 shows the instrumental resolution, obtained by replacing the interface with a fixed mirror. The resolution is roughly inversely proportional to the sampling time of the signal. For a single 1.5 s sampling the resolution is 150 mHz as shown. For multiple samplings some degradation in resolution was observed due to frequency drifts of the acousto-optic driver. For the same reason the resolution does not increase if the sampling time is increased beyond 1.5 s.

Summarizing, we present here a simple heterodyne technique with ultrahigh resolution. Because a spectral range up to 1 GHz can be covered, Fourier transform heterodyne spectroscopy is applicable to a wide variety of fields of research. The ultrahigh resolution makes the technique also suitable to measure extremely small Doppler shifts: the resolution of 150 mHz corresponds to a speed $v \approx 50$ nm/s. It is limited, however, only to *coherent* scattering processes, and the reported resolution is *relative*, not absolute.

References

1. H.Z. Cummins and H.L. Swinney, *Progress in Optics*, Vol. 8, Chapter 2 (North-Holland, Amsterdam, 1970)
2. See for instance: V.G. Levich, *Physiochemical Hydrodynamics* (Prentice-Hall, New Jersey, 1962); R. Loudon, in *Surface Excitations*, Ed. V.M. Agranovich and R. Loudon (Elsevier, Amsterdam, 1984) 589
3. Eric Mazur and Doo Soo Chung, Physica A *to be published* (December, 1987)

High Frequency Modulation Spectroscopy

T.F. Gallagher, C.B. Carlisle, G. Janik, H. Riris, and L.G. Wang

Department of Physics, University of Virginia,
Charlottesville, VA 22901, USA

Frequency modulation (FM) spectroscopy, first developed as a laser spectroscopy tool with a single mode cw dye laser by Bjorklund,[1] is a sensitive differential absorption technique. The essential idea is straightforward. If a laser is frequency modulated at a frequency far in excess of its bandwidth, the sidebands produced by the modulation are spectrally distinct from the laser carrier. If the modulated beam impinges on a square law detector, each of the sidebands beats with the carrier to produce a signal at the modulation frequency. However, the two beat notes are exactly out of phase and cancel. A small absorption of one of the sidebands destroys the cancellation, and a beat note at the modulation frequency appears. The sensitivity derives from the fact that there is no laser noise at the modulation frequency, since it far exceeds the laser bandwidth.

As a sensitive probe of absorption, FM spectroscopy is a promising method for several applications. Among these are sensitive long path measurements of trace gases in the atmosphere and the spectroscopy of nonradiating systems. Examples of the latter are vibrationally excited molecules and collisionally quenched systems. In many of the interesting cases the absorptions of interest have large widths, several GHz, due to Doppler and pressure broadening. Thus for FM spectroscopy to be effective, the modulation frequencies must be at least this high. A benefit of having higher modulation frequencies is that lasers of inherently larger bandwidth, such as pulsed dye lasers, can be used in FM spectroscopy.

A quick survey of commercially available modulators reveals that, above 1 GHz, none are available. In fact there is a good reason for this; useful electro-optic materials have indices of refraction for radio frequency and optical waves which differ by a factor of three, and the resulting phase velocity mismatch generally precludes the use of modulator crystals long enough to be effective modulators. To circumvent this problem we have used the dispersive properties of a waveguide near cutoff to increase the phase velocity of the radio frequency wave to match that of the optical wave, and we have made efficient resonant cavity modulators which operate at frequencies as high as 16 GHz.[2]

Using these modulators we have explored the use of pulsed lasers for FM spectroscopy, the attractions being the large wavelength range, the inherent time resolution, ruggedness, and the possibility of non-linear extension to other spectral ranges. In fact all of these have been realized. With a dye laser of 3 GHz bandwith and 5 ns pulse duration it has been possible to observe absorptions as small as 10^{-4}.[3] A component equally as critical as the high speed modulator is the fast photodetector, which was supplied to us by Hewlett-Packard.[4]

In recent experiments with a multimode cw laser, with an envelope of modes 6 GHz wide, we have found that the fast H-P diode could not withstand the requisite average power, and we had to resort to a different technique, two tone FM spectroscopy. Typically we modulate the laser at two frequencies, 10 GHz $\pm$ 6 Mhz, and detect a signal at 12 Mhz. This allows us to modulate at high frequencies, to detect broad absorptions, while using robust low frequency detectors.[5] This is especially critical in the infrared, where fast detectors are nonexistent. This technique is, in fact, quite sensitive; absorptions as small as 10^{-5} have been observed.[5]

It may be that the two tone technique will be most useful with GaAlAs and lead salt ir diode lasers, for it helps to alleviate two problems, the lack of fast ir detectors and the inherent amplitude modulation produced by modulating the diode laser injection current, the easiest way of frequency modulating such lasers.[6] Preliminary studies with both types of ir diode laser show sensitivities to absorptions smaller than 10^{-5}.[7,8]

This work has been supported by the Electric Power Research Institute.

References

1. G.C. Bjorklund: Opt. Lett. 5, 15 (1980)
2. N.H. Tran, T.F. Gallagher, J.P. Watjen, G. Janik, C.B. Carlisle: Appl. Opt. 24, 4282 (1985)
3. N.H. Tran, R. Kachru, P. Pillet, H.B. van Linden van den Heuvell, T.F. Gallagher, J.P. Watjen: Appl. Opt. 23, 1353 (1984)
4. S.Y. Wang, D.M. Bloom, D.M. Collins: Appl. Phys. Lett. 42, 190 (1983)
5. G.R. Janik, C.B. Carlisle, T.F. Gallagher: J. Opt. Soc. Am. B 3, 1070 (1986)
6. W. Lenth: Opt. Lett. 8, 575 (1983)
7. D.E. Cooper and J.P. Watjen: Opt. Lett. 11, 606 (1986)
8. L.G. Wang, H. Riris, C.B. Carlisle and T.F. Gallagher: Unpublished

394

Laser Sideband Spectroscopy of Molecules

G. Magerl[1], W. Schupita[1], J.M. Frye[1], R.H. Schwendeman[2], D. Peterson[2], Shin-Chu Hsu[2], Yit-Tsong Chen[3], and T. Oka[3]

[1]Technische Universität Wien, A-1040 Wien, Austria
[2]Michigan State University, East Lansing, MI 48824, USA
[3]The University of Chicago, Chicago, IL 60637, USA

Laser sideband spectroscopy is the spectroscopic use of frequency tunable laser sidebands generated by electrooptic mixing of laser and microwave radiation in a nonlinear crystal. With an input of typically 10 W each of laser and microwave power we can generate a few milliwatts of sideband power. The experimental setup has been described previously by MAGERL et al. /1/, and it has been used both at CO laser and CO_2 laser wavelengths. The advantages of the laser sideband method are twofold: First, absolute frequency determination is quite easy and accurate. It simply means to calculate sum or difference of Lamb-dip stabilized laser frequency and of synthesized microwave frequency. Second, because intensities are sufficient for nonlinear saturation we are able to take advantage of this frequency accuracy.

At Michigan State University we built a sideband system around a CO laser. Due to low laser power (a few hundred milliwatts) and due to problems with laser frequency stabilization we had to restrict ourselves to linear absorption spectroscopy to begin with. In the ν_3 fundamental of HCOOH we observed a sample of 10 rovibrational transitions. The spectrum around 1763 cm^{-1} is shown in Fig.1. A comparison of measured line positions with the calculations of WEBER and JOHNS /2/ gave a mean deviation of -6.3 MHz. This indicates the present accuracy in the 5 μm wavelength range.

At CO_2 laser wavelengths, however, we have sufficient laser output and accurate frequency stabilization on line center. Thus we are able to perform Lamb dip spectroscopy, as shown in Fig.2. The trace is part of the ν_5 $^RQ_0(J)$-branch of H_3SiI where the rotational structure is clearly resolved despite line separations well below the Doppler width. In addition, the quadrupole hyperfine structure of each line could be resolved which in fact made it possible to unambiguously assign the J-values and to calculate the quadrupole coupling constant in the excited state. A comparison of calculated splittings (FRYE et al. /3/) and of experimental data taken with the Vienna spectrometer leads to a mean deviation of ±20 kHz, nicely reflecting the present accuracy in the 10 μm wavelength range.

To demonstrate the full experimental potential of sideband spectroscopy we present samples of Stark- and Zeeman-spectra of C_2H_4 taken in Chicago: Figure 3 shows (a) the Stark splitting of $\nu_7, 0_{0,0} \leftarrow 1_{1,0}$ at an electric field strength of 46.82kV/cm and (b) the Zeeman spectrum of the same transition at a magnetic

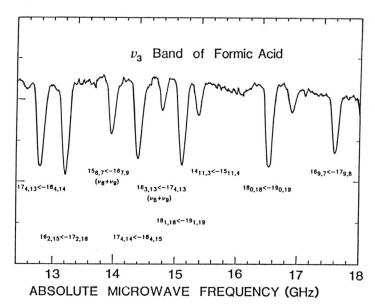

Fig.1 Doppler limited absorption spectrum taken with
CO sideband laser around 1763 cm^{-1}

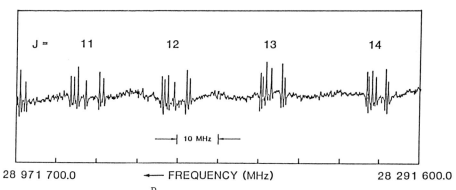

Fig.2 Part of the $\nu 5$ $^R Q_0(J)$-branch of SiH$_3$I with quadrupole
hyperfine splitting resolved

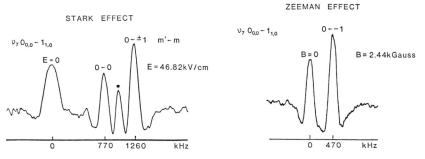

Fig.3 Stark and Zeeman splitting of C$_2$H$_4$ $\nu 7$ $0_{0,0} \leftarrow 1_{1,0}$

field of 0.244 Tesla. Both spectra were taken with the upper sideband of CO_2 laser line 10P(22) at a modulation frequency of 15.630 GHz.

1. G. Magerl et al.: IEEE J.Quantum El., QE-18, 1214 (1982)
2. W. Weber and J.W.C. Johns: J.Mol.Spectrosc., in print
3. J.M. Frye et al.: J.Mol.Spectrosc., in preparation

Sideband Saturation Spectroscopy with a Frequency-Modulated PbSnTe Diode Laser

Y. Ohshima[1], *Y. Matsumoto*[1], *M. Takami*[1], *and K. Kuchitsu*[2]

[1]The Institute of Physical and Chemical Research, Wako,
Saitama 351-01, Japan
[2]Department of Chemistry, Faculty of Science, The University of Tokyo,
Bunkyo-ku, Tokyo 113, Japan

This paper reports successful demonstration of sideband saturation spectroscopy (Doppler-free IR double resonance using modulation sidebands) by using a frequency-modulated PbSnTe diode laser. In previous such experiments [1], sidebands were generated by amplitude modulation of fixed frequency lasers. For applying this technique to atoms and molecules of general spectroscopic interest, it is of crucial importance to use a frequency tunable laser as an IR pump source.

Feasibility of saturating vibration-rotation lines with a tunable diode laser has been well established in IR-RF/MW double resonance [2]: Advantage of using a frequency tunable laser as a pump source has been demonstrated by a large number of RF and MW lines observed in the vibrationally ground and excited states of non-polar molecules of the Td and D3h symmetry. Our primary interest in the present technique is to determine small induced dipole moments in various kinds of non-polar molecules by measuring small Stark splittings.

A block diagram of experimental arrangement is shown in Fig.1. An RF modulation current in the frequency range of 0.15 to 2.0 MHz was coupled into a current source of the diode laser. The laser beam was focused into a pair of Stark electrodes of 2 mm spacing with the polarization of the laser beam perpendicular to the Stark field. A ramp voltage was applied to one of the electrodes for scanning Stark field and 30 kHz square waves to the other for modulation. The saturated IR absorption was detected and lock-in amplified. The sample pressure was maintained at a few mTorr to facilitate saturation of IR absorption lines with a small laser power of the order of 100 microwatts.

The saturation signal was observed first in several low-J lines of the ν_5 band of CF_3H. After optimizing the experimental conditions, the signals were observed in vibration-rotation lines of the ν_4 band of SnH_4. Figure 2 shows an example of the signal observed in the Q(16) E(3) line of $^{120}SnH_4$.

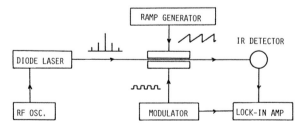

RAMP GENERATOR

IR DETECTOR

DIODE LASER

RF OSC.

MODULATOR

LOCK-IN AMP

Fig.1. Schematic diagram of experimental arrangement

In addition to an ordinary zero-field level crossing signal, two weak signals were observed in the scanning range of the Stark field. The double resonance effect was confirmed by a shift of the resonance voltage when the RF modulation frequency ν_m was changed. The two signals were assigned as those originated from first-order Stark splitting in the ν_4 and ground vibrational state when the sidebands of ν_m frequency difference were in resonance with the $\Delta M = \pm 1$ Stark components of the IR transition. When ν_m was reduced, a second resonance with the sidebands of $2\nu_m$ frequency difference was observed at twice the first resonance voltage. Similar signals were observed in 13 vibration-rotation lines for several isotopic species of SnH_4. A preliminary value for the rotationally-induced dipole moment was determined to be $\theta_z^{xy} = 4.3 \times 10^{-5}$ D in the ground vibrational state. This technique will be applicable to many atomic and molecular systems of fundamental importance.

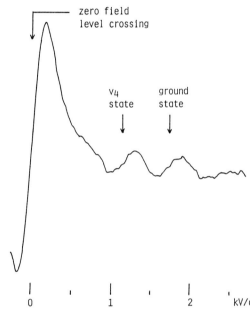

Fig.2. Saturation signal observed in the Q(16) E(3) line of $^{120}SnH_4$ ν_4 band. RF modulation frequency is 0.558 MHz. Stark modulation is 200 V/cm at 30 kHz.

[1]Permanent address: Department of Chemistry, Faculty of Science,
                   The University of Tokyo, Bunkyo-ku, Tokyo 113, Japan
[1] J.Orr and T.Oka; Appl. Phys. 21, 293(1980), and references cited therein.
[2] Y.Ohshima, Y.Matsumoto, M.Takami, and K.Kuchitsu; J. Chem. Phys. 85, 5519(1986), and references cited therein.

A New Method of Signal Recording and Averaging in Diode-Laser Spectroscopy

K. Uehara

Department of Physics, Faculty of Science and Technology,
Keio University, Hiyoshi, Kohoku-ku, Yokohama 223, Japan

1. Introduction

In spectroscopic measurements, signal averaging by repeated frequency sweeps of the light source is a powerful means of improving signal-to-noise ratios. It was recently shown that the operating voltage of a diode laser, when its drive current is held constant during temperature tuning, is a good monitor of the frequency /1/. This implies that signal averaging is possible by recording the spectra as a function of the operating voltage. The present paper reports on the first demonstration of signal averaging in diode-laser spectroscopy by this method.

2. Experimental

An InGaAsP distributed feedback laser(Hitachi HL1341) operating at 1.32 μm is mounted on a copper block, whose temperature is changed by a Peltier element. The dc drive current of the laser is held constant to within ±0.4 μA at 72 mA. The voltage across the laser is amplified and the memory address of a signal averager is advanced linearly with the voltage. A small modulation current at 22 kHz for frequency modulation is superimposed upon the dc current. The dc laser voltage is also used to control the current of the Peltier element to perform repeated automatic temperature sweeps in the desired region. The laser beam transmitted through a 50-cm long absorption cell which is filled with methane at 3 Torr is detected by a Ge photodiode. The photodiode output at the modulation frequency is detected by a lock-in amplifier and accumulated in the signal averager as a function of the laser voltage. When averaging is not necessary, the spectra can be recorded directly on an X-Y recorder.

3. Results

In Fig.1(a) the oscillation wavelength measured by a wavemeter(Advantest TQ-8325) is plotted against the laser voltage for a temperature change between 10°C and 50°C. The curve is smooth and its slope changes by a factor of only 2 between both ends. It means that precise wavelength interpolations by the laser voltage are possible. Figure 1(b) shows a derivative absorption spectrum of methane recorded as a function of the laser voltage by a single sweep. The $R(3)-R(5)$ lines of the $\nu_2 + 2\nu_3$ band are resolved into tetrahedral fine structure components. Repeated measurements showed that the reproducibility of the line positions is better than 20 μV in the laser voltage, which corresponds to 0.003 nm in wavelength.

The results of a signal averaging measurement in a narrow region around the R(4) line are illustrated in Fig.2. The signal-to-noise ratio increases with the number of sweeps, so that the weak lines become clearer by averaging as expected. The signal intensity normalized by the sweep number is nearly constant meaning that the reproducibility of the line positions is good enough for measurements of Doppler-limited resolution.

400

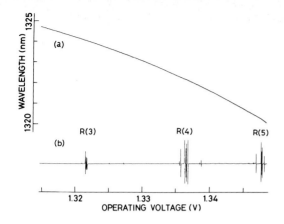

Fig.1 (a)Oscillation wavelength versus the laser voltage for a temperature change between 10 °C (right-hand side) and 50 °C (left-hand side). (b)Derivative absorption spectrum of methane recorded as a function of the laser voltage by a single sweep. Drive current: 72 mA

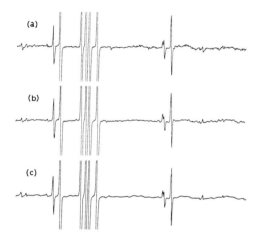

Fig.2 An unaveraged spectrum(a) and averaged spectra(b) and (c) after 4 and 16 sweeps, respectively, in a 0.9-nm wide region around the R(4) line. The signal intensity is normalized by the sweep number. Sweep time: 30 s/sweep

4. Discussions

The present method enables one to monitor the wavelength in real time. As demonstrated in Ref.1, the method can be also applied to multimode diode lasers to which conventional wavemeters can not be used. A new method of wavelength stabilization of diode lasers by controlling the temperature so as to give a constant laser voltage will be reported elsewhere /2/.

References
1. K.Uehara: Opt. Lett. 12, 81(1987)
2. K.Uehara and K.Katakura (in preparation)

Velocity-Modulated Laser Spectroscopy Using GaAlAs Diode Lasers

B. Lindgren, H. Martin, and U. Sassenberg

Institute of Physics, University of Stockholm,
Vanadisvägen 9, S-113 46 Stockholm, Sweden

GaAlAs diode lasers have proven to be useful and inexpensive light sources for molecular laser spectroscopy. As a part of the programme at this laboratory to investigate the characteristics of these lasers when used for high resolution molecular spectroscopy [1,2], we will now report their use in velocity-modulated laser spectroscopy.

This technique was developed by SAYKALLY and coworkers [3] mainly for infrared rotation-vibration studies of molecular ions, but has also been extended to electronic transitions of ions in the visible [4,5]. The ions studied were created in an ac glow discharge. In such a discharge the ion drift velocity along the axis of the discharge tube is reversed every half period of the alternating current. This gives rise to a Doppler shift of the line frequencies each half period in opposite directions. If a tunable laser detects the molecular absorption along the discharge tube, absorption by charged species would be modulated in phase with polarity while that due to neutrals would, at least to a first approximation, be unaffected. This gives an opportunity to detect the ion absorptions by lock-in technique at the frequency of the ac discharge, resulting in first derivative lineshapes.

In practice, neutrals are not unaffected by the ac discharge. The reason is that the concentration of the discharge products, as well as the population densities of all electronic states are also modulated by the discharge current. Unless the two half-periods of the ac discharge are exactly equivalent, absorptions from the neutrals will show up at the frequency of the ac discharge. If one is successful as to exactly match the two half periods of the discharge current, then the neutrals will be completely suppressed in the ion spectrum and will only be seen at twice this modulation frequency. Lock-in technique at this higher frequency can simultaneously with the ion detection be used to detect absorption from neutral species.

The experimental set-up is shown in the block-diagram of Fig. 1. The diode laser (Hitachi HL 7801/7802) can be scanned up to 130 Å by a temperature-regulating device but only in small sections of 1 to 5 Å between the mode jumps. The laser output at room temperature is centered around 7850 Å in the vicinity of which the (2,0) band of the A - X transition of N_2^+ (Meinel bands) would be a good testband. The discharge conditions were rather similar to those given by SAYKALLY et al. [4,5], 4 kV ac at 830 Hz was exciting a mixture of 500 mTorr nitrogen and 9 Torr helium.

As can be seen in Fig. 2 the suppression of the neutral spectrum, mainly consisting of the (7,6) band of the "1st positive system" (B - A) of N_2, is at least 50 times in the lower trace. We estimate that with our experimental conditions absorptions as weak as $1:10^6$ can be detected.

The frequent mode jumps of the diode lasers may of course be regarded as a drawback in their usefulness, but we have found that a selected number of lasers may cover a desired spectral region. Besides the mode jumps there are also regions of mode competition, where the lasers are unreliable. To watch such irregularities it is

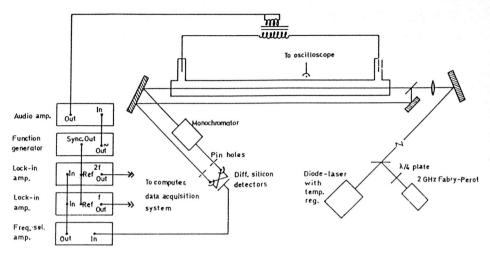

Fig. 1. Block-diagram of a velocity-modulated laser spectrometer

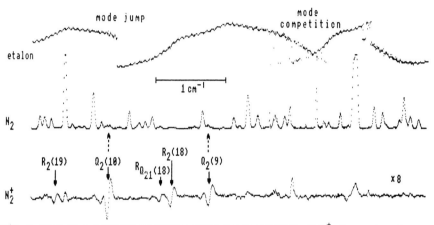

Fig. 2. Part of a velocity-modulated diode laser spectrum of N_2^+ (lower trace) and simultaneous recording of neutral absorption of N_2 (middle trace) and the interferometric control of the laser output (upper trace)

necessary to follow the output of the laser with an interferometer and a reference spectrum. Both mode jumps and mode competition regions can be seen in Fig. 2. In the present case the spectrum of the neutrals also serves as a reference spectrum. We observed that the N_2 measurements given by DIEKE and HEATH [6] had to be reduced by 0.02 cm^{-1} to be in accordance with the N_2^+ measurements of BENESCH et al. [7].

References:
1. T. Gustavsson and H. Martin: Rev. Sci. Instrum. 57, 1132 (1986)
2. T. Gustavsson and H. Martin: Physica Scripta 34, 207 (1986)
3. C.S. Gudeman and R.J. Saykally: Ann. Rev. Phys. Chem. 35, 387 (1984)
4. C.S. Gudeman, C.C. Martner and R.J. Saykally: Chem. Phys. Lett. 122, 108 (1965)
5. M.B. Radunsky and R.J. Saykally: J. Chem. Phys. (in press)
6. G.H. Dieke and D.F. Heath: Johns Hopkins Spectroscopic Report Number 17 (Baltimore 1959)
7. W. Benesch, D. Rivers and J. Moore: J. Opt. Soc. Am. 70, 792 (1980)

403

A High Sensitivity Modulation Method for Atomic Beam Absorption Spectroscopy

C. Salomon, H. Metcalf*, A. Aspect, and J. Dalibard

Laboratoire de Spectroscopie Hertzienne de l'ENS (UA 18) et
Collège de France, 24 rue Lhomond, F-75005 Paris Cedex 05, France

In many experiments on absorption spectroscopy [1-3], modulation techniques are required to extract a weak signal out of large low frequency noise, for instance the technical (amplitude) noise of the laser light. Here we present a very simple (and cheap) technique, which uses optical pumping for modulating the ground state population of cesium atoms. Owing to the relatively small amplitude noise of diode lasers at frequency above $\sim$ 10 kHz we have been able to detect the absorption of as few as 20 atoms on average in our 2 mm^3 interaction volume.

The outline of the apparatus is presented in Fig. 1. A cesium effusive beam is collimated to about 1 mrad. The residual transverse Doppler effect is negligible compared to the natural width of the excited state 6 $P_{3/2}$ (5 MHz). A "chopping" laser diode provides a modulation of the population of the F = 4 ground state. We then detect in phase with this modulation the absorption of a "probe" laser diode by the atomic beam, on the g,F=4$\rightarrow$e,F=5 cycling transition at 852 nm. Note that just outside the oven, all atoms are initially prepumped in the F=4 ground state by a third "prepumping" laser diode.

The chopping diode is square-wave modulated between the g,F=4$\rightarrow$e,F=4 transition and a non resonant frequency 2 GHz away.

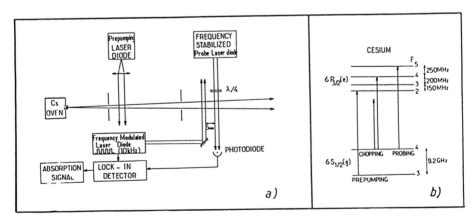

Fig. 1 : a) Experimental set-up. b) Relevant Cs energy levels.

Hence the population in g,F=4 is alternatively nearly 100 %
emptied or unchanged. This is done at about 10 kHz. This
frequency is limited by both the distance between the "chopping"
and "probing" laser beams (5 mm) and the thermal longitudinal
velocity distribution of the atomic beam. The interaction volume
with the Cesium beam, delimited by geometrical optics is 2 mm
(along the laser beam) × 1 mm². The probe laser diode is
stabilized to less than 1 MHz by optical feedback from a F.P.
resonator using the technique of [4].

Absorption spectra are presented in Fig. 2 with about
10000 (a) and 20 (b) atoms in average in the interaction volume.
The atomic density is measured using a hot wire detector and
agrees within a factor 2 with the density calculated from the
oven temperature and the geometry of the experiment.

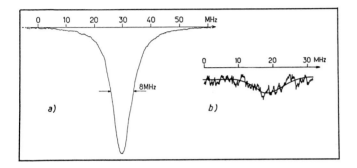

Fig. 2 : Absorption spectra. Probe power 20 μW. a) 10 000 atoms.
τ = 30 ms. b) 20 atoms. τ = 1s. Relative absorption 4 × 10⁻⁶.

References

[1] G.C.BJORKLUND, Opt. Lett. 5, 15 (1980)
[2] J.BIALAS,R.BLATT,W.NEUHAUSER,P.TOSCHEK,Opt. Com. 59,27 (1986)
[3] D.J.WINELAND,W.M.ITANO and J.C.BERGQUIST,Opt. Lett.
    12, 389 (1987).
[4] B.DAHMANI,L.HOLLBERG,R.DRULLINGER,to be published in Opt.
    Lett. (1987) and CLEO,IQEC (1987)
*   Permanent address : Department of Physics, State University
    of New York, Stony Brook, NY 11790 U.S.A.

The Inverse Hook Method
for Measuring Oscillator Strengths

W.A. van Wijngaarden, K.D. Bonin, and W. Happer

Physics Department, Princeton University, Princeton, NJ 08544, USA

We describe an effective new method to measure oscillator strengths for transitions between the excited states of atoms. The oscillator strength is determined by measuring changes in the angular distribution or polarization of fluorescent light emitted by atoms in the initial or final state of the transition of interest, after these atoms have been subject to the a.c. Stark shift of an off-resonant laser pulse. The physics of the situation is very similar to that of the conventional hook method with this difference: the roles of the atoms and the photons have been interchanged. We therefore call this new method The Inverse Hook Method. The inverse hook method is relatively insensitive to the details of the atomic absorption lineshape and also to the temporal and spectral profile of the laser pulse. It yields absolute oscillator strengths and it is especially suitable for measurements of transitions between excited atomic states, including autoionizing states.

High Precision Measurements of Vapor Densities

W.T. Hill, III

Institute for Physical Science and Technology, University of Maryland,
College Park, MD 20742, USA

Abstract

A column-density meter for a highly precise and accurate measurement of
static and transient populations is presented.

Discussion

Anomalous dispersion due to resonant absorption, provides an accurate way
to measure the column density (the product of the number density, N, and
the pathlength, ℓ) and/or the oscillator strength (f) of a medium. The
anomalous dispersion can be probed interferometrically by placing the
absorbing medium in one arm of a low finesse interferometer (e.g., a Mach
Zehnder arrangement), which is illuminated by an appropriate light source.
The distorted interference fringe pattern is then imaged onto photographic
film placed in the exit plane of a stigmatic spectrograph. Traditionally,
$N f \ell$ is estimated by one of the usual hook methods with a precision and accuracy of 5 to 10% when a medium resolution spectrograph (e.g., 1 m in
first order) is used.[1] Recently, it was shown that when the photographic
film is replaced by a photodiode array (25 mm long with 25 µm center-to-
center spacing) it is possible to determine $N f \ell$ to higher precision ($\leqslant$ 1%)
and accuracy ($\leqslant$ 2%).[2]

The basic column-density meter, described in [2] and shown schematically
by the solid traces in Fig. 1, employs a broadband light source and a single
photodiode array. $N f \ell$ is determined numerically by fitting the distorted
fringe profiles (see Fig. 2) to the expression

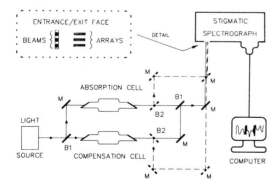

Fig. 1 Schematic of the
column-density meter based
on a Mach Zehnder inter-
ferometer; B1 and B2 represent pairs of matched beam
splitters and M represents
100% reflectors; the dashed
components make simultaneous
measurement of the three
intensities possible when
three diode arrays are
mounted on the spectrograph
(see text)

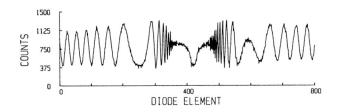

Fig. 2. Typical inter-
ference fringe profile
captured by a photodi-
ode array

$$\cos\theta(\lambda) = \frac{I_0(\lambda) - I_1(\lambda) - I_2(\lambda)}{\sqrt{I_1(\lambda)I_2(\lambda)}} \quad . \tag{1}$$

I_0 describes the fringe pattern shown in Fig. 2, I_1 is the intensity of the light passing through the interferometer arm containing the absorption cell (of length ℓ) and I_2 describes intensity in the other arm. For fringes in the wings of a single absorption line, the phase difference, $\theta(\lambda)$, between the waves is given by

$$\theta(\lambda) = 2\pi\left[\frac{r_0}{4\pi}N\ell\frac{\lambda_0^3}{\lambda-\lambda_0} \pm \epsilon\right]\frac{1}{\lambda} \quad , \tag{2}$$

with ϵ being the pathlength difference between the two arms in the absence of resonant absorption and r_0 the classical electron radius.

The basic density meter, which has been used successfully to measure populations of several atomic vapors (see [2]), has the limitation that I_0, I_1
and I_2 cannot be determined simultaneously. As a result, only static populations can be measured. A more complicated version, which includes the
dashed components in Fig. 1, should be capable of simultaneous measurement
of I_0, I_1 and I_2. Consequently, transient populations could be measured
with high precision and accuracy. Furthermore, to measure very small $N\ell$
values, which produce correspondingly small fringe distortions, a single-
mode tunable laser should replace the broadband light source and the
spectrograph.

This new photometric, fringe profile approach to measuring $N\ell$ is inherently more accurate than the more ususal photographic (hook) methods because the entire fringe profile is used in the determination. The fringe
profile measurement and the numerical reduction require very little input
from the operator and thus yield almost instantaneous $N\ell$ values.

Acknowledgments

This work was supported by the National Science Foundation (grants RII-
8406192 and PHY-8451284) and the Baltimore Gas and Electric Company.

References

1. See M.C.E. Huber and R.J. Sandeman: Rep. Prog. Phys. 49, 397 (1986)
   and references therein.
2. W.T. Hill, III: Appl. Opt. 25, 4476 (1986).

An Atomic Resonance Filter Operating at Fraunhofer Wavelengths

J.A. Gelbwachs

Chemistry and Physics Laboratory, The Aerospace Corporation,
Los Angeles, CA 90009, USA

The atomic resonance filter is an isotropic, wide field-of-view, incoherent, survivable device with an ultranarrow optical bandwidth, namely, ~ 0.01Å in the visible. These properties make it ideal for the detection of a weak laser signal against a continuum background. Applications include open channel laser communications, lidar, and laser radar. We have developed the first atomic filter that exactly matches three intense Fraunhofer lines, i.e., wavelengths at which the solar background is reduced to less than 10% of its continuum value. A high throughput and a fast response time are other important advantages of our filter.

The atomic resonance filter operates by absorbing signal light that corresponds to a transition in an atomic vapor and then emitting light in a different spectral region. Our filter operates at three Fraunhofer lines near 517 nm that arise due to a strong absorption originating from the lowest lying triplet level (metastable level) of Mg vapor in the outer layer of the sun.

Not too surprisingly, our filter is based upon the spectroscopy of atomic Mg. A partial energy level diagram of Mg is shown in Fig. 1. To permit absorption of green light, the Mg vapor is pumped by a laser tuned to the 457 nm intercombination line. Mixing in the triplet level lengthens the effective lifetime ($\tau \sim 6$ msec) so that the transition can be saturated by a modest 4 W/cm^2. The metastable nature of the triplet level contributes to optimum filter performance in three ways. First, it permits pulsed pumping at a 200 Hz rate as well as cw pumping. Secondly, compared to resonance-line pumped active filters, the power extracted from the pumping laser and radiated by the vapor is greatly reduced. Thirdly, filter noise generated by binary collisions is minimized.

Green Fraunhofer wavelength photons are effectively absorbed by the Mg atoms because of the high oscillator strength of the $3p\,^3P^0$ – $4s\,^3S$ transition. To shift the wavelength, a second low-power cw laser with emission at 1.50 μm is provided. Photons from this source excite the Mg atoms to the $4p\,^3P^0$ level. A buffer gas contained within the vapor cell then collisionally transfers population into the nearby $3d\,^3D$ level. The excited vapor than rapidly relaxes to the metastable level with emission at 383 nm, an optimal wavelength for detection by commercial, large area, long-lived PMTs.

By adjusting the IR pump intensity and the buffer gas pressure, the conversion of absorbed green light into UV photons can be made highly

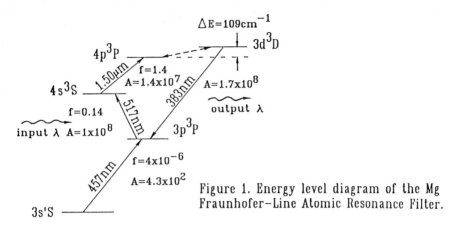

$\Delta E = 109 \text{cm}^{-1}$

$3d^3D$

$4p^3P$

$1.50\mu m$

$f=1.4$

$A=1.4 \times 10^7$

$4s^3S$

$517nm$

$383nm$

$A=1.7 \times 10^8$

output λ

$f=0.14$

input λ $A=1 \times 10^8$

$3p^3P$

$457nm$

$f=4 \times 10^{-6}$

$A=4.3 \times 10^2$

Figure 1. Energy level diagram of the Mg Fraunhofer–Line Atomic Resonance Filter.

$3s'S$

efficient, namely, 90%. This occurs because of the very favorable 15 to 3 degeneracy ratio combined with the faster decay rate of $3d^3D$ level compared to the $4s^3S$ level. Based upon the spontaneous emission rate of the emitting level, a response time of less than 10nsec is predicted.

Development of an Optically Pumped Polarized Deuterium Target

L. Young, R.J. Holt, M.C. Green, and R. Kowalczyk

Argonne National Laboratory, Argonne, IL 60439, USA

The development of a polarized deuterium target for internal use at an electron storage ring is of great interest for fundamental studies in nuclear physics.[1] In order to achieve the maximum allowable target thickness, 10^{14} nuclei/cm^2, consistent with various constraints imposed by the storage ring environment, a flux of 4 x 10^{17} polarized atom/s must be provided. This flux exceeds the capability of conventional atomic beam sources by an order of magnitude.[2] We have been developing an alternative source based upon the spin-exchange optical pumping method in which the flux is limited only by laser power.

In this method, the nuclear polarization of atoms with unpaired electrons (e.g. H,D) is achieved by successive spin-exchange collisions with polarized alkali atoms. The electron spin of the atom of interest is initially polarized by the spin-exchange process. Subsequently the electron spin polarization is transferred to the nucleus via the hyperfine interaction. In our case, for a given alkali atom polarization, approximately 5 spin-exchange collisions ($\sigma_{SE} \sim 10^{-14}$ cm^2) are required to obtain the maximum polarization of the deuteron. [1]

This requirement of several spin exchange scatterings dictates the design of suitable containment vessel in which the "reactants", unpolarized D atoms and polarized alkali atoms, can successfully interact multiple times. The additional constraints imposed by the strict vacuum requirements of the storage ring environment dictates that the interaction occur at relatively low pressure (< 10^{14} atoms/cm^3). As a result of these constraints, two technical· obstacles arise. Firstly, at these low number densities, the walls of the container play a very important role since wall collisions occur at least 100 times more often than spin exchange collisions. [1] Therefore, the surfaces must inhibit depolarization of both the alkali and the D atom as well as D atom recombination. Secondly, the optical pumping photon beam must cover the entire doppler profile (~ 1.0 GHz) of the alkali, since no buffer gas is allowed to induce velocity changing collisions.[3]

To address the first of these obstacles, we have investigated the properties of a dri-film surface [4,5], suggested to us by Swenson and Anderson [6]. At low potassium density, the surface appears to be long lasting (> 100 hrs) and to have a spin relaxation probability/collision for potassium of $\sim 1/500$. The second issue, that of doppler coverage, we have addressed by using two standing wave lasers (CR-599), operating on three adjacent longitudinal modes. In addition, the fold mirror is mounted one a fast PZT (35 kHz) to smear out the laser spectrum.

The polarization of the K, H, or D atoms can be measured by optical detection of magnetic resonance transitions. In this technique, optical

411

pumping of the alkali in a weak B field produces a polarized ensemble with very unequal populations in the magnetic sublevels. When the alkali spin polarization is decreased by applying a radio-frequency field at a Zeeman transition frequency, the transparency of the sample to the optical pumping light is decreased. This can be monitored also as an increase in fluorescence from the sample. By scanning the applied RF frequency and observing the increased fluorescence, the relative populations in the magnetic sublevels can be deduced using a spin temperature model. [7] The polarization of the spin exchange partner, H or D, can also be detected since it communicates with the alkali through spin-exchange. The detection is made simpler by modulating the amplitude of the applied RF and use of a lock-in amplifier to selectively detect only the increased fluoresence occuring at the modulation frequency. The apparatus is shown in Fig. 1.

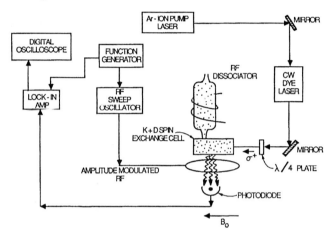

Fig. 1 Schematic diagram of the experimental set-up to observe polarization of K, H, or D in a flow system.

The best result we have obtained to date has been a polarization of 10% with a flux of 5.7×10^{16} H atoms/s. These results were obtained using a dri-film surface at $\sim 300°C$. We appear to be limited by the presence of K on the surface of the spin exchange cell at high alkali densities. We are presently investigating low K-density cells in order to minimize the surface contamination.

This research was supported by the U.S. Department of Energy, Office of Basic Energy Sciences, under contract W-31-109-ENG-38.

References
1. R. J. Holt, Proc. of the Workshop on Nuclear Physics with the Use of Electron Storage Rings, Argonne (1984) ANL-84-50, p. 103.
2. W. Grœbler, ibid 1, p. 223.
3. P. G. Pappas et al., Phys. Rev. Lett. 47, 236 (1986).
4. G. E. Thomas et al., Proc. of The Thirteenth World Conference on Nuclear Target Development, Chalk River, Sept. 1986. To be published in Nucl. Inst. Meth.
5. L. Young et al., Nucl. Inst. Meth. B24/25, 963 (1987).
6. D. R. Swenson and L. W. Anderson, Nucl. Inst. Meth. B12, 157 (1985); L. W. Anderson and D. R. Swenson, private communication.
7. L. W. Anderson and A. T. Ramsey, Phys. Rev. 132, 712 (1963).

Diagnostics of a Cooling Electron Beam by Thomson Scattering of Laser Light

J. Berger[1], P. Blatt[1], P. Hauck[1], W. Meyer[1], R. Neumann[1], C. Habfast[2], H. Poth[2], B. Seligmann[2], and A. Wolf[2]

[1]Physikalisches Institut der Universität Heidelberg,
  Philosophenweg 12, D-6900 Heidelberg, Fed. Rep. of Germany
[2]Kernforschungszentrum Karlsruhe, Institut für Kernphysik,
  D-7500 Karlsruhe, Fed. Rep. of Germany

The method of Thomson scattering frequently used as a diagnostic tool for plasmas /1/ and in few cases for very dense electron beams /2,3,4/, was applied for the first time to a much less dense cooling electron beam. The measurements performed at the electron cooling device for LEAR (Low Energy Antiproton Ring) at CERN demonstrate that it is feasible to determine the absolute energy and the longitudinal velocity distribution (temperature) of an electron beam with a density of less than 10^8 cm^{-3}.

The experimental setup (see Fig.1) consists of the electron cooler /5/, a dye laser pumped by an excimer laser, and an optical analyzing system for the Thomson-scattered photons, equipped with a photon counting electronic device. The linearly polarized light pulses (490nm, 20nsec, 20mJ, 15Hz rep.rate) from the dye laser enter and leave the electron cooling assembly through Brewster windows. The laser beam (Ø2mm) is aligned parallel to the electron beam axis and can be scanned over the cross-section of the electron beam by means of a movable mirror. The laser power is monitored with a photodiode (PD) behind the cooler. Photons backscattered at about 180° from the electron beam (28keV, 2.6A, Ø5cm) are deflected out of the vacuum system, pass a Fabry-Perot interferometer (FPI) for spectral resolution and are focused on a photomutiplier (PM), protected by two wide-band filters against the primary laser light.

The spectra of the Doppler-shifted Thomson photons were measured by tuning the wavelength of the incident laser light, and only photons coincident with the laser pulses were counted. The measurements were performed at different acceleration voltage values of the electron beam (see Fig.2). The shift of the peak position of the spectra produced by the voltage change is in agreement with our calculations /6/, and confirms the correct identification of Thomson-scattered photons. Also, the signal amplitude has the expected order of magnitude. The width of the curves (0.05nm FWHM), dominated by the resolution of the FPI, allows to give an upper bound for the longitudinal electron energy spread ($<10^{-3}$eV). Taking into account the transverse electron energy distribution (0.1 eV) given by the temperature of the cathode (1100 °C), the ratio of longitudinal to transverse electron temperature amounts to $<10^{-2}$. This implies the first direct measurement of the flattened velocity distribution caused by the acceleration process. The absolute electron energy

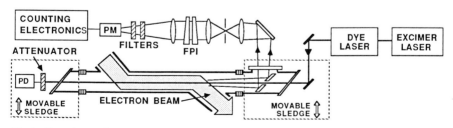

Fig.1: Experimental setup

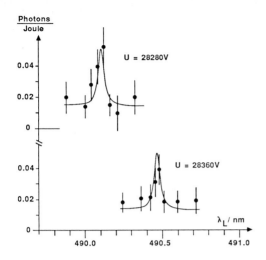

Fig.2: Thomson-scattered photon rate as a function of laser wavelength in the centre of the electron beam for two values of the acceleration voltage U. Least-square fit of Lorentzian lineshape with common width, amplitude and offset. Each point represents an incident laser energy of about 400J accumulated in ≈ 1/2 h.

correlated with the peak position of the signals via the Doppler shift, could be determined in the present experiment with a precision of about 10^{-3}. The measured electron energy indicates a space-charge compensation of about 13% due to positive ions in the rest gas. Data taken by scanning the laser beam across the electron beam are in agreement with the assumption of a parabolic velocity profile and an electron density, uniformly distributed across the beam radius.

A considerable increase of the average laser power is desirable for further diagnostic experiments under comparable conditions in order to get higher scattering rates and thus gain on-line information about the electron beam parameters. This could either be realized by using lasers with high repetition rate, being available now or by means of laser pulse storage techniques /7/.

Acknowledgement: This work was supported by the German Bundesministerium für Forschung und Technologie.

1. D.E. Evans: Physica 82C, 27 (1976)
2. G. Fiocco and E. Thompson: Phys. Rev. Lett. 10, 89 (1963)
3. L.E. Dolotov, O.V. Zyuryukina, A.A. Kolotyrin, A.P. Solov´ev and B.G. Tsikin: Sov. J. Quantum Electron. 12, 624 (1982)
4. V.A. Zhuravlev, O.V. Karpov, V.V. Mikheev, V.E. Muzalevskii and G.D. Petrov: Opt. Spectrosc. 57, 997 (1984)
5. C. Habfast, H. Poth, B. Seligmann, H. Haseroth, C.E. Hill, J.-L. Vallet, and A. Wolf: Status and perspectives of the electron cooling device under construction at CERN, Proc. Third LEAR Workshop on Physics with Cooled Low-Energy Antiprotons in the ACOL Era, Tignes, 1985 (eds. U. Gastaldi, R. Klapisch, J.-M. Richard, and J. Tran Thanh Van) (Editions Frontieres, Gif-sur-Yvette, 1986), p.129
6. C. Habfast, H. Poth, B. Seligmann, A. Wolf, J. Berger, P. Blatt, P. Hauck, W. Meyer and R. Neumann: Submitted to Appl. Phys. B
7. J. Berger, P. Blatt, P. Hauck and R. Neumann: Opt. Commun. 59, 255 (1986)

414

Part XI

Spectroscopic Sources

Diode Pumped Solid-State Laser Oscillators for Spectroscopic Applications

R.L. Byer[1], S. Basu[1], T.Y. Fan[1], W.J. Kozlovsky[1], C.D. Nabors[1], A. Nilsson[1], and G. Huber[2]

[1]Department of Applied Physics, Stanford University,
  Stanford, CA 94305, USA
[2]Institut für Angewandte Physik, Universität Hamburg,
  D-2000 Hamburg 36, Fed. Rep. of Germany

1. Introduction

Solid state laser development has been paced by the improvement of pumping sources. From the helical lamps used to pump the early Ruby lasers, to the linear arc and pulsed flashlamps used to pump Nd:YAG lasers, solid state laser pump sources have improved steadily in power and efficiency. The latest development is pumping solid state lasers with diode lasers and diode laser arrays. The development of high power, efficient, long lived diode lasers promises a revolution in solid state laser technology.

The advantages of diode laser pumping over flashlamp pumping include greater spectral brightness in the pump bands of the ions, greater efficiency compared to flashlamps, longer operating lifetime, and operation without high voltage. At this time the disadvantages are the low power of the diode lasers and the cost per watt of output power. However, the cost of diode lasers is decreasing so rapidly that diode laser pumping should be economically favored over flashlamp pumping for 1 Watt average power Nd:YAG lasers by 1989, and for 10 to 100 Watt Nd:YAG lasers by 1991.

The diode laser is inherently a cw device. The long upper state lifetime of ion levels in solid state laser materials permits energy storage which leads to potential advantages for diode laser pumping of ions. These advantages include Q-switched and modelocked operation; the possibility of extracting high peak and average power from the solid state laser medium in a diffraction limited spatial mode by summing the output of many diode laser pump sources; the ability to operate on many ion wavelengths; and the potential to convert to additional wavelengths by nonlinear optical processes. In addition, the low intrinsic loss of the crystalline medium relative to the semiconductor diode laser medium leads to a potential reduction in the Schawlow Townes linewidth of six orders of magnitude. Diode laser pumping of solid state lasers increases the power spectral brightness of the diode laser pump source without a significant loss in overall efficiency.

Advances in diode laser pumping of solid state lasers has been reviewed recently [1] as has the progress in diode lasers and diode laser arrays[2]. It is worth noting that the first suggestion for diode pumping was by Newman in 1963 who demonstrated fluorescence from the neodymium ion doped in $CaWO_4$[3]. Considerable progress in the 1970's and in the early 1980's coupled with the recent rapid progress in the power and efficiency of diode lasers has led to renewed interest in this approach to solid state laser engineering.

2. Diode Laser Pumped Four Level Lasers

In 1985, Zhou et. al.[4] demonstrated efficient, frequency stable, laser diode pumped Nd:YAG laser operation. The end pumped configuration used a gradient index lens to focus the diode laser radiation into the 5 mm long monolithic Nd:YAG rod. The 2 mW threshold and 25% slope efficiency demonstrated the effectiveness of the end pumping approach. Further, the 10 kHz linewidth showed promise of achieving the expected Shawlow Townes linewidth for this laser of less than 1 Hz-mW. However, spatial hole burning limited single axial mode operation to sub-milliwatt power levels in the standing wave device.

This early work was extended by Sipes who demonstrated 6% wall plug efficiency for diode laser pumping of Nd:YAG[5]. In addition, diode laser pumping of Nd:YLF with internal SHG to generate 532nm was demonstrated by Fan et. al.[6] and by Baer[7]. The lasing ion and the nonlinear material for frequency doubling were combined in diode laser pumped Nd:MgO:LiNbO$_3$. This self-doubling and self-Q-switching laser oscillator operated at 3.6mW threshold with a 39% slope efficiency[8]. Recently, diode laser pumping was extended to Nd:Glass with a surprisingly low threshold of 2.5mW at a slope efficiency of 42%[9]. The low loss of the glass host led to the low threshold in spite of a factor of ten less gain cross section for Nd:Glass compared to Nd:YAG. Nd:Glass operation offered the advantages of broad absorption bands for diode pumping, wide spectral bandwidth for modelocked operation, and high doping levels for side pumping with diode laser arrays.

3. Diode Laser Pumped Three Level Laser Systems

Many of the potential useful ions for near infrared laser operation are three level systems with residual population in the lower level. There was skepticism about whether these laser systems could be pumped by low power diode lasers. Fan and Byer[10] showed theoretically that under proper conditions diode laser pumping was possible. These conditions were realized with the successful diode laser pumping of the 946 nm transition in Nd:YAG[11] which was subsequently frequency doubled to generate blue radiation at 473 nm[12].

The eyesafe requirement of many laser radar applications including global remote wind sensing, led to the search for lasers that operate at wavelengths longer than 1400 nm. Following early work by Duczynski et. al. [13], Fan et. al.[14] successfully demonstrated cw, diode laser pumped, room temperature operation, of the three level laser Tm:Ho:YAG at 2090nm. Diode laser radiation was absorbed at 781nm by theTm ion with subsequent transfer to the holmium ion which oscillated at 2090nm. This transfer laser should allow one absorbed pump photon to generate up to two inverted holmium ions for a potential pump quantum efficiency of two. A threshold of 4.4mW with slope efficiency of 19% were observed in early diode laser pumping experiments. This laser system is more complex than Nd:YAG and requires additional research to optimize the doping concentrations. However, the 8 msec upper level energy storage time of the holmium ion makes possible extraction of high peak power pulses by Q-switching.

4. Q-Switched and Modelocked Operation

The potential to store energy and extract high peak power pulses is a significant advantage for diode laser pumping of crystal and glass lasers. Both Nd:YLF and Nd:YAG have been Q-switched. The cw-pumped laser oscillators were acousto-optically Q-switched at kilohertz repetition rates with peak power levels of greater than 2.5 kW for Nd:YLF and 1.8 kW for Nd:YAG[15]. Q-switched pulse widths of 10 nsec were observed for these short cavity, compact, laser diode pumped oscillators. The higher peak power and larger pulse energy of Nd:YLF compared to Nd:YAG reflected the longer storage time of the Nd:YLF system. The kilowatt peak powers of these lasers allows direct external frequency doubling.

Diode laser pumping of Nd:Glass with active modelocking to achieve pulses of 45psec width was reported by Basu and Byer[16]. The laser oscillator utilized a 3 mm thick Brewster plate of Nd:Glass in an astigmatically compensated three mirror folded resonator similar to the early cw dye laser cavity configuration. The 8 mW diode laser pump threshold of this actively modelocked Nd:Glass laser allowed pumping by a single stripe 30 mW diode laser. Modelocked operation was extended to Nd:YAG which operated at a 90 psec pulsewidth.

5. Single Axial Mode Non-Planar Ring Oscillator

Spatial hole burning in the standing wave monolithic diode laser pumped oscillators led to multi-axial mode operation at powers above the milliwatt level. The desire to maintain the inherent stablility of the monolithic structure and achieve single axial mode operation at high power levels led to the consideration of a ring oscillator. To eliminate spatial hole burning and

417

to assure operation in a single direction, the ring oscillator should contain the three elements of an optical diode in its monolithic structure. These elements are a Faraday rotation, a wave plate, and a polarizer. Kane and Byer[17] proposed and demonstrated the monolithic, unidirectional, single axial mode, non-planar ring oscillator to meet these requirements. The key element in this invention is the breaking of planar symmetry to achieve the combination of Faraday rotation, waveplate and polarizer within the monolithic structure. The nonplanar ring oscillator has operated at up to 163 mW of single axial mode output when argon ion laser pumped, and up to 25 mW of single axial mode power when diode laser pumped. Using single mode fibers to combine the radiation of an NPRO with a single axial mode standing wave oscillator, Kane et. al.[18] have demonstrated that the device operates with free running linewidths of less than 10 kHz.

The advantages of the nonplanar ring oscillator include single frequency operation with frequency stability due to the monolithic construction and immunity to feedback. This last factor is critical in many applications of the NPRO and is an inherent property of the nonplanar ring design[19].

The Nd:YAG nonplanar ring oscillator offers a narrower linewidth and improved frequency stablity compared to the diode laser. For example, the Nd:YAG NPRO with scatter loss of 0.001 cm^{-1} has a projected Schawlow Townes linewidth of 0.3 Hz-mW for a one centimeter diameter ring oscillator. Work is in progress to reduce the technical noise dominated 10 kHz linewidth of the NPRO to the subHertz Shawlow Townes linewidth by using power stable diode pump sources and by using rapid feedback frequency control through an applied magnetic field to offset the slow temperature tuning of -3.1 GHz per degree.

The stable, single axial mode, feedback resistant output of the Nd:YAG nonplanar ring oscillator has allowed efficient SHG by external resonance doubling in LiNbO$_3$[20]. In the early experiments, 15mW of cw 1064nm input was doubled to generated 3mW of single axial mode green output at 15% SHG efficiency. With improvements in the optical coatings, SHG efficiencies approaching 50% are predicted.

6. Amplification

In many applications high peak power radiation is desired but at linewidths that remain single frequency within the Fourier transform limit of the sampling time. We have investigated linear amplification of milliwatt power level sources using a multipass, flashlamp pumped, slab geometry Nd:YAG configuration[21]. The measured 62 dB gain for a 30 microsecond duration, was limited by the onset of superfluorescence. The linear amplification approach to generate high peak power levels offers an alternative to Q-switching. The advantages of linear amplification are the ability to select the pulse width and thus the Fourier Transform linewidth of the output radiation independent of Q-switch dynamics.. The nonplanar ring oscillator, Nd:YAG slab amplifier, and single mode optical fiber for heterodyne mixing was used for a successful demonstration of coherent laser radar at 1064nm[22].

7. Summary

The rapid improvement in diode laser pump sources has led to the recent progress in diode laser pumped solid state lasers. To date electrical efficiencies of greater than 10% have been demonstrated[23]. As diode laser costs decrease with increased production volume, diode laser and diode laser arrray pumped solid state lasers will replace the traditional flashlamp pumped Nd:YAG laser sources. The use of laser diode array pumping of slab geometry lasers will allow efficient, high peak and average power solid state laser sources to be developed.

Perhaps the greatest impact of diode laser pumped solid state lasers will be in spectroscopic applications of the miniature, monolithic devices. The nonplanar ring oscillator offers the potential of sub-Hertz linewidths in a size that matches the recently developed ion traps.It may be possible, one day, to carry in your watch the laser source and single ion that are the basis for a highly accurate miniature clock.

418

This work was supported by U. S. Army Research Office contract DAAG29-84-K-0071 and NASA grant NAG 1-182. W. J. Kozlovsky and C. D. Nabors gratefully acknowledge the support of the Fannie & John Hertz Foundation.

References
1. Robert L. Byer "Diode Pumped Solid State Lasers," to be published Science
2. Peter S. Cross, Gary L. Harnagle, William Streifer, Donald R. Scifres and David F. Welch, "Ultrahigh Power Semiconductor Diode Laser Arrays," to be published Science
3. Roger Newman, Journal of Applied Physics, **34**, 437, (1963)
4. B. K. Zhou, T. J. Kane, G. J. Dixon, and R. L. Byer, Optics Lett., **10**, 62 (1985)
5. D. L. Sipes, Appl. Phys. Lett., **47**, 74 (1985)
6. T. Y. Fan, G. J. Dixon, and R. L. Byer, Optics Lett., **11**, 204 (1986)
7. T. Baer, J. Opt. Soc. Am. B, **3**, 1175 (1986)
8. T. Y. Fan, A. Cordova-Plaza, M. J. F. Diggonet, R. L. Byer, and H. J. Shaw, J. Opt. Soc. Am. B, **3**, 140 (1986)
9. W. J. Kozlovsky, T. Y. Fan and R. L.Byer, Optics. Lett. **11**, 788 (1986)
10. T. Y. Fan and R. L. Byer, IEEE Journ. Quant. Electr., **QE-23**, 605 (1987)
11. T. Y. Fan and R. L. Byer, Continuous wave operation of a room temperature diode laser pumped 946nm Nd:YAG laser," to be published
12. G. J. Dixon, Z. M. Zhang, R. S. F. Chang and N. Djeu paper ThU12-1, CLEO conference, Baltimore, MD, May 1987
13. E. W. Duczynski, G. Huber, and P. Mitzscherlich, in Tunable Solid State Lasers II, ed by A. B. Budgor, L. Esterowitz and L. G. DeShazer (Springer, Berlin 1986) p 282
14. T. Y. Fan, G. Huber, R. L. Byer, P. Mitzscherlich, Continuous wave operation at 2.1 um of a diode laser pumped, Tm-sensitized, Ho:YAG laser at 300K," to be published
15. W. Grossman, Lightwave Electronics Corporation, Mnt. View, CA (private commun)
16. S. Basu and R. L. Byer, "Diode Laser Pumped and Modelocked Nd:Glass Laser," paper WN3 presented at CLEO, Baltimore, MD, May 1987
17. T. J. Kane and R. L. Byer, Optics Lett., **10**, 65 (1985)
18. T. J. Kane, A. C. Nilsson, and R. L. Byer, Optics Lett., **12**, 175 (1987)
19. W. R. Trutna, Jr., D. K. Donald, and Moshe Nazarathy, Optics Lett., **12**, 248 (1987)
20. W. J. Kozlovsky, C. D. Nabors and R. L. Byer, "Second Harmonic Generation of a cw Diode-Laser Pumped Nd:YAG Laser using and Externally Resonant Cavity," to be published
21. T. J. Kane, W. J. Kozlovsky, and Robert L. Byer, Optics Letts., **11**, 21 (1987)
22. T. J. Kane, W. J. Kozlovsky, R. L. Byer, and C. E. Byvik, Optics Letts., **12**, 239 (1987)
23. J. Berger, D. F. Welch, D. R. Scifres, W. Streifer, and P. S. Cross, "370mW, 1.06um TEM_{00} output from a Nd:YAG laser Rod End Pumped by a Monolithic Diode Array," paper ThT10-1 presented at the CLEO conference, Baltimore, MD, May 1987

Spectroscopy of the 3 μm Laser Transitions in YAlO₃:Er

W. Lüthy, M. Stalder, and H.P. Weber

Institute of Applied Physics, University of Bern,
Sidlerstraße 5, CH-3012 Bern, Switzerland

The spectroscopic properties of the 3 μm emission of a YAlO₃:Er laser have been investigated. Erbium crystals with dopant concentrations of 10 %, 20 % and 50 % have been used in the experiments. The emission of YAlO₃ crystals is strongly dependent on the crystal orientation with respect to the polarization of the laser light. The polarization dependence of the most prominent emission lines has been measured for laser crystals oriented with their axes along the crystallographic a- and b- directions. An example is shown in Fig. 1.

The results show that the use of YAlO₃ instead of YAG allows a coarse selection of the laser lines by introducing a polarizing element into the laser resonator. With additional selectively absorbing filters in the laser resonator it is possible to obtain room-temperature laser action on nine lines of the $^4I_{11/2} \rightarrow {}^4I_{13/2}$ manifold in YAlO₃:Er. The wavelengths and the tentative assignments are shown in Fig. 2. The wavelengths of the lines range from 2.71 μm up to 2.92 μm. Six of them have been observed for the first time in a laser operated at 300 K. The bandwidth of the laser lines has been measured with a resolution of 0.15 cm⁻¹ using a grating spectrometer in combination with an image converter. A resulting spectrum is shown in Fig. 3.

In a free-running laser, the bandwidth ranges from 0.5 nm at 2.73 μm up to 2 nm at 2.92 μm. Introducing losses in the center of the 2.92 μm line forces laser action to occur in the wings with a total bandwidth of up to 3.5 nm.

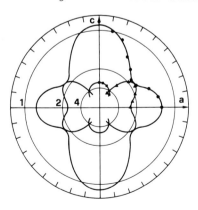

Fig. 1: Polarization dependence of the laser threshold in a YAlO₃:Er (10 %) b-rod. The arrows a and c indicate polarizations with the E-vector ∥ to the a- and c-axes respectively. The radius is inversely proportional to the lasing threshold. The lines are labelled with ▪ for 2.92 μm, ▲ for 2.763 μm and ● for 2.73 μm

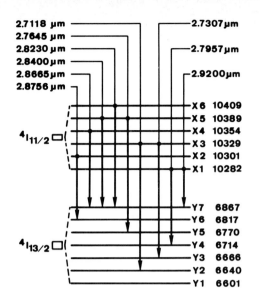

2.7118 µm	2.7307µm
2.7645 µm	
2.8230 µm	2.7957µm
2.8400µm	
2.8665µm	2.9200µm
2.8756 µm	

$^4I_{11/2}$

X6	10409
X5	10389
X4	10354
X3	10329
X2	10301
X1	10282

$^4I_{13/2}$

Y7	6867
Y6	6817
Y5	6770
Y4	6714
Y3	6666
Y2	6640
Y1	6601

Fig. 2: Schematic representation of the $^4I_{11/2}$ -> $^4I_{13/2}$ manifolds with the observed laser wavelengths and the tentative assignments. The 3 lines known from [1,2] are shown at the right hand side of the Fig.

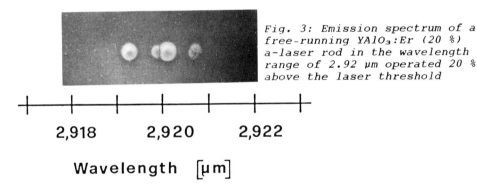

Fig. 3: Emission spectrum of a free-running YAlO$_3$:Er (20 %) a-laser rod in the wavelength range of 2.92 µm operated 20 % above the laser threshold

2,918   2,920   2,922

Wavelength [µm]

[1] A.A. Kaminskii, V.A. Fedorov, A.O. Ivanov, I.V. Mochalov, L.I. Krutova, Sov. Phys. Dokl. 27 (9) 725-727 (1982)

[2] A.A. Kaminskii, Sov. Phys. Dokl. 27 (12) 1039-1041 (1982)

421

Fibre Lasers

A.I. Ferguson, M.W. Phillips, D.C. Hanna, and A.C. Tropper

Department of Physics, University of Southampton,
Southampton, S095NH, United Kingdom

Fibre lasers were first demonstrated in 1963 when a multimode fibre was
transversely pumped by an incoherent light source [1]. There has been a
recent revival of interest in fibre lasers because of the development of
techniques for fabricating single-mode fibres doped with impurities but
with very low losses characteristic of silica fibre used for optical communications. The fabrication technique can be applied to many dopants
including most of the lanthanide series and some transition metal ions.
The attractive features of fibre lasers include the low threshold, the
freedom from thermal distortion problems, the availability of broad gain
profiles so that tuning may be achieved and ultimately so that short pulse
generation may be achieved in a mode-locked system. Thus in many ways
fibre lasers present features available in dye lasers but with additional
advantages such as the ability to Q-switch, long term stability of the
laser medium, the compact configuration, the absence of cooling requirements and the ability to diode pump. Furthermore the confined mode volume
provides low threshold and high gain enabling continuous-wave three-level
laser operation.

Two fibre fabrication techniques have been successfully implemented.
These are the modified chemical vapour deposition (MCVD) technique [2] and
a solution doping technique [3]. The MCVD technique has been used for most
of the lanthanide series and allows dopant levels up to several hundred
parts per million. The loss in the transmission windows of MCVD doped
fibre is as low as 1dB/km. The solution-doping technique is a modification
to the MCVD process in which the dopant ions are added to the fibre preform
from an aqueous solution. The cladding is prepared as in the MCVD process
but the core layers are deposited at a lower temperature than that required
in the MCVD technique, thus forming a porous soot. The preform is then
immersed in an aqueous solution of the rare-earth halides for about an hour
and then dehydrated. Dopant ions including Nd, Ho, Eu, Er, Yb, Dy have
been incorporated with concentrations of up to 4300 parts per million.

A typical fibre laser configuration is shown in Fig. 1. The pump laser
is coupled into the fibre through the input mirror by means of a microscope objective. The input mirror is highly transmitting at the pump wavelength and highly reflecting at the laser wavelength. A variety of pump
lasers have been used including argon ion lasers, dye lasers and GaAlAs
diode lasers. The input end of the fibre is cleaved and butted against the
input mirror. The light emitted from the end of the fibre is collected by
a microscope objective and a stable cavity is formed by a plane output
coupler. The single-mode nature of the fibre at the laser wavelength
ensures single transverse mode output. The collimated region between the
intracavity microscope objective and the output coupler enables tuning
elements and modulators to be inserted into the cavity with low loss.

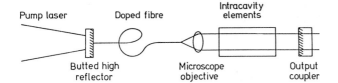

Figure 1 : Schematic diagram of a typical fibre laser cavity configuration

An alternative arrangement which has been used to provide low threshold pump power and high efficiency is to butt the output coupler to the fibre end. With this low loss configuration thresholds of less than 1 mW and slope efficiencies in excess of 35% have been achieved when pumped by a diode laser.

Fibre lasers make useful tunable devices in the near infrared region of the spectrum. The large linewidths produced by the host glass together with the wide range of dopant ions means that fibre lasers can be operated over a significant part of the spectrum. A Nd doped device has been tuned between 0.900 μm and 0.945 μm and between 1.070 μm and 1.145 μm [4] and an Er device has been tuned between 1.528 μm and 1.554 μm [5]. The tuning elements which have been used include birefringent filters and diffraction gratings. In the case of a low-doped Nd fibre laser (300 ppm) the free-running linewidth has been reduced from 500 GHz to 1.5 GHz by use of a 3-plate birefringent filter and two solid etalons (finesse of 20 and thickness of 0.2 mm and 1.0 mm). An integrated fibre grating coupled to the same fibre has produced a bandwidth of 16 GHz. Integrated fibre gratings have also been used with Er doped fibre providing a bandwidth of 5GHz.

Single frequency operation of a fibre laser has recently been reported by using a short length (~ 5 cm) of solution doped fibre [6]. The high dopant level of this fibre ensures good pump absorption over this short length. The relatively large mode spacing together with the use of an integrated fibre grating allows single frequency operation. The resulting linewidth of the unstabilised laser at 1.08 μm is about 1.5 MHz.

Fibre lasers are well suited to Q-switched operation because of the long upper state lifetime. Pulse durations of less than 200 nsec with peak powers in excess of 10W have been produced by a Nd doped fibre laser at 1.09 μm by acousto-optic Q-switching. Pulses of a similar power but of shorter duration (75 nsec) have been generated by a similar laser system operating at 0.9 μm [7].

We have investigated the mode-locked performance of a Nd doped fibre using an acousto-optic modulator. It was found that etalon effects between the fibre end and the output mirror severely limited mode-locked perform-ance. These effects have been reduced by terminating the fibre end in a cell containing index-matching fluid and in this way pulses of less than 300 psec have been achieved.   This pulse duration is still significantly longer than expected and an investigation of dispersion compensation and further reduction in etalon effects is underway.

Diode laser pumping of fibre lasers is attractive for a number of reasons, including the fact that the pump can be easily modulated and thus produce modulated output from the fibre laser. When the laser is driven at a modulation frequency close to the relaxation oscillation frequency many

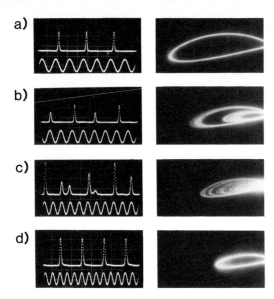

Figure 2 :   Chaotic behaviour : real time output is shown on the left
together with the modulation current, and the corresponding phase
space plot (dI/dt vs I) is shown on the right.

    a)   Stable period doubling   b)   Period quadrupling
    c)   Strange attractor and chaos   d)   Stable period tripling

nonlinear dynamical phenomena are easily demonstrated [8]. These include
resonant driving of relaxation oscillations, hysteresis of the amplitude -
frequency response curve, period doubling and optical chaos. Examples of
some of these phenomena are shown in Fig.2 where real-time oscilloscope
traces are compared with the equivalent phase-space plots. This clearly
shows the existence of a strange attractor between regions of stable period
doubling and period tripling, a hallmark of deterministic chaos.

Fibre lasers provide a convenient tunable source in the near infrared
region. Single frequency, Q-switched and mode-locked operation have been
achieved. Modulated pumping of fibre lasers has demonstrated a rich
variety of nonlinear dynamics including hysteresis, period doubling and
chaos.

Acknowledgement

This work has been funded by SERC/DTI under a JOERS programme. We are
grateful to the Fibre Group, Department of Electronics and Computer
Science, University of Southampton, for the provision of doped fibre.

References

1.   C.G. Young, Appl. Phys. Lett. 2, 151 (1963).

2.   S.B. Poole, D.N. Payne and M.E. Fermann, Electron. Lett. 21, 737
     (1985).

3. J.E. Townsend, S.B. Poole and D.N. Payne, Electron. Lett. 23, 329 (1987).

4. I.P. Alcock, A.I. Ferguson, D.C. Hanna and A.C. Tropper, Optics Lett. 11, 709 (1986).

5. L. Reekie, R.J. Mears, S.B. Poole and D.N. Payne, J. Lightwave Tech. LT-4, 956 (1986).

6. I.M. Jauncey, L. Reekie and D.N. Payne, Postdeadline paper, Conference on Lasers and Electro-Optics, Baltimore (1987).

7. I.P.Alcock, A.C. Tropper, A.I. Ferguson and D.C. Hanna, Electron. Lett. 22, 84 (1986).

8. M.W. Phillips, H. Gong, A.I. Ferguson and D.C. Hanna, Opt. Commun. 61, 215 (1987).

Recent Developments in Nonlinear Optical Materials

R.C. Eckardt[1], Y.X. Fan[1], M.M. Fejer[1], W.J. Kozlovsky[1], C.N. Nabors[1], R.L. Byer[1], R.K. Route[2], and R.S. Feigelson[2]

[1]Department of Applied Physics, Stanford University,
Stanford, CA 94305, USA
[2]Center for Materials Research, Stanford University,
Stanford, CA 94305, USA

1. Introduction

Advances in nonlinear frequency conversion involve developments in laser pump sources, and nonlinear optical materials and techniques. There is a renewed interest in the development of nonlinear optical materials that has resulted from advances in the other areas, and from applications requiring tunable coherent radiation. Recent materials developments have included improvements in available size and optical quality of established nonlinear materials and the development of new materials that extend the conditions under which nonlinear frequency conversion can be used. There is a large amount of earlier materials work that is the subject of reviews [1]. The current effort in nonlinear optical materials development is also extensive, and the discussion here is limited to bulk inorganic materials used in second order nonlinear frequency conversion processes. Silver gallium selenide, lithium niobate and barium borate are discussed to provide examples.

There are about two dozen nonlinear optical materials that are commercially available. In choosing a nonlinear material or selecting new materials for development, it is necessary to consider an number of factors. Linear as well as nonlinear optical properties are important. A material must be transmitting, it must be phasematchable, and it must have good crystalline quality. Different figures of merit have been developed for various conditions. The quantity (d_{eff}^2/n^3) is a simple figure of merit for comparing nonlinearity. Here d_{eff} is the effective nonlinear coefficient for the material for a specified phasematching geometry and direction of propagation, and n is the refractive index. A high nonlinear coefficient alone is not a satisfactory criterion for an acceptable material. The intensity threshold for optical damage must be considered, and nonlinear absorption and intensity dependent changes in the refractive index can also limit nonlinear frequency conversion. Chemical stability and mechanical strength may be of concern. Thermal properties are important for high average power applications. Other properties that may be of concern are birefringent walkoff of pump and generated beams or temporal walkoff of short pulses due to group velocity mismatch. Material losses caused by absorption or scattering are important in high finesse resonant cavity techniques. And perhaps most important is the availability of the nonlinear optical material.

2. Silver Gallium Selenide

Nonlinear materials for applications between 4 and 10 μm in the infrared have been difficult to obtain. Materials which were available for use in this range had problems with low damage threshold, poor optical quality, or very limited availability. The use of AgGaSe$_2$ for nonlinear frequency conversion was first proposed in 1972 [2], and it is only recently that high optical quality crystals of large size have been produced. It can be phasematched for second harmonic generation for fundamental wavelengths between 3.1 and 13 μm. The transparency range extends from 0.8 to 17 μm with some multiphonon absorption at wavelengths longer than 13 μm. The nonlinear optical coefficient is $d_{36} \approx 43 \times 10^{-12}$ m/V. AgGaSe$_2$ has performed well in second harmonic generation and sum generation of the third harmonic of carbon dioxide laser radiation. AgGaSe$_2$ is now being grown in high quality boules yielding angle phasematched crystals of lengths up to 3.5 cm, and it is becoming available as a commercial product.

Some material development problems remain for silver gallium selenide. Post growth heat treatment in the presence of excess Ag$_2$Se is required to reduce the scattering in as-grown

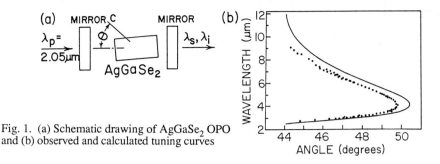

(a) MIRROR C   MIRROR   (b)

$\lambda_p = 2.05\mu m$ AgGaSe$_2$ λ_s, λ_i

Fig. 1. (a) Schematic drawing of AgGaSe$_2$ OPO and (b) observed and calculated tuning curves

WAVELENGTH (μm)
ANGLE (degrees)

material. Adjustment of the heat treatment process is being investigated for higher yields and more complete elimination of the scattering. Scattering loss coefficients as low as 0.01 cm^{-1} have been achieved but values of 0.03 cm^{-1} are more typical in material now being produced. The residual scattering centers are small voids. Development of a surface treatment procedure that would increase damage threshold would be beneficial. The surface damage threshold is 13 MW/cm^2 for pulse durations of approximately 50 ns. The damage threshold for bulk optical damage is 10 times higher than that for surface damage. The bulk material also has excellent average power capabilities; average intensities of 200 kW/cm^2 have been transmitted in the bulk material without damage.

Infrared parametric oscillation has been demonstrated in AgGaSe$_2$ using a 2.05-μm, Q-switched holmium laser as a pump source [3]. This application of the material illustrates the broad spectral range over which it is useful. The output was continuously tunable between 2.65 and 9.0 μm. The 2-cm long parametric oscillator crystal was angle tuned between 45° and 50°. The experimental setup and observed wavelength tuning curve are shown in Fig. 1. Threshold for parametric oscillation was 4 mJ which was approximately one-fourth of the pump energy at which crystal surface damage was observed. The peak efficiency was 16% of the pump energy converted into signal and idler. The tuning range was limited by the mirror reflectivity in the singly resonant oscillator. Improved performance is possible with the 3.5-cm-long crystals now available.

3. Lithium Niobate
Even though lithium niobate has been used as a nonlinear optical material since the mid 1960s, development is continuing to improve and extend its properties. It has a relatively high nonlinear optical coefficient $d_{31} = 6.57 \times 10^{-12}$ m/V, compared to $d_{36}(KDP) = 0.41 \times 10^{-12}$ m/V, and it has high transmission in the near infrared, and good physical properties. It is grown by the Czochralski technique which yields large crystals of good quality. LiNbO$_3$ will 90° phasematch for harmonic generation of fundamental wavelengths near 1.06 μm, thus providing greater angular acceptance and eliminating the problem of birefringent walkoff. Photorefractive damage of LiNbO$_3$, however, has limited its use at high intensity in the visible. Recently two methods for reducing the photorefractive damage have been demonstrated. Doping the material with 5 mole percent of MgO substantially reduces the photorefractive damage that occurs at high intensity [4]. In the second method the material is returned to stoichiometric composition by lithium in-diffusion [5]. This raises the phasematching temperature for fundamental wavelengths near 1.06 μm to 234 C which is well above the annealing point. Preliminary observations have shown that even below the annealing temperature photorefractive damage cleans up after a few seconds. In addition the lithium-diffused stoichiometric material has excellent optical uniformity.

The use of MgO:LiNbO$_3$ has been demonstrated in an external resonant cavity second harmonic generator which provided stable harmonic conversion of a diode-pumped Nd:YAG laser output [6]. The external resonant cavity (Fig. 2) was formed by depositing highly reflecting dielectric coatings on the polished spherical surfaces of the nonlinear crystal. Temperature-tuned 90° phasematching was used, and the resonant frequency of the monolithic cavity was tuned electro-optically by applying a voltage transversely across the MgO:LiNbO$_3$

427

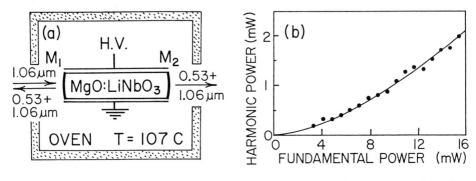

Fig. 2. External resonant cavity second harmonic generation with MgO:LiNbO₃; (a) schematic of setup and (b) observed harmonic power dependence on incident fundamental power

crystal. Feedback control was obtained by deriving the applied voltage from the from the magnitude of the reflected pump wave. The harmonic output was a stable single longitudinal and lowest order transverse mode. A conversion efficiency of 13% was obtained with 15-mW pump power. The measured finesse of the harmonic cavity was 450, indicating that losses for a round trip in the 25-mm crystal were 0.8% when the 0.6% transmission of the coatings was taken into account. The very low transmission loss, good optical quality, and increased resistance to photorefractive damage of the $MgO:LiNbO_3$ material were important properties in this application. Optimization of mirror reflectivities could eliminate the back reflected and transmitted fundamental, increasing harmonic conversion. Preliminary indications are that the lithium-diffused stoichiometric $LiNbO_3$ will perform even better.

4. Barium Borate

The use of barium borate (BaB_2O_4 or BBO) as a nonlinear optical material is a recent development [7]. BBO is particularly useful for generation of ultraviolet radiation as short as 200 nm by sum frequency and harmonic generation [8]. The second through fifth harmonics of 1.06-μm Nd laser radiation can be generated with high efficiency in BBO, and the material appears to have potential for high average power applications [9]. BBO is highly transmitting between 200 nm and 2.2 μm. The nonlinear coefficient is $d_{22} = \pm 1.7 \times 10^{-12}$ m/V. The material is being grown at the Fujian Institute of Research on the Structure of Matter in the Peoples Republic of China, and some is exported for sale. Several other groups have started investigation of the growth of this material.

The melting point of BaB_2O_4 is 1095 ± 3 C. The crystal has two solid phases. The low temperature phase which occurs when crystallization takes place below 925 ± 2 C is non-centrosymmetric with large second order nonlinear susceptibility. A top seeded solution growth technique is required to produce quality crystals of the low temperature phase. The crystals tend to grow in flat boules which limits the length of material that can be harvested in the phasematched direction. Growth of BBO, particularly inclusion free material, is difficult.

Several crystals of excellent quality supplied by the Fujian Institute of Research on the Structure of Matter were used to generate harmonics of Q-switched Nd:YAG laser output. The performance of BaB_2O_4 in the generation of second, third and fourth harmonics compared favorably with that of KDP and ADP crystals. The setup for cascaded harmonic generation to 213 nm is shown in Fig. 3. When pumped with a 540-mJ unstable resonator Q-switched Nd:YAG laser, 5th harmonic energy of 20 mJ was produced. This value could be improved by avoiding excessive depletion of the fundamental which was summed with the fourth harmonic. The 5th harmonic at 213 nm is at a wavelength which can be phasematched for cascaded conversion to the 15th harmonic by 3rd harmonic generation in Ne gas [10].

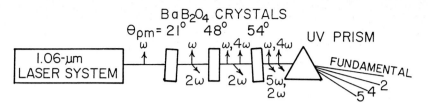

Fig. 3. Schematic representation of setup for fifth harmonic generation of Nd laser radiation in BaB_2O_4

Summary

A limited number of examples were used to illustrate some of the recent developments in nonlinear optical materials. There is a large amount of work on other materials that has not been mentioned. The examples given here are representative of materials development for new spectral regions and for a variety of nonlinear conversion techniques. With one laser and these three materials, it would be possible to generate continuously tunable coherent radiation from 200 nm to 17 μm. The problems of material growth are not trivial, but with continued effort considerable progress is being made.

This work was supported by U. S. Army Research Office contract DAAG29-84-K-0071 and NASA grant NAG 1-182. W. J. Kozlovsky and C. D. Nabors gratefully acknowledge support of the of the Fannie and John Hertz Foundation.

References

1. See for example S. Singe: In *Handbook of Laser Science and Technology,* vol. III: Part 1, ed. by M. J. Weber (CRC Press, Boca Raton, Florida, 1986) p. 3

2. G. D. Boyd, H. M. Kasper, J. H. McFee, and F. G. Storz: IEEE J. Quantum Electron. **QE-8**, 900 (1972)

3. R. C. Eckardt, Y. X. Fan, R. L. Byer, C. L. Marquardt, M. E. Storm, and L. Esterowitz: Appl. Phys. Lett. **49**, 608 (1986)

4. Y. S. Luh, M. M. Fejer, R. S. Feigelson and R. L. Byer: In *Conference on Lasers and Electro-Optics Technical Digest Series 1987, Vol. 14*, (Optical Society of America, Washington, DC, 1987) p. 50

5. D. A. Bryan, R. Gerson and H. E. Tomaschke: Appl. Phys. Lett. **44**, 847 (1984); G.-G. Zhong, J. Jian, Z.-K. Wu, in *Proceeding of the 11th International Quantum Electronic Conference,* 80 CH 1561-0 (IEEE New York, 1980) p. 631

6. W. J. Kozlovsky and R. L. Byer: In *Conference on Lasers and Electro-Optics Technical Digest Series 1987, Vol. 14* (Optical Society of America, Washington, DC, 1987) p. 346; W. J. Kozlovsky, C. D. Nabors, and R. L. Byer: to be published.

7. Chen Chuantian, Wu Bochang, Jiang Aidong and You Guiming: Scienta Sinica **27**, 235 (1985)

8. K. Kato: IEEE J. Quantum Electron. **QE-22**, 1013 (1986); W. L. Glab and J. P. Hessler, "Efficient generation of 200 nm light in β-BaB_2O_4," post deadline paper PD5-1, XV International Quantum Electronics Conference, Baltimore, 26 April - 1 May 1987

9. C.-T. Chen, Y. X. Fan, R. C. Eckardt, and R. L. Byer: Proc. SPIE **681**, 12 (1986)

10. R. Wallenstein, C.-T. Chen, and Y. X. Fan: private communication

Single-Mode Operation of Pulsed Dye Lasers

B. Burghardt, W. Mückenheim, and D. Basting

Lambda Physik, P.O. Box 2663, D-3400 Göttingen, Fed. Rep. of Germany

In large domains of laser spectroscopy narrow-bandwidth probes are an imperative requirement, with single-mode lasers being the ideal instruments. cw-lasers satisfying this desire are commercially available but, if not complemented by pulsed amplifiers, they deliver only low output power, and the costs of such systems are almost inversely proportional to their bandwidth. Pulsed dye lasers, usually pumped by excimer or Nd:YAG lasers, are already suited for many applications and, therefore, present in every spectroscopy laboratory. However, with the exception of a few sophisticated devices they oscillate on several modes.

This drawback can be overcome by reducing the cavity length (Littman design /1/) such that even a rather broad spectral distribution would not cover more than one longitudinal mode, or by reducing the bandwidth via extended pulse duration, which is the appropriate approach for a Hänsch-type /2/ oscillator. This way, being subject of the present paper, offers the unique opportunity of automatic ASE-suppression by utilizing the grating twice /3/. While in the past this way was merely of academic interest, because pump lasers of sufficient pulse duration were not easily accessible, meanwhile a suitable excimer laser, emitting pulses of 300 ns duration at 308 nm /4/, has been made commercially available (Lambda Physik, EMG 602).

For an oscillator equipped with an intracavity etalon, the advantage gained by extended pulse duration can be estimated under some simplifying conditions as follows. The single-pass transmission function of the etalon is, for large finesse F, given by

$$T(\nu) = \left(1 + \left(\frac{2F}{\pi} \sin \frac{\nu - \nu^*}{F \Delta \nu_0} \pi \right)^2\right)^{-1} \qquad (1)$$

with ν^* a frequency of maximum transmission and $\Delta \nu_0$ the width of the single-pass transmission function. With no regard to amplification and loss processes or hole burning in the laser cavity but assuming a constant amplitude $T(\nu^*) = 1$ and negligible divergence of the radiation, the halfwidth $\Delta \nu_N$ of the transmission function after N cavity roundtrips is determined by

$$T(\nu^* + \Delta \nu_N/2)^{2N} = 1/2 . \qquad (2)$$

Expanding the sine and taking only the linear term we get

$$\Delta \nu_N / \Delta \nu_0 = \sqrt{2^{1/2N} - 1} \qquad (3)$$

which, for large N, can be written

$$\Delta \nu_N / \Delta \nu_0 = \sqrt{\ln 2/2} \ / \sqrt{N}. \qquad (4)$$

More interesting for practical applications is the time-averaged bandwidth $\langle \Delta \nu_N \rangle$. According to (3) it can be calculated from

$$\langle \Delta \nu_N \rangle / \Delta \nu_0 = N^{-1} \sum_{k=1}^{N} \sqrt{2^{1/2k} - 1} \tag{5}$$

which, again for large N, may be approximated by means of (4) by

$$\langle \Delta \nu_N \rangle / \Delta \nu_0 = \sqrt{\ln 2/2} \; N^{-1} \sum_{k=1}^{N} k^{-1/2}. \tag{6}$$

Applying the condition $N \gg 1$ for a last time, the sum in (6) can be approximated by $2\sqrt{N} - \sqrt{2} \approx 2\sqrt{N}$ yielding

$$\langle \Delta \nu_N \rangle / \Delta \nu_0 = \sqrt{2\ln 2} / \sqrt{N}. \tag{7}$$

Thus, the final as well as the time-averaged bandwidth reduces in proportion to the inverse square root of the number of cavity roundtrips.

The length of common dye laser oscillators lies between 30 and 40 cm. Therefore, the distance between longitudinal modes ranges from 0.4 to 0.5 GHz. When equipped with an intracavity etalon and pumped by 20 ns pulses, a bandwidth of about 1 GHz is obtained. Hence, two or three modes will usually oscillate. An increase to 200 ns pulse duration is obviously sufficient, to reduce the bandwidth by a factor of 3, permitting only one single longitudinal mode to oscillate.

Due experiments have been performed by using a Hänsch-type dye laser of 2.3 ns roundtrip time and 440 MHz mode separation. The single pass transmission function determined by the intracavity etalon is about $\Delta \nu_0 = 2$ GHz. The dye laser was pumped either by an ordinary excimer laser of 18 to 20 ns pulsewidth or by a precursor of the EMG 602, delivering pulses of 220 ns duration at 308 nm. In both cases the duration of the dye laser output was reduced, namely to 16, 18 and 170 ns respectively. In addition, the time-averaged bandwidth over the first 5 ns was measured by means of a streak camera, when the ordinary excimer laser was used. The results are given in Fig. 1. It is obvious that during the first 20 ns several modes can oscillate ($\Delta \nu_8 = 0.8$ GHz), while a pulse duration of 200 ns or more guarantees single mode operation ($\Delta \nu_{75} = 0.3$ GHz), if the dye laser frequency is chosen such that one mode lies near the center of the transmission function. This condition, however, can be satisfied, e.g., by adjusting the cavity length of the dye laser via a piezo-driven mirror mount.

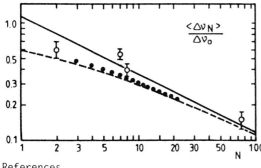

Fig. 1 Normalized average bandwidth $\langle \Delta \nu_N \rangle / \Delta \nu_0$ versus number of roundtrips N. The error bars do not include a +/- 20 % uncertainty in $\Delta \nu_0$. Theoretical curves:
●●●●● Eq. (5),
----- Eq. (6),
——— Eq. (7).

References

/1/ M.G. Littman and H.J. Metcalf, Appl. Opt. 17 (1978) 2224
/2/ T.W. Hänsch, Appl. Opt. 11 (1972) 895
/3/ H. Bücher, U.S. Patent 4399540
/4/ P. Klopotek, U. Brinkmann, D. Basting and W. Mückenheim, CLEO '87, Baltimore, Digest of Technical Papers, p. 326

A Room Temperature Tunable Color Center Laser with a LiF:F$_2^-$ Crystal

S.-H. Liu*, F.-G. Wang, Z. Lin, Y. Taira, K. Shimizu, and H. Takuma

Institute for Laser Science, University of Electro-Communications, Chofu-shi, Tokyo 182, Japan

Color center lasers (CCL) using various types of active centers produced in different host crystals have been extensively studied, and have acquired increasing importance in spectroscopic applications especially in the infared region, where dye lasers are not readily available. However, the wavelength range covered by CCLs is still quite limited. Moreover, many of them require low temperature operation, which is a great inconvenience for practical applications[1].

Among CCL materials, one which is produced by electron-beam or gamma-ray irradiation of LiF is especially interesting because of its capacity for stable room temperature operation and long life. This type of color center was assigned as F_2^- by Gusev et al.[2]. Since then, study of this type of laser has been geographically limited, being active mainly in the Soviet Union and China. On the other hand, Mollenauer studied electron-beam ir-radiated NaF and found that the F_2^+ center produced by irradiation disap-pears producing a new stable center having red-shifted absorption and emis-sion bands[3]. He carried out a detailed study of this type of color cen-ter, which is designated as $(F_2^+)^*$, and concluded that it is probably an impurity-stabilized F_2^+ center. The question whether these two are practi-cally the same or not, together with the fact that it is useful in various applications[4], motivated us to carry out the present study.

LiF crystals, grown by the Czochralski method in an atmosphere which is free from oxygen and water vapor, were irradiated by γ-rays to produce F_2^- cen-ters. Each piece of crystal, has a 1cm X 1cm cross section and is 2cm in length. Both parallel end surfaces are polished optically flat. Four pieces of high optical quality were chosen and irradiated by γ-rays. Four values of dosage were chosen for comparison: 1×10^7, 1×10^8, 1.5×10^8 and 2×10^8rad, respectively.

The experimental setup for the observation of laser oscillation is shown schematically in Fig.1. The pumping source was a Q-switched Nd:YAG laser whose output energy was 200mJ in each pulse of 10ns duration and 1-10Hz repetition rate. Special care was taken to avoid any inclusion of the scat-tered pumping pulse in measuring the output energy, and the adjustment of the resonator was made to obtain optimum coherent output by observing the second harmonics of the output generated by a piece of KDP crystal which was inserted only for adjustment. The threshold pumping energy was less than 1mJ for all the higher dosage crystals, while no oscillation was obtained on the lowest dosage one.

The tuning range was 1132 to 1146 nm as shown in Fig.2. The maximum monochromatic output was 15mJ at the band center (1.14 μm) for a pumping pulse energy of 187mJ, giving a reasonably high efficiency of 8%. All the tested crystals were stable against repetitive pumping, and no sign of deterioration was observed after more than 10^6 shots.

----

*Permanent address: South China Normal University, Guangzhou, PRC.

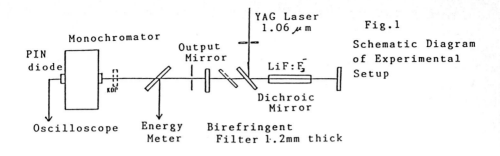

Fig.1

Schematic Diagram
of Experimental
Setup

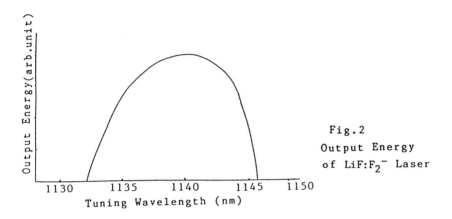

Fig.2
Output Energy
of LiF:F$_2^-$ Laser

   The absorption coefficient at the pumping wavelength was proportional to
the dosage.   In the observed pumping intensity dependence of the output intensity, the high gain but high absorption character of the crystal was apparent.   The high absorption is a common feature of the γ-ray irradiated
crystals in which various types of imperfections are produced.   These
results indicate that a dosage of $(1-3) \times 10^8$rad is about optimum for CCLs.
   Infrared absorption spectra of the present samples showed a dosage-
independent impurity band, which was peaked at 3.4 µm. This observation may
be in favor of Mollenauer's assumption that OH produced by irradiation may
be one of the stabilizing impurities[3].
   In any case, the present study reconfirmed that a CCL using a   γ-ray
irradiated LiF crystal is a versatile coherent source for spectroscopy
around 1.14 µm.

References
1.L.F.Mollenauer: Tunable Lasers , eds. L.F.Mollenauer and J.C.White,
   (Springer, Berlin, 1987).
2.Yu.L.Gusev, S.I.Marennikov and V.P.Chebotayev, Appl. Phys. 14, 121(1977).
3.L.F.Mollenauer, Optics Letters 5, 188(1980).
4.V.P.Chebotayev, S.I.Marennikov and V.A.Smirnov, Appl. Phys. B31,193(1983).

Recent Progress in Semiconductor Lasers

R. Lang

Opto-Electronics Research Labs, NEC Corporation,
4-1-1, Miyazakidai, Miyamaeku, Kawasaki 213, Japan

1. Introduction

GaAs/AlGaAs and InGaAsP lasers are now manufactured in quantities far exceeding any other lasers. While their primary applications are to optical video and audio recording systems as well as optical fiber communication systems, some of them have sufficient qualities for spectroscopic applications. In the following a brief review is presented on the latest developments in semiconductor lasers with emphasis on the frequency stabilized InGaAsP distributed feedback (DFB) lasers and visible-light-emitting AlGaInAs/GaAs lasers.

2. InGaAsP/InP DFB Lasers

Much effort has been devoted in 1980's towards development of diode lasers in 1.2 to 1.6 μm wavelength range, which corresponds to the lowest loss window of an ultra-low loss silica fiber. An $In_xGa_{1-x}As_yP_{1-y}$ light emitting layer, in combination with an InP substrate and InP cladding layers, is adopted for this wavelength range. As an example of high quality InGaAsP/InP lasers, double-channel planar-buried heterostructure (DC-PBH) lasers with Fabry-Perot cavity[1] exhibit typical threshold current below 25 mA, differential quantum efficiency around 50 % and maximum cw output power of 35 - 50 mW at wavelengths of 1.2, 1.3 and 1.55 μm.

These Fabry-Perot cavity lasers tend to oscillate in multi-longitudinal modes even at cw operation, despite expected homogeneity in frequency and space of the gain saturation. Moreover, number of the oscillating modes increases, when they are modulated with current signals at rates exceeding a few 100 MHz, causing serious pulse broadening during transmission through a dispersive optical fiber. To overcome this problem, distributed feedback lasers have been developed, which incorporate grating along a waveguide layer for frequency selective feedback.[2] In spite of the complication of the structure and the fabrication processes, low threshold current and high differential efficiency comparable to those of Fabry-Perot lasers have been attained.

In a DFB cavity with an ideal uniform grating, in the absence of reflections from the end facets, there are two maximum feedback frequencies located symmetrically with respect to the Bragg frequency. Because of this degeneracy, single mode oscillation cannot be obtained with high yield. Introduction of phase shift, amounting to about λ/4, at the center portion of the grating can remove this degeneracy.[3] Depicted in Fig.1 is an example of phase shifted DFB laser with DC-PBH structure.[4] Both facets are antireflection coated with SiN to reduce the reflectivity

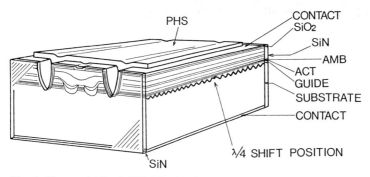

Fig.1 Phase-shifted DFB DC-PBH laser

to below 1%. Almost all phase-shifted DFB DC-PBH lasers oscillate in a single longitudinal mode. The improved frequency selection also results in high main mode-to-side mode intensity ratio amounting to 35 dB at the output power level of 40 mW.

Shown in Fig.2 are examples of observed spectral linewidth of these devices plotted against the injection current normalized with the threshold current Ith.[5] Up to about 2.5 - 3 times the threshold the linewidth decreases inversely proportional to the relative pumping, as expected from the modified Schawlow-Townes formula for semiconductor laser linewidth.[6] However, it tends to increase again at higher current level. The mechanism of this rebroadening is yet unclear. The minimum linewidth attained is 3 to 6 MHz in 1.3 μm lasers, while it is 6 to 10 MHz in 1.5 μm devices.

The semiconductor laser linewidth can be reduced by effectively extending the cavity length employing an external mirror or grating. In Fig. 2, over an order of magnitude narrowing is seen to be achieved by external mirror placed 31cm away from the diode laser facet. Reduction to 1kHz has been reported using an external mirror.[7] A disadvantage in the external mirror scheme is that the laser oscillation tends to be unstable in the coupled cavity operation.[8] An alternative is to employ electronic negative feedback to compensate the carrier density fluctuations due to

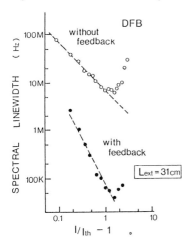

Fig.2 Linewidth vs. normalized pumping with and without an external mirror

spontaneous emission. Reduction to 330kHz, by factor of 1/15 compared with free running state, has been achieved[9], which is even narrower than that expected through the original Schawlow-Townes expression.

3. AlGaInAs/GaAs Visible Lasers

Another exciting development in semiconductor laser technology is towards extension of the emission wavelength into the visible(<700 nm) range. The development is motivated by the desire to have very small size, low power consumption laser source to replace gas lasers, such as He-Ne lasers. A promising material system to cover this range is $(Al_xGa_{1-x})_{0.5}In_{0.5}P/(Al_yGa_{1-y})_{0.5}In_{0.5}P$ III-V compound lattice matched to GaAs substrate. With this system, by varying the compositional fraction x from 0 to 0.4, room temperature lasing from 690 nm (red) down to 580 nm (yellow) is expected to be attainable. Growth of high quality crystal in this system has become feasible using Metal Organic Vapor Phase Epitaxy (MOVPE).

CW room temperature operation has been attained in this material system with x=0 and y=0.4 [10]. The lasing wavelength was 689.7 nm at 25°C. Depicted in Fig.3 is the schematic cross section of this laser. It has a current confining stripe structure formed by the current blocking n-GaAs layers near the top of the figure, but there is no intentional optical waveguiding structure along the active layer. The threshold current is typically around 70 mA, and the spectrum contains several longitudinal modes. These properties are likely to improve with introduction of transverse mode stabilizing structure with refractive index waveguiding which is currently under development.

The reliability of the GaInP/GaAlInP lasers is quite promising. Room temperature life testing results have been reported in which these devices were operating over 2000 hours without any appreciable sign of degradation.[11]

With the same active layer composition, lasing at 678 nm at 20°C[12] and at 679 nm at 24°C[13] have also been reported. The difference in the lasing wavelengths is presumed to be due to the difference in the degrees of disorder in the Ga and In atom configurations in the group III element sublattice of the crystal, which depends on the crystal growth conditions.[14]

The lasing wavelength can be decreased by increasing the Al fraction x in the active layer. With the increase, high quality crystal growth

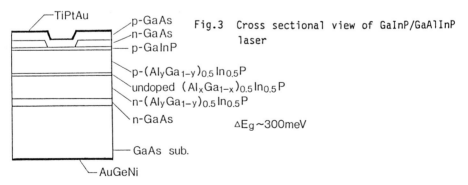

Fig.3 Cross sectional view of GaInP/GaAlInP laser

becomes increasingly more difficult; the threshold current increases, eventually prohibiting room temperature cw operation. However, promising progress is being made towards extension into shorter wavelengths. Room temperature cw oscillation at 661.7 nm has been attained with x=0.1 and y=0.5. [15] CW operation at 0C has been attained at 621 nm for x=0.18 and y=0.6 [16], shorter than the He-Ne laser wavelength at 633 nm. A similar laser oscillated cw at 583.6 nm at liquid nitrogen temperature, which is the shortest cw oscillation wavelength ever attained with a diode laser.[17]

The author gratefully acknowledges Kohroh Kobayashi, Tohru Suzuki and Ikuo Mito for their assistance in preparing this manuscript.

References
1. I.Mito et al., IEEE J. Light Wave Tech., LT-1, 195 (1983).
2. Y.Suematsu et al., IEEE J. Light Wave Tech., LT-1, 161 (1983).
3. H.A.Haus and C.V.Shank, IEEE J. Quantum Electron., QE-12, 532 (1976).
4. M.Yamaguchi et al., Tech. Digest of OFC/IOOC'87, TUC4 (1987).
5. I. Mito et al., Tech. Digest of 12th European Conference on Optical Comunication, vol. II, 67 (1986).
6. C.H.Henry, IEEE J. Quantum Electron., QE-18, 259 (1982).
7. R.Wyatt, Tech. Digest 8th Conf. on Opt. Fiber Commun., TuP2, Feb. 1985
8. R.Lang and K.Kobayashi, IEEE J. Quantum Electron., QE-16, 347 (1980).
9. M.Ohtsu and S.Kotajima, IEEE J. Quantum Electron., QE-21, 1905 (1985).
10. K.Kobayashi et al., Electron. Lett., 21, 931 (1985).
11. A.Gomyo et al., Electron. Lett., 23, 85 (1987).
12. M.Ikeda et al., Appl. Phys. Lett., 48, 89 (1986).
13. M.Ishikawa et al., Appl. Phys. Lett., 48, 207 (1986).
14. A.Gomyo et al., J. Crystal Growth, 77, 367 (1986).
15. K.Kobayashi et al., Electron. Lett., 21, 1162 (1985).
16. S.Kawata et al., Electron. Lett., 22, 1265 (1986).
17. I.Hino et al., Appl. Phys. Lett., 48, 557 (1986).

Spectroscopy with Ultrashort Electrical Pulses *

*D. Grischkowsky, C.-C. Chi, I.N. Duling III, W.J. Gallagher, M.B. Ketchen, and R. Sprik***

IBM Watson Research Center, Yorktown Heights, NY 10598, USA

Recently optoelectronic techniques have been used to generate and detect subpicosecond electrical pulses on coplanar transmission lines.[1,2] The frequency bandwidth of these short electrical pulses ranges up to 1 THz and covers an important part of the far infrared energy spectrum from 0 to 30 cm^{-1}, in which can be found the gap frequencies of superconductors, magnetic excitations, and the far infrared modes of lattices and molecules. With the proper generation geometry, these pulses propagate as a single mode excitation of the transmission line, and the pulse reshaping is determined by the frequency dependent dielectric response of the transmission line materials. These features, plus the fact that the earlier observations showed that the subpicosecond pulses broadened to only 2.6 psec after propagating 8 mm on the transmission line, allow for the following spectroscopic applications of these guided wave electrical pulses. In this paper we will discuss the study of far infrared absorption in superconductors and some magnetic modes in rare-earth garnets.

1. Generation of Ultrashort Electrical Pulses

The ultrashort electrical pulses were generated by photoconductively shorting a charged coplanar transmission line as shown in Fig. 1a, with 70 fsec laser pulses. The resulting electrical pulses were then measured by a fast photoconductive switch, driven by a time delayed beam of the same 70 fsec laser pulses, which connected the transmission line to an electrical probe. In order to measure propagation effects, the excitation point (sliding contact) was moved variable distances away from the sampling photoconductive switch. Consideration of the sliding contact excitation site shows that, to first order, charge is simply transferred from one line to the other creating a symmetrical field distribution with respect to the 2 lines. During the excitation process a current flow is induced between the lines. Localized charge accumulations of opposite sign build up on the segments of the two metal lines under the laser excitation spot, creating a dipolar field distribution, and thereby matching the field pattern of the propagating TEM mode. Consequently, by Fourier analysis of the initial and propagated pulses, the dielectric properties of the transmission line can be completely characterized.

---

*This work was partially supported by the U.S. Office of Naval Research.

** Permanent address: Natuurkundig Laboratorium, Universiteit van Amsterdam, Valckenierstraat 65, 1018 XE Amsterdam, The Netherlands.

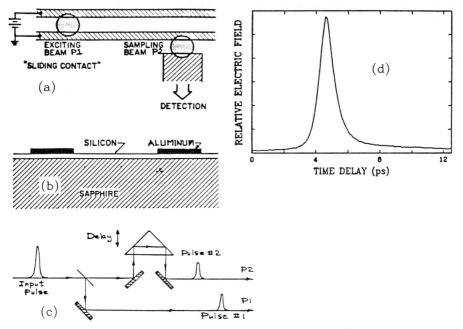

Figure 1. (a) Experimental geometry. (b) Cross-section of SOS wafer. (c) Optical delay line. (d) Measured ultrashort electrical pulse.

Typically, the 20-mm-long transmission line had a design impedance of 100 Ω and consisted of two parallel 5 μm wide, 0.5 μm thick, aluminum lines separated from each other by 10 μm. The measured dc resistance of a single 5 μm line was 200 Ω. The transmission line was fabricated on an undoped silicon on sapphire (SOS) wafer illustrated in cross-section in Fig. 1b. This wafer had been heavily implanted with O^+ ions to ensure the required short carrier lifetime. The laser source is a compensated, colliding-pulse, passively mode-locked dye laser producing 70 fsec pulses at a 100 MHz repetition rate. As depicted in Fig. 1c, the time delay between the exciting and sampling beams was mechanically scanned by moving an air-spaced retro-reflector with a computer controlled stepper motor synchronized with a multichannel analyzer. A measured electrical pulse with an excellent signal-to-noise-ratio is shown in Fig. 1d. For this result, the spatial separation between the exciting and sampling beams was approximately 50 μm, while the laser spot diameters were 10 μm.

2. Superconducting Transmission Lines

We will now describe our observations made on a superconducting transmission line. The geometry of the line is described in Fig. 1, except that it was fabricated out of 900 Å thick niobium metal which becomes superconducting at 9 K. The transmission line was He vapor cooled and temperature regulated in an optical dewar. This arrangement allowed good optical access to the photoconductive switches without resorting to the use of superfluid He. The opto-electronic sampling technique is well suited to operation inside a dewar since all fast electrical signals (0-1 THz) are confined to the sample and only slow signals (0-1 KHz) must leave the dewar.

The input and propagated pulses are shown in Figs. 2a,b for a temperature of 2.6 K. After propagating 3 mm the pulse energy degraded by only 5% and its FWHM broadened to 1.4 psec, but more significantly a strong ringing developed on the trailing edge. This ringing is due to the strong dispersion at frequencies approaching the superconducting energy gap frequency of Nb at 0.7 THz and is the first observation of this phenomenon which was predicted by the superconducting transmission calculations of KAUTZ [3] based on the MATTIS-BARDEEN [4] theory for the complex conductivity of superconductors. The amplitude spectra and the resulting absorption curves were calculated from these experimental results and are shown in Figs. 2c,d. A strong absorption is expected for frequencies above the energy gap of the superconducting niobium, since these frequencies see a normal (resistive) transmission line. At room temperature the line resistance is 500 Ω/mm. The energy gap of niobium at 2.6 K is approximately 3 meV which corresponds to a frequency of 0.72 THz. In Fig. 2d a strong absorption is seen at 0.7 THz and is consequently attributed to pair breaking in the superconducting niobium.

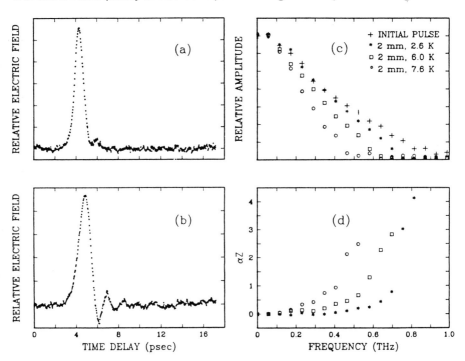

Figure 2. (a) Initial electrical pulse. (b) Electrical pulse after 2 mm propagation on Nb superconducting transmission line at 2.6 K. (c) Fourier transforms of initial and propagated pulses at different temperatures. (d) Absorption spectra of transmission line vs temperature.

The same procedure was followed to obtain the absorption spectra for two additional temperatures approaching the superconducting transition temperature. The closer the niobium is to becoming normal, the smaller the energy gap and the more the pulse will be attenuated. This is supported by the observed movement of the onset of absorption to lower frequencies as the temperature is raised. The predicted position of the absorption edge from the theory is 0.72, 0.64, and 0.49 THz for 2.6, 6.0, and 7.6 K respectively. These values fall above those obtained from the data indicating that only the beginning of the absorption edge is observed, which is further evidence of a strong normal state absorption. With increasing temperature there is an increase in absorption below the pair-breaking frequency due to energy being absorbed by the increasing density of thermally excited quasiparticles. The absorption below the pair-breaking frequency can also be seen to be increasing with frequency. The onset of strong attenuation at the gap frequency and the increase in absorption below the gap with increasing temperature and frequency are both in qualitative agreement with the theoretical expectations.

3. Magnetic Garnets

In order to demonstrate the sensitivity of the technique to determine the absorption of materials in close contact with the transmission line, we studied the magnetic modes in erbium iron garnet (ErIG). The measurements were made by covering a section of an Al transmission line with garnet powder and monitoring the subsequent changes in the propagated pulse shape. Figure 3a. shows propagated electrical pulse without any powder sample on the transmission line at 5 K. When the line was covered with ErIG the pulseshape dramatically changed to that shown in Fig. 3b, where strong oscillation is now seen for the longer time delays. This time domain measurement is converted to the frequency domain in Fig. 3c, by Fourier transforming the two propagated pulseshapes. The resulting spectra show that when the sample is in place, strong resonances appear at the frequencies 10.0 cm^{-1}, and 4.3 cm^{-1} at T = 5 K. A crude estimate of the spectral resolution of the technique based on a total delay time scan of 45 psec is $\simeq 0.7$ cm^{-1}. The measured linewidth, especially for the higher frequency mode, is $\simeq 1.5$ cm^{-1}, clearly broader than the spectral resolution.

Rare-earth iron garnets display a number of well known sharp resonances in the far infrared range of the spectrum relevant for the current experiments. These garnets have two magnetic modes which originate from the exchange resonance between the iron and the rare-earth sublattices in the ferrimagnetic ordered material. The anisotropy in the coupling results in a low frequency mode of ferrimagnetic nature. At somewhat higher frequency are two resonances in the rare-earth garnets associated with the exchange field splitting induced by the iron lattices. These resonances have all been observed and extensively studied in YbIG. The exchange resonance has been observed for the first time by SIEVERS and TINKHAM [5,6] in YbIG using a far-infrared monochromator. Using a spectrometer with an extended spectral range, RICHARDS [7] observed the exchange resonance and in addition the predicted ferrimagnetic resonance in YbIG. ErIG is expected to behave analogous to YbIG and has been studied in some detail by SIEVERS and TINKHAM.[6] In particular for ErIG the exchange resonance was observed at 10.0

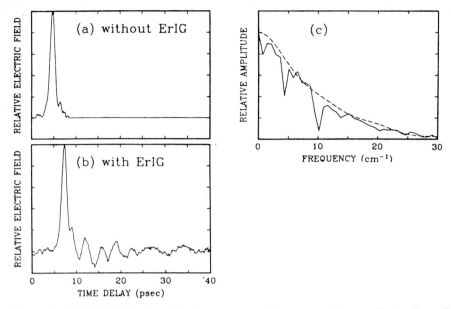

Figure 3. (a) Electrical pulse after 4 mm propagation on an Al transmission line. (b) Electrical pulses after 4 mm propagation on the line covered with EIG garnet powder. (c) Dashed line is spectrum of propagated pulse without ErIG. Solid line is spectrum with ErIG.

cm^{-1} (T = 2 K), which is consistent with our measurement of 9.6 cm^{-1} (T = 2.5 K). SIEVERS and TINKHAM [6] could not observe the ferrimagnetic resonance in ErIG because it was outside their detection range. To our knowledge the ferrimagnetic resonance in ErIG which we observe at 4.3 cm^{-1} has not been reported previously.

4. Comparison with Traditional Far-Infrared Spectroscopy

This new technique has a number of advantages and disadvantages in comparison with the traditional c.w. far infrared spectroscopic techniques based on a mercury arc light source and a monochrometer or interferometer. A c.w. light source has typically $\simeq 10^{-10}$ W available in a bandwidth of 1 cm^{-1} [8], and an infrared interferometer has a spectral resolution better than 0.1 cm^{-1}. The accessible spectral range of such a system is usually from a few cm^{-1} to many hundreds of cm^{-1}. The generated 1 psec electrical pulse (10 mV amplitude) at a 100 MHz repetition rate generates an average c.w. power of 10^{-10} W. This power is distributed in amplitude of the spectral components as shown in fig. 3c. From this spectrum we calculate that for the frequency of 10 cm^{-1}, $\simeq 3 \times 10^{-12}$ W is available within a 1 cm^{-1} bandwidth. However, the effective power is enhanced a factor $\simeq 250$ due to the fact that the actual measurement occurs only during the 45 psec scantime. This makes the effective available infrared power significantly more than the c.w. sources. Another advantage is that similar to an infrared interferometer, the

442

optoelectronic technique detects the electric field and not the intensity. This feature makes the technique sensitive to changes in phase and amplitude of the interfering spectral components resulting in an excellent signal-to-noise ratio with modest infrared power. The current spectral resolution of $\simeq 0.7$ cm^{-1} can be increased simply by extending the delay time scan beyond 45 psec. Currently the spectral range of the pulse technique is between 1-30 cm^{-1}, covering frequencies from the microwave range up to the more readily accessible far infrared frequencies. Because the electrical pulses propagate as a guided wave along the transmission line, the electric and magnetic fields are strongly localized at the surface. This feature makes the technique quite suitable for studying surface excitations. Furthermore, the required amount of sample is very limited since only the area near the transmission lines has to be covered. Perhaps the most attractive feature is the possibility of time resolved far infrared spectroscopy on a picosecond timescale, especially when the available far infrared power can be enhanced to a level where the population of the magnetic levels is strongly influenced. For this case, one can consider applying the technique to observe spin echos from magnetic excitations in the far infrared. The power of the pulse can be increased significantly by amplifying the exciting laser pulses. Preliminary attempts show that electrical pulses of the order of 1 V can be produced, thereby increasing the available power by four orders of magnitude. These unique features of the optoelectronic technique clearly offer new possibilities for the study of far infrared transitions.

References

1. M.B. Ketchen, D. Grischkowsky, T.C. Chen, C.-C. Chi, I.N. Duling, III, N.J. Halas, J.-M. Halbout, J.A. Kash, and G.P. Li, Appl.Phys.Lett. **48**, 751 (1986).
2. W.J. Gallagher, C.-C. Chi, I.N. Duling, III, D. Grischkowsky, N.J. Halas, M.B. Ketchen, and A.W. Kleinsasser, Appl.Phys.Lett. **50**, 350 (1987).
3. R.L. Kautz, J. Appl. Phys. 49, 308 (1978).
4. D.C. Mattis and J. Bardeen, Phys. Rev. 111, 412 (1958).
5. A.J. Sievers, III, and M. Tinkham, Phys.Rev. **124**, 321 (1961).
6. A.J. Sievers, III, and M. Tinkham, J.Appl.Phys. **34**, 1235 (1963).
7. P.L. Richards, J.Appl.Phys. **34**, 1237 (1963).
8. M. Tinkham, J.Appl.Phys., sup. **33**, 1248 (1962).

Part XII

VUV Spectroscopy

Generation and Application
of Coherent Tunable VUV Radiation at 60 to 200 nm

G. Hilber, A. Lago, and R. Wallenstein

Fakultät für Physik, Universität Bielefeld,
D-4800 Bielefeld, Fed. Rep. of Germany

Third- and fifth-order sum- and difference frequency conversion of pulsed dye laser radiation generates coherent tunable radiation in the vacuum ultraviolet at wavelengths λ_{VUV}= 60 - 200 nm. The generated VUV light is of narrow spectral width (ΔE = 0.01 - 1 cm^{-1}) and high spectral intensity (0.03 - 10^4W; 10^8 - $3 \cdot 10^{13}$ photons/pulse). Because of their spectral brightness these VUV light sources are a powerful tool for VUV spectroscopy of atoms, molecules and solid states.

INTRODUCTION

Nonlinear frequency mixing in gases is a well established method for the generation of optical radiation in the spectral region of the vacuum ultraviolet (VUV) at wavelengths λ_{VUV}= 100 - 200 nm and in the extreme ultraviolet (XUV) at λ_{XUV}<100 nm.

The principles of frequency mixing and the experimental progress towards the generation of coherent VUV radiation have been reviewed by W. Jamroz and B.P. Stoicheff[1], by C.R. Vidal[2] and by R. Hilbig et al.[3]

In this contribution we discuss a few examples of more recent investigations of third- and fifth order sum- and difference frequency mixing of dye laser radiation in rare gases. These frequency conversions are simple, reliable methods for the production of tunable coherent VUV and XUV radiation of narrow spectral width and high spectral intensity. The results of current investigations are very promising for a further, considerable increase of the pulse power and for an extension of the tuning range to even shorter wavelengths.

Nonresonant third-order sum- and difference frequency mixing ($\omega_{VUV}= 2\omega_1\pm\omega_2$) in the rare gases Xe and Kr and in Hg vapor generated broadly tunable VUV radiation in the wavelength range λ_{VUV}= 110 - 200 nm.[4-11] Frequency tripling in Ar and Ne produced XUV light in spectral regions between 72 and 105 nm.[12,13]

In these experiments laser pulse powers of 1 to 5 MW provided conversion efficiencies of typically 10^{-5} to 10^{-6}. The power of the generated VUV light pulses was in the range of 1 to 20 W (0.3 - $6\cdot10^{10}$ photons/pulse). Main limitations on the efficiency are caused by dielectric gas-breakdown in the focus of the laser light and by nonlinear intensity dependent changes of the refractive index[2,14] which destroy the phase-matching.

The VUV intensity generated by nonresonant conversion methods is sufficient for most investigations in linear (absorption or fluorescence) spectroscopy. Other applications (like multiphoton excitation and ionization or photodissociation) require more powerful VUV light pulses.

The pulse power can be increased by several orders of magnitude using resonant conversion methods.[15]

Tuning the laser frequency, for example, to a two-photon resonance the induced polarization is resonantly enhanced. This enhancement provides conversion efficiencies of $\eta > 10^{-4}$ even at input powers of only a few kilowatts.

The two-photon resonant conversion is usually of the type ω_{VUV} = $2\omega_1\pm\omega_2$ where ω_1 is tuned to a two-photon transition, and ω_2 is a variable frequency.

In the past this type of frequency conversion has been investigated in detail in metal vapors (like Sr[16,17], Mg[19-20],Cs[21], Ba[22], Hg[23-25] and Zn[26]) and in the rare gases Xe and Kr.[10,27-30]

For the experimental realization of the frequency mixing rare gases are very advantageous. Enclosed in a simple glas or metal cell (equipped with appropriate windows) these gases provide a nonlinear medium of homogeneous, easily variable density. These gases are thus an appropriate medium for the construction of a reliable VUV light source.

Very promising for this purpose is the resonant frequency conversion in Kr.[31] Because of the high excitation energy of the lowest two-photon resonance 4p-5p[5/2,2] the sum- and difference frequency ($\omega_{VUV} = 2\omega_R \pm \omega_T$) of the two-photon resonant radiation ($\lambda_R = 216.6$ nm) and of tunable dye laser light ($\lambda_T = 217-900$ nm) generates radiation in the region of the XUV ($\lambda_{XUV} = 72.3-96.7$ nm) and of the VUV ($\lambda_{VUV} = 123-216$ nm), respectively.

Tuning ω_T for example in the range $\lambda_T = 219-364$ nm the sum frequency generated light at $\lambda_{XUV} = 72.5-83.5$ nm. In agreement with theoretical predictions the conversion efficiency η is almost constant within this spectral range. At input powers $P_R = 14$ kW and $P_T = 400$ kW the pulse power of the XUV exceeded $P_{XUV} = 20$ W. Absorptions in the Kr gas reduced, however, the power of the detected XUV light to about 5 W (effective efficiency $\eta = 1.2 \cdot 10^{-5}$). With laser light at $\lambda_T = 272-737$ nm the difference frequency generates continously tunable radiation at $\lambda_{VUV} = 127-180$ nm. In this range the conversion efficiency increases with wavelength by more than one order of magnitude. At $\lambda_{VUV} = 135$ nm, for example, input powers $P_R = 0.2$ MW and $P_T = 1.2$ MW generate VUV light with $P_{VUV} = 250$ W ($\eta = 1.8 \cdot 10^{-4}$). At $\lambda_{VUV} = 175$nm, a lower input ($P_R = 80$ kW, $P_T = 560$ kW) produced VUV light pulses of $P_{VUV} = 1.8$ kW ($\eta = 2.8 \cdot 10^{-3}$). This spectral variation of η is in agreement with the calculated wavelength dependence of the nonlinear susceptibility and of the gas pressure required for optimum VUV output.

In these experiments the UV radiation of the wavelength $\lambda_R = 216.6$ nm (resonant with the two-photon transition 4p-5p [5/2/2,2]) was generated by sum frequency mixing in KDP ($\omega_R = \omega_{UV} + \omega_{IR}$) of frequency doubled dye laser radiation ($\omega_{UV} = 2\omega_L$ with $\lambda_L = 544$ nm) and of the fundamental (ω_{IR}) of the Nd:YAG laser. In principle, radiation at $\lambda_R = 216.6$ nm can be generated also by doubling the frequency of a blue laser ($\lambda_L = 433$ nm) In the past the only crystal suited for this second harmonic generation was the deuterated KB5 crystal.[32] Because of the low conversion efficiency of 2 to $4 \cdot 10^{-2}$ the generated pulse powers are typically 60 to 120 kW.

Considerable higher efficiencies (of about 15 percent) are now obtainable with the new nonlinear optical material Barium-β-borate (BBO).[33-36] Besides the high conversion efficiency this material generates the second harmonic of radiation at wavelengths as short as 410 nm. Therefore not only the 4p-5p [5/2,2] transition can be used for the two-photon resonant enhancement of $X^{(3)}$ but also the resonances 4p-5p[3/2,2] and 4p-5p[1/2,0] which require UV radiation at λ_{UV}= 214.7 nm and 212.5 nm, respectively. The use of the transition 4p-5p[1/2,0] is of particular advantage since in this case the resonant enhancement of $X^{(3)}$ is considerably larger than the one obtained with the transition 4p-5p[5/2,2] which has been used in previous investigations.

If ω_R is tuned to the different two-photon transitions the calculated ratios[15] of the output P_{XUV} of the resonant third harmonic ω_{XUV}= $3\omega_R$ are, for example, R_1= 8.7 and R_2= 10.6 where R_1= P_{XUV}(5p[1/2,0])/P_{XUV}(5p[5/2,2]) and R_2= P_{XUV}(5p[5/2,2])/P_{XUV} (5p[3/2,2]).

In the experiment the frequency of a Stilbene 3 dye laser was doubled in a suitable BBO-crystal. The frequency tripling was achieved by focusing the UV laser light into a pulsed Kr jet (lens: f= 200 mm). The measurements - performed with the same (low) UV power level of about 40 kW - provide values of R_1= 6.0 and R_2= 7.7. These values are in good agreement with the theoretical estimations if saturation is taken into account.

The resonant frequency conversion with UV laser radiation ω_R - generated by the efficient frequency doubling in BBO - offers several advantages. First, the frequency ω_R can be tuned to the two photon transition, which provides the largest resonant enhancement of $X^{(3)}$. Second, the spectral width of the UV light is determined by the line width of the dye laser radiation and thus narrowband radiation is easily obtained (the spectral width of the radiation at ω_R= ω_{UV}+ ω_{IR} is usually limited by the rather large line width of the Nd:YAG laser light). Finally, the wavelength (λ_L= 425 nm) of the radiation resonant with the transition 4p-5p[1/2,0] is located in the center of the tuning range of Stilbene 3 which is efficiency excited by the third harmonič of a Nd:YAG laser as well as by the radiation of excimer lasers.

Considering these new advantageous possibilities (together with the results of a general theoretical treatment of the frequency mixing in gases)[37] the two-photon resonant conversion in Kr might become a standard method for the generation of broadly tunable coherent radiation in the spectral region of the VUV and of the XUV.

FIFTH-ORDER FREQUENCY MIXING

In the experiments decribed so far third order frequency mixing of dye laser radiation (λ_L= 216-800 nm) generated continuously tunable VUV in the wavelength range of 72 to 200 nm. In principle an extension of this tuning range to shorter wavelengths is possible by conversion processes of higher order.[28,38] Sum-frequency mixing of fifth order, for example, should produce coherent VUV at wavelengths λ_{XUV}= 42-72 nm.

In the past fifth harmonic generation has been investigated with powerful fixed frequency solid state (Nd-YAG or Nd-Glass) and gas (XeCl, KrF) lasers.[29,38-40] In one of these experiments[40] input powers of more than 300 MW (mode-locked Nd-YAG fourth Harmonic, λ= 266.1 nm) provide, for example, conversion efficiencies of 10^{-5} to 10^{-6}. Since the pulse power of most dye laser systems is lower by one or two orders of magnitude nonresonant fifth-order frequency mixing of this radiation would produce intensities below a useful level.

A considerable increase of the generated VUV power is expected, however, from resonant six-wave mixing. In fact resonant fifth-order conversion has been demonstrated in Ar.[3] In these investigations the UV radiation (λ_{UV}= 318.9 nm) of a frequency doubled dye laser (λ_L= 637.8 nm) is resonant with the four-photon Ar-transition $3p-9p$ $[1/2,0]$. Simultaneously $3\omega_{UV}$ is close to the transition frequency ω_{res} of the first Ar resonance ($3p^1S_0$ - $4s[3/2,1]$). The energy difference $\Delta E= 3\omega_1 - \omega_{res}$ of 91 cm^{-1} is sufficiently small so that the fifth-order conversions ω_{XUV} = $5\omega_{UV}$ and $\omega_{XUV}=4\omega_{UV}+\omega_L$ are not only four-photon but also (almost) three-photon resonant. The conversion ω_{XUV}= $3\omega_{UV}+ 2\omega_L$ is five- and nearly three-photon resonant.

Tunable radiation is generated by ω_{XUV}= $4\omega_{UV}+ \omega_V$. With the radiation of a second dye laser (λ_V= 216 - 800 nm) the six-wave mixing ω_{XUV}= $4\omega_{UV}+ \omega_V$ should cover the whole range of 58 to 72nm.

In the experiment λ_V was tuned, for example, in the range of 550 - 580 nm (dye: Rhodamine 6G) and of 275 - 290 nm (the frequency doubled dye laser radiation). In this way the generated XUV was tunable in the wavelength regions λ_{XUV}= 69.85 - 70.1 nm and λ_{XUV}= 62.0 - 62.6 nm, respectively.

Of special interest is of course the XUV pulse power obtainable by this method. Measurements of the dependence of the power

450

P_F of the fifth harmonic on the laser power P_L (P_L= o.2 - 1.5 MW) confirmed that P_F is proportional to P_L^5. The value of P_F could be estimated by comparison with the known pulse power of the VUV generated by fourwave mixing processes. The measured ratio of P_F and of the tripled power P_T is about one hundred. Since for the laser pulse power P_{UV}= 1.5 MW (used in the present investigation) P_T is on the order of 10 W the power P_F should be about 0.1 W or $3 \cdot 10^8$ photons/pulse.

Analogous to the results in Ar six-wave mixing in Ne should generate VUV at even shorter wavelengths. In the case of Ne the UV radiation of a frequency doubled dye laser (λ_{UV}= 282.1nm) and of the sum frequency $\omega_{UV}' = \omega_{UV} + \omega_{IR}$ (λ_{UV}'= 223.0 nm) is four-photon resonant with the 2p-6p$[1/2,0]$ transition ($\omega(2p-6p) = 3\omega_{UV}'$ + ω_{UV}). The frequency $3\omega_{UV}'$ is close to ω_{res} of the resonance transition $2p^1S_o \rightarrow 3s[3/2,1]$. The difference $\Delta E = 3\omega_1 - \omega_{res}$ is only 69.8 cm^{-1}. The (almost) three- and four-photon resonant conversion $\omega_{XUV} = 3\omega_{UV}' + \omega_{UV} + \omega_V$ (with λ_V= 216 - 800 nm) should thus be well suited for the production of tunable XUV in the range λ_{XUV}= 46.2 - 54.8 nm.

At present the resonant six wave mixing in Ar and Ne is subject of detailed investigations. The results obtained so far indicate that the mixing schemes described above are appropriate for the desired extension of the tuning range of the VUV light generated by nonlinear frequency up conversion.

APPLICATIONS

The discussed experimental results demonstrate that nonlinear optical frequency conversion produces widely tunable VUV radiation. Because of the narrow spectral width and the high intensity the VUV light is a powerful tool for high resolution spectroscopy of atoms and molecules. This has been demonstrated, for example, by absorption spectroscopy (carried out on CO[41,42], H$_2$[43,44] and N$_2$[45], by excitation spectroscopy (in CO[8,46], NO[47,48], H$_2$[44,49], Xe$_2$[50,51], Kr$_2$ and Ar$_2$[52] or by state selective resonant-excitation-ionization spectroscopy - first demonstrated in CO and NO[53]- which made possible state-selective photoionization and photodissociation spectroscopy of the H$_2$ molecule from excited states[54].

With narrow band L$_\alpha$ -radiation it was possible to measure the spatial velocity distribution of atomic hydrogen generated by the photodissociation of Hl[55]. Other applications of L_α radiation included the resonant photodissociation of H and D[56], the investigation of the H atom in an external electric[57] or strong magnetic field[58] and the hydrogen-atom photofragment spectroscopy[59].

The powerful VUV generated by resonant frequency mixing is suited for applications which require very intense laser radiation like the multiphoton excitation of atoms[25] and molecules, photodissociation studies of molecules[59] or plasma diagnostics.

Today the number of spectroscopic applications of coherent laser-generated VUV light is still small. Because of the simple way of generation and the excellent spectral properties there is no doubt that in the future this radiation will be very useful for a large variety of spectroscopic applications.

RFERENCES

1 W. Jamroz and B.P. Stoicheff, in Progress in Optics, E.Wolf, ed., North Holland, Amsterdam 1983, vol.20,pp.326-380.
2 C.R.Vidal, in Tunable Lasers (I.F.Mollenauer and J.C.White, eds., Springer Verlag, Heidelberg 1984)
3 R.Hilbig, G.Hilber, A.Lago, B.Wolff, and R. Wallenstein, Comments At.Mol.Phys., part D18, 157 (1986)
4 G.C.Bjorklund, IEEE J.Quantum Electron. QE-11, 287 (1975).
5 R.Mahon, T.J. McIlrath and D.W. Koopman; Appl.Phys.Lett.33, 305 (1978)
6 D.Cotter, Optics Commun. 31, 397 (1979).
7 R.Wallenstein, Optics Commun. 33, 119 (1980)
8 R.Hilbig and R.Wallenstein, IEEE J.Quantum Electron. QE-17, 1566 (1981)
9 R.Hilbig, PhD thesis, 1984 (to be published in Appl.Phys.).
10 R.Hilbig, G.Hilber, A.Timmermann, and R.Wallenstein; "Laser Techniques in the Extreme Ultraviolet", in Proc AIP Conf. vol. 119,p. 1 (1984)
11 R.Hilbig and R. Wallenstein; Appl. Optics 21,913 (1982).
12 R.Hilbig and R.Wallenstein; Optics Commun. 44, 283 (1983).
13 R.Hilbig, A.Lago and R.Wallenstein, Optics Commun. 49, 297 (1984)
14 H.Langer, H.Puell, and H.Röhr; Opt. Commun.34,137 (1980).
15 D.C.Hanna, M.A.Yuratich, and D.Cotter, Nonlinear Optics of Free Atoms and Molecules; Berlin: Springer Verlag, 1979.
16 R.T.Hodgson, P.P.Sorokin, and J.J.Wynne; Phys.Rev.Lett.32, 343 (1974).
17 H.Scheingraber, H.Puell, and C.R.Vidal; Phys.Rev.A 18,2585 (1978).
18 D.M.Bloom, J.T.Yardley, J.F.Young, and S.E.Harris; Appl.Phys. Lett. 427 (1974).
19 S.C.Wallace and G.Zdaziuk; Appl.Phys.Lett.28, 449 (1976).
20 H.Junginger, H.B.Puell, H.Scheingraber, and G.R.Vidal, IEEE J.Quantum Electron. QE-16, 1132 (1980).
21 K.M.Lang, J.F.Ward, and B.J.Orr, Phys.Rev.A 9,2440 (1974).
22 J.Heinrich and W.Behmenburg; Appl.Phys.23, 333 (1980).

23 F.S.Tomkins and R.Mahon; Opt.Lett.6, 179 (1981),IEEE J.Quantum Electron. QE-18, 913 (1982).
24 J.Bokor, R.R.Freeeman, R.L.Panock, and J.C.White, Opt.Lett. 6, 182 (1981);
M.Jopson, R.R.Freeman, and J.Bokor; Proc.XIIth IQEC, App. Phys. B 28,203 (1982); see also "Laser Techniques in the Extreme Ultraviolet" in Proc. AIP Conf.,vol.90.
25 R.Hilbig and R.Wallenstein, IEEE J.Quantum Electron. QE-19, 1759 (1983).
26 W.Jamroz, P.E.LaRocque, and B.P.Stoicheff; Opt.Lett.7, 617 (1982).
27 R.Hilbig and R.Wallenstein, IEEE J.Quantum Electron, QE-19, 194 (1983).
28 J.Bokor, P.H.Bucksbaum and R.R. Freeman; Opt.Lett.8,217 (1983).
29 K.D.Bonin and T.J.McIlrath, JOSA B2, 527 (1984).
30 A.Lago, Ph.D.Thesis, Bielefeld 1987.
31 G.Hilber, A.Lago, and R.Wallenstein, JOSA B (inpress).
32 J.A.Paisner, M.L.Spaeth, D.C.Gerstenberger, and I.W.Rudermann, Appl.Phys.Lett. 476 (1978).
33 C.Chen, B.Wu, A.Jiang, and G.You, Sci. Sinica (Ser.B) vol.28, pp. 235-243 (1985).
34 C.Chen, Y.X.Fan, R.C.Eckardt, and R.L.Byer, in 'Digest of Conference on Lasers and Electro-Optics' (Opt.Soc. of America, Washington, D.C.(1986) p. 322
35 K.Kato, IEEE J.Quantum Electron. QE-22,1013 (1986).
36 K.Miyazaki, H.Sakai, and T.Sato, Opt.Lett.11, 797 (1986).
37 A.Lago, G.Hilber, and R.Wallenstein, Phys.Rev.A (in press).
38 J.Reintjes, Appl.Opt. 19, 3889 (1980) and references therein.
39 J.Reintjes, L.L.Tankersley and R.Christensen; Opt.Commun.39, 355 (1981).
40 J.Reintjes, R.C.Eckardt, C.Y.Shen, N.E.Karangelen, R.C.Elton, and R.A.Andrews; Phys.Rev.Lett.37, 1540 (1976).
41 A.C.Provorov, B.P.Stoicheff and S.C.Wallace, J.Chem.Phys.67, 5393 (1977).
42 J.C.Miller, R.N.Compton and R.W.Cooper, J.Chem.Phys.76,3967 (1982).
43 M.Rothschild, H.Egger, R.T.Hawkins, J.Bokor, H.Pummer and C.K. Rhodes, Phys.Rev.A23,206 (1981).
44 E.E.Marinero, C.T.Rettner and R.N.Zare, Chem.Phys.Lett.95, 486 (1983).
45 P.R.Herman and B.P.Stoicheff, Opt.Lett.10, 502 (1985).
46 P.Klopotek and C.R.Vidal, Can.J.Phys.62,1426 (1984).
47 J.R.Banic, R.H.Lipson, T.Efthimiopoulos, and B.P.Stoicheff, Opt.Lett.6,461 (1981).
48 H.Scheingraber and C.R.Vidal, JOSA B2, 343 (1985).
49 F.J.Northrup, J.C.Polanyi, S.C.Wallace and J.M.Williamson, Chem.Phys.Lett.105,34 (1984).
50 R.H.Lipson, P.E. LaRoque, and B.P.Stoicheff, Opt.Lett.9,402 (1984)
51 R.H.Lipson, P.E.LaRoque, and B.P.Stoicheff, J.Chem.Phys.82, 4470 (1985).
52 B.P.Stoicheff and A.A.Madej (see contribution at this conference)
53 H.Zacharias, H.Rottke, and K.H.Welge, Opt.Commun 35, 185 (1980)
54 H.Rottke and K.H.Welge, Chem.Phys.Lett.99, 456 (1983);
J. Physique 46, (1-127) (1985)
55 R.Schmiedl., H.Dugan, W.Meier, and K.H.Welge, Z.Phys.A304, 137 (1982).

56 H.Zacharias, H.Rottke, J.Danon and K.H.Welge, Opt.Commun.<u>37</u>, 15 (1981).
57 H.Rottke and K.H.Welge, Phys.Rev.A <u>33</u>, 301 (1986).
58 A.Holle, J.Main, G.Wiebusch, and K.H.Welge (see contribution at this conference).
59 H.J.Krautwald, L.Schnieder, K.H.Welge, and M.N.R.Ashfold, Faraday Discuss.Chem.Soc.<u>82</u>,7 (1986).

Third-Harmonic Generation in Both Positively and Negatively Dispersive Xe

P.R. Blazewicz and J.C. Miller

Chemical Physics Section, Oak Ridge National Laboratory,*
P.O. Box X, Oak Ridge, TN 37831, USA

The multiphoton ionization (MPI) spectra of rare gases at pressures above about 10^{-2} Torr are strongly influenced by the presence of third-harmonic (TH) light which is generated in the focal region concurrently with any multiphoton absorption [1,2]. The presence of this additional frequency at 3ω and its intensity dependence on wavelength, pressure, focal length, or the addition of buffer gas must be accounted for in order to fully understand the MPI spectrum. Most dramatic is the complete suppression at modest pressures of resonances in the MPI which are both one- and three-photon allowed. The disappearance of resonances is due to the interference between the two possible excitation pathways involving either one TH photon or three laser photons [1,2]. Conversely, MPI is an excellent way to study TH generation, providing an in-situ probe of the wavelength dependence and relative intensity of the vacuum ultraviolet (VUV) light without the need for MgF_2 windows or vacuum monochromators.

Two-color MPI studies provide another dimension for the study of TH generation. In these experiments, one laser is responsible for VUV generation and a second, probe laser ionizes via a fourth-photon resonance. Using a weak probe laser, we have previously investigated the cancellation effect on the Xe $6s[3/2]_{J=1}^0$ state and determined the role of dimers in the MPI spectrum [2]. In these experiments the probe laser was too weak to fundamentally alter the TH generation.

In the present work, we have investigated modifications of TH generation by a second, <u>strong</u> laser. We report the observation of <u>tunable</u> TH light in the positively dispersive region on the low energy side of the Xe $6s'[1/2]_1^0$ state induced upon introduction of a second, strong laser field. We also report observations of strong, systematic increases and decreases in ionization signals generated to the blue of the 6s' level. Our results are interpreted in terms of phase-matching changes induced by the second laser which affect TH production and subsequent ionization of Xe atoms. In addition, we present theoretical expressions developed by PAYNE and GARRETT [3] for the effect of the second laser field on TH production and MPI on both the positively and negatively dispersive sides of the three-photon resonances. Similar expressions were recently published independently by TEWARI and AGARWAL [4].

Typically, the initial pumping with three photons at ω_1 may fall in the positively dispersive region ($\Delta k > 0$) just to lower energy than the

---
*Operated by Martin Marietta Energy Systems, Inc. under contract DE-AC05-84OR21400 with the U.S. Department of Energy.

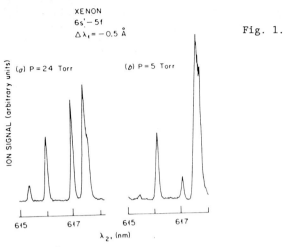

Fig. 1.

$6s'[1/2]^0_{J=1}$ resonance of Xe or the negatively dispersive region ($\Delta k<0$) to the blue. The second laser is then tuned to strongly couple to a particular J level in the 5f manifold of states at the fourth photon level. Figure 1 shows MPI spectra at two pressures obtained by scanning the second laser through the 5f levels while ω_1 is tuned 0.5 Å to the red of the 6s resonance. The first and third peaks represent excitation to J=4 levels by three photons of frequency ω_1 and one photon of ω_2. The second and fourth peaks arise from pumping of J=2 levels in the same way but in addition can be produced by one TH photon plus one photon at frequency ω_2. The change in relative intensity of these two pairs of peaks reflects differing amount of TH light due to different phase-matching conditions at the two pressures. In addition, TH light generated in positively dispersive Xe has been directly observed in a second ionization cell.

When laser one is tuned to the negatively dispersive region to the blue of the 6s' state, normal TH generation occurs. In this case, the addition of the second laser can either enhance or reduce the TH production. Figure 2 illustrates this effect. The broad envelope is due to ionization by a TH photon plus visible photons. Depending on details of

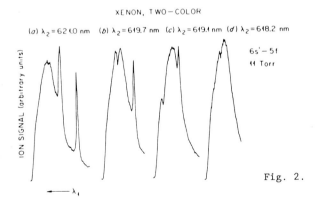

Fig. 2.

the phase-matching, coupling to the 5f levels with the second laser can lead to either peaks or dips in the ionization signal. Enhancement occurs when $b\Delta k < (b\Delta k)_{optimum}$ and deenhancement when $b\Delta k \geq (b\Delta k)_{optimum}$. The above relationship appears to be quite generally true and peaks can be converted to dips or visa versa by changing the phase-matching in several different ways. These include changes in the laser two frequency (Fig. 2), changes in pressure, addition of buffer gas or use of a different focal length lens. This quite general experimental observation is not fully understood theoretically [3]. Further details of this work may be found elsewhere [5,6].

1. J. C. Miller, R. N. Compton, M. G. Payne, W. R. Garrett: Phys. Rev. Lett. 45, 114 (1980); J. C. Miller, R. N. Compton: Phys. Rev. A25, 2056 (1982).
2. R. N. Compton, J. C. Miller: J. Opt. Soc. Am. B2, 355 (1985).
3. M. G. Payne, W. R. Garrett: private communication.
4. S. P. Tewari, G. S. Agarwal: Phys. Rev. Lett. 56, 1811 (1986).
5. P. R. Blazewicz, M. G. Payne, W. R. Garrett, J. C. Miller: Phys. Rev. A34, 5171 (1986).
6. P. R. Blazewicz, J. C. Miller: to be published.

Laser Depletion Spectroscopy of Core-Excited Levels

S.E. Harris and J.K. Spong

Edward L. Ginzton Laboratory, Stanford University,
Stanford, CA 94305, USA

1. Introduction

For several years now we have been developing techniques for using the properties of high lying metastable levels for high resolution spectroscopy of
core-excited levels [1]. At Tegernsee [2], we discussed the use of anti-
Stokes scattering to obtain the isotopic shift of the He $2s^1S$ level in ^{3}He
and ^{4}He. At Jasper [3], we described the application of a hollow cathode
based tunable anti-Stokes source to the high resolution spectroscopy of the
$3p^6$ shell of neutral potassium. Our paper at Interlaken [4] reported the
transfer of energy from the metastable Li($1s2s2p$ ^4P) level to the radiating
Li($1s2p^2$ ^2P) level, thereby interconnecting the quartet and doublet manifolds. Most recently in Maui [5], we discussed the properties of a subclass
of metastable core-excited levels which have been termed quasi-metastable
[1]. These levels which lie near the bottom of the core-excited manifolds
of the alkali atoms and alkali-like ions, autoionize sufficiently slowly to
have radiative yields of about 50%, and typically dominate the neutral XUV
emission spectrum of these atoms and ions. They form the basis of the laser
depletion spectroscopy technique [6,7] which we describe here.

2. Depletion Spectroscopy

Figure 1 shows a partial energy level diagram of neutral rubidium. In Rb,
the lowest quasi-metastable level is the even parity $4p^55s5p$ $^4S_{3/2}$ level
which radiates at 82.4 nm [8] to the valence level $4p^65p$ $^2P_{3/2}$.

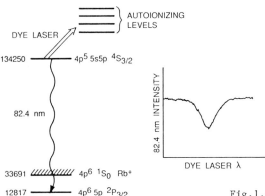

167032 $4p^5 5s$ $^3P_2$ Rb$^+$

AUTOIONIZING LEVELS

DYE LASER

134250 — $4p^5 5s5p$ $^4S_{3/2}$

82.4 nm

82.4 nm INTENSITY

DYE LASER λ

33691 $4p^6$ $^1S_0$ Rb$^+$

12817 — $4p^6 5p$ $^2P_{3/2}$

0 cm^{-1} — $4p^6 5s$ $^2S_{1/2}$

Fig.1. Energy level diagram of Rb
showing depletion technique

458

In this technique, we use laser produced x-rays to impulsively excite this quasi-metastable level and following this excitation, monitor the flourescence at 82.4 nm. A tunable dye laser is used to transfer the radiating quasi-metastable atoms to other levels within the core-excited manifold. As the wavelength of the dye laser is scanned, a level is encountered, and the quasi-metastable population is transferred to it, depleting the quasi-metastable flourescence at 82.4 nm. The shape and position of the depleted signal, as a function of dye laser wavelength, determine the position and linewidth of the autoionizing level. Because the visible dye laser is the resolving instrument, the resolution of this technique is determined by its bandwidth; relative to the position of the quasi-metastable level, the energies of the accessed levels may be determined to sub-wavenumber accuracy.

The quasi-metastable level is excited using laser produced plasma excitation, shown schematically in Fig. 2. A 350 mj, 7 ns pulse of 1064 nm Nd:YAG radiation is focused onto a solid tantalum target, forming a plasma which then radiates a blackbody-like continuum of soft x-rays into the surrounding Rb vapor. These x-rays photoionize p-shell electrons of neutral Rb, leaving it in the excited ion configuration Rb^+ $4p^5 5s$. The ions then undergo charge transfer collisions with the ground state neutral atoms, forming quasi-metastable atoms and ground state ions. The quasi-metastables flouresce at 82.4 nm, and the radiation is collected by a wide slit spectrometer. The transfer beam is the output of a dye laser pumped by the same Nd:YAG laser which provided the 1064 nm plasma beam. The dye laser traverses the excited Rb vapor along the line of sight of the spectrometer, depleting the population in the viewed region.

Laser produced plasma excitation is important to the depletion technique because it provides large count rates on the 82.4 nm emission line, allowing analog detection and a good signal to noise ratio. In addition, because the x-ray excitation pulse is short relative to the flourescent lifetime of the quasi-metastable level, transfer of the population out of the level can take place after the pumping pulse has ended, giving good contrast to the depletion signals. These two features allow rapid data collection; scanning is typically done at 1 cm^{-1}/sec.

Figure 3 shows the signal level at 82.4 nm as a function of dye laser wavelength in the tuning range of Coumarin 500 dye. Two depletion signals are evident in the quasi-metastable flourescence at 505.0 nm and 499.9 nm, corresponding to absolute energies of 154044 and 154253 cm^{-1}, respectively.

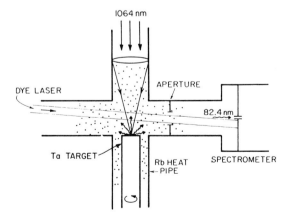

Fig. 2. Schematic of experimental setup

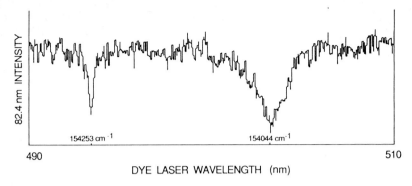

Fig. 3. Experimental trace for the dye laser wavelength region 490-510 nm

3. Sub-Doppler Linewidth Measurements

For rapidly autoionizing levels, such as that located at 154044 cm^{-1} in Fig.3, the halfwidth of the depletion signal is the Lorentz width of the autoionizing level, and therefore it is the Fourier transform of the natural autoionizing lifetime. In such cases, the lifetime of the level can be measured directly from the depletion curve. For narrower, longer lived states however, such as that located at 154253 cm^{-1} in Fig. 3, Doppler broadening, hyperfine structure and the laser probe width may determine the halfwidth of the depletion signal. In this case, a saturation technique is employed which makes use of the fact that the transition probability in the far wings of a convolved Doppler-Lorentz (Voigt) profile varies as the oscillator strength and Lorentz width of the transition, and is independent of the Doppler-hyperfine-probe laser linewidth. The reason for this is that the wings of a Lorentzian lineshape drop off more slowly than those of a Doppler lineshape, so that at sufficiently large detunings,the Voigt profile is essentially Lorentzian in shape.

Figure 4a shows a depletion signal using a weak probe laser of energy E, which is insufficient to saturate the transition shown. The area of the depletion signal, A_{unsat}, is measured. In Fig. 4b, the laser energy has

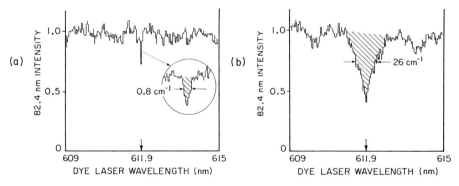

Fig. 4. a) Unsaturated depletion curve, laser energy E = 6.8 x 10^{-6} J/cm^2
b) Saturated depletion curve, E = 1.2 x 10^{-2} J/cm^2
Shaded regions are A_{unsat} and A_{sat}, respectively.

been increased to γE, large enough to strongly saturate the transition even at detunings several Doppler widths from line center. One measures the depletion area, A_{sat}, and the depth, D, of the saturated signal. The Lorentz width, $\delta\nu_L$, is given by

$$\delta\nu_L = (A_{sat})^2 / (2D\gamma A_{unsat}). \tag{1}$$

Using this technique, the lifetime of the transition shown in Fig. 4 was measured to be 41 ps. This saturation method is capable of measuring level lifetimes as long as the pulse length of the probe laser, and therefore 10-100x narrower than the Doppler width of these lines. No knowledge of the transition oscillator strength, absolute probe laser intensity or probe shape is required.

4. Extreme Ultraviolet Flourescence

The ability to measure sub-Doppler linewidths has led to the identification of a number of core-excited levels whose autoionizing times were sufficiently long that large branching ratios to XUV radiation were expected. This was confirmed by pumping the $4p^54d(^3D)5s$ $^2D_{3/2}$ $4p^55s6s$ $^4P_{3/2}$, and $4p^55s6s$ $^2P_{1/2}$ levels located at 148904, 151310 and 152314 cm^{-1} respectively. Branching ratios to XUV radiation of 3%, 100% and 35%, occurring at 77.2, 76.2 and 75.6 nm, respectively, were observed from these levels. The favorable gain cross sections and long autoionizing lifetimes of these states make them attractive candidates for XUV laser systems.

5. Acknowledgements

The authors thank R. Buffa, A. Imamoglu, J.D. Kmetec and J.F. Young for help and discussions. The work described here was supported by the Army Research Office and the Air Force Office of Scientific Research.

6. References

1. S.E.Harris, J.F.Young: JOSA B, 4, 547 (1987)
2. S.E.Harris, J.F.Young, W.R.Green, R.W.Falcone, J.Lukasik, J.C.White, J.R.Willison, M.D.Wright, G.A.Zdasiuk: In Laser Spectroscopy IV, ed. by H.Walther and K.W.Rothe (Springer, Berlin, Heidelberg 1979)
3. S.E.Harris, R.W.Falcone, M.Gross, R.Normadin, K.D.Pedrotti, J.E.Rothenberg, J.C.Wang, J.R.Willison, J.F.Young: In Laser Spectroscopy V, ed. by A.R.W.McKellar, T.Oka, and B.P.Stoicheff (Springer,Berlin, Heidelberg 1981)
4. S.E.Harris, J.F.Young, R.G.Caro, R.W.Falcone, D.E.Holmgren, D.J.Walker, J.C.Wang, J.E.Rothenberg, J.R.Willison: In Laser Spectroscopy VI, ed. by H.P.Weber and P.Luthy (Springer,Berlin, Heidelberg 1983
5. S.E.Harris, J.F.Young, A.J.Mendelsohn, D.E.Holmgren, K.D.Pedrotti, and D.P.Dimiduk: In Laser Spectroscopy VII, ed. by T.W.Hansch and Y.R.Shen (Springer,Berlin, Heidelberg 1985)
6. J.K.Spong, J.D.Kmetec, S.C.Wallace, J.F.Young, S.E.Harris: Phys.Rev.Lett. (1987, to be published)
7. K.D.Pedrotti: Opt.Commun. (1987, to be published)
8. J.Reader: Phys.Rev. A (to be published)

Laser Spectroscopy of Atoms and Molecules Below 100 nm

W.E. Ernst*, *T.P. Softley+*, *L.M. Tashiro*, and *R.N. Zare*

Department of Chemistry, Stanford University, Stanford, CA 94305, USA

1. Introduction

For most atoms and molecules single photon excitation from the ground state to Rydberg states requires working in the vacuum ultraviolet (VUV). In recent years frequency upconversion of high power dye lasers has extended the application of laser spectroscopy into the VUV and even the windowless region below 104 nm, the extreme vacuum ultraviolet (XUV) [1]. In this work we generate XUV radiation by tripling the frequency of a frequency doubled dye laser in a free jet of the rare gases Ar and Xe [2]. In order to avoid unwanted UV multiphoton absorption processes in the sample under investigation we separate the third harmonic XUV beam from the collinearly propagating high power fundamental UV light in a dichroic beam splitter set-up. The XUV is focussed into an atomic or molecular beam for studying the photoexcitation, ionization or fragmentation of the species.

2. XUV Laser Spectrometer

Figure 1 shows the complete spectrometer set-up. A pulsed dye laser operating with various dyes is pumped by a Nd-YAG laser of 6 ns pulse length at 10 Hz repetition rate. The dye laser output is frequency doubled in a KDP crystal. The generated UV radiation of about 10mJ/pulse is focussed into a pulsed free jet of rare gas atoms which serves as a gaseous nonlinear medium [2]. Frequency tripling in a Xe jet yields about 10^8 photons per pulse in the 93 nm region. The XUV and the UV fundamental are incident on a beam splitter consisting of a coated quartz plate. The light is s-polarized with the angle of incidence being 70°. With a quarter wavelength layer of MgF_2 the reflection is about 45% in the XUV and 1% in the UV [3]. The light passes a second beam splitter with the result that the fundamental UV intensity is reduced by four orders of magnitude and about 20% of the generated XUV can be used for spectroscopy. The first beam splitter has a torroidal surface focussing the XUV on to the sample, which is in our case a pulsed atomic or molecular beam. An electron multiplier detects the XUV and can be used for absorption measurements. Visible or UV fluorescence from the sample is collected and focussed on to a photomultiplier. Generated ions are detected in a 50 cm time-of-flight mass spectrometer.

3. Spectroscopic Applications

In a first application of the XUV laser spectrometer we have studied the Stark effect of Xe autoionizing Rydberg states in the vicinity of the second ionization limit around 92.5 nm. These states autoionize and Xe^+ ions are detected in

---

* Present address: Institut für Molekülphysik, Freie Universität Berlin, Arnimallee 14, D-1000 Berlin 33, W. Germany.
+ Present address: University Chemical Laboratory, Lensfield Road, Cambridge, CB2 1EW, England.

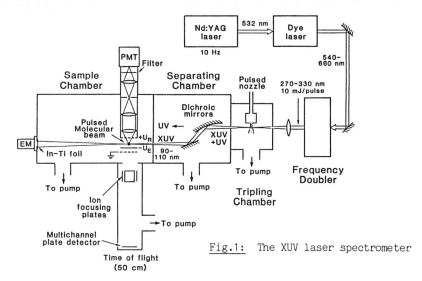

Fig.1: The XUV laser spectrometer

the time-of-flight mass spectrometer. Using different repeller and extractor plate potentials we varied the electric field strength in the XUV-Xe-atom interaction region between 30 V/cm and 2000 V/cm and recorded Rydberg series from n=14 to the ionization limit. In addition to the ns and nd series, np lines appear in the spectrum already at low field strengths and at higher fields hydrogen like Stark multiplets start to be observed [4].

In addition we studied the autoionization structure in the photoionization efficiency curve of O_2 between 90 and 95 nm at 0.001 nm resolution. Again, ions were detected. The rotational temperature in the beam was about 10 K. The linewidth of the rotational structure appeared to be determined by autoionization broadening. In another application a cold N_2 beam was generated and the N_2 b $^1\Pi_u$ - X $^1\Sigma_g^+$ spectrum around 98 nm was recorded by monitoring the N_2^+ ion yield. The b $^1\Pi_u$ state lies below the ionization limit and the ion yield was due to a 1 + 1 resonance enhanced multiphoton ionization, i.e. the absorption of an XUV and a UV photon which comes from the residual UV fundamental radiation. The UV intensity, however, was too low for inducing any UV multiphoton absorption which was carefully checked. The XUV spectrometer can also be used for the study of the photodissociation dynamics of small molecules. As an example, we used XUV radiation between 90.3 and 92.5 nm to excite CO_2 molecules into high Rydberg states below the ionization threshold. The excited molecules dissociate into CO and O with the fragment CO being in highly excited triplet states. Visible fluorescence from these CO states and the total XUV absorption were simultaneously recorded as a function of the exciting XUV wavelength yielding information about the dissociation process [5].

The work was funded in part by ONR under N00014-78C-0403 and NSF PHY 85-06668. WEE was supported by a Heisenberg Fellowship of the Deutsche Forschungsgemeinschaft and TPS by a Harkness Fellowship of the Commonwealth Fund.

References

1. R. Wallenstein, see contribution in this volume
2. C.T. Rettner, E.E. Marinero, R.N. Zare, A.H. Kung, J.Phys.Chem.88, 4459(1984)
3. R.W. Falcone and J. Bokor, Opt.Lett. 8, 1 (1983)
4. W.E. Ernst,T.P. Softley, R.N. Zare, to be published
5. T.P. Softley, W.E. Ernst, L.M. Tashiro, R.N. Zare, Chem.Phys. (in press)

Internal State Selected Population and Velocity Distribution of Hydrogen Desorbing from Pd(100)

L. Schröter[1], *G. Ahlers*[1], *H. Zacharias*[1], *and R. David*[2]

[1]Fakultät für Physik, Universität Bielefeld,
 D-4800 Bielefeld, Fed. Rep. of Germany
[2]IGV, Kernforschungsanlage, D-5700 Jülich, Fed. Rep. of Germany

State selective investigations of molecule-surface dynamics have attracted much interest recently. While several groups are studying the scattering of molecules from surfaces by tunable laser spectroscopy [1], only a few experiments have been carried out on molecular desorption [2]. Recently, the rotational state selective detection of molecular hydrogen has been achieved with high sensitivity, which subsequently has been used for first experiments on the desorption of hydrogen from Cu(111), Cu(100) [3] and poly-Pd [4]. The permeation of hydrogen through these metals facilitates continuous desorption experiments, removing experimental constraints associated with measurements in consecutive adsorption-desorption cycles.

The D_2 molecules were detected by resonantly enhanced two-photon ionization. The rovibrational state selectivity was achieved by single photon excitation in the B $^1\Sigma_u^+$ ← X $^1\Sigma_g^+$ Lyman bands with a VUV laser tunable around 106 nm. This tunable VUV light was generated by frequency tripling a tunable UV laser in Xe. At input intensities of ~ 1 MW at λ ~ 319 nm about 0.5 W of VUV radiation was available in the center of the UHV chamber. Within the tuning range of Xe we could excite part of the (4-0) and (7-1) Lyman bands. The excited molecules were ionized by a second UV laser photon. D_2^+ ions were measured by a MCP detector.

Rotational temperatures T_{rot} in the vibrational ground state determined at various surface temperatures (375 k $\leq T_s \leq$ 740 k) are shown in Fig.1. T_{rot} is slowly increasing with T_s, however, always less than T_s with a tendency of leveling-off at higher surface temperatures. This rotational cooling, previously observed in molecular surface scattering [1] and by us in desorption from sulfur covered poly-PD [4], was for the first time observed for a clean single crystal surface. The relative vibrational state population shows Arrhenius behaviour. From the slope of an Arrhenius plot an activation energy of about E_a(v=1) ~ 4.2 kcal/mol for clean Pd(100) is observed.

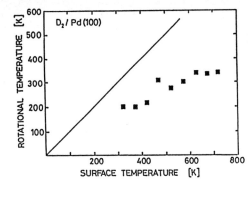

Fig. 1:

Rotational temperature of $D_2(v" = 0)$ as a function of the surface temperature

Fig. 2:

Kinetic energy $\langle E \rangle/2k$ of desorbing hydrogen in specific internal states. $T_s = 405$ k, clean Pd(100).

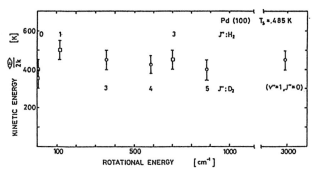

Velocity distributions of D_2 molecules in a specific internal state can be measured in the same set-up. After resonant ionization, which provides the state selection, the generated D_2^+ ions drift in an electric field free region to the detector with exact the same velocity as the formerly neutral D_2 molecules [5]. Measuring the time-of-flight from ionization to detection directly gives the velocity of the desorbing D_2 molecules. At a surface temperature of about 485 K the mean kinetic energy $\langle E/2k \rangle$ obtained is shown in Fig.2 for various rotational states of H_2 and D_2. A slight decrease of velocity ($\sim$ 10-20%) was observed, similar to a recent measurement of Ertl et al. [6] in the elastic scattering component of NO from Ge.

[1] J.A. Barker, D.J. Auerbach, Surf. Sci. Rep., 4, 1 (1985)

[2] G. Comsa, R. David, Surf. Sci. Rep. 5, 145 (1985)

[3] G.D. Kubiak, G.O. Sitz, R.N. Zare, J. Chem. Phys. 83, 2538 (1985)

[4] H. Zacharias, R. David, Chem. Phys. Lett. 115, 205 (1985);

    L. Schröter, H. Zacharias, R. David, Appl. Phys. A 41, 95 (1986)

[5] J. Häger, Y.R. Shen, H. Walther, Phys. Rev. A 31, 1962 (1985)

[6] A. Mödl et al., Phys. Rev. Lett. 57, 384 (1986)

Laser Spectroscopy of the Diamagnetic Hydrogen Atom in the Chaotic Regime

A. Holle, J. Main, G. Wiebusch, H. Rottke, and K.H. Welge

Fakultät für Physik, Universität Bielefeld,
D-4800 Bielefeld, Fed. Rep. of Germany

Atomic diamagnetism is still an essentially open problem of fundamental importance. Since the discovery of atomic quasi-Landau resonances by Garton and Tomkins (1) it has been a subject of intense experimental and theoretical research (2). With its purely Coulombic field, the hydrogen atom has served as basis for virtually all theoretical work (3). The general interest arises from the non-separability of the Schrödinger equation, even with the simplest Hamiltonian, containing the Coulomb and diamagnetic term only ($\gamma = B/(2.35 \times 10^5 T)$):

$$H = \frac{1}{2}p^2 - 1/r + \frac{1}{8}\gamma^2 \rho^2 \quad \text{(a.u.)}.$$

Experiments with the diamagnetic H-atom have recently been performed by us for the first time. The results have been compared with exact quantum mechanical calculations in the energy range below the ionization threshold (4). Measurements in the energy range around threshold have revealed the existence of a multitude of further quasi-Landau resonance types (5, 6).

Briefly hydrogen atoms are excited to even parity final states with m=0 by resonant two-photon absorption through the Paschen-Back resolved 2p state:

$$H(1s,m=0) + h\nu(\lambda=121.6 \text{ nm}) \rightarrow H(2p,m=0),$$
$$H(2p,m=0) + h\nu(\lambda \approx 365 \text{nm}) \rightarrow H^*(m=0).$$

The vacuum-ultraviolet laserlight for the first excitation step and ultraviolet laserlight for the second excitation step was pulsed and linearly polarized parallel to the magnetic field. The bandwidth was 3 GHz and 1 GHz, respectively. Electrons produced by photo- or fieldionization were detected with a surface barrier diode. While taking the spectrum in fig. 1a the vuv was fixed to a Δ m=0 Lyman-α transistion and the uv was scanned through an energy range around threshold. The discovered new resonances, i.e. modulations of the energy spectrum, were found by Fourier

466

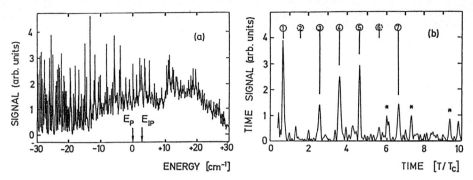

Fig.1: (a) Excitation ionization spectrum of H-atom Balmer series in a static homogeneous magnetic field of 5.96 T, (b) Fourier transformation of (a), plotted is the absolute value squared, T_c is the cyclotron period. (From ref.7)

transforming the experimental data into a time spectrum (fig.1b). The resonance types ① to ⑦ were explained by classical trajectory calculation (6). A detailed discussion of shape, energy dependence and quantization of the corresponding trajectories is given in ref.7. As an example fig.2 shows 21 trajectories at a magnetic field of 6 Tesla in the energy range from -40 cm^{-1} to $+60$ cm^{-1} in their ρ-z projection. This trajectory type explains quantitatively the position of peak ④ at 3.6 T_c.

The method of Fourier transforming an absorption spectrum measured at constant magnetic field over a defined energy range, however, bears a principle problem: The resonances are energy dependent. This leads to a broadening of the Fourier peaks, as the Fourier integral averages the dependence. Strongly energy dependent resonance types vanish or interfere in an undefined way, if the energy range is too large. But a shortening of the measured range again leads to a broadening of the Fourier peaks, because of the smaller width of the finite Fourier integral. Thus, this

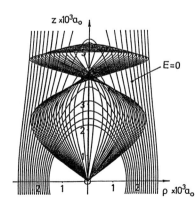

Fig.2: Calculated closed classical trajectories of electron motion at energies from -40 cm^{-1} to $+60$ cm^{-1}, drawn in increments of 5 cm^{-1}. Projection on (ρ, z) coordinates. Equidistant (10 cm^{-1}) potential lines, E=0: field-free ionization potential.

467

method allows to discover only those resonance types which depend only little on energy. Indeed, the new types discovered recently are characterized by being nearly independant of energy over the measured range (7).

This principle problem, arising from the energy dependance of the quasi-Landau resonances, can be solved by taking the scaling properties of the classical Hamiltonian into account. Introducing semiclassical quantization of the trajectories the resonances appear equidistant in $\gamma^{-1/3}$ if the scaled energy $E_{scal} = E\gamma^{-2/3}$ is kept constant. In other words, the classical system does not depend on binding energy E and magnetic field strength γ independantly, it only depends on a special ratio of both. Fourier transformation of such scaled spectra yields sharp structures, corresponding to classical trajectories (for a proper devivation see (7)).

For a propper detection of energy dependant resonance types (e.g. the ones denoted by asterix in fig.1b), it is necessary to measure at constant scaled energy. Fig.3a shows a scaled spectrum at E_{scal} = -0.45. During measurement, the magnetic field was scanned with a rate proportional to $\gamma^{-1/3}$ and the laser was stepped synchroniously to satisfy the scaling relation. The Fourier transformation of the scaled spectrum is given in fig.3b. It shows clear structures and each peak can be explained by a classical trajectory. For the four most prominent resonance types the shapes of the corresponding trajectories are given in the figure. The $n\gamma^{-1/3}$-values for peaks and trajectories agree to better than 0.5%.

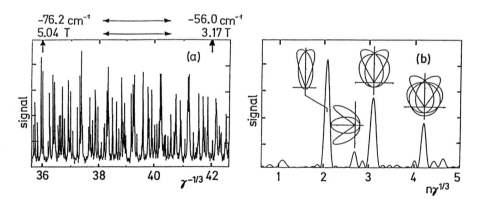

Fig.3:   (a) Scaled excitation spectrum at E_{scal} = -0.45.   (b) Fourier transformation of (a), plotted is the absolute value squared, for the peaks at $n\gamma^{1/3}$ = 2.10, 2.70, 3.13, 4.13 the corresponding trajectories are shown in ρ-z projection.

In future measurements, where the scaled energy shall be increased in small steps, we will study the energy development of these peaks to energies above threshold and hope to get a deeper understanding of the correspondance between classical and quantum mechanics and classical and quantum chaos for the diamagnetic H-atom.

REFERENCES

(1) W.R.S. Garton and F.S. Tomkins, Astrophys. J. 158, 839 (1969)
(2) J.C. Gay, in "Progress in Atomic Spectroscopy, Part C", edited by H.J. Beyer, and H. Kleinpoppen, (Plenum, New York, 1984), pp. 177-246; C.W. Clark, K.T. Lu, and A.F. Starace, in "Progress in Atomic Spectroscopy, Part C", edited by H.J. Beyer, and H. Kleinpoppen, (Plenum, New York, 1984), pp. 247-320
(3) G. Wunner, U. Woelk, I. Zech, G. Zeller, T. Ertl, F. Geyer, W. Schweitzer, and H. Ruder, Phys. Rev. Lett. 57, 3261 (1986); D. Wintgen, and H. Friedrich, Phys. Rev. Lett. 57, 571 (1986); M.L. Du, J.B. Delos, Phys. Rev. Lett. 58, 1731 (1987); D. Delande, J.C. Gay Phys. Rev. Lett. 57, 2006 (1986); D. Wintgen, Phys. Rev. Lett. 58, 1589, (1987);
(4) A. Holle, G. Wiebusch, J. Main, K.H. Welge, G. Zeller, G. Wunner, T. Ertl, and H. Ruder, Z. Phys. D 5, 279 (1987); D. Wintgen, A. Holle, G. Wiebusch, J. Main, H. Friedrich, and K.H. Welge, J. Phys. B 19, 1557 (1986);
(5) A. Holle, G. Wiebusch, J. Main, B. Hager, H. Rottke, and K.H. Welge, Phys. Rev. Lett. 56, 2594 (1986)
(6) J. Main, G. Wiebusch, A. Holle, and K.H. Welge, Phys. Rev. Lett. 57, 2789 (1986)
(7) J. Main, A. Holle, G. Wiebusch, and K.H. Welge, Z. Phys. D (1987), accepted for publication

Index of Contributors

Proceedings of highly esteemed series of Conferences

published in

Springer Series in Optical Sciences

Volume 7

Laser Spectroscopy III

Proceedings of the Third International Conferences, Jackson Lake Lodge, Wyoming, USA, July 4–8, 1977

J.L. Hall, J.L. Carlsten (Eds.)

1977. 296 figures. XI, 468 pages.
ISBN 3-540-08543-2

Volume 21

Laser Spectroscopy IV

Proceedings of the Fourth International Conference Rottach-Egern, Federal Republic of Germany, June 11–15, 1979

H. Walther, K. W. Rothe (Eds.)

1979. 411 figures, 19 tables. XIII, 652 pages.
ISBN 3-540-09766-X

Volume 30

Laser Spectroscopy V

Proceedings of the Fifth International Conference
Jasper Park Lodge, Alberta, Canada, June 29–July 3, 1981

A. R. W. McKellar, T. Oka, B. P. Stoicheff (Eds.)

1981. 319 figures. XI, 495 pages.
ISBN 3-540-10914-5

Springer-Verlag
Berlin Heidelberg New York
London Paris Tokyo

Volume 40

Laser Spectroscopy VI

Proceedings of the Sixth International Conference
Interlaken, Switzerland, June 27–July 1, 1983

H. P. Weber, W. Lüthy (Eds.)

1983. 258 figures. XVII, 442 pages.
ISBN 3-540-12957-X

Contents: Photons in Spectroscopy. – Spectroscopy of Elementary Systems. – Coherent Processes. – Novel Spectroscopy. – High Selectivity Spectroscopy. – High Resolution Spectroscopy. – Cooling and Trapping. – Collisions and Thermal Effects on Spectroscopy. – Atomic Spectroscopy. – Rydberg-State Spectroscopy. – Molecular Spectroscopy. – Transient Spectroscopy. – Surface Spectroscopy. – NL-Spectroscopy. – Raman and CARS. – Double Resonance and Multiphoton Processes. – XUV – VUV Generation. – New Laser Sources and Detectors. – Index of Contributors.

Volume 49

Laser Spectroscopy VII

Proceedings of the Seventh International Conference
Maui, Hawaii, USA, June 24–28, 1985

T. W. Hänsch, Y. R. Shen (Eds.)

1985. 298 figures. XV, 419 pages.
ISBN 3-540-15894-4

The emphasis in these proceedings is on new developments in basic principles and phenomena, coherent sources and techniques, and various applications of laser spectroscopy. There are sections on laser cooling, trapping and manipulation of atoms and ions, applications to basic physics, Rydberg states, atomic spectroscopy, molecular and ion spectroscopy, VUV and X-ray sources and spectroscopy, nonlinear optics and wave-mixing spectroscopy, quantum optics, squeezed states and chaos, coherent transient effects, surfaces and clusters, ultrashort pulses and applications to solids, spectroscopic sources and techniques, and miscellaneous applications.

Springer

Ultrafast Phenomena V

Proceedings of the Fifth OSA Topical Meeting
Snowmass, Colorado,
June 16–19, 1986

G. R. Fleming, A. E. Siegman (Eds.)

1986. 427 figures. Approx. 576 pages. (Springer
Series in Chemical Physics, Volume 46).
ISBN 3-540-17077-4

Contents: Mode Locking and Ultrashort Pulse
Generation. – Ultrafast Optical Generation and
Measurement Techniques. – Electrooptic
Sampling Techniques. – Nonlinear Optics and
Continuum Generation. – Applications to
Semiconductors, Quantum Wells, and Solid
State Physics. – Chemical Reaction Dynamics.
– Dynamics of Biological Processes. – Energy
Transfer and Relaxation. – Coherent Spectro-
scopic Techniques. – Index of Contributors.

Ultrafast Phenomena IV

Proceedings of the Fourth International
Conference
Monterey, California, June 11–15, 1984

D. H. Auston, K. B. Eisenthal (Eds.)

1984. 370 figures. XVI, 509 pages. (Springer
Series in Chemical Physics, Volume 38).
ISBN 3-540-13834-X

Contents: Part I: Generation and Measure-
ment Techniques. – Part II: Solid State Physics
and Nonlinear Optics. – Part III: Coherent
Pulse Propagation. – Part IV: Stimulated
Scattering. – Part V: Transient Laser Photoche-
mistry. – Part VI: Molecular Energy Redistri-
bution, Transfer, and Relaxation. – Part VII:
Electronics and Opto-Electronics. – Part VIII:
Photochemistry and Photophysics of Proteins,
Chlorophyll, Visual Pigments, and Other
Biological Systems. – Index of Contributors.

Springer-Verlag
Berlin Heidelberg New York
London Paris Tokyo

Picosecond Phenomena III

Proceedings of the Third International Confer-
ence on Picosecond Phenomena, Garmisch-
Partenkirchen, Federal Republic of Germany,
June 16–18, 1982

K. B. Eisenthal, R. M. Hochstrasser, W. Kaiser,
A. Laubereau (Eds.)

1982. 288 figures. XIII, 401 pages. (Springer
Series in Chemical Physics, Volume 23).
ISBN 3-540-11912-4

Contents: Advances in the Generation of
Ultrashort Light Pulses. – Ultrashort Measuring Techniques. – Advances in Optoelectronics. – Relaxation Phenomena in Molecular
Physics. – Picosecond Chemical Processes.
Ultrashort Processes in Biology. – Applications
in Solid-State Physics. – Index of Contributors.

Picosecond Phenomena II

Proceedings of the Second International
Conference on Picosecond Phenomena, Cape
Cod, Massachusetts, USA,
June 18–20, 1980

R. Hochstrasser, W. Kaiser, C. V. Shank (Eds.)

1980. 252 figures. 17 tables. XII, 382 pages.
(Springer Series in Chemical Physics,
Volume 14). ISBN 3-540-10403-8

Contents: Advances in the Generation of Pico-
second Pulses. – Advances in Optoelectronics.
– Picosecond Studies of Molecular Motion. –
Picosecond Relaxation Phenomena. – Pico-
second Chemical Processes. – Applications in
Solid State Physics. – Ultrashort Processes/
Biology. – Spectroscopic Techniques. – Index
of Contributors.

Picosecond Phenomena I

Proceedings of the First International Confer-
ence on Picosecond Phenomena, Hilton Head,
South Carolina, USA,
May 24–26, 1978

C. V. Shank, E. P. Ippen, S. L. Shapiro (Eds.)

1978. *Out of print*